T0320503

STAR-DISK INTERACTION IN YOUNG STARS

IAU SYMPOSIUM No. 243

COVER ILLUSTRATION: 3D MHD simulations of magnetospheric accretion.

Three-dimensional MHD simulations of accretion onto a young star with a complex magnetic field. The magnetic configuration is represented by a superposition of dipole and quadrupole fields. The quadrupole field at the surface of the star (at the magnetic pole) is about twice as large compared to the dipole field. Their magnetic moments are aligned and are shown by the $\mu-$vector which is inclined relative to the rotation axis of the star Ω at an angle $\Theta = 30°$. Simulations show that the disk is disrupted by the dipole component of the field, but the quadrupole component governs the flow closer to the star. The color background shows slices of density distribution (dark-blue: lowest density, red: highest density). Red lines are closed magnetic field lines of the magnetically-dominated magnetosphere, while yellow lines show magnetic field lines affected by the disk matter. From: Long, Romanova & Lovelace (2007).

INTERNATIONAL ASTRONOMICAL UNION

UNION ASTRONOMIQUE INTERNATIONALE

STAR-DISK INTERACTION IN YOUNG STARS

PROCEEDINGS OF THE 243th SYMPOSIUM OF THE INTERNATIONAL ASTRONOMICAL UNION HELD IN GRENOBLE, FRANCE MAY 21–25, 2007

Edited by

JÉRÔME BOUVIER

Laboratoire d'Astrophysique, Grenoble, France

and

IMMO APPENZELLER

ZAH, Landessternwarte, Heidelberg, Germany

Shaftesbury Road, Cambridge CB2 8EA, United Kingdom

One Liberty Plaza, 20th Floor, New York, NY 10006, USA

477 Williamstown Road, Port Melbourne, VIC 3207, Australia

314–321, 3rd Floor, Plot 3, Splendor Forum, Jasola District Centre, New Delhi – 110025, India

103 Penang Road, #05–06/07, Visioncrest Commercial, Singapore 238467

Cambridge University Press is part of Cambridge University Press & Assessment, a department of the University of Cambridge.

We share the University's mission to contribute to society through the pursuit of education, learning and research at the highest international levels of excellence.

www.cambridge.org
Information on this title: www.cambridge.org/9780521874656

First published 2007

A catalogue record for this publication is available from the British Library

ISBN 978-0-521-87465-6 Hardback
ISSN 1743-9213

Table of Contents

Part 4. MAGNETOSPHERIC EJECTION

Part 5. STAR-DISK MAGNETOSPHERIC COUPLING

Part 6. COMPANIONS, PLANETS, AND EFFECTS OF THE STELLAR MASS

Part 7. CONCLUSION

From left to right: Jérôme Bouvier, Claude Bertout, Gibor Basri, Immo Appenzeller

Preface

Disk accretion and jet outflows are intimately associated with the formation of stars and planets. One central issue raised by recent observational studies is the origin of the physical connection between accretion and wind/jet processes. It has become clear that the physical connection takes place within 1 AU of the central star, in a region where the interaction between the star and the inner disk is still poorly understood.

IAU Symposium 243 "Star-Disk Interaction in Young Stars" was intended to review the observational constraints available on the physical processes thought to be at work at the star-disk interface, to confront the predictions of the latest numerical and analytical MHD models of star-disk-jet systems with observations, and to explore the consequences of these processes for stellar angular momentum evolution and inner disk structure. Indeed, understanding the structure and evolution of the star-disk interaction region in young stars is critical to our understanding of the star and planetary system formation process.

The conference was held from May 21 to 25, 2007 in Grenoble, and brought together nearly 150 specialists from 21 countries, theorists, numericists and observers, to discuss a variety of topics, including magnetic fields in young stars, signatures of magnetospheric accretion and inner disk truncation, wind and jet ejection processes, and star-disk magnetospheric coupling in low-mass young stars, as well as in intermediate mass pre-main sequence stars and young brown dwarfs.

The motivations for organizing this Symposium were several. While the topics of star-disk interaction was regularly (albeit usually only briefly) touched upon at a number of conferences on star formation, no former conference had yet been held specifically dedicated to this central topics. The spectacular advances made in this field from observations and theory over the last few years convinced us that the time was ripe to provide the community with an opportunity to present and discuss recent (and sometimes debated) results. Last but not least, we wanted to dedicate this Symposium to Claude Bertout, in recognition of his pioneering contributions to our understanding of the physics of young stars, but also to acknowledge his patient efforts to educate a new generation of students and young researchers in France into the star formation arena.

It is a great pleasure to acknowledge the financial support of our sponsors listed on the next page of these Proceedings, the active support of the members of the SOC for defining an exciting science program, and the equally efficient support of the members of the LOC to overcome the difficulties inherent to the organization of such a large conference. Finally, we would like to thank the speakers and participants who contributed so much to the success of this Symposium by the quality of their presentations and for the lively discussions which took place throughout this conference.

Jérôme Bouvier & Immo Appenzeller,
Grenoble, September 5, 2007

THE ORGANIZING COMMITTEE

Scientific

S. Alencar (Brazil)
G. Basri (USA)
S. Cabrit (France)
M. Jardine (UK)
O. Regev (Israel)
K. Shibata (Japan)
I. Appenzeller (Germany)
J. Bouvier (chair, France)
S. Edwards (USA)
R. Keppens (Belgium)
B. Reipurth (USA)

Local

A. Blanc-Senet
J. Bouvier
F. Malbet
T. Montmerle
F. Bouillet
J. Ferreira
F. Ménard
M.-H. Sztefek

Acknowledgements

The symposium is sponsored and supported by the IAU Division VI (Interstellar Matter), and by the IAU Commissions No. 25 (Stellar Photometry and Polarimetry), No. 27 (Variable Stars), No. 35 (Stellar Constitution), and No. 36 (Theory of Stellar Atmospheres).

Financial support by the
International Astronomical Union,
Centre National de la Recherche Scientifique,
Laboratoire d'Astrophysique de Grenoble,
Programme National de Physique Stellaire,
Observatoire des Sciences de l'Univers de Grenoble,
Region Rhône-Alpes,
Ville de Grenoble,
Grenoble Alpes Metropole,
Université Joseph Fourier Grenoble,
and
Centre National d'Études Spatiales
is gratefully acknowledged.

Photo Credit : Jean Bérezné

Editors' Note

This volume contains the invited and contributed papers presented at the IAU Symposium 243. The posters are available on the Symposium web site (http://www.iaus243.org).

J. Bouvier & I. Appenzeller
September 5, 2007

Participants

Alicia **Aarnio**, Vanderbilt University, Nashville, USA alicia.n.aarnio@vanderbilt.edu
Vanessa **Agra-Amboage**, Laboratoire d'Astrophysique de Grenoble, Grenoble, France amboage@obs.ujf-grenoble.fr
Suzanne **Aigrain**, University of Exeter, UK suz@astro.ex.ac.uk
Evelyne **Alecian**, Royal Military College of Canada, Kingston, Canada evelyne.alecian@obspm.fr
Silvia **Alencar**, Universidade Federal de Minas Gerais, Brazil silvia@fisica.ufmg.br
Jean-Jacques **Aly**, Service d'Astrophysique, CEA Saclay, France jean-jacques.aly@cea.fr
Simone **Antoniucci**, INAF-Osservatorio Astronomico di Roma, Italy antoniucci@oa-roma.inaf.it
Immo **Appenzeller**, Landessternwarte Heidelberg, Germany immo@appenzeller.net
David **Ardila**, Spitzer Science Center, Caltech, Pasadena, USA ardila@ipac.caltech.edu
Costanza **Argiroffi**, Universita di Palermo, Italy argi@astropa.unipa.it
Marc **Audard**, ISDC & Geneva Observatory, Switzerland Marc.Audard@obs.unige.ch
Jean-Charles **Augereau**, Laboratoire d'Astrophysique de Grenoble, France augereau@obs.ujf-grenoble.fr
Carla **Baldovin**, ISDC & Geneva Observatory, Switzerland carlabaldovin@gmail.com
Nairn **Baliber**, Las Cumbres Observatory & UCSB, Santa Barbara, USA baliber@lcogt.net
Dinshaw **Balsara**, University of Notre Dame, Notre Dame, USA dbalsara@nd.edu
Mary **Barsony**, Space Science Institute, Baltimore, USA barsony@spacescience.org
Jeffrey **Bary**, University of Virginia, Charlottesville, USA jbary@virginia.edu
Gibor **Basri**, University of California, Berkeley, USA basri@berkeley.edu
Pierre **Bastien**, Université de Montréal, Canada bastien@astro.umontreal.ca
Tracy **Beck**, Gemini Observatory, USA tbeck@gemini.edu
Myriam **Benisty**, Laboratoire d'Astrophysique de Grenoble, France Myriam.Benisty@obs.ujf-grenoble.fr
Jean-Philippe **Berger**, Laboratoire d'Astrophysique de Grenoble, France Jean-Philippe.Berger@obs.ujf-grenoble.fr
Claude **Bertout**, Observatoire de Paris, France Claude.Bertout@obspm.fr
Nicolas **Bessolaz**, Laboratoire d'Astrophysique de Grenoble, France Nicolas.Bessolaz@obs.ujf-grenoble.fr
Rosaria **Bonito**, INAF, Osservatorio Astronomico di Palermo, Italy sbonito@astropa.unipa.it
Jerome **Bouvier**, Laboratoire d'Astrophysique de Grenoble, France jbouvier@obs.ujf-grenoble.fr
Herve **Bouy**, IAC & UC Berkeley, Berkeley, USA hbouy@astro.berkeley.edu
Sean **Brittain**, Clemson University, Clemson, USA sbritt@clemson.edu
Sylvie **Cabrit**, Observatoire de Paris, France sylvie.cabrit@obspm.fr
Marilena **Caramazza**, INAF, Osservatorio Astronomico di Palermo, Italy mcarama@astropa.unipa.it
Alessio **Caratti o Garatti**, Thüringer Landessternwarte, Tautenburg, Germany caratti@tls-tautenburg.de
John **Carr**, Naval Research Laboratory, Washington D.C., USA carr@nrl.navy.mil
Gael **Chauvin**, Laboratoire d'Astrophysique de Grenoble, France gchauvin@obs.ujf-grenoble.fr
Andrea **Ciardi**, Observatoire de Paris, France a.ciardi@imperial.ac.uk
Deirdre **Coffey**, Osservatorio Astrofisico di Arcetri, Italy dac@arcetri.astro.it
Celine **Combet**, Laboratoire d'Astrophysique de Grenoble, France ccombet@obs.ujf-grenoble.fr
Kevin **Covey**, Harvard-Smithsonian Center for Astrophysics, Cambridge, USA kcovey@cfa.harvard.edu
Rachel **Curran**, Dublin Institute for Advanced Studies, Ireland rcurran@cp.dias.ie
Fabio **De Colle**, Dublin Institute for Advanced Studies, Ireland fdc@cp.dias.ie
Vasily **Demichev**, Space Research Institute, Moscow, Russia demichev@iki.rssi.ru
Suzan **Edwards**, Smith College, Northampton, USA sedwards@smith.edu
Jochen **Eislöffel**, Thüringer Landessternwarte, Tautenburg, Germany jochen@tls-tautenburg.de
Davide **Fedele**, European Southern Observatory, Garching, Germany dfedele@eso.org
Christian **Fendt**, Max Planck Institut für Astronomie, Heidelberg, Germany fendt@mpia.de
Matilde **Fernandez**, Instituto de Astrofisica de Andalucia, Granada, Spain matilde@iaa.es
Jonathan **Ferreira**, Laboratoire d'Astrophysique de Grenoble, France Jonathan.Ferreira@obs.ujf-grenoble.fr
Ettore **Flaccomio**, INAF, Osservatorio Astronomico di Palermo, Italy ettoref@astropa.unipa.it
Laure **Fouchet**, ETHZ - Institute of Astronomy, Zurich, Switzerland fouchet@phys.ethz.ch
Jorge **Gameiro**, Centro de Astrofisica, Universidade Porto, Portugal jgameiro@astro.up.pt
Rebeca **Garcia Lopez**, INAF, Observatory of Rome, Italy garcia@tls-tautenburg.de
Tommaso **Gatti**, Osservatorio Astrofisico di Arcetri, Italy tomcat@arcetri.astro.it
Adrian **Glauser**, Paul Scherrer Institut, Villigen, Switzerland adrian.glauser@psi.ch
Matthieu **Gounelle**, Museum of Natural History, Paris, France gounelle@mnhn.fr
Jose **Gracia**, Dublin Institute for Advanced Studies, Ireland jgracia@cp.dias.ie
Jane **Gregorio-Hetem**, University of Sao Paulo, Brazil jane@astro.iag.usp.br
Scott **Gregory**, University of St Andrews, UK sg64@st-andrews.ac.uk
Nicolas **Grosso**, Observatoire Astronomique de Strasbourg, France Nicolas.Grosso@obs.ujf-grenoble.fr
Mario Giuseppe **Guarcello**, Universita di Palermo, Italy mguarce@astropa.unipa.it
Manuel **Guedel**, Paul Scherrer Institut, Villigen, Switzerland guedel@astro.phys.ethz.ch
Hans **Günther**, Hamburger Sternwarte, Hamburg, Germany moritz.guenther@hs.uni-hamburg.de
Sylvain **Guieu**, IPAC, Caltech, Pasadena, USA guieu@ipac.caltech.edu
Marcelo **Guimaraes**, Universidade Federal de Minas Gerais, Brazil mmg@fisica.ufmg.br
Tim **Harries**, University of Exeter, UK th@astro.ex.ac.uk
Coel **Hellier**, University of Keele, UK ch@astro.keele.ac.uk
Patrick **Hennebelle**, Ecole Normale Suprieure, Paris, France patrick.hennebelle@ens.fr
William **Herbst**, Wesleyan University, Middletown, Connecticut, USA wherbst@wesleyan.edu
Gregory **Herczeg**, Caltech, Pasadena, USA gregoryh@astro.caltech.edu
Naomi **Hirano**, Academia Sinica, IAA, Taipei, China hirano@asiaa.sinica.edu.tw
Susan **Hojnacki**, Rochester Institute of Technology, USA smhpci@cis.rit.edu
Subhon **Ibadov**, IoA Tajik Academy of Science, Tajik Republic ibadovsu@yandex.ru
Shu-ichiro **Inutsuka**, Kyoto University, Japan inutsuka@tap.scphys.kyoto-u.ac.jp
Andrea **Isella**, INAF, Osservatorio Astrofisico di Arcetri, Italy isella@arcetri.astro.it
Moira **Jardine**, University of St Andrews, UK mmj@st-andrews.ac.uk
Vera **Jatenco-Pereira**, University of Sao Paulo, Brazil jatenco@astro.iag.usp.br
Christopher **Johns-Krull**, Rice University, Houston, USA cmj@rice.edu
Isabelle **Joncour**, Universite Joseph Fourier, Grenoble, France ijoncour@obs.ujf-grenoble.fr
Joel **Kastner**, Rochester Institute of Technology, USA jhk@cis.rit.edu
Toshiaki **Kawamichi**, Kyoto University, Japan kawamiti@kwasan.kyoto-u.ac.jp
Rony **Keppens**, Centre for Plasma Astrophysics, Leuven, Belgium Rony.Keppens@wis.kuleuven.be
Stefan **Kraus**, Max-Planck-Institute for Radioastronomy, Bonn, Germany skraus@mpifr-bonn.mpg.de
Aleksandra **Kravtsova**, Sternberg Astronomical Institute, Moscow, Russia kravts@sai.msu.ru
Volodymyr **Kryvdyk**, Univesity of Kyiv, Ukraine kryvdyk@univ.kiev.ua
Akshay **Kulkarni**, Cornell University, Ithaca, USA akshay@astro.cornell.edu
Ryuichi **Kurosawa**, University of Nevada, Las Vegas, USA rk@physics.unlv.edu
Sergey **Lamzin**, Sternberg Astronomical Institute, Moscow, Russia lamzin@sai.msu.ru
Natália **Landin**, Universidade Federal de Minas Gerais, Brazil nlandin@fisica.ufmg.br
Jarron **Leisenring**, University of Virginia, Charlottesville, USA jml2u@virginia.edu
Sebastien **Leygnac**, Dublin Institute for Advanced Studies, Ireland sleygnac@cp.dias.ie
Min **Long**, Cornell University, Ithaca, USA Long@astro.cornell.edu
Tigran **Magakian**, Byurakan Observatory, Byurakan, Armenia tigmag@sci.am
Fabien **Malbet**, Laboratoire d'Astrophysique de Grenoble, France Fabien.Malbet@obs.ujf-grenoble.fr
Claire **Martin-Zaidi**, CEA Saclay, Saclay, France claire.martin-zaidi@cea.fr

Robert **Mathieu**, University of Wisconsin, Madison, Wisconsin, USA mathieu@astro.wisc.edu
Titos **Matsakos**, University of Turin, Italy matsakos@ph.unito.it
Sean **Matt**, University of Virginia, Charlottesville, USA spm5x@virginia.edu
Owen **Matthews**, Paul Scherrer Institut, Switzerland owen.matthews@psi.ch
Leonid **Matveyenko**, Space Research Institute, Moscow, Russia lmatveen@iki.rssi.ru
Giuseppina **Micela**, INAF, Osservatorio Astronomico di Palermo, Italy giusi@astropa.unipa.it
Subhanjoy **Mohanty**, Harvard University, Cambridge, USA smohanty@cfa.harvard.edu
Jean-Louis **Monin**, Laboratoire d'Astrophysique de Grenoble, France Jean-Louis.Monin@obs.ujf-grenoble.fr
Thierry **Montmerle**, Laboratoire d'Astrophysique de Grenoble, France montmerle@obs.ujf-grenoble.fr
Gareth **Murphy**, Laboratoire Astrophysique de Grenoble, France gmurphy@obs.ujf-grenoble.fr
Antonella **Natta**, INAF, Osservatorio di Arcetri, Italy natta@arcetri.astro.it
Elena **Nikogossian**, Byurakan Observatory, Byurakan, Armenia elena@bao.sci.am
Elena **Nokhrina**, Moscow Institute of Physics and Technology, Russia nokhrinaelena@gmail.com
Johan **Olofsson**, Laboratoire d'Astrophysique de Grenoble, France olofsson@obs.ujf-grenoble.fr
Salvatore **Orlando**, INAF, Osservatorio Astronomico di Palermo, Italy orlando@astropa.unipa.it
Antonio **Pedrosa**, Center for Astrophysics of Oporto, Portugal apedrosa@multimeios.pt
Giovanni **Peres**, Universita di Palermo, Italy peres@astropa.unipa.it
Christophe **Pinte**, University of Exeter, UK pinte@astro.ex.ac.uk
Giovanni **Pinzon**, Universidad Antonio Narino, Bogota, Colombia gpinzon@on.br
Linda **Podio**, INAF, Osservatorio Astrofisico di Arcetri, Italy lindapod@arcetri.astro.it
Frederic **Pont**, Observatoire de Genève, Switzerland frederic.pont@obs.unige.ch
Loredana **Prisinzano**, INAF, Osservatorio Astronomico di Palermo, Italy loredana@astropa.inaf.it
Samira **Rajabi**, Centro de Astrofisica Universidade do Porto, Portugal samira@astro.up.pt
Tom **Ray**, Dublin Institute for Advanced Studies, Ireland tr@cp.dias.ie
Luisa **Rebull**, Spitzer Science Center, Caltech/JPL, Pasadena, USA rebull@ipac.caltech.edu
Oded **Regev**, Technion, Haifa, Israel regev@physics.technion.ac.il
Terrence **Rettig**, University of Notre Dame, Notre Dame, USA trettig@nd.edu
Veronica **Roccatagliata**, Max Planck Institut fuer Astronomie, Heidelberg, Germany roccata@mpia.de
Marina **Romanova**, Cornell University, Ithaca, USA romanova@astro.cornell.edu
Giuseppe **Sacco**, INAF, Osservatorio Astronomico di Palermo, Italy sacco@astropa.unipa.it
Jürgen **Schmitt**, Hamburger Sternwarte, Hamburg, Germany jschmitt@hs.uni-hamburg.de
Salvatore **Sciortino**, INAF, Osservatorio Astronomico di Palermo, Italy sciorti@astropa.unipa.it
Rogel Mari **Sese**, University of Tsukuba, Japan rmdsese@yahoo.com
Frank **Shu**, University of California San Diego, USA shufrnk5@netscape.net
Stephen **Skinner**, University of Colorado, Boulder, USA skinners@casa.colorado.edu
Daniil **Smirnov**, Sternberg Astronomical Institute, Moscow, Russia danila@sai.msu.ru
Noam **Soker**, Technion, Haifa, Israel soker@physics.technion.ac.il
John **Stauffer**, Spitzer Science Center, Caltech, Pasadena, USA stauffer@ipac.caltech.edu
Chantal **Stehlé**, Observatoire de Paris, France chantal.stehle@obspm.fr
Beate **Stelzer**, INAF, Osservatorio Astronomico di Palermo, Italy stelzer@astropa.unipa.it
Caroline **Terquem**, Institut d'Astrophysique de Paris, France terquem@iap.fr
Ovidiu **Tesileanu**, Universita di Torino, Italy tesilean@ph.unito.it
Paola **Testa**, MIT, Boston , USA testa@space.mit.edu
David **Tilley**, University of Notre Dame, USA dtilley@nd.edu
Jeff **Valenti**, STScI, Baltimore, USA valenti@stsci.edu
Aline **Vidotto**, University of Sao Paulo, Brazil aline@astro.iag.usp.br
Emma **Whelan**, Dublin Institute for Advanced Studies, Ireland ewhelan@cp.dias.ie
Tatiana **Yelenina**, INAF, Osservatorio Astronomico di Palermo, Italy elenina@astropa.inaf.it
Ruslan **Yudin**, Pulkovo Observatory, Pulkovo, Russia ruslan61@hotmail.com
Claudio **Zanni**, Laboratoire d'Astrophysique de Grenoble, France Claudio.Zanni@obs.ujf-grenoble.fr

Star-Disk Interaction in Young Stars
Proceedings IAU Symposium No. 243, 2007
J. Bouvier & I. Appenzeller, eds.

doi:10.1017/S1743921307009362

The accretion disk paradigm for young stars

Claude Bertout
Institut d'Astrophysique, 98bis, Boulevard Arago, 75014 Paris, France

Abstract. Accretion and magnetic fields play major roles in several of the many models put forward to explain the properties of T Tauri stars since their discovery by Alfred Joy in the 1940s. Early investigators already recognized in the 1950s that a source of energy external to the star was needed to account for the emission properties of these stars in the optical range.

The opening of new spectral windows from the infrared to the ultraviolet in the 1970s and 1980s showed that the excess emission of T Tauri stars and related objects extends into all wavelength domains, while evidence of outflow and/or infall in their circumstellar medium was accumulating.

Although the disk hypothesis had been put forward by Merle Walker as early as 1972 to explain properties of YY Orionis stars and although Lynden-Bell and Pringle worked out the accretion disk model and applied it specifically to T Tauri stars in 1974, the prevailing model for young stellar objects until the mid-1980s assumed that they experienced extreme solar-type activity. It then took until the late 1980s before the indirect evidence of disks presented by several teams of researchers became so compelling that a paradigm shift occurred, leading to the current consensual picture.

I briefly review the various models proposed for explaining the properties of young stellar objects, from their discovery to the direct observations of circumstellar disks that have so elegantly confirmed the nature of young stars. I will go on to discuss more modern issues concerning their accretion disk properties and conclude with some results obtained in a recent attempt to better understand the evolution of Taurus-Auriga young stellar objects.

Keywords. Stars: formation, stars: pre-main sequence.

---

1. Introduction

Like many topics in astrophysical research, the idea that stars might be evolving bodies that form from diffuse gas was first dreamt of by great ancient minds. The 18th century scientist Pierre-Simon de Laplace already thought that the Sun and planets coalesced from a flattened and rotating gaseous nebula under the action of gravity (Marquis de Laplace 1798). The idea was reborn in the mid-20th century when the theory of stellar evolution emerged and in less than one generation, evolved into a full-fledged, highly successful research area that grew very quickly and branched into several more specialized subfields. This symposium, devoted to one of these subfields of active topical research, investigates the role of both a magnetic field and accretion in shaping the properties of young stellar objects.

It is amusing, but also revealing of the scientific approach, that these two physical processes were already invoked in the late 1940s for explaining some properties of young stars even *before* it became apparent that they were young. There is only a relatively small number of basic physical phenomena that can be called upon to explain the properties of celestial bodies. If one proceeds by elimination in a logical and careful way to restrict the range of possible mechanisms that could explain an observed situation, there is a good chance that one will correctly identify a plausible physical basis for any phenomenon. Laplace's conclusion that gravity was at work in shaping the primitive solar nebula proceeded from such a line of thought.

The devil, however, is in the details. There is a long path and many dead ends from such an initial intuition to a consensual paradigm explaining a wide range of observational properties. The next section recalls some of the steps that researchers have gone through over the last 60 years to establish a working paradigm for young stars; because of space limitations, I will focus mainly on T Tauri stars (TTSs hereafter). I then briefly discuss some of the current open issues concerning disk properties.

2. Historical notes

Alfred H. Joy (1882–1973) is remembered for his observational contributions to the field of stellar spectroscopy, beginning with the Mount Wilson radial velocity and spectroscopic parallax program, which led in 1935 to the first comprehensive Hertzsprung-Russell diagram with 4179 stars (Adams *et al.* 1935), followed by his pioneering studies of long-period regular and semi-regular variables (e.g., Joy 1952, 1954) and his involvement with late-type erratic variables (e.g., the first observation of a flare in UV Cet by Joy & Humason 1949; Joy 1958). However, as George Herbig (1974) pointed out, "It is difficult to be certain which was Joy's most far-reaching contribution, but the T Tauri stars are his most famous monument. In the early 1950s it was becoming painfully clear that given the sources of stellar energy, star formation must be a continuing thing, presumably going on before our eyes [...] Several years passed before it came to be recognized generally that the link [between stars and the interstellar clouds] existed, described in that 1945 investigation by Joy. [...] Twenty years later, the dimensions of that discovery are just beginning to be appreciated".

The paper that George Herbig refers to is, of course, the now classical work that defined the class of "T Tauri Variable Stars" (Joy 1945). This work was soon followed by additional discoveries of TTSs in Taurus (Joy 1949) and Ophiuchus (Struve & Rudkjøbing 1949). The criteria given by Joy to define the class of TTSs were

(*a*) "irregular light variations of about 3 mag,

(*b*) spectral type F5-G5 with emission lines resembling the solar chromosphere,

(*c*) low luminosity,

(*d*) association with dark or bright nebulosity".

Seventeen years later, Herbig (1962) gave a purely spectroscopic definition of the class, using the following criteria:

(*a*) "The hydrogen lines and the lines of CaII are in emission.

(*b*) The fluorescent FeI emission lines $\lambda\lambda$ 4063, 4132 are present (they have been found only in T Tauri stars).

(*c*) the [SiII] emission lines $\lambda\lambda$ 4068, 4076 are usually but not always present. Probably [SII] $\lambda\lambda$ 6717, 6731 and [OI] $\lambda\lambda$ 6300, 6363 are also characteristic.

(*d*) Recent results [...] suggest that the presence of strong LiIλ6707 absorption, in those stars in which an absorption spectrum can be seen at all, may constitute another primary criterion".

Herbig also notes that the emission lines are usually superimposed on a continuous spectrum ranging from an approximately normal absorption spectrum in the range F to M to pure continuous emission, with all intermediary stages in between. Herbig's spectroscopic characteristics define what we now call the classical T Tauri stars (CTTSs hereafter). When all present in the same stars, these criteria signal one of the most active stars of the class. Less active CTTSs do not show FeI fluorescent emission or forbidden emission, but all have strong Balmer and CaII H and K line emission, as well as strong LiI absorption, a feature now known to be the main signature of their youth.

The "veiling" continuous emission that obliterates the late-type absorption spectrum of CTTSs had already been noted by Joy and was considered as early as 1950 as a strong hint that a non-stellar emission component contributes to the blue part of the CTTS spectra (Greenstein 1950). In fact, Greenstein anticipated much of the controversy that would develop around CTTSs in the following decades when he wrote that "The veiling continuum and the emission lines must be produced in a circumstellar fringe or extended chromosphere".

2.1. *1940s and 1950s: first TTS models*

The first model for TTSs that emerged shortly after their discovery was motivated by their association with dark nebulae. These stars, postulated e.g. Struve & Rudkjøbing (1949), might be normal field stars that were crossing and interacting with interstellar clouds. However, it quickly became clear that the kinetic energy of the gas falling onto the star was not high enough to account for the observed line and excess continuum luminosity; assuming a differential velocity of 10 km/s between gas cloud and star and a cloud density of 10^{-18} g/cm^3, Greenstein (1950) derived a kinetic luminosity of a few $10^{-5} L_{\odot}$, whereas the observed excess luminosity can be greater than $L_{\odot}$ in extreme CTTSs. Even though the value of a typical CTTS luminosity excess was unknown in the early 1950s, it was already clear that it was orders of magnitude larger than what the accretion of interstellar cloud gas could provide. Faced with this problem, Greenstein (1950) thus proposed a variant of this model in which the (then newly discovered) magnetic field of the interstellar matter falling onto the star acted to compress and heat the infalling matter and led to the propagation of Alfvén waves in the stellar surroundings that could be responsible for the widths of several hundreds of km/s observed in CTTS Balmer lines. Greenstein (1950) went on to speculate that "the incoming material apparently does not itself carry sufficient energy to the star, but it seems well suited as a source of the high kinetic temperature and as a trigger mechanism for the release of excess energy from star spots, in the form of flares bright enough to affect the total spectrum of the object".

It is therefore clear that the two physical processes, accretion and magnetism – the topic of today's symposium because we have strong evidence that they are responsible for young star activity – had already been identified as such as early as in 1950. Unfortunately, as data started to accumulate, these promising physical findings were clouded by what appeared to be contradictory observations, and it took 40 years to disentangle the various issues and understand "the nature of the objects of Joy"†.

What was less evident in the late 1940s and early 1950s was the evolutionary status of TTSs. Although Ambartsumian (1947, 1954) introduced the concept of stellar OB and T associations and proposed that they were the sites of ongoing star formation shortly after TTSs were discovered, these ideas were long in gaining mainstream acceptance in the community, presumably due to the difficulties of communication between the Eastern the Western worlds after World War II. The first paradigm shift, from stars passing by chance through interstellar clouds to stars being born in interstellar clouds, took place around 1955 and was strongly influenced by Ambartsumian's ideas.

The state of research on TTSs in the mid-50s is summarized in the proceedings of the IAU Symposium No.3 "*Non-stable stars*" that took place in Dublin in September 1955. There, Herbig (1957) discussed the nature of TTSs, whose pre-main sequence status appears to be by then an accepted fact, but it is their relationship to the newly discovered Herbig-Haro objects (Herbig 1951; Haro 1952) that takes the center stage in Herbig's

† This cute expression is the title of a paper by Rydgren *et al.* (1976) that fails to disclose the true nature of these objects but remains a strong contribution to our knowledge of their infrared properties.

talk, which gives arguments based in part on the association of Burnham's nebula with T Tau itself, supporting the view that the nebulous Herbig-Haro objects might be the precursors of TTSs. Another subject of discussion during the symposium was the relationship between flare stars and TTSs. This interest was motivated by Guillermo Haro having discovered a number of flare stars (which he called flash stars to distinguish them from UV Cet stars) in Taurus and Orion. In his talk, Haro (1957) concluded that TTSs and UV Cet stars must have a similar nature. Alfred Joy, on the other hand, reviewed the properties of the three classes of late-type variables, the TTSs, the UV Cet stars, and the first examples of cataclysmic variables and wisely concluded: "Somewhat similar effects are noted in the spectral changes of the three groups of stars, but the underlying cause of the strange and unexpected behaviour, which may be quite different in the three types, invites both observations and analysis" (Joy 1957). In contrast, the somewhat similar spectral properties of TTSs, flare stars, Herbig-Haro objects and cataclysmic variables appeared so unusual to Ambartsumian that he felt compelled to propose that they were due to the liberation of large amounts of energy connected with nuclear processes but different from the already known thermo-nuclear reactions. He went on to add: "It may be that the material being brought up from the interior contains pre-stellar matter of high density. It may represent matter that is in an altogether peculiar state, thus far unknown to us. [...] the rapidly growing knowledge of T Tauri stars permits us to pass on to the study of the laws of physical processes of a new type, which govern a number of phenomena taking place in these stars" (Ambartsumian 1957). Clearly, understanding the origin of the energy input needed to explain the exotic properties of TTSs had already become a true challenge for astrophysicists, and some bold ideas were emerging. Reacting to this suggestion, Burbidge & Burbidge (1955) emphasized that the energy sources responsible for flare activity must have different natures in different classes of stars and that no new physical mechanism was needed to understand them. They wrote "To sum up, we wish to divide the stars showing flaring into two main categories: (a) The T Tauri and related objects in which MHD processes outside the star are responsible. (b) The dMe flare stars in which internal MHD processes are responsible". By the end of the 1950s, then, the main physical question concerning these stars was on the table: were the radiative energy losses observed in TTSs covered by the star itself or by some external energy source?

2.2. *1960s: evidence for T Tauri winds*

The first major review of the T Tauri phenomenon by Herbig (1962) and Leonard Kuhi's thesis work (Kuhi 1964) were the milestones of the first half of the 1960s and a major evolution in ideas on the envelopes of T Tauri stars. Herbig reviewed the scarce high-resolution spectroscopic data and concluded that "mass ejection by T Tauri stars [...] seems to be such a common trait of the family that it must find some natural explanation in the contractional process", while Kuhi convincingly demonstrated that the H_α line profiles of several bright stars of the class could be modeled by a spherical wind. Derived mass-loss rates were in the $10^{-8}-10^{-7} M_\odot/\mathrm{yr}$, and this work became the first of numerous attempts to model the line profiles of young stars. Another result obtained by Kuhi attracted much less attention, although it was crucial. Namely, Kuhi (1966) found that the UV excess of TTSs is correlated with H_α intensity, making it apparent that the UV excess was due to Balmer continuum emission. This finding opened the door to a model where, by analogy with the Sun, one postulates that the T Tauri chromospheres could possibly be denser than in the Sun and could thus give rise to the UV excess.

Meanwhile, a few doors away from Herbig's office at Lick Observatory, Merle F. Walker was studying faint peculiar stars in Orion that showed particularly strong UV excesses.

He had been using a Lallemand electronic camera to obtain flux-calibrated spectrograms of a number of these objects, now known as YY Ori stars after their prototype, and had noticed the presence of red-displaced, highly variable absorption components in the Balmer lines of these objects, indicating infall velocity of up to 400 km/s. In these spectra the Balmer jump is clearly in emission, and the overall line spectra look very much like those of usual TTSs. Walker (1963) suggested that "the presence of the excess continuous emission may have some connection with this possible infall of material", a prophetic statement that, unfortunately, no one paid attention to.

In the late 1960s, another major discovery was made by Eugenio E. Mendoza V. (1966, 1968) when he found that several TTSs and related objects had strong infrared excesses up to 5μm. Mendoza, who was using Johnson's photometric equipment at the Lunar and Planetary Laboratory, also confirmed the UV excess and was the first to present quasi-simultaneous UBVRIJHKLM spectral energy distributions for a few bright young stars (T Tau, SU Aur, RW Aur, V380 Ori, and R Mon). By the end of the 1960s, major pieces of the T Tauri puzzle were known: their optical line spectra, their UV and IR excesses, their "veiling", their outflows and sometimes inflows. The energetics problem had become much more acute when the IR properties of young stars were discovered, but by then this question was set aside as specialists were focusing on trying to understand the mass motions in the surroundings of young stars.

2.3. *1970s: pioneers and the great confusion*

The 1970s were a very contrasted decade as far as T Tauri research was concerned. On one hand, there were a few pioneers who anticipated the nature of TTSs, but their work remained largely unnoticed. On the other hand, there was the rest of us, who were more and more puzzled and confused by the data that we kept accumulating.

Meet the pioneers: observer Merle F. Walker and theoreticians Donald Lynden-Bell and Jim Pringle. Walker (1972) postulated a circumstellar disk for explaining the spectroscopic properties of YY Ori stars. He noted "That we might expect the infall to occur in a preferential plane during the last stages of the collapse of the prestellar cloud is suggested by analogy with the solar system, where the coplanar orbits of the planets indicate that the presolar nebula must have collapsed to a plane during the formation of the system". It was, to the best of my knowledge, the first time that a circumstellar disk was proposed as an explanation for specific observed properties of young stars. In Walker's model, the UV and blue continua are formed in a shock front where the infalling matter reaches the stellar surface, and the Balmer line emission originates in the disk with the red-displaced absorption components being formed in that part of the disk seen in projection onto the star by the observer. While Walker also discussed the possibility of a spherically symmetric envelope around these stars, he clearly favors the disk model in particular because "the shock front does not cover the entire surface of the star, so that the stellar absorption spectrum will be visible, if it is sufficiently luminous compared to the blue continuum, regardless of the optical thickness of the infalling matter or the material at the shock front". Because envelope modeling efforts started in the late 70s, it was not yet apparent to most researchers that the low visual extinction of most TTSs poses major difficulties for any model involving a spherically symmetric, dusty envelope. Here, Walker was providing an elegant solution to a problem that he was first to foresee, but unfortunately that solution had been forgotten a few years later when the problem actually did appear.

Lynden-Bell & Pringle (1974) built a formidable theoretical foundation to the empirical disk picture envisioned by Walker. Their accretion disk model is such a powerful physical and mathematical construct that it underlies most of the work on circumstellar disks done ever since. While they applied their work specifically to TTSs, the community of young star specialists hardly noticed that Lynden-Bell and Pringle were offering an answer to many of their questions. Instead, it is in its application to interacting binaries that the accretion disk model, which had been independently worked out in that context by Shakura & Sunyaev (1973), met with immediate success.

In 1976, researchers met for the IAU Symposium No. 75 entitled "Star formation", which was held in Geneva. Some emphasis was given during that conference to results of computations of protostellar collapse in spherical symmetry, since several groups had attempted to reproduce the pioneering numerical computations by Larson (1969). Relating these computations to observations of YY Ori and TTSs appeared difficult, although there were some hints that the IR spectral energy distribution of some bright TTSs bore some resemblance to theoretical spectra of the collapsing protostar models computed at the end of the accretion phase (cf. Larson 1977, and references therein). An amazingly upbeat review of young stellar object (YSO hereafter) properties was given on that occasion by Steve Strom (1977), who proposed in particular that most properties of TTSs could be explained by a dense gaseous spherical envelope at temperature $\approx 2 \cdot 10^4$ K and that Herbig-Haro objects were reflection nebulae illuminated by embedded stars. Strom mentioned briefly a new model for TTSs that was being worked out by Roger Ulrich (1976), who was able to reproduce the Type III P Cygni line profiles often observed in the H_α line of TTSs in a model involving line emission from a shock front at the stellar surface. The shock wave is formed by the matter from an infalling and rotating protostellar envelope when it impacts the star.

The discussion following Strom's talk demonstrates the state of confusion that was prevalent among other observers of TTSs, with the contradictory evidence of infall and outflow obscuring most other issues. In his summary talk, Herbig (1977b) called for decisive observations and concluded: "After all, when the historians of science look back on our times with the perspective of the years, all that we do today will certainly be seen to have been either wrong, or irrelevant, or obvious". Thirty years later, this opinion appears to have been overly pessimistic; while some of the ideas discussed during that symposium were clearly wrong, none was irrelevant or particularly obvious. What remains stunning even today, however, is that during the Geneva symposium only two invited speakers mentioned, and only in passing, the Lynden-Bell & Pringle (1974) accretion disk model; these reviewers were the two theoreticians Leon Mestel and Richard Larson.

We should not close this brief review of the 1970s without mentioning two extremely important pieces of work; Herbig (1977a) discussed observational evidence for FU Ori outbursts and their relationship to TTSs, while Cohen & Kuhi (1979) presented their *"magnum opus"*, an exhaustive discussion of TTS spectroscopic properties.

2.4. *1980s: extension of the knowledge base and paradigm shift*

The first half of the 1980s saw an unprecedented increase in the available data on YSOs, owing to the launch of several orbiting observatories (IUE, IRAS, and the *Einstein* X-ray Observatory), as well as a strong increase in the efficiency of detectors used for ground-based observations in various wavelength domains. These technological advances led to a succession of discoveries, which are cited below in approximate chronological order.

• Snell *et al.* (1980) discovered the first instance of bipolar CO outflows, which turned out to be a ubiquitous phenomenon in starforming regions.

• Observations with the IUE satellite showed that the UV excess of TTSs extends to the far-UV and that the UV line spectrum is dominated by Mg II and includes transition region lines from ions such as Si II and C IV (e.g, Giampapa *et al.* 1981).

• The *Einstein* X-ray Observatory revealed that typical TTS X-ray luminosities were within the 10^{30-31} erg/s range (e.g., Feigelson & Decampli 1981), which ruled out spherical collapse as a possibility for explaining CTTS properties. Repeated observations also demonstrated the strong variability of TTSs in X-rays (Montmerle *et al.* 1983).

• The discovery of highly collimated jets shed new light on the nature of Herbig-Haro objects (Mundt & Fried 1983).

• The association of Herbig-Haro objects with bipolar molecular outflows, discovered by Edwards & Snell (1984), was crucial for understanding the nature of outflows emanating from YSOs.

• Mid-IR ground-based observations led to a new characterization of YSOs based on their IR properties (Lada & Wilking 1984), thus paving the way for the work by Frank Shu and collaborators that led them to propose an evolutionary sequence from embedded protostellar sources to main sequence stars.

• IRAS observations showed that the IR excess of TTSs extends to the far-IR and that their spectral energy distribution (SED hereafter) is typically given by $\lambda F_\lambda \propto \lambda^{-1}$ (Rucinski 1985).

Meanwhile, theory was also moving forwards on several fronts. A first line of investigation was the deep chromosphere model advocated by Herbig and Kuhi. Detailed atmosphere models that included deep chromospheres were investigated, e.g., by Cram (1979) and Calvet *et al.* (1984). These models assumed that the chromospheric temperature minimum is located at a continuum optical depth $\tau_c \approx 0.1$–0.3 instead of $\tau_c \approx 10^{-3}$–10^{-4} as in the Sun. Such a model accounts for the bulk properties of TTS emission line spectra and Balmer emission, but is unable to reproduce (a) the observed Balmer decrement, which indicates that the H_α line originates from a more extended region, (b) the Balmer line shapes, which require high velocities of several hundred km/s that are not present in the chromospheric region, and (c) the observed IR excess in the 1 μm region, which remains optically thin in this atmospheric model where the main opacity source is the H^- ion. Furthermore, this model ignores the energetics issue, which was emerging again after William De Campli (1981) discussed the energetic constraints involved in generating a strong T Tauri wind. He showed that the wind luminosity of a thermally driven wind with mass-loss rate $\approx 10^{-8} M_\odot$/yr exceeds the stellar luminosity. De Campli went on to show that Alfvén-wave driven winds are more efficient and may account for moderately strong T Tauri winds. Detailed Alfvén-wave driven wind models were computed by De Campli (1981), Hartmann *et al.* (1982), and Lago (1984). Magnetic fields were also invoked to explain the photometric variability of TTSs, as Appenzeller & Dearborn (1984) found from computations of pre-main sequence stellar structure models, that variable magnetic fields can cause visual brightness variations that are comparable in range (up to 3.5 mag) to what is observed in TTSs. It had therefore become evident in 1985 that a promising TTS model would combine a magnetically active star with a deep chromosphere and an Alfvén-wave driven wind. However, a model involving only the magnetic field had serious shortcomings in that it involved strong constraints on the stellar magnetic field strength and was unable to reproduce either the observed IR excess or the spectroscopic evidence for infall in some stars. Indeed, such a model was never developed because TTS research instead experienced a sudden paradigm shift with the comeback of the accretion disk model.

The return of the disk model for TTSs was announced by several precursor works. First, Martin Cohen (1983) suggested that the CTTS HL Tau was surrounded by a dusty disk. HL Tau has a "flat" SED in the IR (when plotted as $\log \lambda F_\lambda$ vs. $\log \lambda$), it is a TTS with both a 3.1 μm ice feature and 10 μm silicate absorption, and it displays a record linear polarization of 13%. Cohen suggested that all these properties represent strong evidence of a circumstellar disk. Then, Appenzeller *et al.* (1984) showed that the forbidden lines of TTSs are often blue-displaced and can be interpreted as being formed in collimated outflows partially occulted by optically thick, dusty "screens", i.e., circumstellar disks. This first spectroscopic evidence of optically thick disks around TTSs was particularly convincing because the interpretation of optically thin forbidden line emission is much easier than that of permitted lines such as H_α. Finally, Hartmann & Kenyon (1985) showed that FU Orionis outbursts can be explained as episodes of strong mass accretion from a disk. Emission from the hot, optically thick accretion disk dominates the system light at its maximum.

The IAU Symposium No. 122 "*Circumstellar matter*" held in Heidelberg in June 1986 can be seen as the moment when the disk hypothesis for TTSs re-emerged successfully on the public scene. Frank Shu presented there his ongoing work on the 4 stages of pre-main sequence evolution (Shu & Adams 1987). Mundt (1987) discussed the large body of data concerning molecular outflows and optical jets, while Cohen (1987) discussed the possible role of disks in bipolar outflows observed in many different types of stars. Walter (1987) presented the properties of X-ray selected, "naked" TTSs that were to become known as weak-emission lines TTSs (WTTSs), i.e., TTSs that lack circumstellar disks. Meanwhile Finkenzeller & Basri (1987) reported evidence of inflowing material in moderately active CTTSs. Finally, Bertout (1987) compared observed SEDs of TTSs with models of stars interacting with accretion disks.

The second half of the 1980s was marked by a wide exploration by several teams of the various consequences of the new paradigm, until accumulated evidence in favor of the accretion disk picture became so convincing that a consensus emerged concerning the presence of accretion disks around TTSs. The success of the model can be attributed not only to the fact that it solves many problems but also that it obeys *lex parsimoniæ*, also called Occam's razor rule. In other words, one hypothesis, the presence of a disk and its interaction with the star, accounts for most exotic properties of young stars. We briefly list in the following some influential works that led to the current consensus on disks and magnetospheric accretion. First, Frank Shu, Charlie Lada, and their collaborators developed an empirical/theoretical classification of YSOs based on the evolution of their SEDs and provided a global, physically consistent scheme for star formation (Adams *et al.* 1987; Shu *et al.* 1987). Rydgren & Zak (1987) showed that the fraction of TTS system luminosity due to the IR-excess component is typically in the 0.4–0.8 range and that the average mid-infrared spectral slope in their sample is $\lambda F_\lambda \propto \lambda^{-3/4}$. Kenyon & Hartmann (1987), in an effort to explain this mid-IR slope, which is shallower than the law $\lambda F_\lambda \propto \lambda^{-4/3}$ expected for classical accretion disks, considered flaring, passive disks as an alternative to accretion disks. They also worked out detailed models of FU Ori disks (Hartmann & Kenyon 1987). Edwards *et al.* (1987) extended the Appenzeller *et al.* (1984) study of forbidden line emission in TTSs and refined the forbidden line profile computation, confirming the need for an optically thick disk to explain these observations. Bertout *et al.* (1988) and Basri & Bertout (1989) used quasi-simultaneous observations of TTSs in the UV, optical, and near-IR spectral ranges to compute detailed models of SEDs and derived mass accretion rates in the range 10^{-9}–$10^{-7} M_\odot$/yr in the framework of the boundary layer model for the star-disk interaction. They also suggested, based on the observed rotational modulation of its UV and optical flux, that DF Tau was experiencing magnetospheric accretion. Adams *et al.* (1988) constructed models for TTSs with flat IR-spectra and were the first to suggest that gravitational instabilities might play a role in the accretion process. The first estimates of disk lifetimes were provided by Strom *et al.* (1989), while Cabrit *et al.* (1990) provided the first evidence for a connection between disk accretion and ejection of matter in a wind. Finally,

Koenigl (1991) adapted the Ghosh & Lamb (1979) magnetospheric accretion model to TTSs.

From 1990 on, the accretion disk paradigm was widely accepted and became the framework of "normal" T Tauri research. The body of theoretical results that has accumulated since then is very impressive and has led in particular to sophisticated 2D models of accretion disks (e.g., D'Alessio *et al.* 1998; Malbet *et al.* 2001; Dullemond *et al.* 2002; Pinte *et al.* 2006) that take into account in a self-consistent way such important processes as the vertical stratification of gas and dust in the disk, as well as some of the details of the magnetospheric accretion process (see the invited papers by S. Alencar, T. Harries, D. Ardila, and M. Romanova in this volume). Meanwhile, models of magnetized accretion-ejection structures, which were long limited to self-similar solutions due to the complexity of the MHD equations to be solved, are now becoming more realistic so that comparison with observations appears possible (see the reviews by S. Edwards, S. Cabrit, and F. Shu in this volume). On the observational side, one must of course mention that the presence of circumstellar disks in CTTSs has been confirmed by images from the *Hubble* Space Telescope and ground-based adaptive optics observations (e.g., Burrows *et al.* 1996; Close *et al.* 1997; Stapelfeldt *et al.* 1998, 2003). Today, near-IR interferometry allows us to probe the inner disk structure (see the review of F. Malbet in this volume). The next section provides a brief overview of some issues concerning basic disk properties that, in the opinion of this writer, remain open.

3. Remarks on some open issues and a conclusion

Basic properties of T Tauri disks that are still currently a matter of debate are the disk masses and mass-accretion rates. A number of methods have been used over the years to derive the mass-accretion rates from the optical spectra of large samples of CTTSs (e.g., Valenti *et al.* 1993; Hartigan *et al.* 1995; Gullbring *et al.* 1998). These estimates show that the stellar accretion rates of CTTSs are relatively well-known relative to each other, provided that they are determined using the same assumptions and methods; but rates determined with different methods can differ by a factor of 10. Furthermore there is an uncertainty of typically a factor 3 on individual mass-accretion rates. The origin of these differences is understood fairly well (see Gullbring *et al.* 1998). The assumptions leading to these different sets of accretion rates differ mainly in the visual extinction values. While their relative validity can be discussed, none of the assumptions made by the various groups appears unreasonable. It should be emphasized, however, that the low mass-accretion rates derived by Gullbring *et al.* (1998) are lower limits to the true rates. These authors use the J flux to estimate the photospheric luminosity and the visual extinction values that they adopt are particularly low; both assumption result in a low estimate for the accretion luminosity. A main reason for adopting these low mass-accretion rates is because they provide for a consistent picture of disk evolution with the relatively low disk masses that were determined, e.g., by Beckwith & Sargent (1991); a disk of mass $10^{-2} M_\odot$ provides an adequate reservoir for feeding a star during 10^6 years at a rate of $10^{-8} M_\odot/\mathrm{yr}$.

Disk masses, however, are far from being well known and could be vastly underestimated. If there is appreciable grain growth and settling in the disk during the CTTS phase, resulting in the formation of a population of "sand and pebbles", as observations seem to indicate (cf. Natta *et al.* 2007), more mass than previously envisioned might be needed to reach the observed millimeter flux (Hartmann *et al.* 2006). There are other, more indirect lines of evidence that point in the direction of underestimated disk masses. It has been shown that the mass accretion rate depends on the stellar mass as $\dot{M} \propto M_*^{2.1}$, albeit with a large scatter (Muzerolle *et al.* 2003; Natta *et al.* 2005). As discussed by Alexander & Armitage (2006) and Hartmann *et al.* (2006), a similar relationship between disk and stellar masses would be helpful for understanding the dependency of the mass accretion rate on the stellar mass. Another hint of higher disk masses comes from evolutionary aspects. In the current scheme of star formation, it is usually taken for granted that CTTSs evolve into WTTSs when their disk are fully accreted or dissipate for any other reason. The fact that CTTSs and WTTSs share the same region of the Hertzsprung-Russell diagram is usually interpreted as evidence of a large scatter in the properties of circumstellar accretion disks. While hard evidence for such an evolution was lacking until now, recent results provide support for this evolutionary scheme. Bertout *et al.* (2007) used the individual parallaxes computed for members of the Taurus-Auriga moving group by Bertout & Genova (2006) to reassess the ages and masses of moving group members. They find that the age distributions of CTTSs and WTTSs are significantly different. Assuming that accretion of disk matter by the star proceeds at a constant rate and terminates when the disk is fully accreted, they find empirical evidence for evolution from CTTS to WTTS, provided that the disk mass is $\propto M_*^{2.85}$, leading to a disk lifetime of $\tau_d = 4 \cdot 10^6 (M_*/M_\odot)^{0.75}$. This surprising result implies very low masses for disks surrounding brown dwarfs *unless* the masses of CTTS disks have been underestimated. If confirmed, this result raises the question of the possible role of gravitational instabilities in driving the accretion process in these stars.

During this brief review of 60 years of T Tauri research, I have noted several instances of unfortunate "memory loss" when results obtained by one generation of scientists were forgotten, or ignored, by the next generation. Another example, observed in several recent papers, concerns the physical reason that led to the accretion disk hypothesis in the first place. The main physical problem that the disk hypothesis solves concerns the energetics of TTSs; the disk offers a reservoir of potential energy external to the star that can be tapped to power the phenomena associated with these stars (emission excesses at all wavelengths, jet and wind). A passive (non-accreting) disk only reprocesses the luminosity of the star and does not bring any gain as far as energetics of the system are concerned. While a fraction of the stellar luminosity is absorbed by the disk and re-radiated in the direction of the observer because of the system's anisotropy, this additional IR flux (compared to the same star but without a circumstellar disk) corresponds to at most 25% of the stellar luminosity in a flat disk (Adams & Shu 1986) and up to 45% in a flaring disk (Kenyon & Hartmann 1987), but there is no external energy source supplementing the star's in this model. TTSs with "classical" characteristics in the optical and UV spectral ranges (emission lines, veiling, UV excess, etc.) must therefore be surrounded by actively accreting disks rather than by passive disks.

The move from one generation of researchers to the next occurs on a timescale of one decade in modern research. Given the current pressure on young researchers to publish quickly and profusely, it is understandable that they have little time to find their own way in the maze of literature pertaining to previous research in their field, so as to gain a broad and unbiased view of their entire field of research before adding their own findings toward an even fuller picture. Above all other goals, I hope that this short review will offer some help in this formidable task.

Acknowledgements

I am deeply indebted to the scientific organizing committee for the dedication of this conference and for their kind and careful attention to avoiding the embarrassing moments that are often associated with such an honor. I commend the committee for organizing a clearly focused and very timely conference and gratefully acknowledge IAU support.

References

Adams, F. C., Lada, C. J., & Shu, F. H. 1987, ApJ, 312, 788
Adams, F. C. & Shu, F. H. 1986, ApJ, 308, 836
Adams, F. C., Shu, F. H., & Lada, C. J. 1988, ApJ, 326, 865
Adams, W. S., Joy, A. H., Humason, M. L.-S., & Brayton, A. M. 1935, The spectroscopic absolute magnitudes and parallaxes of 4179 stars ([Chicago, 1935]), 11
Alexander, R. D. & Armitage, P. J. 2006, ApJ, 639, L83
Ambartsumian, V. A. 1947, Stellar Evolution and Astrophysics (Acad. Sci. Armenian S.S.R., Erevan)
Ambartsumian, V. A. 1954, Memoires of the Societe Royale des Sciences de Liege, 1, 293
Ambartsumian, V. A. 1957, in IAU Symposium, Vol. 3, Non-stable stars, ed. G. H. Herbig, 177
Appenzeller, I. & Dearborn, D. S. P. 1984, ApJ, 278, 689
Appenzeller, I., Oestreicher, R., & Jankovics, I. 1984, A&A, 141, 108
Basri, G. & Bertout, C. 1989, ApJ, 341, 340
Beckwith, S. V. W. & Sargent, A. I. 1991, ApJ, 381, 250
Bertout, C. 1987, in IAU Symposium, Vol. 122, Circumstellar Matter, ed. I. Appenzeller & C. Jordan, 23–27
Bertout, C., Basri, G., & Bouvier, J. 1988, ApJ, 330, 350
Bertout, C. & Genova, F. 2006, A&A, 460, 499
Bertout, C., Siess, L., & Cabrit, S. 2007, A&A
Burbidge, G. R. & Burbidge, E. M. 1955, The Observatory, 75, 212
Burrows, C. J., Stapelfeldt, K. R., Watson, A. M., *et al.* 1996, ApJ, 473, 437
Cabrit, S., Edwards, S., Strom, S. E., & Strom, K. M. 1990, ApJ, 354, 687
Calvet, N., Basri, G., & Kuhi, L. V. 1984, ApJ, 277, 725
Close, L. M., Roddier, F., Northcott, M. J., Roddier, C., & Graves, J. E. 1997, ApJ, 478, 766
Cohen, M. 1983, ApJ, 270, L69
Cohen, M. 1987, in IAU Symposium, Vol. 122, Circumstellar Matter, ed. I. Appenzeller & C. Jordan, 39–50
Cohen, M. & Kuhi, L. V. 1979, ApJS, 41, 743
Cram, L. E. 1979, ApJ, 234, 949
D'Alessio, P., Canto, J., Calvet, N., & Lizano, S. 1998, ApJ, 500, 411
De Campli, W. M. 1981, ApJ, 244, 124
Dullemond, C. P., van Zadelhoff, G. J., & Natta, A. 2002, A&A, 389, 464
Edwards, S., Cabrit, S., Strom, S. E., *et al.* 1987, ApJ, 321, 473
Edwards, S. & Snell, R. L. 1984, ApJ, 281, 237
Feigelson, E. D. & Decampli, W. M. 1981, ApJ, 243, L89
Finkenzeller, U. & Basri, G. 1987, in IAU Symposium, Vol. 122, Circumstellar Matter, ed. I. Appenzeller & C. Jordan, 103–104
Ghosh, P. & Lamb, F. K. 1979, ApJ, 232, 259
Giampapa, M. S., Calvet, N., Imhoff, C. L., & Kuhi, L. V. 1981, ApJ, 251, 113
Greenstein, J. L. 1950, PASP, 62, 156
Gullbring, E., Hartmann, L., Briceno, C., & Calvet, N. 1998, ApJ, 492, 323
Haro, G. 1952, ApJ, 115, 572
Haro, G. 1957, in IAU Symposium, Vol. 3, Non-stable stars, ed. G. H. Herbig, 26
Hartigan, P., Edwards, S., & Ghandour, L. 1995, ApJ, 452, 736
Hartmann, L., Avrett, E., & Edwards, S. 1982, ApJ, 261, 279
Hartmann, L., D'Alessio, P., Calvet, N., & Muzerolle, J. 2006, ApJ, 648, 484
Hartmann, L. & Kenyon, S. J. 1985, ApJ, 299, 462
Hartmann, L. & Kenyon, S. J. 1987, ApJ, 312, 243
Herbig, G. 1974, QJRAS, 15, 526
Herbig, G. H. 1951, ApJ, 113, 697
Herbig, G. H. 1957, in IAU Symposium, Vol. 3, Non-stable stars, ed. G. H. Herbig, 3
Herbig, G. H. 1962, Advances in Astronomy and Astrophysics, 1, 47
Herbig, G. H. 1977a, ApJ, 217, 693

Herbig, G. H. 1977b, in IAU Symposium, Vol. 75, Star Formation, ed. T. de Jong & A. Maeder, 283
Joy, A. H. 1945, ApJ, 102, 168
Joy, A. H. 1949, ApJ, 110, 424
Joy, A. H. 1952, ApJ, 115, 25
Joy, A. H. 1954, ApJS, 1, 39
Joy, A. H. 1957, in IAU Symposium, Vol. 3, Non-stable stars, ed. G. H. Herbig, 31
Joy, A. H. 1958, PASP, 70, 505
Joy, A. H. & Humason, M. L. 1949, PASP, 61, 133
Kenyon, S. J. & Hartmann, L. 1987, ApJ, 323, 714
Koenigl, A. 1991, ApJ, 370, L39
Kuhi, L. V. 1964, ApJ, 140, 1409
Kuhi, L. V. 1966, PASP, 78, 430
Lada, C. J. & Wilking, B. A. 1984, ApJ, 287, 610
Lago, M. T. V. T. 1984, MNRAS, 210, 323
Larson, R. B. 1969, MNRAS, 145, 271
Larson, R. B. 1977, in IAU Symposium, Vol. 75, Star Formation, ed. T. de Jong & A. Maeder, 249–267
Lynden-Bell, D. & Pringle, J. E. 1974, MNRAS, 168, 603
Malbet, F., Lachaume, R., & Monin, J.-L. 2001, A&A, 379, 515
Marquis de Laplace, P.-S. 1798, Exposition du systeme du monde (Paris : V. Courcier, 1798; 551 p. ; in 4.; DCC.4.448)
Mendoza V., E. E. 1966, ApJ, 143, 1010
Mendoza V., E. E. 1968, ApJ, 151, 977
Montmerle, T., Koch-Miramond, L., Falgarone, E., & Grindlay, J. E. 1983, ApJ, 269, 182
Mundt, R. 1987, in IAU Symposium, Vol. 122, Circumstellar Matter, ed. I. Appenzeller & C. Jordan, 147–158
Mundt, R. & Fried, J. W. 1983, ApJ, 274, L83
Muzerolle, J., Hillenbrand, L., Calvet, N., Briceño, C., & Hartmann, L. 2003, ApJ, 592, 266
Natta, A., Testi, L., Calvet, N., *et al.* 2007, in Protostars and Planets V, ed. B. Reipurth, D. Jewitt, & K. Keil, 767–781
Natta, A., Testi, L., Randich, S., & Muzerolle, J. 2005, Memorie della Societa Astronomica Italiana, 76, 343
Pinte, C., Ménard, F., Duchêne, G., & Bastien, P. 2006, A&A, 459, 797
Rucinski, S. M. 1985, AJ, 90, 2321
Rydgren, A. E., Strom, S. E., & Strom, K. M. 1976, ApJS, 30, 307
Rydgren, A. E. & Zak, D. S. 1987, PASP, 99, 141
Shakura, N. I. & Sunyaev, R. A. 1973, A&A, 24, 337
Shu, F. H. & Adams, F. C. 1987, in IAU Symposium, Vol. 122, Circumstellar Matter, ed. I. Appenzeller & C. Jordan, 7–22
Shu, F. H., Adams, F. C., & Lizano, S. 1987, ARA&A, 25, 23
Snell, R. L., Loren, R. B., & Plambeck, R. L. 1980, ApJ, 239, L17
Stapelfeldt, K. R., Krist, J. E., Menard, F., *et al.* 1998, ApJ, 502, L65
Stapelfeldt, K. R., Ménard, F., Watson, A. M., *et al.* 2003, ApJ, 589, 410
Strom, K. M., Strom, S. E., Edwards, S., Cabrit, S., & Skrutskie, M. F. 1989, AJ, 97, 1451
Strom, S. E. 1977, in IAU Symposium, Vol. 75, Star Formation, ed. T. de Jong & A. Maeder, 179–197
Struve, O. & Rudkjøbing, M. 1949, ApJ, 109, 92
Ulrich, R. K. 1976, ApJ, 210, 377
Valenti, J. A., Basri, G., & Johns, C. M. 1993, AJ, 106, 2024
Walker, M. F. 1963, AJ, 68, 298
Walker, M. F. 1972, ApJ, 175, 89
Walter, F. M. 1987, in IAU Symposium, Vol. 122, Circumstellar Matter, ed. I. Appenzeller & C. Jordan, 107

Star-Disk Interaction in Young Stars
Proceedings IAU Symposium No. 243, 2007
J. Bouvier & I. Appenzeller, eds.

doi:10.1017/S1743921307009374

T Tauri stars: from mystery to magnetospheric accretion

Gibor Basri
Department of Astronomy, University of California, Berkeley, CA 94720, USA
email: Basri@berkeley.edu

Abstract. This is a selective historical overview of the progess in understanding T Tauri spectra. Originally they were understood to be very young, but the physical conditions (or even geometry) of the material on the star and in its surroundings were mysterious. The origin and meaning of the emission lines was largely unknown. Today we have a detailed consensus of what is happening near and on these newly forming stars. They are very magnetically active, and the stellar field is strong and extensive enough to control both the final accretion onto the star and the launching of outflows which solve the angular momentum problem during formation. Much of this consensus has emerged from spectral information, but much remains to be learned. I highlight some of the seminal breakthroughs that have led to the current picture. There are very complex and time-variable components to the entire physical system that constitutes a T Tauri star, and spectral information at various wavelengths and resolutions is crucial to making further progress.

Keywords. Accretion disks, line: formation, line: profiles, stars: magnetic fields, stars: formation, stars: activity

---

The T Tauri stars provide us with a detailed glimpse into the processes that led to the formation of our own Sun and planetary system. Rather than speculating on our origins, we are able to make detailed observations of the physical processes at work. While imaging is very informative in providing context and morphological information at large scales, spectroscopy provides our chance to delve right into the detailed processes and dynamics at the heart of the star formation process. The spectra of T Tauri stars were enigmatic right from the first observations, suggesting that these are no ordinary stars, but rather unique forces and conditions are in play. Although they share some similarities with other types of stars, a number of remarkable spectral characteristics made it clear that once we can untangle the meaning of their spectra, we will gain insight into crucial new processes in the environments and lives of stars.

Bertout (this volume) provides an excellent overview of the whole journey from a mysterious new class of objects to our current understanding of the essential nature of T Tauri stars (TTS). In this overview, I concentrate on the interpretation of spectra, both low and high resolution, and in the time domain. My career spans the time between confusion on the basic physical scenarios which might explain these spectra and excellent consensus that we are looking at young magnetically active stars surrounded by accretion disks, in which magnetic fields play a crucial role in simultaneous infall and outflow regions surrounding the star. I make no pretense at a comprehensive review, but give a somewhat personal tour through some of the highlights of this journey in the spectral domain. I tend to cite historical work rather than the latest results. This should be useful for workers new to the field in providing a context for current work. My basic message is the need for simultaneous diagnostics which probe different parts of the same system, and the need to include the time domain when studying these highly variable and complex systems.

1. Basic Characteristics of T Tauri Spectra

1.1. *Low Resolution: Chromospheres and Accretion*

At low resolution, the basic set of optical spectra that characterize the T Tauri stars are illustrated in Figure 1, taken from Bertout (1989). They illustrate the basic points about TTS spectra. Firstly, it was recognized from early on (Joy 1945) that many of the emission lines can also be seen in the solar chromosphere. Dumont *et al.* (1973) took the earlier suggestion of Herbig and pushed the chromosphere to high optical depth (also adjusting its temperature) and computed Hα line fluxes and profiles. While the fluxes could be reproduced (albeit with arbitrary large heating requirements), the profiles had strong (unobserved) central reversals. The chromospheric hypothesis was explored further by a number of authors, and studied with more modern methodology by Calvet, Basri & Kuhi (1984). These authors concluded that a number of TTS features could be reproduced by extremely strong chromospheres (without explaining how such strong chromospheres could be actually supported). These included line fluxes in CaII, MgII, and some iron lines, as well as some of the blue continuum veiling. There were concerns that the stronger absorption lines had bright emission cores (not observed) in such models, but NLTE effects had not been properly treated. It was clear, however, that certain spectral characteristics could never be reproduced simply by a very strong chromosphere (both strength and breadth as well as profile shape are issues). The Hα lines of the more active TTS, in particular, have to arise in a dynamic region off the star. Models which might work for Hα did great violence to other diagnostics (MgII, for example) compared with observed characterics. These conclusions were reinforced by Herbig (1985).

Nonetheless, it is actually true that TTS are extreme examples of stellar magnetic activity. Finkenzeller & Basri (1987) provide a very graphic example of this by comparing eclipses in software for TTS (spectra divided by inactive standards) with a solar eclipse spectral image. Ultraviolet and X-ray comparisons also place many TTS at the top of a progression of increasing activity that reaches back to the Sun (eg. Calvet *et al.* 1985). There is a set of TTS which appear to just have strong chromospheres; they do not exhibit diagnostics of disk accretion or winds but are otherwise similar to classic TTS. These are now known as "weak-lined" TTS or WTTS. Although many make the mistake of defining them as having Hα strengths less than a certain equivalent width (typically 10Å), one must be careful of the spectral class if using that method (see White & Basri 2003), who suggest using profile widths to distinguish them if possible). Only the lowest spectrum in Figure 1 is from a WTTS. The first direct proof that WTTS have very strong magnetic fields came from a novel Zeeman method employed by Basri, Marcy & Valenti (1992). These and subsequent measurements (discussed in more depth by Johns-Krull in this volume) provide one of the fundamental requirements for claiming magnetospheric accretion on TTS.

The other spectral characteristic somewhat unique to TTS which points at the presence of accretion is the so-called spectral "veiling". It was noticed very early on, even in low dispersion spectra (Figure 1), that the depth or strength of stellar absorption lines was smaller in some TTS and there are even cases in which no absorption lines are seen at all. Rydgren, Strom & Strom (1976) give a nice summary of the evidence that a bluish continuum light superposed on a normal stellar spectrum could be the cause. They identify the Balmer continuum as in emission in the more heavily veiled cases, and suggest that an optically thin Paschen continuum is responsible for most of the rest. They were still thinking in terms of an ionized wind as the source of the continuum.

Basri & Bertout (1989) correctly identified the veiling as due to the accretion shock from disk accretion, with a small contribution from the disk itself in the far red, increasing

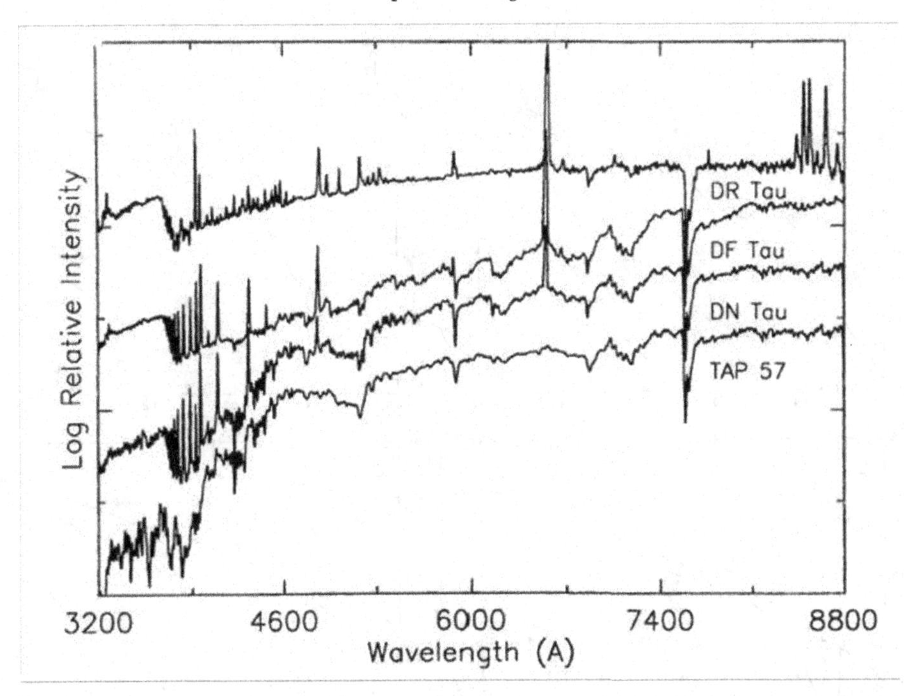

Figure 1. Low Resolution Spectra of TTS. Four sample spectra in the optical covering a range of accretion activity from none to high (moving upward). The underlying stars are all late K spectral type, and the bottom WTTS shows only the effects of a strong chromosphere. As accretion increases, Balmer line and continuum emission is increasingly apparent (with Hα becoming much stronger than the underlying continuum), while stellar absorption features are diluted by continuum veiling. In the upper spectrum, the Ca II infrared triplet has come strongly into emission, while the blue end is dominated by the Ca II H&K lines and a forest of iron emission lines. No absorption features remain, and the continuum is markedly bluer throughout.

to domination in the infrared (Figure 2). They suggest that in fact this veiling continuum is one of the cleanest diagnostics for accretion rates, which has proved to be the case. The methodology and first detailed measurement of the spectral characteristics of the veiling were pioneered by Hartigan *et al.* (1989) and Basri & Batalha (1990). An extensive set of accretion rates were produced by several authors, including Valenti, Basri & Johns (1993), Hartigan, Edwards & Ghandour (1995), and Gullbring *et al.* (1998). Unfortunately, the accuracy of these accretion rates is still problematic to within factors of a few, due to the complex region in which the diagnostic arises.

1.2. *High Resolution: Accretion and Outflow*

Even in the days before the discovery of bipolar molecular outflows or collimated optical jets, it was clear from line profiles that TTS have dynamic flows around them. Although the typical TTS does not exhibit a classical P Cygni profile at Hα (strong blue-shifted absorption below the continuum), many of them do show blue-shifted absorption features superposed on broad and strong emission. The broad strong emission by itself implies dynamic flows off the stellar surface, since its strength relative to the continuum is derived in part from the fact that it arises from a volume substantially larger than the photosphere. Its breadth is also thought to arise at least in part due to strong turbulent and organized flows (Doppler shifts rather than opacity broadening). The blue-shifted absorption has to arise in a cooler region (or more technically, a region with a lower source function) on the outside of the main emission region which is flowing toward the

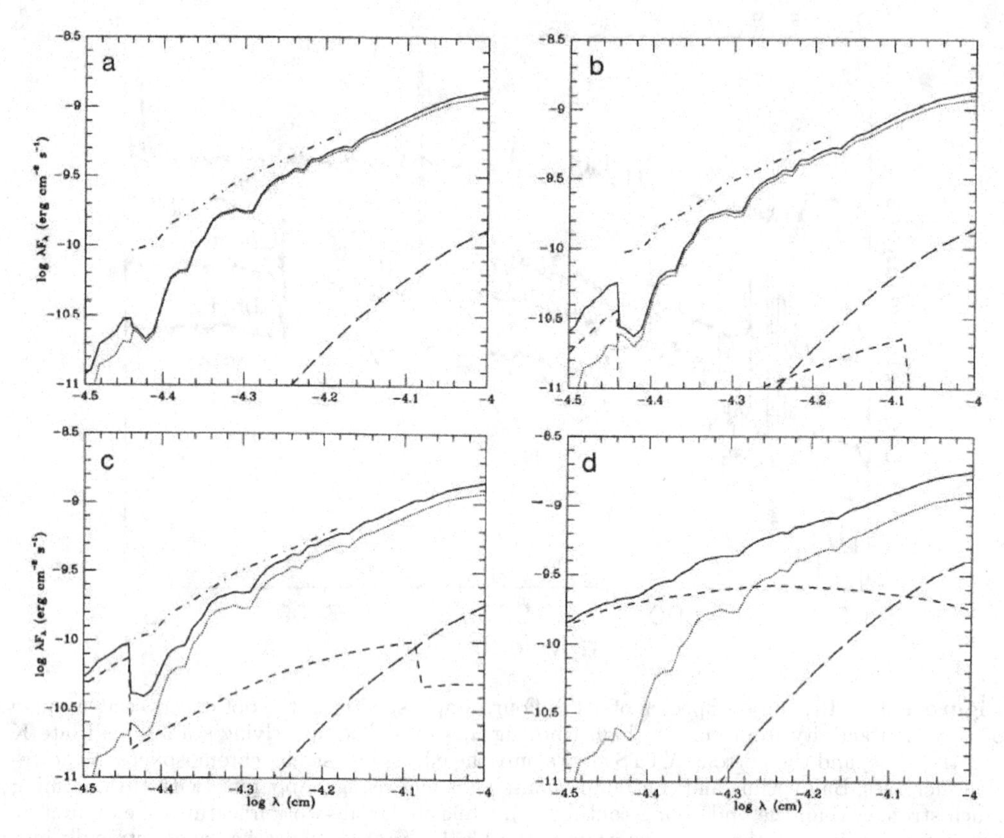

Figure 2. The veiling continuum. The spectral energy distributions arising from four combinations of a star with a disk providing increasing accretion rates from 3×10^{-9} to 2×10^{-7} (from Basri & Bertout 1989). The solid line is the composite spectrum made from the star (dots) and accretion shock emission (dashed). The dash-dot line indicates where the peaks of the Balmer lines would lie if they arose only on the star.

observer. Evidence for winds is often also seen in blue-shifted absorption components in the higher Balmer series, NaD lines, CaII H&K lines, and other strong lines. A sample of Hα profiles is shown in Figure 3.

Less obvious at first, but still noted enough to create a separate subtype of young stars (the YY Ori stars; Walker 1972) is the presence of red-shifted absorption components. These are seen less often in Hα (for technical reasons due to the physical properties of the line formation region and NLTE effects), but can be seen in many of the other lines noted in the paragraph above. The same arguments there apply here but with the sign of the flow reversed – these spectral features must indicate inflow. In the early days when spherical symmetry was often invoked, it was puzzling to find signs of outflow and inflow in the same star. This was an early sign that spherical symmetry does not really apply to TTS, and could have been taken as evidence for disk accretion. When the disk accretion hypothesis began to take hold in the middle 1980s, the first conception was that the disk extended right up to the star, and accretion occurred in a "boundary layer". One of the strong pieces of evidence that magnetospheric accretion is a better picture comes from the actual infall velocities, which are closer to free-fall rather than orbital velocities (Edwards *et al.* 1994). A sample of Hβ profiles from that paper is given in Figure 4.

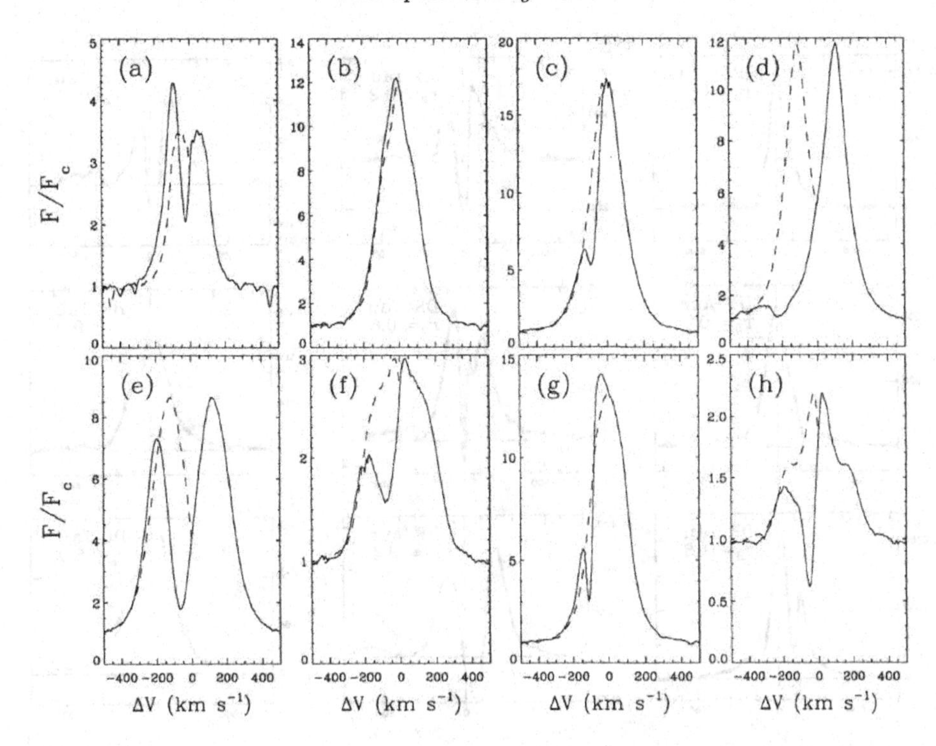

Figure 3. Hα lines in TTS. A representative set of Hα profiles taken from Johns & Basri (1995a). Note the fact that some of the profiles have peaks far above the continuum (difficult to obtain without a large emitting volume). Superposed on the data are the red sides of the profile reflected about the stellar rest velocity (dashed lines). Note the common presence of blue-shifted absorption (though with different manifestations), and the far wings are often symmetric. The stars shown are AA Tau, BP Tau, DF Tau, DR Tau, RW Aur, RY Aur, T Tau, and SU Aur.

Beyond the obvious imprint of dynamical flows, the interpretation of strong emission line profiles in TTS has been rather vexing. The Balmer lines have always been the iconic spectral diagnostics, yet they combine the most difficult aspects of radiative transfer into a single package. They are very optically thick, arise from a high energy lower level which causes NLTE effects to play an important role, form at a temperature which typically requires non-radiative heating to play an important role, arise not only in the stelllar chromosphere but also both accretion and outflow regions which are not spherically or even axisymmetric, and are quite variable (reflecting the complexities of their formation). The result of all this is that it is difficult to properly model them from both a physical and radiative transfer point of view, and even more difficult to obtain unique models which are clearly correct.

An example of this is provided by some early work by Kuhi (1964) and Ulrich (1976). Both authors are able to plausibly reproduce the (small) sample of line profiles they are working with. Kuhi assumes a spherical outflow scenario requiring ballistic ejection from the stellar surface and ionization due to dissipation of stellar magnetic energy (making an early guess that stellar magnetic fields play a role in angular momentum losses). Ulrich assumes an infall model in which interstellar magnetic fields control the flow down to an accretion shock on the star, producing primary and secondary cooling regions which produce profiles that depend fairly strongly on their inclination to the observer. This is an early version of a magnetic accretion model, but the field is not the star's. Remarkably, this model produces profiles which also look like they have blue-shifted

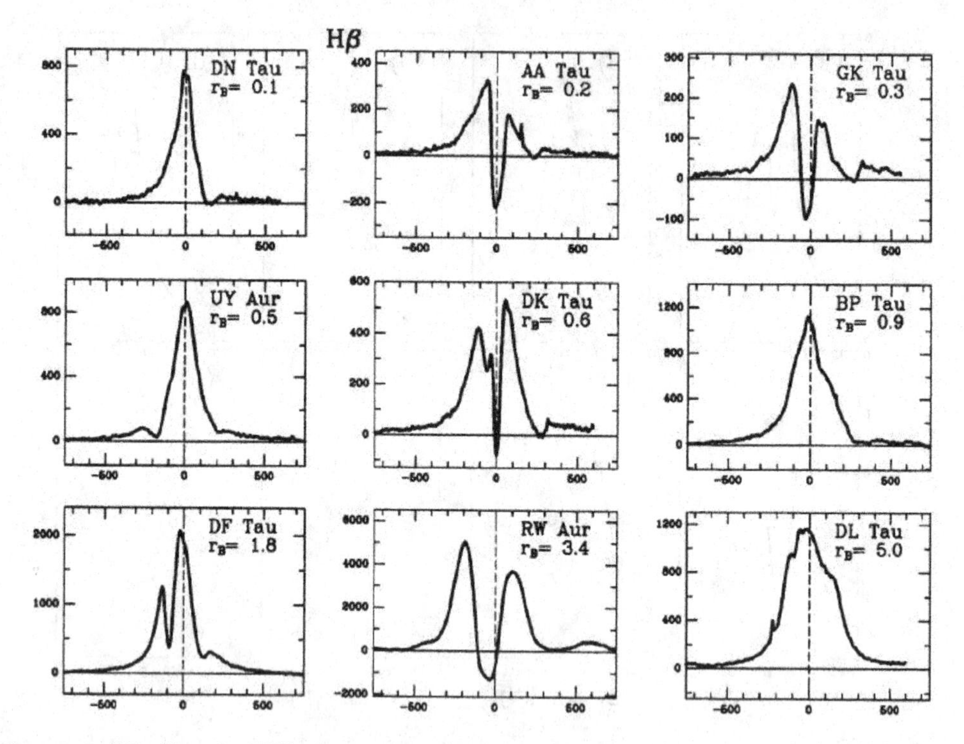

Figure 4. Hβ lines in TTS. A representative set of line profiles taken from Edwards *et al.* (1994). The vertical scale is arbitary and the horizontal scale is in km/s. The amount of blue continuum veiling (a measure of the strength of accretion) is given as a value of r_B for each spectrum.

absorption. Although neither paper passes the muster of modern spectral modeling, they are both plausible and clever efforts for their time, and still make the point that line profiles can be deceiving.

A series of other papers wrestled not only with producing profiles but also plausible physical mechanisms for producing the sort of mass loss implied by these models. That is not that easy, given that neither radiation pressure nor a thermal Parker wind is likely to work in TTS at the required level of mass loss. A suggestion of Alfven-wave driven winds remained on the table. The case for spherical outflow was finally put to rest by Hartmann *et al.* (1990), who showed that more careful calculations of such winds produce classical P Cygni profiles rather than the typical TTS profile. Attention then turned to the magnetospheric accretion hypothesis (see below), while it was realized that other diagnostics provided a better approach to the study of the outflow region(s). The NaD lines have always had obvious wind components (which go below the continuum, and are easier to interpret), while forbidden lines diagnose the wind (or jet) further from the star. Recently, the HeI 10830Å line (which was waiting for good detectors in that spectral region to come along) looks very promising (cf. Edwards, this volume).

2. Diagnostics of Magnetospheric Accretion

Once it became apparent that classical TTS are surrounded by accretion disks, thought turned to how material actually gets to the star, and how the angular momentum problem is solved (since it is clear that TTS rotate far below breakup velocities). The original idea was that the disk extended right to the star, where material lost the other half of its original potential energy in slowing from orbital velocity to the slow stellar surface in a boundary layer (Lynden-Bell & Pringle 1974; Bertout, Basri & Bouvier 1988). The difference between this and the current idea is that in magnetospheric accretion the material is slowed at several stellar radii and more or less freefalls to the stellar surface. This produces higher infall velocity signatures in line profiles, a stronger and hotter accretion shock, and moves the position of the shock from the equator to higher latitudes. Depending on the stellar field configuration, it can also produce asymmetrical shock regions (with respect to the stellar rotation axis), modulation of the accretion signature by stellar rotation, and much greater chances for short timescale variability. Indeed, Bertout, Basri & Bouvier (1988) noted periodic variability in the blue excess from DF Tau and ascribed it to magnetically controlled accretion. One can estimate the extend of the stellar surface which is covered by shocks from the veiling; it is at most a few percent (smaller than simple dipole models predict – likely indicating a more complicated field configuration).

2.1. *Time Variability*

The variability of TTS spectra (and photometry) was one of the early defining characteristics of these objects. Examples of papers examining spectral variability at low dispersion include Aiad *et al.* (1984) who found variability in as little as 2 hours, and even 10 minutes at high dispersion (Mundt & Giampapa 1982). Such short timescales are a direct indication that we are observing phenomena near the star, since it is not possible for substantial profile changes to occur over large-scale regions in such short coordinated intervals. That these changes occur in the outflow absorption features similarly argues that the part of the wind absorption region being sampled cannot arise over a very big part of the disk, and makes it more likely that these variations probe a part of outflow arising from near the magnetospheric boundary. Even more direct evidence of magnetospheric influence can come from the modulation of brightness or profile changes which have the stellar rotation period.

One of the clearest examples of a profile variability study showing that there are 3 components of the Hα profile which arise from the stellar chromosphere, magnetospheric accretion, and an outflow modulated by the stellar field is the study of SU Aur by Giampapa *et al.* (1993) and Johns & Basri (1995b). A collection of more than 100 high resolution spectra found clear periodicity consistent with the stellar rotation period in both infall (in Hβ) and outflow profile components (in Hα and Hβ) as shown in Figure 5. Even more interestingly, these components were out of phase: one side of the system shows stronger inflow and the other side stronger outflow. They are interpreted as caused by a tilted dipole field (with respect to the stellar rotation axis and the disk plane), and the connection with stellar rotation leaves little room for alternatives to the stellar magnetic field as the source of the variations.

These papers also introduced two ways of displaying the information inherent in time series of high resolution profiles. One is to simply display the normalized variance profile along with the average profile, which highlights the velocities at which the profile displays most of its variability. This typically shows that the most variability is associated with the outflow part of the profile, even when an absorption feature is not very strong. It also highlights the infall variability, showing red-shifted variance features in Hα even when there is no overt absorption at all. To study infall, however, it is better to concentrate on Hβ (Edwards *et al.* 1994). That paper made it clear that the infall velocities are high enough that magnetospheric accretion is the preferred interpretation, particularly since the strength of the absorption is correlated with other accretion diagnostics like veiling.

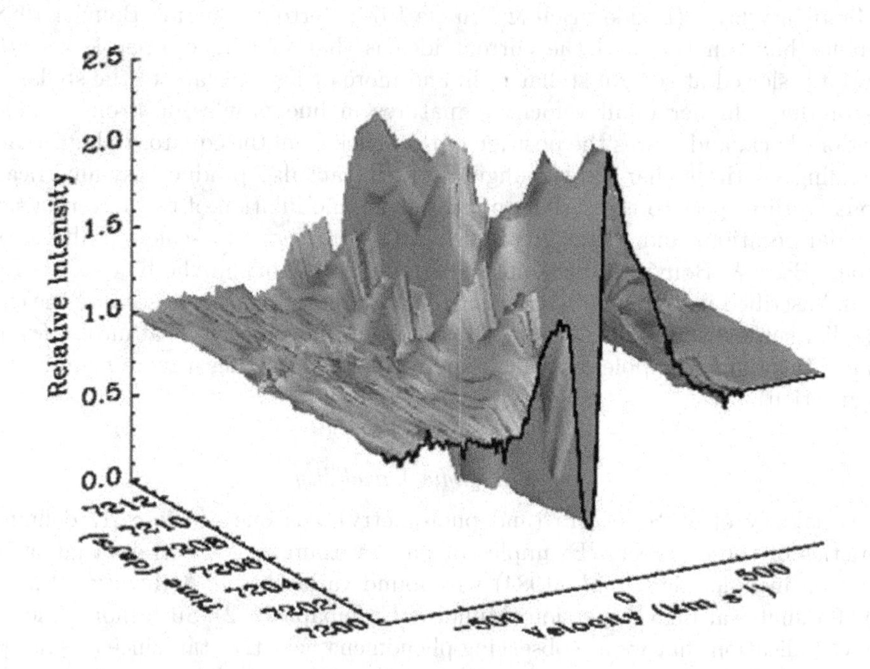

Figure 5. Profile variability in SU Aur. A time-resolved history of the Hα profile in SU Aur. Notice the periodic variation of the blue side of the line, showing an outflow component that is modulated by the stellar magnetic field on the stellar rotation period.

The other means of studying profile variability is using "correlograms" (Johns & Basri 1995a), which show the extent to which variations at any velocity in the profile are correlated with variations at any other velocity. If the variations of different parts of the profile are uncorrelated, one gets a narrow diagonal line whose width shows the extent in velocity of the locality (probably induced by local turbulence). When the whole profile moves together, the correlogram takes on a blocky or square appearance. Both sorts of behavior are observed on TTS, and the use of this sort of diagnostic is just coming into play from a modeling point of view (Kurosawa, Harries & Symington 2006).

2.2. *Profile Modeling*

The modeling of the actual profiles seen in TTS is one of the hardest tasks facing us today. As mentioned at the beginning, there are many reasons for this because of the inherent geometrical complexity of the emitting region, our ignorance of the physical conditions in it (velocities, densities, and heating rates), and details of radiative transfer. A series of papers by Muzerolle and collaborators (Muzerolle, Calvet & Hartmann 1998, 2001; and other papers for higher and lower mass objects) constitutes the primary standard for such work. They use "cartoon" models for the geometry and physical conditions, but a reasonable (although still axisymmetric) approach to the radiative transfer. These calculations show that the magnetospheric hypothesis produces plausible line profiles (Figure 6), and allows estimation of accretion rates from high resolution spectra. The extent to which actual TTS resemble these computed profiles has been discussed by the CfA group who note the areas of agreement, and by Alencar & Basri (2000) who also point out the substantial exceptions. Since these models are quite physically simplistic,

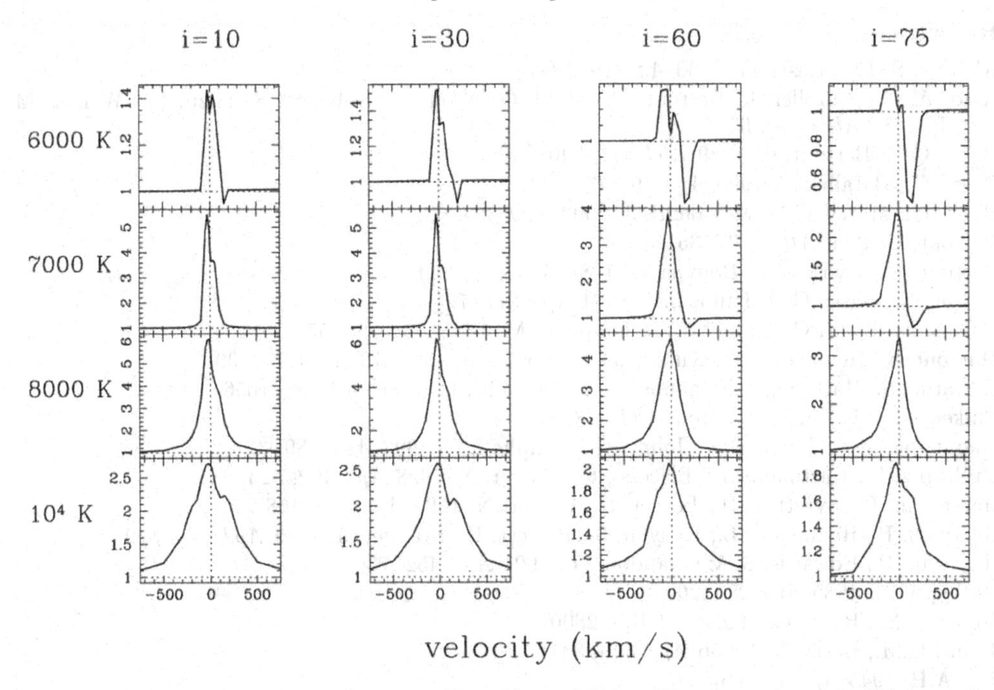

Figure 6. Computed magnetospheric profiles. A set of Hβ profiles from the work of Muzerolle, Calvet & Hartmann (2001). Here the effect of different amounts of heating viewed at different inclinations are shown. The infall absorption feature is more obvious at high inclination, the magnetosphere produces a broader profile if it is hotter, and there is a characteristic asymmetry in the emission profile which is clearer for cooler, more edge-on cases.

the extent of agreement is quite heartening, and there is general consensus that this is the correct paradigm to use.

More recently, there has been an explosion of more sophisticated efforts which use more physically realistic modeling. These compute more realistic magnetic configurations, use good radiative transfer, and work on variability as well. They are the subject of a number of papers in this volume, and I commend the reader to them. The Balmer lines are the most difficult case, although they also span the greatest physical region. It will be important for the future to examine other diagnostics in greater detail, being sure to conduct synoptic observations so that the range of variability and the appearance of a given system at different phases are included in our understanding of these highly complex and ever-changing systems. The reward is a detailed understanding of the processes which are at the basis of the near stellar environment during star formation, and which tell us how the star manages to collect material and dump angular momentum to grow into the final product.

Acknowledgements

I would like to acknowledge the enjoyable and productive collaborations and discussions I have had with Claude Bertout and his students and collaborators over the past 2 decades. These have made my sojourn in the area of star formation much more enjoyable and effective. I would also like to thank my students: Jeff Valenti, Chris Johns-Krull, Subu Mohanty, and David Ardila for choosing to work with me and then making me very proud to have worked with them. I also acknowledge the ongoing support of the National Science Foundation through a series of grants.

References

Alencar, S.H.P., Basri, G. 2000 *AJ* 119, 1881

Aiad, A., Appenzeller, I., Bertout, C., Stahl, O., Wolf, B., Isobe, S., Shimizu, M., Walker, M. F. 1984 *A&A* 130, 67

Basri, G. & Bertout, C. 1989 *ApJ* 341, 340

Basri, G., Batalha, C. 1990 *ApJ* 363, 654

Basri, G., Marcy, G.M. & Valenti, J. 1992 *ApJ* 390, 622

Bertout, C. 1989 *ARAA* 27, 351

Bertout, C., Basri, G. & Bouvier, J. 1988 *ApJ* 330, 350

Calvet, N., Basri, G. & Kuhi, L.V. 1984 *ApJ* 277, 725

Calvet, N., Basri, G., Imhoff, C., Giampapa, M. 1985 *ApJ* 293, 575

Dumont, S., Heidmann, N., Kuhi, L. V., Thomas, R. N. 1973 *A&A* 29, 199

Edwards, S., Hartigan, P., Ghandour, L., Andrulis, C. 1994 *AJ* 108, 1056

Finkenzeller, U., Basri, G. 1987 *ApJ* 318, 823

Giampapa, M.S., Basri, G.S., Johns, C.M., Imhoff, C. 1993 *ApJS* 89, 321

Gullbring, E., Hartmann, L., Briceño, C., Calvet, N. 1998 *ApJ* 492, 323

Hartmann, L., Avrett, E.H., Loeser, R., Calvet, N. 1990 *ApJ* 349, 168

Hartigan, P., Hartmann, L., Kenyon, S., Hewett, R., Stauffer, J. 1989 *ApJS* 70, 899

Hartigan, P., Edwards, S. & Ghandour, L. 1995 *ApJ* 452, 736

Herbig, G.H. 1985 *ApJ* 289, 269

Johns, C.M., Basri, G. 1995a *AJ* 109, 2800

Johns, C.M., Basri, G. 1995b *ApJ* 449, 341

Joy, A.H. 1945 *ApJ* 102, 168

Kurosawa, R., Harries, T.J. & Symington, N.H. 2006 *MNRAS* 370, 580

Kuhi, L.V. 1964 *ApJ* 102, 168

Lynden-Bell, D., Pringle, J.E. 1974 *MNRAS* 168, 603

Mundt, R., Giampapa, M.S. 1982 *ApJ* 256, 156

Muzerolle, J., Calvet, N. & Hartmann, L. 1998 *ApJ* 492, 743

Muzerolle, J., Calvet, N. & Hartmann, L. 2001 *ApJ* 550, 944

Rydgren, A.E., Strom, S.E. & Strom, K.M. 1976 *ApJS* 30, 307

Ulrich, R. 1976 *ApJ* 210, 377

Valenti, J., Basri, G. & Johns, C.M. 1993 *AJ* 106, 2024

Walker, M.F. 1972 *ApJ* 175, 89

White, R., Basri, G. 2003 *ApJ* 582, 1109

Star-Disk Interactions in Young Stars
Proceedings IAU Symposium No. 243, 2007
J. Bouvier & I. Appenzeller, eds.

doi:10.1017/S1743921307009386

What can X-rays tell us about accretion, mass loss and magnetic fields in young stars?

Thierry Montmerle

Laboratoire d'Astrophysique de Grenoble, BP 53X, 38400 St Martin d'Hères, France
email: montmerle@obs.ujf-grenoble.fr

Abstract. Until recently, X-rays from low-mass young stars ($10^5 - 10^6$ yr) were thought to be a universal proxy for magnetic activity, enhanced by 3-4 orders of magnitude with respect to the Sun, but otherwise similar in nature to all low-mass, late-type convective stars (including the Sun itself). However, there is now evidence that other X-ray emission mechanisms are at work in young stars. The most frequently invoked mechanism is accretion shocks along magnetic field lines ("magnetic accretion"). In the case of the more massive A- and B-type stars, and their progenitors the Herbig AeBe stars, other, possibly more exotic mechanisms can operate: star-disk magnetic reconnection, magnetically channeled shocked winds, etc. In any case, magnetic fields, both on small scale (surface activity) and on large scale (dipolar magnetospheres), play a distinctive role in the emission of X-rays by young stars, probably throughout the IMF.

Keywords. accretion disks, line: identification, plasmas, stars: activity, stars: coronae, stars: early-type, stars: magnetic fields, stars: pre–main-sequence, stars: winds, outflows, X-rays: stars

---

1. Introduction: early stellar evolution and magnetic fields

Low-mass stars, and in particular solar-like stars, form as a result of the collapse of an extended protostellar envelope, via the formation of an embedded accretion disk. Such disks live for a few million years, throughout the so-called "classical T Tauri" (CTTS) phase (e.g., Hillenbrand 2006). In the early phases, mass loss is observed to take place in the form of bipolar jets and outflows: this is sometimes called the "accretion-ejection" phenomenon. Although there are significant differences in the proposed theoretical models, it is widely accepted that accretion and ejection are closely coupled via *magnetic fields*, at least out to spatial scales of a few stellar radii, and perhaps even (in some models) throughout the accretion disk ("disk winds") (e.g., Pudritz *et al.* 2007).

In contrast, magnetic fields are not thought to play a major role in the early evolution of massive stars, but there is now evidence for their presence in a significant fraction of O and B stars, and for their influence on radiative winds. The situation of intermediate-mass stars (the so-called "Herbig AeBe" stars, with $M_\star \sim 2 - 3 M_\odot$), is less clear, but in a few cases there is indirect evidence for large-scale magnetic fields.

Direct measurements and modeling of magnetic fields have made spectacular advances in recent years, thanks mainly to observations of the Zeeman effect via spectropolarimetic measurements and Doppler imaging techniques (e.g., Donati *et al.* 2007; Strassmeier & Rice 2006; Yang *et al.* 2007). However, indirect access to magnetic fields has been provided for a long time by X-ray observations (spectra, timing), based on the idea that the only way to confine a hot plasma ($T \sim 10^6 - 10^7$ K, i.e., thermal X-ray energies ~ 0.1 to a few keV) is to trap it in closed magnetic loops (e.g., Feigelson & Montmerle 1999; Güdel 2004; Micela & Favata 2005). The numbers obtained by the various methods for the surface magnetic field intensities are quite similar: $B_\star \approx 0.1 - 1$ kG, i.e., comparable to values obtained in present-day solar active regions.

2. X-ray emission from young stars: processes and environments

Stellar X-rays are thermal in the $\sim$ keV range, and can be produced as the end result of internal structure processes. This is the case for low-mass stars, which have outer convective envelopes. Magnetic fields are currently thought to be generated via the dynamo effect at the bottom of the convective zone (the so-called "tachocline", e.g., Brun & Zahn 2006), and to buoy out to the surface across the convective zone. Reconnections between magnetic loops of opposite polarity, anchored in the photosphere, result in flaring and sudden heating of the photospheric gas to X-ray temperatures. The prototype of this behavior is the Sun itself, as observed in particular by the *Yohkoh* satellite (Peres *et al.* 2004). The X-ray signatures of this stellar "magnetic activity" are: *temporal variability* over a time scale of a few hours (flares: fast rise followed by slow decay corresponding to cooling); frequent 2-temperature spectra, with a *dominant hard component* ($T \sim$ a few 10 MK), and a less important soft component ($T \sim$ a few MK); *"coronal"* plasma densities ($n_e \sim 10^{10} - 10^{11}$ cm^{-3}) (e.g., Wolk *et al.* 2005). The level of X-ray luminosity (expressed in L_X/L_{bol}) is 3-4 orders of magnitude higher than in the Sun, in fast rotating stars (like RS CVn binaries) or in fully convective stars (like dMe stars and T Tauri stars).

Stellar X-rays can also be produced by *shocks* at the photospheric level. This is the case of the winds of massive stars, which have been known for a long time to be radiatively unstable. As a result, myriads of shock waves, with velocities $\sim$ several 100 km.s^{-1}, criss-cross the wind and emit X-rays (e.g., Kudritzki & Puls 2000). The X-ray signatures of radiative winds are: *no (or small but random) temporal variability* over time scales of hours; the overall spectrum is dominated by the weakly absorbed outer layers: consequently, the spectrum is basically *soft* (sub-keV, or equivalent $T \sim$ a few MK), and the plasma densities are comparatively *low* ($n_e \sim 10^9 - 10^{11}$ cm^{-3}) (see, e.g., Owocki & Cohen 1999). As explained below (§ 4), when a magnetic field exists and is sufficiently strong to confine the wind within a large-scale closed magnetosphere, the signature is modified, the most important change being the possibility of *rotational modulation* of the X-ray emission, at the rotational period of the star.

Additional X-ray emission mechanisms are possible, as a result of the interactions between the star and the surrounding medium. The now classical general picture of young low-mass stars is that of a central star postulated to be surrounded by a large-scale magnetospheric structure, linking the star and the disk at the corotation radius R_c (typically $R_c \sim 2 - 3R_\star \approx 0.05$ AU): beyond R_c, the disk is in Keplerian rotation. This opens two new possibilities for X-ray emission: (i) *magnetic interactions* between the star and its circumstellar disk, which will be the topic of § 5; (ii) *shock interactions* between the jet and the surrounding medium (protostellar envelope close to the star, and/or ambient ISM farther away) (see Güdel, this volume). Fig. 1 summarizes the various X-ray emission regions associated with low- and intermediate-mass stars.

3. Low-mass stars: magnetic activity

The X-ray emission from TTS (both "classical", still surrounded by disks, and "weak-line", without disks), is extremely well documented, after over 25 years of X-ray observations of star-forming regions. On the one hand, to date many star-forming regions have been observed by various X-ray satellites, with typical "short" exposures of 30-150 ksec, yielding hundreds of individual T Tauri detections down to the brown dwarf regime. On the other hand, the unique, very long *Chandra* exposure (850 ksec) of the Orion Nebula Cluster (the so-called "Chandra Orion Ultradeep Project", or COUP; PI E. Feigelson), has yielded in a single observation over 1500 detections of TTS (to which are added detections of massive stars and protostars, see below) over a $17' \times 17'$ FOV (Getman

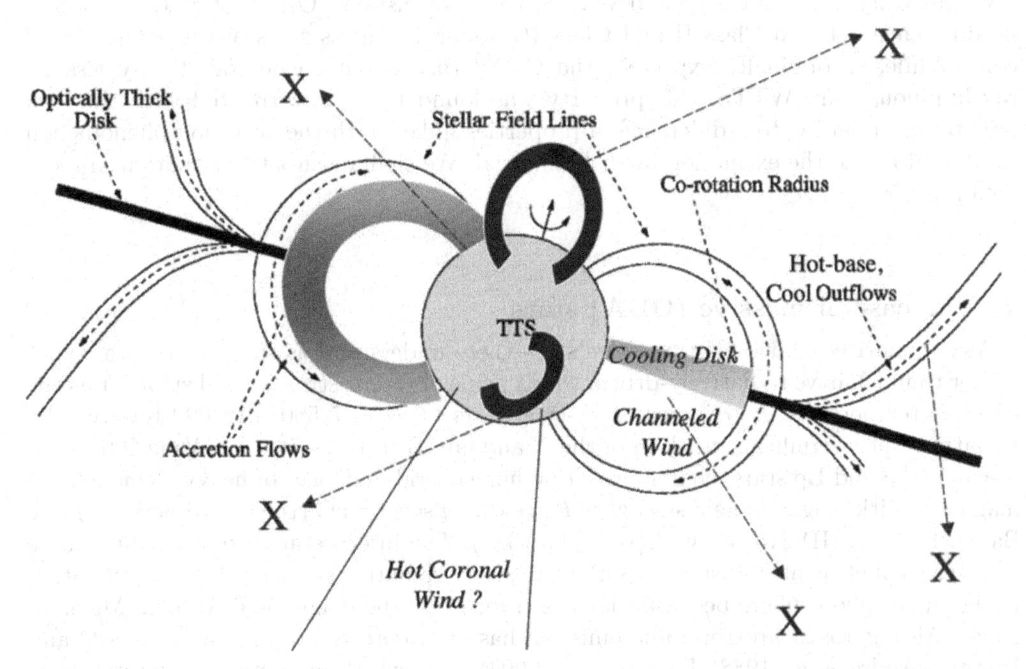

Figure 1. The various X-ray emission regions that may exist in the environment of low- and intermediate-mass stars: magnetic reconnections (on the star or between the star and the disk), and shocks (magnetically channeled winds, accretion on the stellar surface, outflow collisions with the envelope and/or with the interstellar medium, etc.). (Adapted from Stassun 2001)

et al. 2005). Another important observation is the "XMM-Newton extended survey of Taurus" (XEST; PI M. Güdel), a medium exposure ($\sim$ 30 ksec/field), medium angular resolution (a few arcsec), but large spatial extension (25 XMM fields, i.e., a total of $\sim$ 5 square degrees) of the Taurus clouds, yielding over 2400 identifications (mostly 2MASS), of which only $\sim$ 160 are characterized to date as young stars (Güdel *et al.* 2006).

For our purpose here, I briefly comment on three main global results from these satellite observations, restricting the discussion to TTS (for a similar discussion about protostars, see Montmerle 2007). There has been a long-standing debate in the literature about whether there was a statistically meaningful difference in X-rays (luminosities, spectra, etc.) between "classical" and "weak" TTS. The question arose originally simply because the additional extinction of the disk of CTTS (if seen more or less edge-on) would *a priori* make their spectra harder and their luminosities smaller, and this effect was not seen. On the other hand, we now know that there are in fact *three* classes of TTS: (i) *accreting* TTS, which have both evidence for the presence of a disk (mostly from near-IR excess), and for accretion (enhanced Hα, CaII, etc., emission: the usual CTTS criterion; see also § 5); (ii) *non-accreting*, more evolved TTS, which still have a disk but no evidence for accretion (hereafter "DTTS", for "disk" TTS; unofficial designation !); (iii) even more evolved *diskless* TTS, which are equivalent to "weak" TTS (evidence neither for a disk nor for accretion). As a result, comparing CTTS and WTTS implies in reality making a distinction between DTTS and "pure" WTTS, and is not the same as comparing the IR-classified "Class II" sources (= CTTS + DTTS) and "Class III" sources (= WTTS).

With the increased sensitivity (i.e., larger sample, lower mass limits, etc.) of the COUP observations, and using only CTTS defined by accretion-related criteria (Hα, CaII),

Preibisch *et al.* (2005) found a statistically significant difference of a factor ~ 2 in X-ray luminosity between CTTS and WTTS, while for instance Ozawa *et al.* (2005) found no difference between Class II and Class III young low-mass stars in the ρ Oph cloud core. While, as originally expected, the CTTS turned out to be statistically less X-ray luminous than WTTS, this property was found to be uncorrelated with the disk orientation, pointing to a difference in properties linked with the accretion phenomenon itself, and not to the extinction by disk material. We shall discuss this situation in more detail in § 5.

4. The case of massive (OBA) stars

As summarized above (§ 2), massive stars (here understood as stars of spectral types earlier than F), have radiatively-driven winds, which become stronger and stronger as the effective temperature T_{eff} increases. A-type stars ($T_{eff} \sim 7,500 - 10,000$ K) are fully radiative. Yet a significant fraction of the A and late B stars ($\sim 5\%$, see Wade 2005), the so-called Ap and Bp stars, characterized by huge overabundances of heavy elements, are magnetic, with magnetic field strengths $B_\star$ reaching several kG (the record being held by Babcock's star, HD 215441, with $B_\star \sim 11.5$ kG). The interpretation of overabundances is in terms of element diffusion driven outwards by radiation pressure, but accumulating in the upper photosphere because they are trapped by the magnetic field (e.g., Michaud 2004). Also, gyrosynchrotron radio emission has been detected from a number of O and B stars (André *et al.* 1988; Trigilio *et al.* 2004). The widely accepted interpretation is that of fossil fields brought from the ISM during the early formation and evolutionary stages, although recent work suggest the possibility of an internal, non-convective origin (see MacDonald & Mullan 2004, and references therein).

On the other hand, X-ray observations have shown that O and B stars obey a simple correlation: $L_X/L_{bol} \sim 10^{-7}$, over a wide range of luminosities (Berghöfer *et al.* 1997). This correlation has been nicely explained by Owocki & Cohen (1999), in terms of a rather subtle balance between the X-ray emissivity of shocks in the radiatively unstable winds, and extinction as a function of depth in the wind. Yet the COUP observations of a sample of 9 O7 to B3 stars in the vicinity of the Trapezium (the "strong wind" sample of Stelzer *et al.* 2005) have shown a significant excess of X-ray emission over the nominal $L_X/L_{bol} \sim 10^{-7}$ correlation in three stars, as well as three cases of X-ray rotational modulation in the whole sample. The most spectacular case is that of θ^1 Ori C, the most massive star (O7, $M_\star \sim 45\ M_\odot$) of the Trapezium cluster. The COUP observation confirmed (with much better statistics) the earlier *ROSAT* result of X-ray modulation (Gagné *et al.* 1997), with an amplitude of a factor ~ 2, at the rotation period of the star ($P_{rot} = 15.4d$). To explain both the high X-ray luminosity of θ^1 Ori C ($L_X/L_{bol} \sim 10^{-5}$), and its rotational modulation, Babel & Montmerle (1997) proposed the existence of a dipolar magnetic field, strong enough to confine the radiative wind, and to channel it along both hemispheres into an equatorial shock. At this shock X-rays are generated, and absorbed by an equatorial disk formed by the cool, post-shock material, resulting in a rotational modulation of the X-ray emission.

This *predicted* magnetic field was subsequently detected by Donati *et al.* (2002), with an observed value $B_\star \sim 1$ kG, recently confirmed by Wade *et al.* (2006). Fig. 2 is a sketch of the so-called "magnetically channeled wind shock" (MCWS) model as introduced by Babel & Montmerle (1997). More elaborate numerical calculations (e.g., Townsend & Owocki 2005) have now refined this model, and recent high-resolution X-ray spectra of θ^1 Ori C have fully confirmed its validity (Schulz *et al.* 2003).

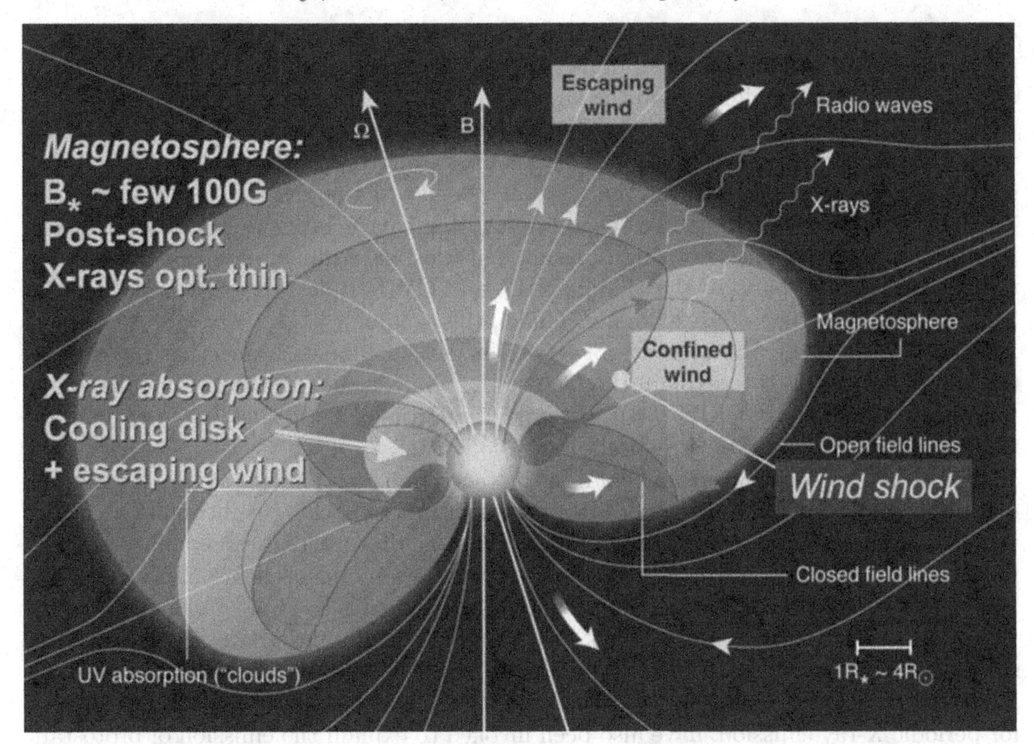

Figure 2. The "magnetically channeled wind shock" (MCWS) model, initially proposed by Babel & Montmerle (1997) to explain the X-ray rotational modulation of θ^1 Ori C. In this model, the radiatively-driven wind of a hot star (O, B, or A star) is confined by a strong magnetic field, and "self-collides" along the equator, heating the post-shock gas to X-ray energies. Then the shocked wind cools in a dense disk. This disk, in turn, absorbs the X-rays, and if the viewing geometry is favorable, causes a rotational modulation of the X-ray flux. (Adapted from Montmerle 2001)

5. Magnetospheric accretion and star-disk interactions

There is widespread support, both observational and theoretical, for "magnetic accretion" in CTTS, i.e., infall of disk material onto the central star, channeled along a dipolar magnetosphere assumed to connecting the star and the inner disk in the vicinity of the corotation radius. However, a number of recent optical observations suggest the existence of a more structured magnetosphere, with discrete "accretion funnels" linking the disk to the star. This departure from cylindrical symmetry may have various natural causes. For instance, "oblique rotator" 3D MHD stationary models (Romanova *et al.* 2004; also this volume) predict the existence of two main symmetrical accretion funnels and a distortion (warp) of the inner disk structure. The existence of such an inner disk warp in AA Tau, which is seen nearly edge-on, was inferred from multicolor photometry by Bouvier *et al.* (2007), modulating the X-ray luminosity (Grosso *et al.* 2007).

What about X-rays in general ? In the preceding sections, we have argued in favor of the widespread interpretation of X-rays from hundreds of young stars in terms of magnetic activity originating in a convective dynamo. However, in a so far handful of cases, the X-rays must be interpreted in terms of emission by shocks from magnetically channeled accretion –reminiscent, in a way, of the MCWS model proposed for some O stars, but with matter being channeled inwards instead of outwards.

The first case of "non-magnetic activity" X-ray emission from CTTS was reported by Kastner *et al.* (2002) for TW Hya. Thanks to its proximity ($d \simeq 60$ pc), TW Hya is one of the brightest CTTS, and high-resolution spectra could be obtained using the *Chandra* gratings. The analysis revealed an unusually high plasma density ($n_e \sim 10^{12}\mathrm{cm}^{-3}$), i.e., an order of magnitude higher than the largest coronal densities, and a very soft spectrum ($T_X = 2 \times 10^6$ K). Kastner *et al.* (2002) showed that, combined with the absence of time variability during the observations, the data could be interpreted in terms of an accretion shock near the stellar surface at the free-fall velocity of the gas ($v_{ff} \sim 200$ km s^{-1}). To date, two more CTTS (BP Tau and V4046 Sgr) have been found to show X-ray accretion spectral signatures of high plasma densities, determined independently of the temperature on the basis of the He-like NeIX and OVII triplet line ratios (Robrade & Schmitt 2006; Günther *et al.* 2006; also Güdel, this volume).

On the other hand, the presence of several accretion funnels connecting the disk to the star is now invoked to explain the factor $\sim 2-3$ deficiency in the X-ray emission of (accreting) CTTS and WTTS (§ 3), in terms of additional "self-shielding" provided by the discrete accretion flows (Preibisch *et al.* 2005).

Another star-disk situation may hold. In the "accretion-ejection" paradigm, corotation, due to magnetic locking, is assumed. However, if magnetic locking is incomplete , i.e., if there is a differential rotation between the star and the disk, there is a possibility of "self-reconnection" within the star-disk magnetic configuration and resulting X-ray emission. It is in this context that Montmerle *et al.* (2000) explained the "triple flare" observed by *ASCA* in the Class I protostar YLW15. However, in spite of repeated observations of this star, and even during the two-week-long exposure of COUP, no other case of periodic X-ray flaring on young stars was found. Star-disk interactions, without explicit evidence for periodic X-ray emission, have also been invoked to explain the emission of protostars in general (Preibisch 2004), and arguments in favor of large magnetic structures linking the star and the inner disk have been presented for some Orion TTS (Favata *et al.* 2005).

6. The mysterious Herbig stars

The so-called Herbig AeBe (HAeBe) stars are the young predecessors of intermediate-mass stars ($M_\star \sim 2-4\ M_\odot$), the future main-sequence A and B stars. They are entirely radiative and have relatively cool effective temperatures ($T_{eff} \sim 5,000-6,000$ K), and therefore are not expected to show any sign of magnetic activity, nor a significant wind. Yet, their detection rate in X-rays is quite high ($\sim 76\%$, Stelzer *et al.* 2006). The X-ray luminosities, known for many years to reach levels in excess of the brightest TTS (e.g., Zinnecker & Preibisch 1994), as well as their soft spectra, preclude, contrary to the A stars, the presence of unresolved low-mass companions as the general explanation of their X-ray emission.

In the presumed absence of magnetic fields, some form of accretion shock can be invoked. However, because the stars are more massive than TTS, their free-fall velocities are larger ($v_{ff} \approx 500-600$ km s^{-1}), implying harder X-rays than observed. The high-resolution (*XMM* RGS) spectrum of AB Aur, the first among HAeBes, with its density-sensitive OVII triplet, does not show evidence for accretion-shock plasma densities (Telleschi *et al.* 2007), although a Ne excess is present in the low-resolution spectra of several HAeBe stars (Swartz *et al.* 2005).

UV observations by *FUSE* may hold the answer. In recent observations of HD 163296, Deleuil *et al.* (2005) found that the line profile of several strongly ionized heavy elements gave evidence for a weak wind ($\dot{M} \sim 7 \times 10^{-9}\ M_\odot$ yr^{-1}, $\mathrm{v_w} \sim 300$ km s^{-1}), but with a much higher emissivity than a normal, freely expanding wind. These authors suggested that, instead of expanding freely, this wind is confined by a large-scale magnetosphere, with a predicted $B_\star \sim 700$ G, in a fashion very similar with the MCWS model of θ^1 Ori

C. Independently, a Zeeman search for magnetic fields in three other HAeBe stars has resulted in one 5σ detection of the same order ($B_\star = 450 \pm 93$ G) (see Hubrig *et al.* 2007).

Thus, the MCWS model appears promising also to explain the X-ray emission of some HAeBe stars, and implies the existence of magnetic fields in a significant fraction of them, which may be the predecessors to the Ap-Bp stars. In that sense, in X-rays HAeBe stars offer more similarities with OBA stars than with CTTS stars, despite the fact that, like CTTS, they are surrounded by circumstellar disks.

7. Conclusions

• Magnetic activity-related X-ray emission, i.e., magnetic reconnection, is by far the most widespread mechanism in convective, low-mass young stars. In a few cases there is indirect evidence for a star-disk reconnection in lieu of the common reconnections between magnetic loops on the star. Thus, as a rule X-rays can be safely taken a proxy for stellar magnetic fields, provided some signatures are checked: hard spectrum, flarelike light curve, coronal densities. Note that the large-scale ($R_c \approx 0.05$ AU) dipolar corotating magnetosphere (possibly oblique to the rotation axis) assumed to mediate accretion and ejection *cannot be detected in X-rays* if it is really in a steady state.

• However, there are a few exceptions (three to date) to the general magnetic activity picture. In these few cases, X-rays come from accretion shocks. The signatures are clearly different from the preceding case: soft spectrum, absence of flares, densities much larger than coronal. High densities are best proven by He-like triplets, resolved by grating spectroscopy on *XMM* or on *Chandra*.

• Conversely, in the more massive stars, the dominant X-ray emission mechanism is shocks pervading their radiatively unstable winds. X-rays are then precious to probe the inner structure of the wind (density and temperature as a function of radius). However, a large fraction of the OB stars (up to $\sim 50\%$ in the ONC, Stelzer *et al.* 2005) show indications of magnetic fields when they are very young. If the magnetic fields are strong enough, then they can confine the wind inside a closed magnetosphere, and the resulting "magnetically channeled" flows from both hemispheres collide and emit shock X-rays, with an X-ray luminosity exceeding that of the standard wind instability mechanism.

• The so-called "Herbig AeBe stars" are commonly referred to as T Tauri stars scaled up in mass, because of the presence of circumstellar disks and/or envelopes. However, from the point of view of X-ray emission, they seem to be more related to massive stars. In particular, at least in some cases of X-ray luminous HAeBe stars, the MCWS model may explain the X/UV emission. For less X-ray luminous HAeBe stars, the presence of a low-mass companion remains the most likely explanation for the X-ray emission.

All in all, we conclude that X-rays from young stars, which are thermal in the $\sim 0.1-10$ keV range covered by *Chandra* and *XMM*, always result from some combination of shocks and magnetic fields. On the one hand, magnetic activity dominates in the vast majority of low-mass stars, while on the other hand wind shocks dominate in a majority of high-mass stars. Although the number of "hybrid" cases (i.e., magnetic fields + shocks) is small, they give important insights into the physics of accretion (CTTS), and into the origin and early evolution of magnetic fields in massive stars (Ap-Bp stars).

References

André, P., *et al.* 1988, *ApJ* 335, 940
Babel, J., & Montmerle, T. 1997, *ApJ* (Letters) 485, L29
Berghöfer, T.W., *et al.* 1997, *A&A* 322, 167

Bouvier, J., *et al.* 2007, *A&A* 463, 1017
Brun, A. S., Zahn, J.-P. 2006, *A&A* 457, 665
Deleuil, M., *et al.* 2005, *A&A* 429, 247
Donati, J.-F., *et al.* 2002, *MNRAS* 333, 55
Donati, J.-F., *et al.* 2007, in *14th Cambridge Workshop on Cool Stars, Stellar Systems, and the Sun*, Ed. G. van Belle, ASP Conf. Ser., in press (astro-ph/0702159)
Favata, F., *et al.* 2005, *ApJ*(Suppl.) 160, 469
Feigelson, E. D., & Montmerle, T. 1999, *ARA&A* 37, 363
Gagné, M., *et al.* 1997, *ApJ* (Letters) 478, L87
Getman, K. V., *et al.* 2005, *ApJS* 160, 319
Grosso, N., *et al.* 2007, *A&A* in press
Güdel, M. 2004, *Astr.Ap.Rev.* 12, 71
Güdel, M., *et al.* 2006, *A&A* in press (astro-ph/0609160)
Günther, H. M., *et al.* 2006, *A&A* 459, L29
Hillenbrand, L. A. 2006, in *A Decade of Discovery: Planets Around Other Stars*, Ed. M. Livio, STScI Symposium Ser., 19, in press
Hubrig, S., *et al.* 2007, *A&A* 463, 1039
Kastner, J. H., *et al.* 2002, *ApJ* 567, 434
Kudritzki, R.-P., & Puls, J. 2000, *ARA&A* 38, 613
MacDonald, J., Mullan, D. J. 2004, *MNRAS* 348, 702
Micela, G., Favata, F. 2005, *Sp. Sci. Rev.* 108, 577
Michaud, G. 2004, in *The A-Star Puzzle*, Eds. J. Zverko *et al.* IAUS 224 (Cambridge, UK: Cambridge University Press), p.173
Montmerle, T. 2001, *Science* 293, 2409
Montmerle, T. 2007, *Mem. Soc. Astr. It.* in press
Montmerle, T., *et al.* 2000, *ApJ* 532, 1097
Owocki, S.P., & Cohen, D.H. 1999, *ApJ* 520, 833
Ozawa, H., Grosso, N., Montmerle, T. 2005, *A&A* 438, 963
Peres, G., Orlando, S., Reale, F. 2004, *ApJ* 612, 472
Preibisch, T. 2004, *A&A* 428, 569
Preibisch, T., *et al.* 2005, *ApJS* 160, 401
Pudritz, R. E., *et al.* 2007, in *Protostars and Planets V*, Eds. B. Reipurth, D. Jewitt, & K. Keil (Tucson: University of Arizona Press), p.277
Robrade, J., Schmitt, J. H. M. M. 2006, *A&A* 449, 737
Romanova, M. M., *et al.* 2004, *ApJ* 610, 920
Schulz, N. S., *et al.* 2003, *ApJ* 595, 365
Stassun, K. G. 2001, in *From Darkness to Light*, Eds. T. Montmerle & P. André, ASP Conf. Ser., 243, p.599
Stelzer, B., Flaccomio, E., Montmerle, T., *et al.* 2005, *ApJS* 160, 557
Stelzer, B., *et al.* 2006, *A&A* 457, 223
Strassmeier, K. G., Rice, J. B. 2006, *A&A* 460, 751
Swartz, D. A., *et al.* 2005, *ApJ* 628, 811
Telleschi, A., *et al.* 2007, *A&A* in press (astro-ph/0610456)
Townsend, R. H. D., Owocki, S. P. 2005, *MNRAS* 357, 251
Trigilio, C., *et al.* 2004, *A&A* 418, 593
Wade, G.A. 2005, in *Element Statification in Stars*, Eds. G. Alecian, O. Richard, & S. Vauclair, EAS Publ. Series, 17, p.227
Wade, G. A., *et al.* 2006, *A&A* 451, 195
Wolk, S. J.; *et al.* 2005, *ApJ* 160, 423
Yang, Hao, Johns-Krull, C. M., Valenti, J. A. 2007, *ApJ* 133, 73
Zinnecker, H., Preibisch, T. 1994, *A&A* 292, 152

Star-Disk Interaction in Young Stars
Proceedings IAU Symposium No. 243, 2007
J. Bouvier & I. Appenzeller, eds.

doi:10.1017/S1743921307009398

Measurements of magnetic fields on T Tauri stars

Christopher M. Johns–Krull
Physics and Astronomy Department, Rice University, Houston, TX 77005, USA
email: cmj@rice.edu

Abstract. Stellar magnetic fields including a strong dipole component are believed to play a critical role in the early evolution of newly formed stars and their circumstellar accretion disks. It is currently believed that the stellar magnetic field truncates the accretion disk several stellar radii above the star. This action forces accreting material to flow along the field lines and accrete onto the star preferentially at high stellar latitudes. It is also thought that the stellar rotation rate becomes locked to the Keplerian velocity near the radius where the disk is truncated. This paper reviews recent efforts to measure the magnetic field properties of low mass pre-main sequence stars, focussing on how the observations compare with the theoretical expectations. A picture is emerging indicating that quite strong fields do indeed cover the majority of the surface on these stars; however, the dipole component of the field appears to be alarmingly small. On the other hand, at least one accretion model which takes into account the non-dipole nature of the magnetic field provides predictions relating various stellar and accretion parameters which are present in the current data.

Keywords. Accretion, accretion disks, stars: formation, stars: magnetic fields, stars: pre–main-sequence

1. Introduction

It is now generally accepted that accretion of circumstellar disk material onto the surface of a classical T Tauri star (CTTS) is controlled by a strong stellar magnetic field (e.g. see review by Bouvier *et al.* 2007). The first detailed magnetospheric accretion model for CTTSs was developed by Uchida & Shibata (1984). This model includes both accretion of disk material onto the star as well as the formation of a shock driven bipolar outflow; however, rotation is ignored. Camenzind (1990) first considered the rotational equilibrium of a CTTS with a kilogauss strength dipolar magnetic field accreting matter from a circumstellar disk. Electric currents in the stellar and disk magnetospheres are found to offset the angular momentum accreted with the disk material, producing an equilibrium rotation rate with the disk truncated close to the corotation radius. A wind is then driven off the disk outside the corotation radius. Variations of this magnetospheric accretion model have been studied analytically or semi-analytically, sometimes without an attendant outflow (Königl 1991; Collier Cameron & Campbell 1993) and sometimes with (Shu *et al.* 1994). In all cases, the field truncates the inner disk at or close to corotation and an equilibrium rotation rate (P_{rot}) is established which depends on the (assumed) dipolar field strength, the stellar mass (M_*), radius (R_*), and the mass accretion rate ($\dot{M}$) in the disk. The relationships published in these papers can be used to predict the stellar field strength on CTTSs for which measurements for the other parameters exist. The predicted field variations from star to star correlate extremely well from study to study, even though the magnitude of the predicted fields can vary substantially from one study to another due to different assumptions regarding the efficiency of the field and disk coupling, ionization state in the disk, and so on (Johns–Krull *et al.* 1999b; Johns–Krull 2007).

Observationally, support for magnetospheric accretion in CTTSs is significant and is reviewed elsewhere in this volume. Despite these successes, open issues remain. Most current theoretical models assume the stellar field is a magnetic dipole with the magnetic axis aligned with the rotation axis. As discussed below, spectropolarimetric measurements are often at odds with this assumption. On the other hand, it is expected that even for complex magnetic geometries, the dipole component of the field should dominate at distance from the star where the interaction with the disk is taking place, so the dipole assumption may not seriously contradict current theory. In the case of the Sun, the dipole component appears to become dominant at $2.5R_{\odot}$ or closer (e.g. Luhmann *et al.* 1998). For expected disk truncation radii of 3 – 10 R_* in CTTSs, this suggests the dipole component will govern the stellar interaction with the disk. Additionally, Gregory *et al.* (2006) show that accretion can occur from a truncated disk even when the stellar field geometry is quite complex; however, no study has considered the torque balance between a star and its disk in the case of a complex stellar field geometry. Another concern is the work of Stassun *et al.* (1999) who find no correlation between rotation period and the presence of an infrared (IR) excess indicative of a circumstellar disk in a sample of 254 stars in Orion. However, IR excess alone is not a good measure of the accretion rate. Muzerolle, Calvet & Hartmann (2001) note that current theory predicts a correlation between rotation period and mass accretion rate which they do not observe. Muzerolle *et al.* (2001) suggest that variations in the stellar magnetic field strength from star to star may account for the lack of correlation. Indeed, there are several stellar and accretion parameters that enter into the equilibrium relationship, and the stellar magnetic field remains the quantity measured for the fewest number of CTTSs. Here, we review magnetic field measurements on TTSs, paying particular attention to how the magnetic field data agrees or not with the predictions of magnetospheric accretion models for young stars. We refer the reader to the contribution by Alecian in this volume for a review of magnetic field measurements on higher mass Herbig Ae/Be stars.

2. Techniques

Virtually all measurements of stellar magnetic fields make use of the Zeeman effect. Typically, one of two general aspects of the Zeeman effect is utilized: (1) Zeeman broadening of magnetically sensitive lines observed in intensity spectra, or (2) circular polarization of magnetically sensitive lines. When an atom is in a magnetic field, different projections of the total orbital angular momentum are no longer degenerate, shifting the energy levels taking part in the transition. In the simple Zeeman effect, a spectral line splits into 3 components: 2 σ components split to either side of the nominal line center and 1 unshifted π component. The wavelength shift of a given σ component is

$$\Delta\lambda = \frac{e}{4\pi m_e c^2}\lambda^2 gB, \tag{2.1}$$

where g is the Landé g-factor of the specific transition, B is the strength of the magnetic field, and λ is the wavelength of the transition. Evaluating the constants, the wavelength shift is

$$\Delta\lambda = 4.67 \times 10^{-7}\ \lambda^2 gB \text{ mÅ}, \tag{2.2}$$

where λ is in Å and B is in kG. One thing to note from this equation is the λ^2 dependence of the Zeeman effect. Compared with the λ^1 dependence of Doppler line broadening mechanisms such as rotation and turbulence, this means that observations in the IR are generally more sensitive to the presence of magnetic fields than optical observations.

The simplest model of the spectrum from a magnetic star assumes that the observed line profile can be expressed as $F(\lambda) = F_B(\lambda) * f + F_Q(\lambda) * (1 - f)$; where F_B is the spectrum formed in magnetic regions, F_Q is the spectrum formed in non-magnetic (quiet) regions, and f is the flux weighted surface filling factor of magnetic regions. The magnetic spectrum, F_B, differs from the spectrum in the quiet region not only due to Zeeman broadening of the line, but also because magnetic fields affect atmospheric structure, causing changes in both line strength and continuum intensity at the surface. Most studies *assume* that the magnetic atmosphere is in fact the same as the quiet atmosphere because there is no theory to predict the structure of the magnetic atmosphere. If the stellar magnetic field is very strong, the splitting of the σ components is a substantial fraction of the line width, and it is easy to see the σ components sticking out on either side of a magnetically sensitive line. In this case, it is relatively straightforward to measure the magnetic field strength, B. Differences in the atmospheres of the magnetic and quiet regions primarily affect the value of f. If the splitting is a small fraction of the intrinsic line width, then the resulting observed profile is only subtly different from the profile produced by a star with no magnetic field, and more complicated modelling is required to be sure all possible non-magnetic sources (e.g. rotation, pressure broadening, turbulence) have been properly constrained.

In cases where the Zeeman broadening is too subtle to detect directly, it is still possible to diagnose the presence of magnetic fields through their effect on the equivalent width of magnetically sensitive lines. For strong lines, the Zeeman effect moves the σ components out of the partially saturated core into the line wings where they can effectively add opacity to the line and increase the equivalent width. The exact amount of equivalent width increase is a complicated function of the line strength and the true Zeeman splitting pattern (Basri *et al.* 1992). This method is primarily sensitive to the product of B multiplied by the filling factor f. Since this method relies on relatively small changes in the line equivalent width, it is very important to be sure other atmospheric parameters which affect equivalent width (particularly temperature) are accurately measured.

Measuring circular polarization in magnetically sensitive lines is perhaps the most direct means of detecting magnetic fields on stellar surfaces, but it is also subject to several limitations. When viewed along the axis of a magnetic field, the Zeeman σ components are circularly polarized, but with opposite helicity; and the π component(s) is(are) absent. The helicity of the σ components reverses as the polarity of the field reverses. Thus, on a star like the Sun that typically displays equal amounts of + and − polarity fields on its surface, the net polarization is very small. If one magnetic polarity does dominate the visible surface of the star, net circular polarization is present in Zeeman sensitive lines, resulting in a wavelength shift between the line observed through right- and left-circular polarizers. The magnitude of the shift represents the surface averaged line of sight component of the magnetic field (which on the Sun is typically less than 4 G even though individual magnetic elements on the solar surface range from $\sim$ 1.5 kG in plage to $\sim$ 3.0 kG in spots). Several polarimetric studies of cool stars have generally failed to detect circular polarization, placing limits on the net magnetic field strength present of 10 − 100 G (e.g. Vogt 1980; Brown & Landstreet 1981; Borra, Edwards & Mayor 1984). The interpretation resulting from these studies is that the late-type stars studied (primarily main sequence and RS CVn types) likely have complicated surface magnetic field topologies which display approximately equal amounts of opposite polarity field which results in no detectable net magnetic field. On the other hand, stars with strong dipole components, such as the magnetic Ap stars, show quite strong circular polarization in their photospheric absorption lines (e.g. Mathys 2004 and references therein). If CTTSs do have strong dipole components, circular polarization should be detectable in photospheric absorption lines.

3. Zeeman broadening measurements

3.1. *The equivalent width method*

TTSs typically have $v \sin i$ values of 10 km s^{-1}, which means that observations in the optical typically cannot detect the actual Zeeman broadening of magnetically sensitive lines because the rotational broadening is too strong. Nevertheless, optical observations can be used with the equivalent width technique to detect stellar fields. Basri *et al.* (1992) were the first to detect a magnetic field on the surface of a TTS. They studied two TTSs showing no evidence for accretion, the so-called weak line or naked TTSs (WTTSs or NTTSs). Basri *et al.* find a value of $Bf = 1.0$ kG on the NTTS Tap 35. This detection is illustrated in Figure 1. Here, the abscissa, $S(1000)$ is the *magnetic sensitivity* of each photospheric line. It is the line equivalent calculated from a photospheric model with a magnetic field of 1000 G divided by the line equivalent width calculated for the same atmosphere but with no magnetic field. The ordinate is the ratio of the observed equivalent widths of the lines in Tap 35 divided by the observed equivalent width in the lines of the inactive main sequence star 61 UMa, which has a similar spectral type to Tap 35 (see Basri *et al.* 1992 for additional details). In addition to Tap 35, Basri *et al.* also observed the NTTS Tap 10, finding only an upper limit of $Bf < 0.7$ kG. Guenther *et al.* (1999) apply the same technique to spectra of 5 TTSs, with apparent significant field detections on two stars; however, these authors analyze their data using models different by several hundred K from the expected effective temperature of their target stars, which can introduce artifacts in equivalent width analyses. In principle, the equivalent width technique can separately measure B and f; however, in practice this is quite difficult and the technique primarily gives a measure of the product Bf (see Basri *et al.* 1992; Guenther *et al.* 1999). While measurements of actual Zeeman broadening as described below can give more detailed information about the magnetic fields on TTSs, that method is biased towards stars with intrinsically narrow line profiles, and hence is generally less usefull when studying rapidly rotating stars. Line blending makes equivalent width measurements more difficult in rapidly rotating stars as well; however, the equivalent width method used on IR lines (where the density of lines is lower in many regions) is likely to be the only way to get robust mean field measurements on high $v\sin i$ TTSs.

3.2. *Zeeman broadening of infrared* Ti I *lines*

As described above, observations in the IR help solve the difficulty in detecting direct Zeeman broadening. There are two principle IR diagnostics that have been utilized for magnetic field measurements in late-type stars. The first is a series of Zeeman sensitive Fe I lines at 1.56 μm, including one with a Landé-g value of 3.00 at 1.5649 μm (e.g. Valenti *et al.* 1995, Rüedi *et al.* 1995). These Fe I lines have a relatively high excitation potential, and as a result are best used to study G and early K type stars. To date, no TTS magnetic field measurements have been made using these lines; however, Guenther & Emerson (1996) demonstrate the suitability of these lines for TTS magnetic field work and present observations of these lines in Tap 35 which give an upper limit of $Bf < 2000$ G, consistent with the result of Basri *et al.* (1992) described above. For later spectral types such as the majority of TTSs with field measurements, lower excitation potential lines are best. There are several Ti I lines near 2.2 μm which are suitable for magnetic field

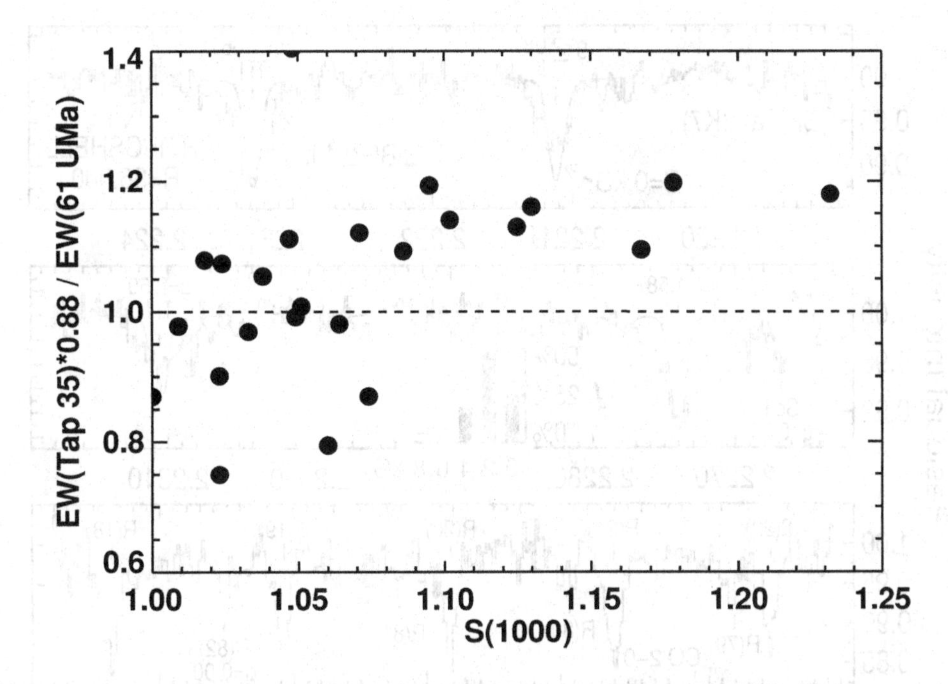

Figure 1. The first field detection on a TTS. Data is taken from Basri *et al.* (1992). The abscissa gives the magnetic sensitivity of each observed line, while the ordinate gives the ratio of the line equivalent width observed in Tap 35 divided by that observed in the inactive reference star 61 UMa. The positive correlation here demonstrates the presence of a magnetic field.

work on late K and M stars. Saar & Linsky (1985) first made use of these lines to study the magnetic field on the dMe flare star AD Leo. By far, observations of these K band Ti I lines have yielded the most information on the magnetic fields of TTSs, starting with the measurement of the magnetic field on BP Tau given by Johns–Krull *et al.* (1999b). These authors found that the broadening of the Ti I lines in BP Tau could not be well fit assuming a single magnetic field component with some value of B and f. Instead, they find that a distribution of magnetic field strengths is required. For example, one fit includes atmospheric components with field strengths of 0, 2, 4, and 6 kG magnetic fields, with individually determined filling factors which sum to 1.0. This distribution of magnetic field strengths can be characterized by the mean field $\bar{B} = \Sigma B_i f_i = 2.6 \pm 0.3$ kG for BP Tau.

Robust Zeeman broadening measurements require Zeeman insensitive lines to constrain nonmagnetic broadening mechanisms. Fortunately, numerous CO lines at 2.31 μm have negligible Landé-g factors, making them an ideal null reference. These CO lines are well fit by synthetic stellar models with only rotational and turbulent broadening. In contrast, the 2.2 μm Ti I line spectra are best fit by a model with a distribution of field strengths as described above (and see Figure 2). A total of 16 TTSs now have magnetic field measurements based on observations of the K band Ti I lines (Johns–Krull *et al.* 1999b; Johns–Krull, Valenti & Saar 2004; Yang, Johns–Krull & Valenti 2005; Johns–Krull 2007). These studies show that strong magnetic fields appear to be ubiquitous on TTSs. The mean magnetic field strength, $\bar{B}$, of most TTSs is ~ 2.5 kG. Thus, on these low surface gravity stars, the magnetic pressure dominates the photospheric gas pressure (see Johns–Krull *et al.* 2004; Johns–Krull 2007).

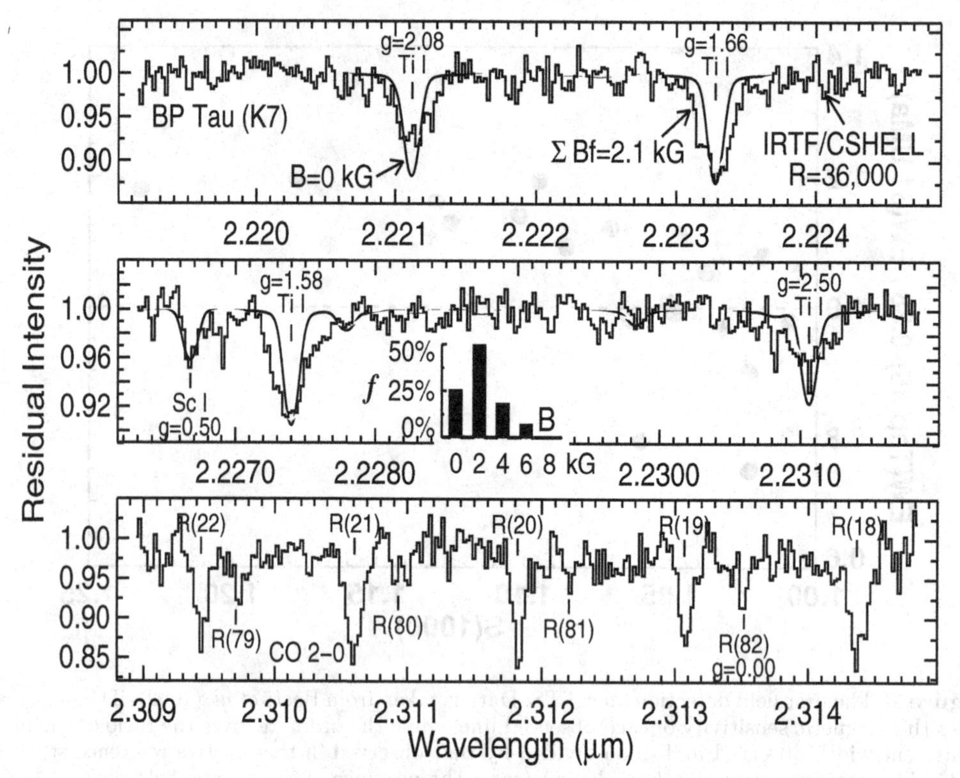

Figure 2. An IRTF/CSHELL spectrum of the K7 CTTS BP Tau (histogram) is compared with synthetic spectra based on magnetic (doubled curve) and nonmagnetic (single curve) models. Zeeman insensitive CO lines are well fit by both models. The Zeeman sensitive Ti I lines are much broader than predicted by the nonmagnetic model. The magnetic model reproduces the observed spectrum, using a distribution of magnetic field strengths (inset histogram) with a mean of 2.1 kG (Johns–Krull 2007) over the entire stellar surface. The effective Landé-g factor is given for each atomic line.

4. Spectropolarimetry and magnetic field topology

Zeeman broadening measurements are sensitive to the distribution of magnetic field strengths, but they have limited sensitivity to magnetic geometry. In contrast, circular polarization measurements for individual spectral lines are sensitive to magnetic geometry, but they provide limited information about field strength. The two techniques complement each other well, as we demonstrate below.

4.1. *The photospheric fields of T Tauri stars*

As mentioned above, existing magnetospheric accretion models assume that intrinsic TTS magnetic fields are dipolar; however, this would be unprecedented for cool stars. Nevertheless, the typical mean field (2.5 kG) measured is similar in magnitude to the dipole field strength required to truncate the accretion disk and enforce disk locking. Higher order multi-polar components of the magnetic field should fall off more rapidly with distance, so if the surface field on the star is dominated by higher order components, even stronger surface fields would be required on the star to produce the required field strength at the inner edge of the disk a few stellar radii from the surface of the star. If we assume for the moment that the mean fields described above are in fact dipolar, we can then ask what net longitudinal magnetic field, B_Z, should be measured using spectropolarimetry? The answer depends on the angle the dipole field axis makes with the line of sight. If the field axis is 90° from the line of sight, $B_Z = 0$. If the dipole axis is aligned with the line of sight, $|B_Z| \sim 0.64 B_e$ where B_e is the equatorial value of the dipole field strength (B_e is the predicted field value tabulated in Johns–Krull *et al.* 1999b and Johns–Krull 2007). The exact value of the coefficient depends a little on the value of the limb darkening coefficient used. Assuming a dipolar field geometry observed at an angle of 45° bewteen the field axis and the line of sight, $|B_Z| \sim 800$ G if the mean field strength on the stellar surface is 2.5 kG.

Overall, there are relatively few measurements of B_Z for TTSs. Until recently, T Tau was the only TTS observed polarimetrically, with a 3σ upper limit of $|B_Z| < 816$ G set by Brown & Landstreet (1981). T Tau has been the focus of more recent study: Smirnov *et al.* (2003) report a detection of a net field of 160 ± 40 G on T Tau which was not confirmed by Smirnov *et al.* (2004) or Daou *et al.* (2006). Johnstone & Penston (1986, 1987) set 3σ upper limits on $|B_Z|$ on 3 TTSs: 494 G (RU Lup), 1110 G (GW Ori), and 2022 G (CoD-34 7151). Donati *et al.* (1997) used the rapid rotation of the diskless NTTS V410 Tau to effectively isolate strips on the stellar surface and detect net circular polarization from the star; however, no field strength was ascribed to these results. In addition, Donati *et al.* do not detect polarization on two other rapidly rotating TTSs. Yang *et al.* (2007) detect a net field of $B_Z = 149 \pm 33$ G on TW Hya on one night of their 6 night monitoring campaign on this star, finding only (3σ) upper limits of ~ 100 G on the other nights. Additional studies also only find upper limits (3σ) of 100–200 G on 4 additional CTTSs (Johns–Krull *et al.* 1999a; Valenti & Johns–Krull 2004). In light of the strong magnetic fields measured using Zeeman broadening techniques, the general absence of polarimetric detections strongly suggest the magnetic fields on TTSs are not dipolar, at least at the stellar surface. Again, as higher order terms will fall off more rapidly with distance, it is expected that the dipole component of the field will indeed dominate at distances of several stellar radii. However, measuring the fields at these distances is quite difficult. The only direct field measurement above the surface of a TTSs is the recent detection of circular polarization in the line profiles of FU Ori (Donati *et al.* 2005). Here, the fields detected are likely in the accretion disk, and the measured fields may not be anchored in the star at all.

4.2. *Magnetic fields in accretion shocks on CTTSs*

Johns–Krull *et al.* (1999a) made the surprising discovery of circular polarization in emission line diagnostics that form predominantly in the accretion shock at the surface of CTTSs. This circular polarization signal is strongest in the narrow component of the He I 5876 Å emission line, but it is also present in the Ca II infrared triplet lines (e.g. Yang *et al.* 2005). Valenti & Johns–Krull (2004) detect He I polarization in four CTTSs: AA Tau, BP Tau, DF Tau, and DK Tau. Symington *et al.* (2005) also detect He I polarization at greater than the 3σ level in three stars (BP Tau, DF Tau, and DN Tau) in their survey of seven CTTSs, and Yang *et al.* (2005) detect polarization in this line in the CTTS TW Hya. All these stars are characterized by He I emission lines which have strong narrow components (NCs) to their line profiles (see Edwards *et al.* 1994 or Alencar & Basri 2000 for a discussion of NC and broad component, BC, emission in CTTSs). Smirnov *et al.* (2004) reported detections of circular polarization in the He I 5876 Å emission line of T Tau on all 3 nights they observed the star, though with significant variability from one night to the next (field measurements range from +350 G to +1100 G with no uncertainty estimates). T Tau's He I line is dominated by BC emission. Daou et al. (2006) observed T Tau on 2 nights, finding field values in the He I line of -29 ± 116 G on one night and

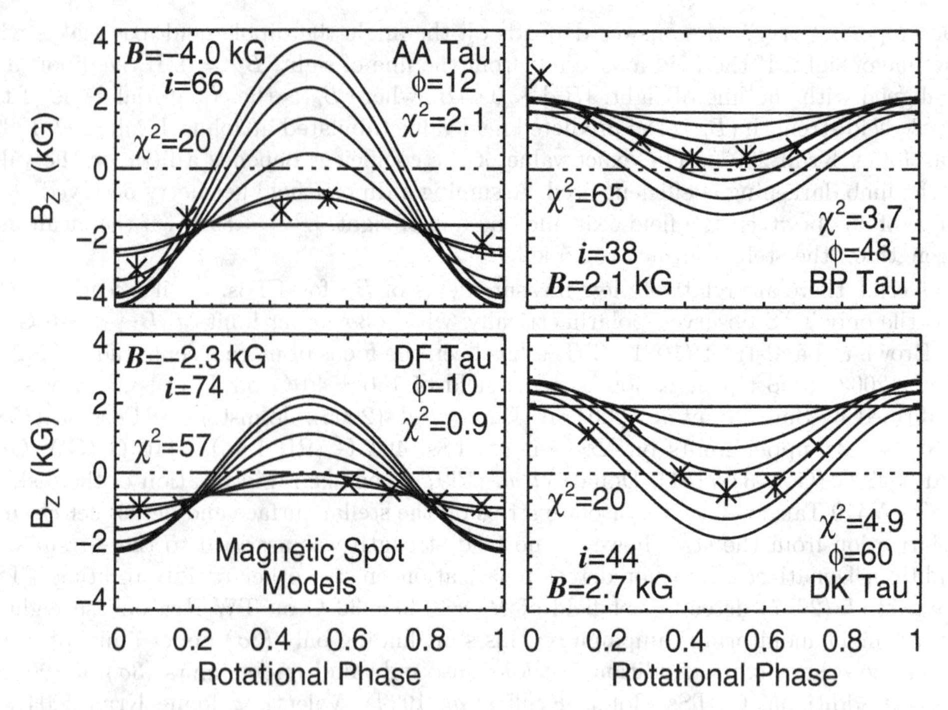

Figure 3. Crosses (×) with vertical error bars indicate the net longitudinal magnetic field (B_z), measured on six consecutive nights using the He I 5876 Å accretion diagnostic. The family of curves show predicted B_z values for a simple model with radial magnetic field lines concentrated in a spot a latitude ϕ. Magnetic field strengths are constrained by independent Zeeman broadening measurements. Reduced χ^2 is 1-5 for the best fitting model, shown as a long dashed curve for each star. Reduced χ^2 is 20-60 for a model that assumes $B_z = 0$.

-43 ± 300 G on the second. TW Hya's He I line has a significant broad component, but Yang *et al.* (2005) do not report any polarization in this part of the line.

The NC of the He I emission is commonly associated with the accretion shock itself at the stellar surface, whereas the BC may have contributions from the magnetospheric accretion flow and/or a hot wind component (e.g. Beristain, Edwards & Kwan 2001). Since the BC of the He I emission line forms over a large, extended volume, its magnetic field strength should be weaker than at the stellar surface. In addition field line curvature may enhance polarization cancellation in the BC. As a result, circular polarization in the BC of the He I 5876 Å emission is predicted to be less than in the NC. Therefore, the result of Smirnov *et al.* (2004) for T Tau is quite surprising. Additional observations of T Tau and other CTTSs dominated by BC emission are needed to confirm the polarization detections. Such observations can strongly constrain the formation region of the BC emission. For example, it is difficult to see how formation of this component over an extend region such as in a hot wind can produce significant polarization characteristic of magnetic field strengths in excess of 1000 G.

The polarization measured in the He I lines is observed to be variable. Figure 3, taken from Valenti & Johns–Krull (2004), shows measurements of B_Z determined from the He I line on 6 consecutive nights for 4 CTTSs. The field values in the He I line vary smoothly on rotational timescales, suggesting that uniformly oriented magnetic field lines in accretion regions sweep out a cone in the sky as the star rotates. Rotational modulation implies a lack of symmetry about the rotation axis in the accretion or the magnetic field or both. For example, the inner edge of the disk could have a concentration

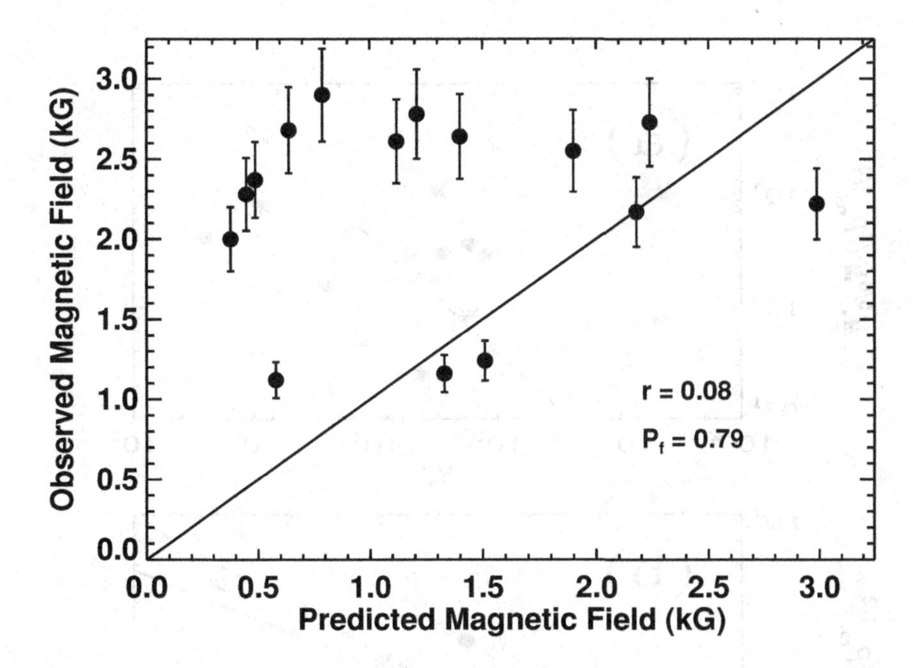

Figure 4. Measured mean magnetic fields as diagnosed by K band Ti I line profiles versus predicted fields using magnetospheric accretion models which assume disk-locking in CTTSs (taken from Johns–Krull 2007).

of gas that corotates with the star, preferentially illuminating one sector of a symmetric magnetosphere. Alternatively, a single large scale magnetic loop could draw material from just one sector of a symmetric disk. Many variants are possible. Figure 3 shows one interpretation of the He I polarization data. Predicted values of B_Z are shown for a simple model consisting of a single magnetic spot at latitude ϕ that rotates with the star. The magnetic field is assumed to be radial with a strength equal to the measured values of $\bar{B}$. Inclination of the rotation axis is constrained by measured $v \sin i$ and rotation period, except that inclination (i) is allowed to float when it exceeds 60° because $v \sin i$ measurements cannot distinguish between these possibilities. Predicted variations in B_Z are plotted for spot latitudes ranging from 0° to 90° in 15° increments. The best fitting model is the heavy curve. The corresponding spot latitude and reduced χ^2 are given on the right side of each panel. Large values of χ^2 on the left side of each panel rule out the hypothesis that no polarization signal is present. In all four cases, this simple magnetic spot model reproduces the observed He I time series. Similar results are found by Symington *et al.* (2005).

5. Confronting theory with observations

At first glance, it might appear that magnetic field measurements on TTSs are generally in good agreement with theoretical expectations. IR Zeeman broadening measurements indicate mean fields on several TTS of $\sim$ 2 kG, similar in value to those predicted by magnetospheric accretion models (Johns–Krull *et al.* 1999b; Johns–Krull 2007). However, in detail the field observations do not agree with some aspects of the theory. This is shown in Figure 4 where the measured mean magnetic field strengths are plotted

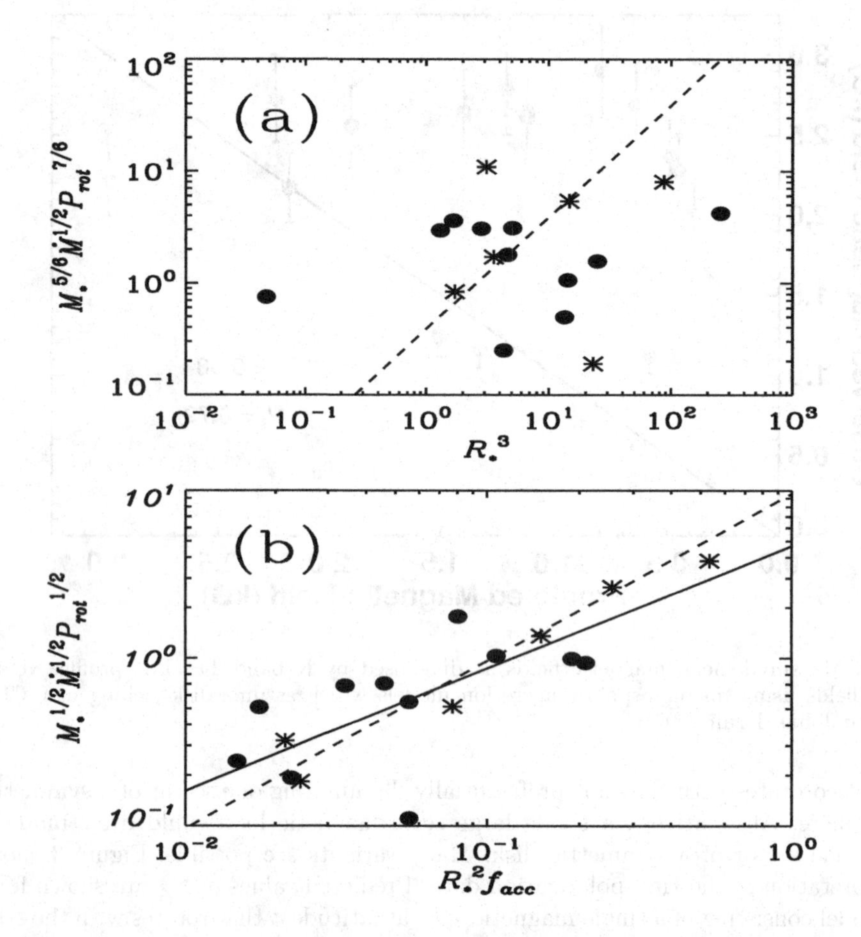

Figure 5. (a) The top panel shows the quantity $(M_*/M_\odot)^{5/6}(\dot{M}/1 \times 10^{-7} M_\odot yr^{-1})^{1/2} P_{rot}^{7/6}$ versus $(R_*/R_\odot)^3$ for the sample of stars from Valenti *et al.* (1993). Single CTTSs are shown in solid circles while CTTSs in binary systems are shown in asterisks. The dashed line shows the best fit line whose slope (1.0) is predicted by equation dipolar magnetospheric accretion models. (b) The bottom panel shows the quantity $(M_*/M_\odot)^{1/2}(\dot{M}/1 \times 10^{-7} M_\odot yr^{-1})^{1/2} P_{rot}^{1/2}$ versus $(R_*/R_\odot)^2 f_{acc}$ for the same sample of stars. This expression is derived without the assumption of a dipolar field, which requires the additional parameter f_{acc} which is the filling factor of accreting zones on each star. Shown in the solid line is the best fit line to the data, and shown in the dashed line is best fit line whose slope (1.0) is predicted non-dipolar accretion models.

versus predicted field strengths. Clearly, the measured field strengths show no correlation with the predicted ones. The field topology measurements give some indication to why there may be a lack of correlation: the magnetic fields on TTSs are not dipolar. On the other hand, the smoothly varying polarization detected in the He I accretion shock lines suggests that the region where the disk interacts with the stellar field is dominated by a simple magnetic field geometry. Since the dipole component of the field falls off the least rapidly with distance, it may well be that the stellar field at the disk truncation radius is dominated by the dipole component. The disk material then loads onto these field lines and accretes onto the star, landing at the surface in those regions which contribute the the large scale dipolar field. Perhaps then, the correct correlation to look for is between the predicted fields and the dipole component, or the predicted fields and the field in the He I region? Currently, there are not enough reliable measurements of either of these field diagnostics to look for such a correlation.

One potential test though is to look for other correlations predicted by magnetospheric accretion theory. As shown in Figure 4, the fields on TTSs are found to all be rather uniform in strength. Assuming the field is in fact constant from one TTS to the next, Johns–Krull & Gafford (2002) looked for correlations among the other stellar and accretion parameters relevant in the models, finding little evidence for the predicted correlations (Figure 5a). On the other hand, Johns–Krull & Gafford (2002) showed how the models of Ostriker & Shu (1995) could be extended to take into account non-dipole field geometries. Once this is done, current data do reveal the predicted correlations (Figure 5b), suggesting magnetospheric accretion theory is basically correct as currently formulated.

6. Conclusion

The current magnetic field measurements show that the strong majority of TTSs are covered by kilogauss magnetic fields. The observations also suggest these fields manifest themselves in a complicated surface topology and that the dipole component of the field is likely small on TTSs. Despite this surface complication, fields measured in the accretion shock on CTTSs suggest that the disk interacts with a primarily dipolar field geometry several stellar radii above the star. However, it is likely the strength of this dipole component is substantially weaker than current models require. Additional high precision spectropolarimetry is required to determine the true dipole component of the field (e.g. Donati *et al.* 2007), and new theoretical studies of magnetospheric accretion with realistic field geometries are urgently needed. Gregory *et al.* (2006) have made a first attempt in this direction, but additional work, including calculations of the torque balance are still needed.

Acknowledgements

I wish to acknowledge partial support from the NASA Origins of Solar Systems program through grant numbers NAG5-13103 and NNG06GD85G made to Rice University.

References

Alencar, S. H. P., & Basri, G. 2000, *AJ* 119, 1881
Basri, G., Marcy, G. W., & Valenti, J. A. 1992, *ApJ* 390, 622
Beristain, G., Edwards, S., & Kwan, J. 2001, *ApJ* 551, 1037
Borra, E. F., Edwards, G., & Mayor, M. 1984, *ApJ* 284, 211
Bouvier, J., Alencar, S. H. P., Harries, T. J., Johns-Krull, C. M., & Romanova, M. M. 2007, *Protostars and Planets V* 479
Brown, D. N. & Landstreet, J. D. 1981, *ApJ* 246, 899
Camenzind, M. 1990, *Rev. Mod. Ast.* 3, 234
Collier Cameron, A. & Campbell, C. G. 1993, *A&A* 274, 309
Daou, A. G., Johns-Krull, C. M., & Valenti, J. A. 2006, *AJ* 131, 520
Donati, J.-F., Paletou, F., Bouvier, J., & Ferreira, J. 2005, *Nature* 438, 466
Donati, J.-F., Semel, M., Carter, B. D., Rees, D. E., & Collier Cameron, A. 1997, *MNRAS* 291, 658
Donati, J.-F., *et al.* 2007, *ArXiv Astrophysics e-prints* arXiv:astro-ph/0702159
Edwards, S., Hartigan, P., Ghandour, L., & Andrulis, C. 1994, *AJ* 108, 1056
Gregory, S. G., Jardine, M., Simpson, I., & Donati, J.-F. 2006, *MNRAS* 371, 999

Guenther, E. W., & Emerson, J. P. 1996, *A&A* 309, 777
Guenther, E. W., Lehmann, H., Emerson, J. P., & Staude, J. 1999, *A&A* 341, 768
Johns–Krull, C. M. 2007, *ApJ* in press
Johns–Krull, C. M., & Gafford, A. D. 2002, *ApJ* 583, 685
Johns–Krull, C. M., Valenti, J. A., & Koresko, C. 1999b, *ApJ* 516, 900
Johns-Krull, C. M., Valenti, J. A., Hatzes, A. P., & Kanaan, A. 1999a, *ApJL* 510, L41
Johns–Krull, C. M., Valenti, J. A., & Saar, S. H. 2004, *ApJ* 617, 1204
Johnstone, R. M. & Penston, M. V. 1986, *MNRAS* 219, 927
Johnstone, R. M. & Penston, M. V. 1987, *MNRAS* 227, 797
Königl, A. 1991, *ApJL* 370, L39
Luhmann, J. G., Gosling, J. T., Hoeksema, J. T., & Zhao, X. 1998, *JGR* 103, 6585
Mathys, G. 2004, *The A-Star Puzzle: Proc. IAUS* 224, 225
Muzerolle, J., Calvet, N., & Hartmann, L. 2001, *ApJ* 550, 944
Ostriker, E. C., & Shu, F. H. 1995, *ApJ* 447, 813
Rüedi, I., Solanki, S. K., & Livingston, W. 1995, *A&A* 302, 543
Saar, S. H. & Linsky, J. L. 1985, *ApJL* 299, L47
Shu, F. H., Najita, J., Ostriker, E., Wilkin, F., Ruden, S., & Lizano, S. 1994, *ApJ* 429, 781
Smirnov, D. A., Fabrika, S. N., Lamzin, S. A., & Valyavin, G. G. 2003, *A&A* 401, 1057
Smirnov, D. A., Lamzin, S. A., Fabrika, S. N., & Chuntonov, G. A. 2004, *Astronomy Letters* 30, 456
Stassun, K. G., Mathieu, R. D., Mazeh, T., & Vrba, F. J. 1999, *AJ* 117, 2941
Symington, N. H., Harries, T. J., Kurosawa, R., & Naylor, T. 2005, *MNRAS* 358, 977
Valenti, J. A., Basri, G., & Johns, C. M. 1993, *AJ* 106, 2024
Valenti, J. A. & Johns–Krull, C. M. 2004, *Ap&SS* 292, 619
Valenti, J. A., Marcy, G. W., & Basri, G. 1995, *ApJ* 439, 939
Vogt, S. S. 1980, *ApJ* 240, 567
Uchida, Y. & Shibata, K. 1984, *PASJ* 36, 105
Yang, H., Johns–Krull, C. M., & Valenti, J. A. 2005, *ApJ* 635, 466
Yang, H., Johns–Krull, C. M., & Valenti, J. A. 2007, *AJ* 133, 73

Discussion

SKINNER: I wonder if you have any information about how the fields on these stars evolve with age?

JOHNS–KRULL: My graduate student, Hao Yang, is working on this problem. We have measurements of several stars in the TW Hya association, which is about 10 Myr old compared to 1.5 – 3 Myr for Taurus which is where most of the field measurements I showed you come from. Right now, it looks like the fields in the TWA stars are stronger on average; however, these stars are smaller on average due to their older age, so it looks like magnetic flux might be conserved. These results are very preliminary at this point.

LAMZIN: I don't really believe the distinction between NC and BC emission in the He I line, and the results of Smirnov *et al.* show that the entire He I line is polarized in T Tau, indicating it forms very close to the star.

JOHNS–KRULL: I agree that these results are very interesting, and I think additional observations are needed to see how general they are.

Star-Disk Interaction in Young Stars
Proceedings IAU Symposium No. 243, 2007
J. Bouvier & I. Appenzeller, eds.

doi:10.1017/S1743921307009404

Magnetism, rotation and accretion in Herbig Ae-Be stars

E. Alecian[1,2], G.A. Wade[1], C. Catala[2], C. Folsom[1], J. Grunhut[1], J.-F. Donati[3], P. Petit[3], S. Bagnulo[4], T. Boehm[3], J.-C. Bouret[5] and J.D. Landstreet[6]

[1]Dept. of Physics, Royal Military College of Canada
email: evelyne.alecian@rmc.ca
[2]Observatoire de Paris, LESIA, France
[3]Laboratoire d'Astrophysique, Observatoire Midi-Pyrénées, France
[4]European Southern Observatory, Casilla 19001, Santiago 19, Chile
[5]Laboratoire d'Astrophysique de Marseille, France
[6]Dept. of Physics & Astronomy, University of Western Ontario, Canada

Abstract. Studies of stellar magnetism at the pre-main sequence phase can provide important new insights into the detailed physics of the late stages of star formation, and into the observed properties of main sequence stars. This is especially true at intermediate stellar masses, where magnetic fields are strong and globally organised, and therefore most amenable to direct study. This talk reviews recent high-precision ESPaDOnS observations of pre-main sequence Herbig Ae-Be stars, which are yielding qualitatively new information about intermediate-mass stars: the origin and evolution of their magnetic fields, the role of magnetic fields in generating their spectroscopic activity and in mediating accretion in their late formative stages, and the factors influencing their rotational angular momentum.

Keywords. Stars : pre-main sequence, stars : magnetism, stars : rotation, stars : accretion

---

1. Introduction

1.1. *Magnetism and rotation in the main sequence A and B stars*

Between about 1.5 and 10 $M_\odot$, at spectral types A and B, about 5 % of main sequence (MS) stars have magnetic fields with characteristic strengths of about 1kG. Such stars also show important chemical peculiarities and are thus usually called the magnetic chemically peculiar Ap/Bp stars. The strength of the magnetic fields of these stars cannot be explained by an envelope dynamo as in the Sun. Until now, the most reliable hypothesis has been to assume a fossil origin for these magnetic fields. This hypothesis implies that the stellar magnetic fields are relics from the field present in the parental interstellar cloud. Its also implies that magnetic fields can (at least partially) survive the violent phenomena accompanying the birth of stars, and can also remain throughout their evolution and until at least the end of the MS, without regeneration.

According to the fossil field model, we should observe magnetic fields in some pre-main sequence (PMS) stars of intermediate mass, the so-called Herbig Ae/Be stars. However no magnetic field was observed up to recently in these stars (except HD 104237, Donati *et al.* 1997). Can we obtain some observational evidence of the presence of magnetic fields during the PMS phase of evolution, as predicted by the fossil field hypothesis? If some Herbig Ae/Be stars are discovered to have magnetic fields, is the fraction of magnetic to non-magnetic Herbig stars the same as the fraction for main sequence stars? Is the magnetic field in Herbig stars strong enough to explain the strength of that of Ap/Bp stars?

Chemical peculiarities and magnetism are not the only characteristic properties observed in the Ap/Bp stars. Most magnetic MS stars have rotation periods (typically of a few days) that are several times longer than the rotation periods of non-magnetic MS stars (a few hours to one day). It is usually believed that magnetic braking, in particular during PMS evolution, when the star can exchange angular momentum with its massive accretion disk, is responsible for this low rotation (Stępień 2000; Stępień & Landstreet 2002). An alternative involves a rapid dissipation of the magnetic field during the early stages of PMS evolution for the fastest rotators, due to strong turbulence induced by rotational shear developed under the surface of the stars, as the convection do in the solar-type stars (see e.g. Lignières *et al.* 1996). In this scenario, only slow rotators could retain their initial magnetic fields, and evolve as magnetic stars to the main sequence. So the question to be addressed is the following: does the magnetic field control the rotation of the star, or else does the rotation of the star control the magnetic field? We propose that this question can be answered by studying rotation and magnetic fields in Herbig Ae/Be stars.

1.2. *The Herbig Ae/Be stars*

The Herbig Ae/Be stars are intermediate-mass pre-main sequence stars, and therefore the evolutionary progenitors of the MS A and B stars. They are distinguished from the classical Be stars by their IR emission and the association with nebulae, characteristics which are due to their young age.

They display many observational phenomena often associated with magnetic activity. First, high ionised lines are observed in the spectra of some stars (e.g. Bouret *et al.* 1997; Roberge *et al.* 2001), and X-ray emission have been detected, coming from some Herbig stars (e.g. Hamaguchi *et al.* 2005). In active cool stars, many of these phenomena are produced in hot chromospheres or coronae. Some authors mentioned rotational modulation of resonance lines which they speculate may be due to rotation modulation of winds structured by magnetic field (Praderie *et al.* 1986; Catala *et al.* 1989, 1991, 1999).

In the literature we find many clues of the presence of circumstellar disks around these stars, from spectroscopic data showing strong emission, and also from photometric data (e.g. Mannings & Sargent 1997, 2000). Recently, using coronagraphic data and interferometric data, some authors have also found direct evidence of circumstellar disks around these stars (Grady *et al.* 1999, 2000; Eisner *et al.* 2003). A careful study of these disks shows that they have similar properties to the disk of their low mass counterpart (Natta *et al.* 2001), the T Tauri stars, whose emission lines are explained by magnetospheric accretion models (Königl 1991; Muzerolle *et al.* 1998, 2001). Finally, Muzerolle *et al.* (2004) have sucessfully applied their magnetospheric accretion model to Herbig stars to explain the emission lines in their spectra.

For all these reasons we suspect that the Herbig stars may host large-scale magnetic fields that should be detectable with current instrumentation. However, many authors tried to detect such fields without much success (Catala *et al.* 1993, 1999; Donati *et al.* 1997; Hubrig *et al.* 2004; Wade *et al.* 2007). But in 2005, a new high-resolution spectropolarimeter, ESPaDOnS, has been installed at the Canada-France-Hawaii Telescope (CFHT). We therefore decided to proceed to survey many Herbig stars in order to investigate rotation and magnetism in the pre-main sequence stars of intermediate mass.

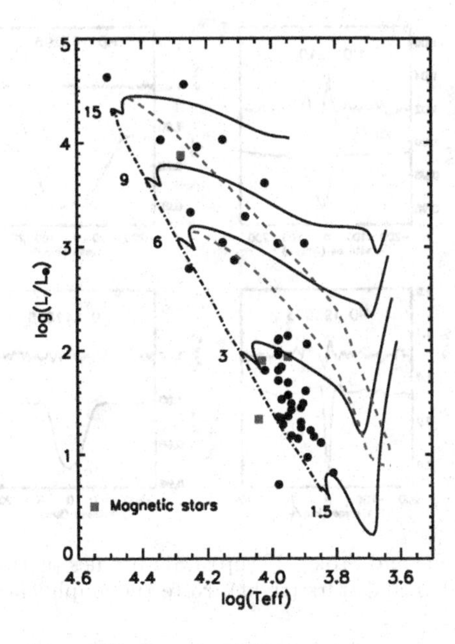

Figure 1. Herbig stars plotted in a HR diagram. The red squares are the magnetic Herbig stars. The PMS evolutionary tracks are plotted for different masses (full lines). The birthline for 10^{-5} and 10^{-4} $M_{\odot}.\mathrm{yr}^{-1}$ mass accretion rates are plotted as dashed lines (Palla & Stahler 1993). The dash-dotted line is the ZAMS.

2. Observations and reduction

2.1. *Our sample*

We have selected Herbig stars in the catalogues from Thé *et al.* (1994) and Vieira *et al.* (2003) with a visual magnitude brighter than 12. Our sample contain 55 stars which have masses ranging from 1.5 $M_{\odot}$ to 15 $M_{\odot}$ with all ages between the birthline and the zero-age main sequence (ZAMS). In Fig. 1 are plotted the stars of our sample in an HR diagram, as well as the evolutionary tracks computed with the CESAM code (Morel 1997), and the birthlines computed by Palla & Stahler (1993) with two mass accretion rates during the protostellar phase : 10^{-5} $M_{\odot}.\mathrm{yr}^{-1}$ and 10^{-4} $M_{\odot}.\mathrm{yr}^{-1}$.

2.2. *Observations and reduction*

Our data were obtained using the high resolution spectropolarimeter ESPaDOnS installed on the 3.6 m Canada-France-Hawaii Telescope (Donati *et al.* in prep.) during many scientific runs.

We used this instrument in polarimetric mode, generating spectra of 65000 resolution. Each exposure was divided in 4 sub-exposures of equal time in order to compute the optimal extraction of the polarisation spectra (Donati *et al.* 1997; Donati *et al.* 2007, in prep.). We recorded only circular polarisation, as the Zeeman signature expected in linear polarisation is about one order of magnitude lower than circular polarisation. The data were reduced using the "Libre ESpRIT" package especially developed for ESPaDOnS, and installed at CFHT (Donati *et al.* 1997; Donati *et al.* 2007, in prep.). After reduction, we obtained the intensity Stokes I and the circular polarisation Stokes V spectra of the stars observed.

We then applied the Least Squares Deconvolution procedure to all spectra (Donati *et al.* 1997), in order to increase our signal to noise ratio. This method assumes that all lines of the intensity spectrum have a profile of similar shape. Hence, this supposes that

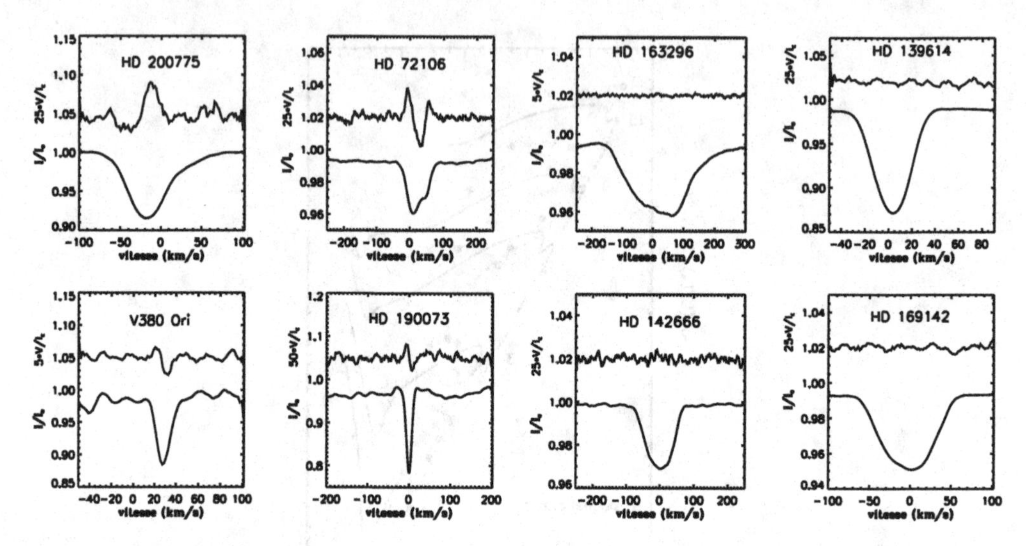

Figure 2. Stokes I (bottom) and Stokes V (up) LSD profiles plotted for the 4 magnetic stars (left) and 4 undetected stars (right). Note the amplification factor in V.

all lines are broadened in the same way. We can therefore consider that the observed spectrum is a convolution between a profile (which is the same for all lines) and a mask including all lines of the spectrum. We therefore apply a deconvolution to the observed spectrum using the pre-computed mask, in order to obtain the average photospheric profiles of Stokes I and V. In this procedure, each line is weighted by its signal to noise ratio, its depth in the unbroadened model and its Landé factor. For each star we used a mask computed using "extract stellar" line lists obtained from the Vienna Atomic Line Database (VALD)†, with effective temperatures and $\log g$ suitable for each star (Wade *et al.* 2007). We excluded from this mask hydrogen Balmer lines, strong resonance lines, lines whose Landé factor is unknown and emission lines. The results of this procedure are the mean Stokes I and Stokes V LSD profiles (Fig. 2).

3. Results

3.1. *Discovery of magnetic fields in Herbig stars*

Thanks to the high performance of the instrument ESPaDOnS and to the LSD method, *we have discovered four new magnetic Herbig Ae/Be stars* (Wade *et al.* 2005; Catala *et al.* 2006; Alecian *et al.* 2007). In Fig. 2 we plotted the Stokes I and V profiles of each newly discovered magnetic Herbig Ae/Be star : HD 200775, HD 72106, V380 Ori and HD 190073, and of four stars in which a magnetic field has not been detected (the undetected stars, hereafter). Contrary to the undetected stars, the Stokes V profiles of the magnetic stars are not null and display a strong Zeeman signature, of the same width of the photospheric I profile, characteristic of the presence of a magnetic field in the stars.

These 4 detections among our sample of 55 stars lead to the conclusion that *7% of Herbig Ae/Be stars are magnetic.* The projection of the distribution of magnetic and non-magnetic main sequence A and B stars on the pre-main sequence phase, assuming a fossil field hypothesis, predict that between 5 and 10 % of Herbig stars should be

† http://www.astro.univie.ac.at/∼vald/

Table 1. Fundamental, geometrical and magnetic parameters of the magnetic Herbig Ae/Be stars. References : 1 : Alecian *et al.* (2007), 2 : Folsom *et al.* (2007), in prep., 3 : Alecian *et al.* (2007), in prep., 4 : Catala *et al.* (2006).

Star	Sp.T.	$v \sin i$ (km.s^{-1})	age (Myr)	P (d)	$B_{\rm P}$ (kG)	β (°)	i (°)	$\frac{d_{\rm dip}}{R_*}$	$B_{\rm P(ZAMS)}$ (kG)	Ref.
HD 200775	B2	26	0.1	4.328	1	78	13	0.1	3.6	1
HD 72106	A0	41	10	0.63995	1.5	58	23	0	1.5	2
V380 Ori†	A2	9.8	2.8	[7.6, 9.8]	1.4	[90, 85]	[36, 49]	0	2.4	3
HD 190073‡	A2	8.5	1.5		[0.1, 1]	[0, 90]	[0, 90]		[0.3, 3]	4

† Work in progress. We need more data of V380 Ori to choose between the 7.6 and the 9.8 periods. Therefore two solutions are possible for β and i

‡ Although we have observed HD 90073 over more that 2 years, no variation of the Stokes V profile has yet been detected.

magnetic, which is consistent with our observations. We therefore bring a new strong argument in favour of the fossil field hypothesis (Wade *et al.* 2007).

3.2. *Topology and intensity of the magnetic fields*

3.2.1. *The magnetic Herbig stars*

We determined the topology and the intensity of the magnetic fields of the four magnetic Herbig stars in order to compare them to the magnetic MS A and B stars. With this aim, we used the oblique rotator model described by Stift (1975). We consider a dipole placed at a distance $d_{\rm dip}$ on the magnetic axis of a spherical rotating star with a magnetic intensity at the pole $B_{\rm P}$. The rotation axis of the star is inclined at an angle i with respect to the line of sight and makes an angle β with the magnetic axis.

According to Landi degl'Innocenti (1973), in the weak field approximation, the Stokes V profile is proportional to the magnetic field projected onto the line of sight and integrated over the surface of the star (B_ℓ, the longitudinal magnetic field, hereafter). As the star rotates, the visible magnetic field changes, resulting in variation of B_ℓ. Therefore the Stokes V profile changes with the rotation phase.

In order to determine the geometrical and magnetic parameters i, β, $B_{\rm P}$ and $d_{\rm dip}$, as well as the rotation period P of the star, we observed the stars at different rotation phases and fit simultaneously all the Stokes V profiles observed for each star. With this aim we calculated a grid of V profiles, using the oblique rotator model, for each date of observations, varying the five parameters. Then, for each star, we applied a χ^2 minimisation to find the best model which matches simultaneously all the V profiles observed. Fig. 3 shows the result of our fitting procedure for one star : HD 200775. The synthetic Stokes V are superimposed on the observed ones (Alecian *et al.* 2007).

The values of the geometrical and magnetic parameters are summarized in Table 1 for each stars. In the case of HD 190073, the topology and the intensity of its magnetic field are not constrained, because, during 2 years observations, the Stokes V profile has not been observed to vary. There are 3 possible explanations for this : the inclination i is very small, the obliquity angle β is very small, or the rotation period of the star is very long. More observations will allow us to discard two of these solutions. However the stability of the magnetic field over more than 2 years and the shape of the Stokes V profiles lead us to the conclusion that this star hosts a large-scale fossil magnetic field (Catala *et al.* 2006).

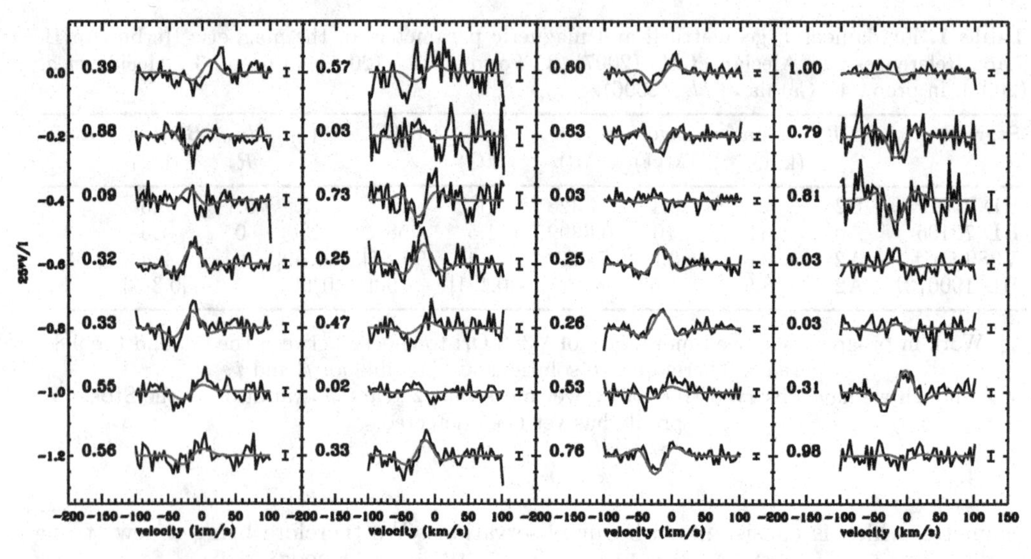

Figure 3. Stokes V profiles of HD 200775 superimposed to the synthetic ones, corresponding to our best fit. The rotation phase and the error bars are indicated on the left and on the right of each profiles, respectively (Alecian *et al.* 2007).

The success of our fitting procedure for the 3 stars HD 200775, HD 72106 and V380 ori, as well as our discussion on HD 190073, lead to the conclusion that the *magnetic Herbig Ae/Be stars host globally dipolar magnetic fields*, similar to the Ap/Bp stars.

Assuming the conservation of magnetic flux during the PMS evolution, and using the radius of the stars and their predicted radius on the ZAMS for the same mass, we can estimate the magnetic intensity at their surface they will have when they will reach the ZAMS (see Table 1). We found intensities ranging from 300 G to 3.6 kG, which is very close to what is observed in the Ap/Bp stars. Hence we bring 2 more new strong arguments in favour of the fossil field hypothesis.

3.2.2. *The undetected stars*

In the case of the other 51 stars no magnetic field was detected. However, a non-detection does not mean that the star does not host a magnetic field. The star may host a magnetic field which is too weak to be detected with our method, or we may observed it at a rotation phase where the Stokes V profile was too weak to be detected. However, for two of these stars (HD 139614, HD 169142), with $v \sin i$ similar to the magnetic Herbig stars, we obtained data during many successive nights, with a 3σ uncertainty on the measured B_ℓ less than 50 G, which is typical of the uncertainties on B_ℓ obtained in the magnetic Herbig stars. Therefore, if these stars would host a magnetic field similar to the magnetic Herbig stars, we would have detected it. This implies that if these stars host a magnetic field, their magnetic intensities are much lower than those of the magnetic Herbig stars. We will consider only these two well-constrained stars in the following, because more work needs to be done to give conclusions on the magnetic intensity of the other stars.

4. Consequences on magnetospheric accretion in Herbig stars

We can use these intensities to test if magnetospheric accretion can occur in these stars. Following Wade *et al.* (2007) we used the models of Königl (1991), Shu *et al.*

(1994) and Collier Cameron & Campbell (1993), which consider a dipole magnetic field aligned with the rotation axis, and coupled with the star up to the corotation radius or a fraction of the corotation radius. The matter from the disk is therefore funnelled along the magnetic lines to the surface of the star. We used the equations, given by these authors, of the polar magnetic field as a function of the parameters of the stars to calculate the minimum intensity required to trigger magnetospheric accretion in these stars (see also Johns-Krull *et al.* 1999). We considered a mass accretion rate of 10^{-8} $M_{\odot}.\mathrm{yr}^{-1}$, which is the typical value in Herbig stars (e.g. Blondel *et al.* 1993)

In the case of the magnetic stars we found for the models of Königl (1991) and Shu *et al.* (1994), that only one star over four has a strong enough magnetic intensity to cause magnetospheric accretion, whereas with the model of Collier Cameron & Campbell (1993), the four stars can have magnetospheric accretion.

In the case of the two well constrained undetected stars, considering that they could host a magnetic field with intensity lower than few 100 G, the minimum field intensity required is too high to have magnetospheric accretion.

Therefore, according to the models that we considered, and taking into account our observations, magnetospheric accretion cannot occur in all Herbig stars.

5. Conclusions

We used the new spectropolarimeter ESPaDOnS installed at the CFHT to proceed in a survey of the Herbig stars, in order to investigate their rotation and magnetic field. We discovered four magnetic stars whose field topology is similar to the MS magnetic A-B stars. We also show that the magnetic intensities of these fields and the proportion of magnetic Herbig stars can explain the magnetic intensity and the proportion of magnetic fields among the MS stars, in the context of fossil field model. We therefore bring fundamental arguments in favour of this hypothesis.

The four magnetic Herbig stars are slow rotators ($v \sin i < 41$ km.s^{-1}) which supports the view that magnetic Herbig Ae/Be stars are the progenitors of the magnetic Ap/Bp stars. Among these magnetic stars two have very low $v \sin i$ (< 10 km.s^{-1}) and are very young (age$<$ 2.8 Myr). Assuming these stars are true slow rotators, this implies that there exists a braking mechanism which acts very early during the PMS evolution of the intermediate mass stars. We could think that this braking mechanism has a magnetic origin, although among the undetected stars we also observe some stars with small $v \sin i$ (~ 15 km.s^{-1}). The nature of the braking mechanism requires further study.

Finally it has been proposed that all Herbig stars should undergo magnetospheric accretion, as the spectra of all stars show similar emission. However, we calculated the minimum polar magnetic intensity of the magnetic Herbig stars and of two well-constrained undetected stars to get magnetospheric accretion, using three different models which have been developed for the T Tauri stars (Königl 1991; Collier Cameron & Campbell 1993; Shu *et al.* 1994). Considering the intensity of the magnetic stars, as well as the maximum magnetic intensity of the two well-constrained undetected stars, if they host a magnetic field, our first conclusion is that magnetospheric accretion cannot occur in all the Herbig stars (a statistical study taking into account all the undetected stars is in progress in order to confirm this result). Therefore, either the models that we used are not well adapted to the Herbig stars, or the emission lines are not only produced by magnetospheric accretion. A thorough observational and theoretical study of the emission in the spectra, as well as the surroundings of the Herbig Ae/Be stars is necessary to better understand the interaction of these stars with their surroundings.

References

Alecian, E., Catala, C., Donati, J.-F., Petit, P., Wade, G. A., Landstreet, J. D., Böm, T., Bouret, J.-C., Bagnulo, S., Folsom, C. & Silvester, J., 2007, *MNRAS*, submitted

Blondel, P. F. C., Talavera, A. & Djie, H. R. E. T. A., 1993, *A&A*, 268, 624

Bouret, J.-C., Catala, C. & Simon, T., 1997, *A&A*, 328,606

Catala, C., Alecian, E., Donati, J.-F., Wade, G.A., Landstreet, J.D., Böm, T., Bouret, J.-C., Bagnulo, S., Folsom, C. & Silvester, J., 2006, *A&A*, 462, 293

Catala, C., Bohm, T., Donati, J.-F. & Semel, M., 1993, *A&A*, 278, 187

Catala, C., Czarny, J., Felenbok, P., Talavera, A. & Thé, P. S., 1991, *A&A*, 244, 166

Catala, C., Donati, J. F., Böhm, T., Landstreet, J., Henrichs, H. F., Unruh, Y., Hao, J. & Collier Cameron, A., *et al.*, 1999, *A&A*, 345, 884

Catala, C., Simon, T., Praderie, F., Talavera, A., Thé, P. S. & Tjin A Djie, H. R. E., 1989, *A&A*, 221, 273

Collier Cameron, A. & Campbell, C. G., 1993, *A&A*, 274, 309

Donati, J.-F., Semel, M., Carter, B. D., Rees, D. E. & Collier Cameron, A., 1997, *MNRAS*, 291, 658

Eisner, J. A., Lane, B. F., Akeson, R. L., Hillenbrand, L. A. & Sargent, A. I., 2003, *ApJ*, 588, 360

Grady, C. A., Devine, D., Woodgate, B., Kimble, R., Bruhweiler, F. C., Boggess, A., Linsky, J. L., Plait, P., Clampin, M. & Kalas, P., 2000, *ApJ*, 544, 895

Grady, C. A., Woodgate, B., Bruhweiler, F. C., Boggess, A., Plait, P., Lindler, D. J., Clampin, M. & Kalas, P., 1999, *ApJL*, 523, L151

Hamaguchi, K., Yamauchi, S. & Koyama, K., 2005, *ApJ*, 618, 360

Hubrig, S., Schöller, M. & Yudin, R. V., 2004, *A&A*, 428, L1

Johns-Krull, C. M., Valenti, J. A. & Koresko, C., 1999, *ApJ*, 516, 900

Königl, A., 1991, *ApJL*, 370, L39

Landi degl'Innocenti, E. & Landi degl'Innocenti, M., 1973, *Sol. Phys.*, 31, 299

Lignières, F., Catala, C. & Mangeney, A., 1996, *A&A*, 314, 465

Mannings, V. & Sargent, A. I., 1997, *ApJ*, 490, 792

Mannings, V. & Sargent, A. I., 2000, *ApJ*, 529, 391

Morel, P., 1997, *A&AS*, 124, 597

Muzerolle, J., Calvet, N. & Hartmann, L., 1998, *ApJ*, 492, 743

Muzerolle, J., Calvet, N. & Hartmann, L., 2001, *ApJ*, 550, 944

Muzerolle, J., D'Alessio, P., Calvet, N. & Hartmann, L., 2004, *ApJ*, 617, 406

Natta, A., Prusti, T., Neri, R., Wooden, D., Grinin, V. P. & Mannings, V., 2001, *A&A*, 371, 186

Palla, F. & Stahler, S. W., 1993, *ApJ*, 418, 414

Praderie, F., Catala, C., Simon, T. & Boesgaard, A. M., 1986, *ApJ*, 303, 311

Roberge, A., Lecavelier des Etangs, A., Grady, C. A., Vidal-Madjar, A., Bouret, J.-C., Feldman, P. D., Deleuil, M., Andre, M., Boggess, A., Bruhweiler, F. C., Ferlet, R. & Woodgate, B., 2001, *ApJL*, 551, L97

Shu, F., Najita, J., Ostriker, E., Wilkin, F., Ruden, S. & Lizano, S., , *ApJ*, 429, 781

Stępień, K., 2000, *A&A*, 353, 227

Stępień, K. & Landstreet, J. D., 2002, *A&A*, 384, 554

Stift, M. J., 1975, *MNRAS*, 172, 133

Thé, P. S., de Winter, D. & Perez, M. R., 1994, *A&AS*, 104, 315

Vieira, S. L. A., Corradi, W. J. B., Alencar, S. H. P., Mendes, L. T. S., Torres, C. A. O., Quast, G. R., Guimarães, M. M. & da Silva, L., 2003, *AJ*, 126, 2971

Wade, G. A., Bagnulo, S., Drouin, D., Landstreet, J. D. & Monin, D., 2007, *MNRAS*, 376, 1145

Wade, G. A., Drouin, D., Bagnulo, S., Landstreet, J. D., Mason, E., Silvester, J., Alecian, E., Böhm, T., Bouret, J.-C., Catala, C. & Donati, J.-F., 2005, *A&A*, 442, L31

Star-Disk Interaction in Young Stars
Proceedings IAU Symposium No. 243, 2007
J. Bouvier & I. Appenzeller, eds.

doi:10.1017/S1743921307009416

Magnetic field at the inner disk edge

Moira Jardine[1], Scott G. Gregory[1] and Jean-François Donati[2]

[1]SUPA, School of Physics and Astronomy, North Haugh, St Andrews, UK, KY16 9SS
email: mmj@st-andrews.ac.uk, sg64@st-andrews.ac.uk

[2]Laboratoire dAstrophysique, Observatoire Midi-Pyrénées, 14 Av. E. Belin, F-31400 Toulouse, France
email: donati@ast.obs-mip.fr

Abstract. Our present understanding of the coronal structure of T Tauri stars is fragmentary and observations in different wavelength regimes often appear to give contradictory results. X-ray data suggest the presence of magnetic loops on a variety of scales, from compact loops of size less than a stellar radius, up to very large loops of up to 10 stellar radii which may connect to the disk. While some stars show a clear rotational modulation in X-rays, implying distinct bright and dark regions, many do not. This picture is complicated by the accretion process itself, which also contributes to the X-ray emission. The location of the inner edge of the accretion disk and the nature of the magnetic field there are still hotly-contested issues. Accretion indicators often suggest the presence of discrete accretion funnels. This has implications for the structure of the corona, as does the presence of an outflowing wind. All of these factors are linked to the structure of the magnetic field, which we are now beginning to unravel through Zeeman-Doppler imaging. In this review I will describe the present state of our understanding of the magnetic structure of T Tauri coronae and the impact this has during such an early evolutionary stage.

Keywords. Stars:coronae

---

1. Introduction

T Tauri stars represent a fascinating phase of stellar evolution. It is during this period, when stars of intermediate to low mass contract onto the main sequence, that planets may form in the disks around these stars. Two outstanding questions remain however about this phase of stellar evolution. The first is the rate at which material can accrete from this disk onto the star. This influences the lifetime of the disk and so sets the challengingly brief timescale for planet formation. The second question is the mechanism by which these stars lose angular momentum. Most T Tauri stars are fairly slow rotators, despite their contraction and the possible gain in angular momentum associated with accretion of material from a surrounding disk.

The nature of the stellar magnetic field is central to both of these questions. The stellar magnetic field disrupts the inner edge of the accretion disk and channels the accreting material onto the star in *accretion funnels.* This process not only governs the torques that are exchanged between the disk and the star but also, by determining the inner extent of the accretion disk, it may influence the minimum orbital radius of exoplanets (Romanova & Lovelace 2006). Only a certain fraction of the stellar field, however, intersects the accretion disk and contributes to this process. Some stellar field lines may be closed and never intersect the disk. These field lines will contribute to the X-ray emission of the star and may provide the "coronal component" of the X-ray spectrum. The remaining field lines are open and can carry the stellar wind, thus allowing the star to lose angular momentum.

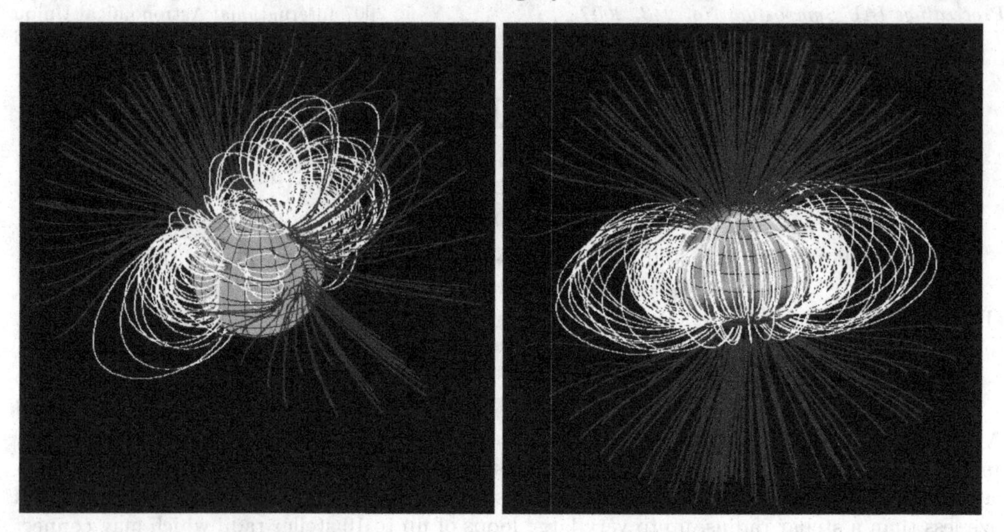

Figure 1. Field structures for the marginally pre-main sequence 1 $M_\odot$ star AB Dor (left) and the fully-convective main sequence 0.3 $M_\odot$ star V374 Peg (right) extrapolated from Zeeman-Doppler surface magnetic maps (Donati *et al.* 1999; Donati *et al.* 2006). White denotes closed field lines, while blue denotes open field lines.

2. Observational indicators of magnetic structure

The last few years have seen great advances in our knowledge of the surface magnetic fields of T Tauri stars. Zeeman-Broadening measurements can now be used to measure mean fields on the stellar surface (Johns-Krull *et al.* 2004; Valenti & Johns-Krull 2004; Yang *et al.* 2005). This gives distributions of local field strengths on the stellar surface that may be as high as 6kG. The results are similar on both accreting and non-accreting T Tauri stars and show little correlation with either the stellar rotation rate or Rossby number.

A complementary approach is to use circular polarisation measurements either in photospheric lines (presumably formed across the whole stellar surface) or the HeI 5876Å emission line believed to form in the accretion shock (Yang *et al.* 2007). Polarisation measurements taken at a range of rotation phases can then be used to provide information on the structure of the magnetic field both across the surface and in the accretion shock. The low circular polarisation measured in photospheric lines rules out the presence of a global dipole, suggesting perhaps a complex field that may be locally very intense, but which is organised into small enough scales that the individual polarisations cancel out. In comparison, the strong, rotationally modulated polarisation in the Helium emission lines suggests that the field associated with accretion is much simpler with few reversals in sign. This is no doubt to be expected, since it is the largest-scale field lines that will interact with the disk and these are likely to have the simplest structure.

The nature of the star-disk interaction will clearly depend on the structure of the whole coronal field, and in particular the degree to which the complexity that seems apparent at the stellar surface may have died away by the time the coronal field reaches the inner edge of the disk. This is illustrated in Fig. 1 which shows, for two very different stars, coronal field structures extrapolated from Zeeman-Doppler maps of the surface field. This is done using the *Potential Field Source Surface* method developed many years ago for modelling the solar corona (Altschuler & Newkirk Jr. 1969). This assumes that at some height above the stellar surface (known as the source surface) the outward pressure of the hot coronal gas forces the magnetic loops open and allows the gas to escape to

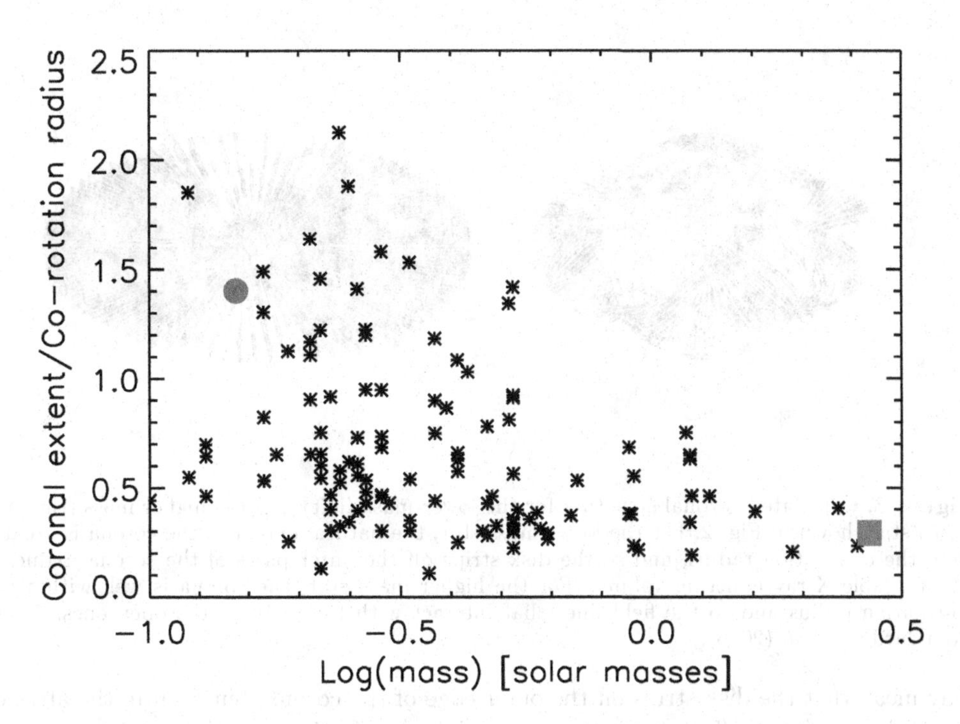

Figure 2. Calculated coronal extent (in units of the Keplerian co-rotation radius) as a function of stellar mass for stars in the COUP sample. Two stars (marked with a circle and square) have been selected as examples. Their coronal structure is shown more fully in Fig. 3. Taken from Jardine *et al.* (2006).

form the stellar wind. Inside the source surface, the field is potential and so along each field line the gas settles into hydrostatic balance (Jardine *et al.* 2002a,b). As can be seen from Fig. 1, the magnetic fields on stars with different internal structures can be very different. The higher-mass star (AB Dor) has a radiative core and a dynamo that produces a field that is complex at the stellar surface although on the largest scale it resembles a tilted dipole. The lower-mass star (V374 Peg), is fully convective and has a field structure that resembles an aligned dipole. On the largest scales, such fields are both simple and would interact with an accretion disk (if one were present) to produce perhaps only one accretion funnel in each hemisphere. An accretion disk that penetrated closer to the star, however, would intersect a much more complex field in the first case, possibly producing many more accretion funnels.

These results suggest that the growth of the radiative core in T Tauri stars may be associated with the onset of a different type of dynamo activity and hence a different coronal field structure. Some indication of this may be apparent in the COUP sample of stars. Those that are expected to have a radiative core appear to have a lower X-ray luminosity than their fully-convective counterparts (Rebull *et al.* 2006).

The nature of the star-disk interaction is also likely to be influenced by the coronal structure. Recent results by Jardine *et al.* (2006) suggest that the nature of the star-disk interaction may be fundamentally different in higher and lower mass stars. If we consider pre-main sequence stars of progressively lower mass, their pressure scale heights Λ will *increase* (since for a polytrope $R \propto M_*^{-1/3}$ and so $\Lambda \propto M_*^{-5/3}$) while their co-rotation radii will *decrease*. Thus the lower the stellar mass, the more likely it is that the co-rotation radius will be within the corona (see Figs. 2 and 3). If a disk is present, this

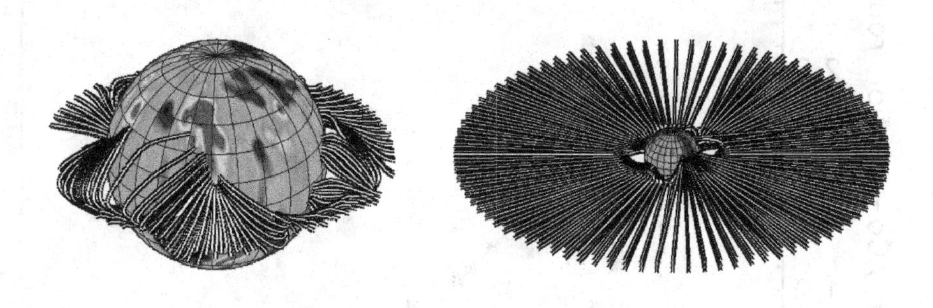

Figure 3. Calculated coronal structure for the lower-mass (left) and the higher mass (right) T Tauri star shown in Fig. 2. For the lower mass star, the natural extent of the corona is greater than the co-rotation radius, and so the disk strips off the outer parts of the corona, reducing the available X-ray emitting volume. For the higher mass star, the corona is well within the co-rotation radius and so the field lines that interact with the disk are the open ones. Taken from Jardine *et al.* (2006).

may mean that the disk strips off the outer edge of the corona, but even in the absence of a disk, centrifugal effects may do the same job. As Fig. 4 shows, this not only explains the drop in X-ray emission measures towards lower masses seen in the COUP dataset (Getman *et al.* 2005; Preibisch *et al.* 2005), but it also suggests that for the higher-mass stars, whose coronae may not extend as far as the co-rotation radius, any disk that exists at the co-rotation radius may interact not with the closed X-ray emitting field lines of the star but rather with the open, wind-bearing lines (see Fig. 3). A further calculation of the associated accretion rates as a function of stellar mass also reproduced the accepted relationship (Gregory *et al.* 2006b).

3. Implications for angular momentum loss

The importance of the open field lines (i.e. those with one footpoint on the star and the other either attached to the disk, or to the interstellar field) has become apparent from several independent lines of research. It has also emerged out of MHD modelling of accretion onto a stellar magnetic field which additionally suggests that the open field lines may themselves carry much of the accretion torque (von Rekowski & Brandenburg 2004; Long *et al.* 2007). So how much of the stellar field is in the form of open field? One clue to this may come from the detection of rotational modulation of X-ray emission from the COUP stars (Flaccomio *et al.* 2005). This suggests that the X-ray coronae are compact, with discrete bright and dark regions. Gregory *et al.* (2006a) have modelled the structure of T Tauri coronae and found significant rotational modulation, with X-ray periods that are typically equal to the optical period (or one half of it). The dark regions are sites of open field where the stellar wind escapes. Some preliminary measurements of such wind properties are now being made.

At the same time as observations of stellar magnetic fields have suggested a complex structure, modelling of accretion inflows and wind outflows have also raised questions about the applicability of "disk locking" models. In the original theory of disk locking a dipolar magnetic field connects to the disk at the truncation radius (Ghosh & Lamb 1978;

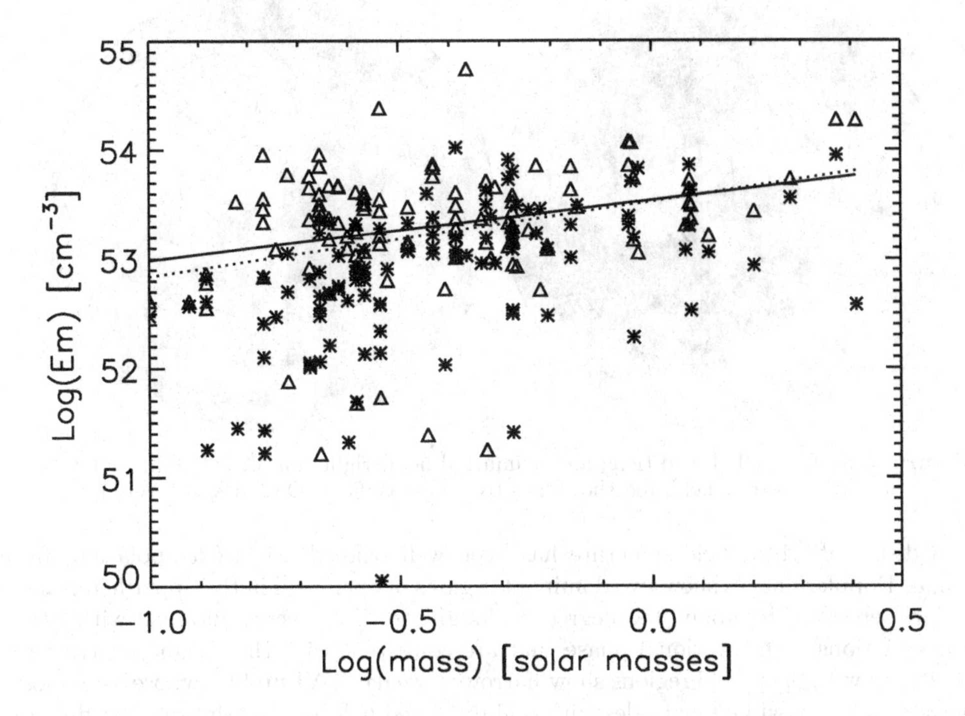

Figure 4. Emission measure as a function of stellar mass for stars in the COUP sample. Crosses show the observed values, while triangles show the calculated values. The solid line is the best fit to the calculated values, while the dotted line is the best fit to the observed values. Taken from Jardine *et al.* (2006).

Königl 1991). Strong fields are however needed in order to ensure that the truncation radius is close to the co-rotation radius, thus ensuring that the star is close to torque balance. This raises problems for the model since field strengths derived from circular polarisation measurements are of order 100G, not the 1000G values needed by disk-locking theory. In addition, the field lines that connect beyond the co-rotation radius and provide the spin down torque will be opened up by the shearing effect of the Keplerian velocity of the disk, and so will be unable to exert a torque on the star. The X-wind model Shu *et al.* (1994) provides a mechanism for allowing accretion to occur whilst the angular momentum of the accreting material is removed from the system by a wind launched from a region close to the corotation radius. Observations of T Tauri jets however suggest they carry much more angular momentum than is present at this radius. Such a large amount of angular momentum must come from further out in the disk in the form of a disk wind.

An accretion powered stellar wind on the other hand carries away the angular momentum imparted to the star by the accretion process. Its effectiveness (relative to a disk wind) lies in the large lever arm that is associated with its large Alfvén radius. This wind is powered by the energy of the accretion flow itself. Indeed, a significant fraction of this energy is needed to drive the wind and a consistent model for this energy transfer process remains the greatest challenge for this theory. Observational signatures of such hot winds are now becoming accessible (Dupree *et al.* 2005). A very promising technique for examining the nature of such winds is the use of co-ordinated observations, using both FUSE data and Zeeman-Doppler imaging to determine the magnetic field structure. This has been done for AB Dor, a star in the post T Tauri phase which no longer

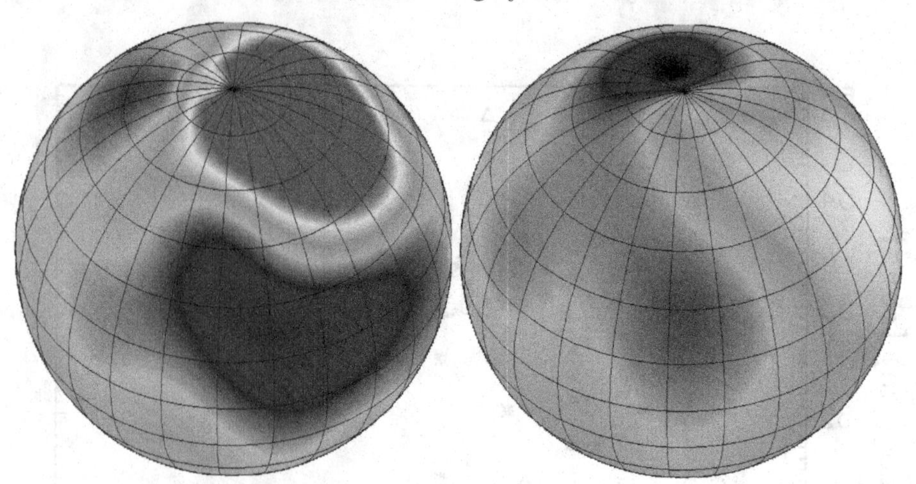

Figure 5. Surface radial field (left) and azimuthal field (right) for V2129 Oph. Blue shows negative polarity field, red shows positive polarity field (Donati *et al.* 2007).

has a disk, but whose field structure has been well studied. Field extrapolations from Zeeman-Doppler images show two dominant regions of open field in the upper hemisphere that are separated by about 180 degrees in longitude. UV spectra obtained with FUSE show variations with rotational phase that are consistent with this magnetic structure. Longitudes with open field regions show narrower, weaker OVI profiles whose asymmetry suggests outflows, while longitudes with mainly closed field regions show higher flux and the presence of compact active regions at mid-latitudes and of height 1.3-1.4 stellar radii (Dupree *et al.* 2006).

4. What can we learn from time-dependent variations?

While this is a promising technique for studying the structure of both the wind regions and the X-ray bright regions of T Tauri coronae, it provides only an instantaneous view of the system. This is a limitation shared by many T Tauri observations. A complementary approach is the type of long-term study that has been undertaken of AA Tau, which accretes at a moderate rate of only $10^{-9}\mathrm{M}_{\odot}\mathrm{yr}^{-1}$. Over the last 10 years there have been four campaigns monitoring the photometric and spectral diagnostics of magnetospheric accretion on timescales of days to years (for a review see Bouvier *et al.* 2007). The system is seen at a high inclination of about 75 degrees and the photometric brightness variations are consistent with a warp of the inner disk that is situated at 8.8 $R_\star$ (approximately at the co-rotation radius). This warp may be caused by a large-scale magnetic field inclined to the stellar rotation axis (Romanova *et al.* 2003; Long *et al.* 2007), and is consistent with He I observations (Valenti & Johns-Krull 2004). For AA Tau, Hβ line profiles also show redshifted absorption components that appear near the middle of photometric eclipse, suggesting a spatial relationship between the disk warp and the funnel flows. The general characteristics of this variability are reproduced by non-symmetric radiative transfer models (Symington *et al.* 2005) although the predicted degree of variability is too high. On a timescale of weeks the change in the redshifted absorption component of Hα can be interpreted as due to field inflation (Lynden-Bell & Boily 1994; Goodson *et al.* 1997) as field lines that link to the disk are sheared by the Keplerian flow until they expand and ultimately reconnect. On longer timescales, however, AA Tau's V-band eclipses change shape over several years, perhaps suggesting that the inner disk warp

Figure 6. Calculated coronal field extrapolated from the surface radial field map of V2129 Oph shown in Fig. 5. The smaller field lines have been selected to show the field structure near the stellar surface.

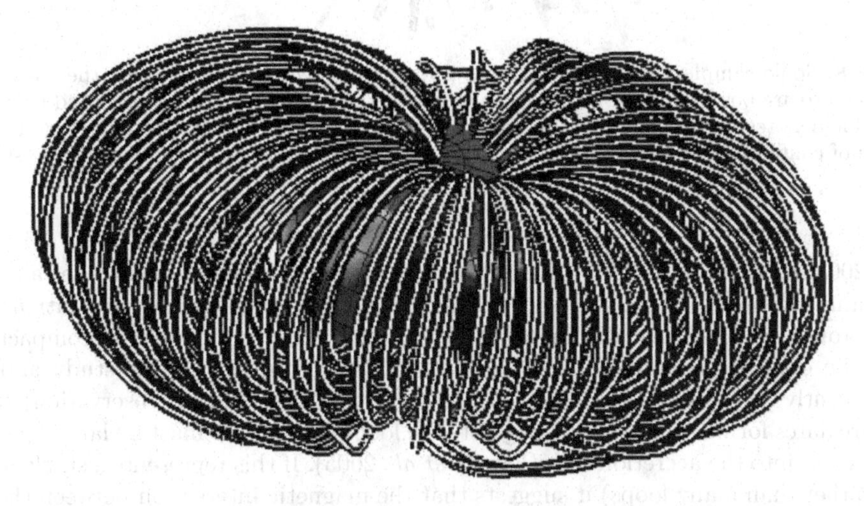

Figure 7. Calculated coronal field extrapolated from the surface radial field map of V2129 Oph shown in Fig. 5. The larger field lines have been selected to show the global structure.

changes on that timescale. This may be pointing towards a magnetic cycle, or longer timescale variation in the accretion process.

Perhaps one of the most promising approaches to studying the variable accretion processes in T Tauri stars is the use of multi-wavelength campaigns. This has been done for the Coronet cluster using simultaneous X-ray, radio, near-infrared and optical monitoring (Forbrich *et al.* 2007). This is the first attempt to examine the correlation between radio and X-ray variability in class 0 and class 1 protostars. Preliminary results suggest that the variability in the radio and X-ray wavebands is uncorrelated, but that a relationship does exist between the luminosity at X-ray and radio wavelengths (Benz & Guedel 1994).

In addition to the ubiquitous flaring-like variability seen in the X-ray emission of the COUP stars, there is also a clear rotational modulation in a small but significant number of these stars. Such modulation has previously been difficult to detect because it requires continuous observation over several rotation periods (Güdel *et al.* 1995; Marino

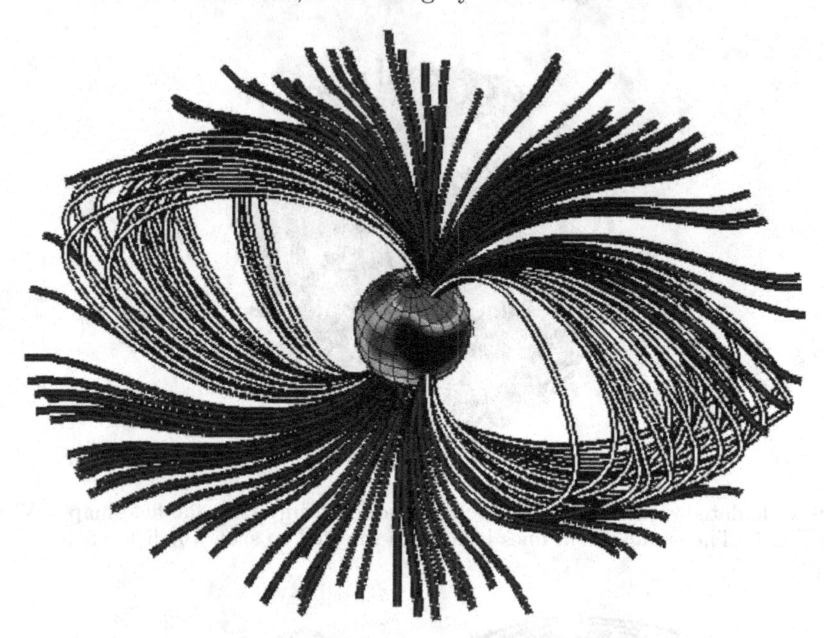

Figure 8. Some sample field lines that pass through the accretion disk inside the co-rotation radius and so are possibly capable of carrying an accretion flow. White denotes closed field lines, while blue denotes open field lines. We note, however, that these closed field lines would not be capable of confining the $\sim 10^7$ K plasma of the stellar corona and so are unlikely to be stable.

et al. 2003; Hussain *et al.* 2005). This rotational modulation shows clearly that although these stars appear to lie in the "saturated" part of the rotation-activity relation, they have coronae in which the dominant X-ray emission structures must be compact and cannot be distributed homogenously in longitude. Both in the COUP study and also more recently in the DROXO (Deep Rho Ophiuchi XMM-Newton Observation) study, there are flares for which modelling suggests the loops responsible must be large - perhaps reaching out into the accretion disk (Favata *et al.* 2005). If this represents a single flaring loop (rather than many loops) it suggests that the magnetic interaction between the star and the disk may be a violent one. Most flares from T Tauri stars are at much lower energies however and show a distribution in energy that is broadly compatible with what is observed on the Sun, which is consistent with energy release from a corona dominated by nanoflares (Caramazza *et al.* 2007).

The question still remains however of the contribution to the X-ray emission that may be made by the accretion process itself. Material falling onto a T Tauri star at the free-fall speed v_{ff} will be shocked to a temperature $T \propto v_{ff}^2 \propto M_\star/R_\star$. For typical values, this gives T= 2 – 4MK which would contribute to the soft X-ray spectrum. A few well studied stars such as TW Hya (Stelzer & Schmitt 2004; Kastner *et al.* 2002) and more recently MP Muscae do indeed show the high densities and soft excess that would be expected from accretion, although this is not true of all T Tauri stars (Argiroffi *et al.* 2007). Of course not all of the infall will occur at the free-fall speed (as is apparent from modelling of accretion infalls (Gregory *et al.* 2006b) and there may be some absorption by the accretion column itself (Gregory *et al.* 2007). Modelling of the X-ray emission from the shock promises to help untangle the different contributions to the overall X-ray spectrum (Günther *et al.* 2007).

5. Zeeman-doppler imaging of T Tauri stars

Our understanding of the nature of the magnetic field that links T Tauri stars to their disks has now entered a new phase with the availability of surface magnetic maps obtained using Zeeman-Doppler imaging. The first of these stars to be observed in this way is V2129 Oph, a young, moderately accreting T Tauri star aged around 1.5 - 3.0 My (Donati *et al.* 2007). With a derived period of 6.5 d, a mass of $1.35 \pm 0.15 M_\odot$ and radius of 2.4 ± 0.3 $R_\odot$ it is fairly typical of many stars of this class. A map of the surface brightness shows a clear polar spot, in addition to many lower-latitude spots. Maps of the surface magnetic field derived from the photospheric lines show a complex topology with many regions of mixed polarity. Maps produced using the accretion-sensitive Ca IR triplet or HeI 5876Å line show a simpler structure with one dominant region of positive polarity at the same rotation phase as the greatest emission in the HeI line. This suggests that the coronal field at low to moderate heights is likely to be complex, while the larger-scale field is much simpler. It is this larger-scale field that interacts with the disk and so is responsible for the single dominant region of accretion in the visible hemisphere.

Fig. 5 shows the surface radial and azimuthal magnetic fields. The strongest component of the field is a 1.2 kG (tilted) octupole, while the dipole component is much weaker with a strength of only 0.35 kG. This can be seen most clearly by extrapolating the field and selecting out field lines of different heights. Fig. 6 shows the octupolar structure clearly, while Fig. 7 shows the tilted dipole. It is this dipolar field that survives out to the co-rotation radius and so is likely to interact with the disk. Thus while the closed-field corona (which will be bright in X-rays) is highly structured, the accreting field (shown in Fig. 8) is much simpler, in accordance with the behaviour of the HeI line. It seems, therefore, from this first example that the physical locations of the X-ray bright coronal regions and the accreting regions are quite distinct. Given the complexity of the small-scale field, it would be unlikely that a correlation would be found between the rotationally-induced optical and X-ray emission variation (Stassun *et al.* 2007).

6. Conclusions

Our understanding of the processes by which T Tauri stars accrete material from their disks while remaining at moderate rotation rates has advanced enormously over the past few years. The old models of stars with strong dipolar fields that lock the star to its disk have been replaced by (or perhaps only challenged by) new and more dynamic models. In these pictures, the star and the disk are not in torque balance. The accretion flows, which form into discrete funnels, transport angular momentum onto the star. Through some as yet unknown mechanism this process of accretion allows the star to drive a powerful wind that carries away enough angular momentum to keep the star rotating slowly.

The picture is more complicated than we might have thought some years ago, but observations of both the X-ray structure of T Tauri coronae and also of the magnetic field structure point to a degree of complexity inconsistent with a dipolar field. The very recent Zeeman-Doppler maps of V2129 Oph show a field that is simple on large scales, leading to a single accretion funnel in the observable hemisphere, while displaying significant complexity on the smaller scales on which the X-ray bright regions will be organised. There are also distinct regions of open field (see Fig. 9) which will be capable of carrying a hot stellar wind. More studies of this type, exploring stars at a range of evolutionary stages and so with a range of internal structures, would be extremely valuable.

The complexity seen in such "snapshots" of stars are also seen in the variability of the accretion process which appears to change on timecales from hours to years. Multiwavelength studies promise to be enormously valuable in studying the relationship between

Figure 9. Calculated coronal field extrapolated from the surface radial field map of V2129 Oph shown in Fig. 5. White denotes closed field lines capable of confining hot, X-ray bright gas, while blue denotes open, wind-bearing field lines.

the closed-field, X-ray bright regions of the stellar corona, the accreting regions and the open-field wind-bearing regions. It appears that only by understanding the relationship between these different components of the coronal structure can we hope to understand the rotational evolution of T Tauri stars.

References

Altschuler, M. D. & Newkirk Jr., G. 1969, Solar Phys., 9, 131

Argiroffi, C., Maggio, A., & Peres, G. 2007, A&A, 465, L5

Benz, A. O. & Guedel, M. 1994, A&A, 285, 621

Bouvier, J., Alencar, S. H. P., Harries, T. J., Johns-Krull, C. M., & Romanova, M. M. 2007, in Protostars and Planets V, ed. B. Reipurth, D. Jewitt, & K. Keil, 479–494

Caramazza, M., Flaccomio, E., Micela, G., *et al.* 2007, ArXiv e-prints, 706

Donati, J.-F., Collier Cameron, A., Hussain, G., & Semel, M. 1999, MNRAS, 302, 437

Donati, J.-F., Forveille, T., Cameron, A. C., *et al.* 2006, Science, 311, 633

Donati, J.-F., Jardine, M. M., Gregory, S., *et al.* 2007, MNRAS, submitted

Dupree, A. K., Ake, T. B., Brickhouse, N. S., Hussain, G. A. J., & Jardine, M. 2006, in ASP Conf. Ser. 348: Astrophysics in the Far Ultraviolet: Five Years of Discovery with FUSE, ed. G. Sonneborn, H. W. Moos, & B.-G. Andersson, 168–+

Dupree, A. K., Brickhouse, N. S., Smith, G. H., & Strader, J. 2005, Ap. Lett., 625, L131

Favata, F., Flaccomio, E., Reale, F., *et al.* 2005, ApJS, 160, 469

Flaccomio, E., Micela, G., Sciortino, S., *et al.* 2005, ApJS, 160, 450
Forbrich, J., Preibisch, T., Menten, K. M., *et al.* 2007, A&A, 464, 1003
Getman, K. V., Flaccomio, E., Broos, P. S., *et al.* 2005, ApJS, 160, 319
Ghosh, P. & Lamb, F. 1978, ApJ, 223, L83
Goodson, A. P., Winglee, R. M., & Boehm, K.-H. 1997, ApJ, 489, 199
Gregory, S. G., Jardine, M., Cameron, A. C., & Donati, J.-F. 2006a, MNRAS, 373, 827
Gregory, S. G., Jardine, M., Simpson, I., & Donati, J.-F. 2006b, MNRAS, 371, 999
Gregory, S. G., Wood, K., & Jardine, M. 2007, ArXiv e-prints, 704
Güdel, M., Schmitt, J., Benz, A., & Elias II, N. 1995, A&A, 301, 201
Günther, H. M., Schmitt, J. H. M. M., Robrade, J., & Liefke, C. 2007, A&A, 466, 1111
Hussain, G., Brickhouse, N., Dupree, A., *et al.* 2005, ApJ, 621, 999
Jardine, M., Cameron, A. C., Donati, J.-F., Gregory, S. G., & Wood, K. 2006, MNRAS, 367, 917
Jardine, M., Collier Cameron, A., & Donati, J.-F. 2002a, MNRAS, 333, 339
Jardine, M., Wood, K., Collier Cameron, A., Donati, J.-F., & Mackay, D. H. 2002b, MNRAS, 336, 1364
Johns-Krull, C. M., Valenti, J. A., & Saar, S. H. 2004, ApJ, 617, 1204
Kastner, J. H., Huenemoerder, D. P., Schulz, N. S., Canizares, C. R., & Weintraub, D. A. 2002, ApJ, 567, 434
Königl, A. 1991, ApJ, 370, L39
Long, M., Romanova, M. M., & Lovelace, R. V. E. 2007, MNRAS, 374, 436
Lynden-Bell, D. & Boily, C. 1994, MNRAS, 267, 146
Marino, A., Micela, G., Peres, G., & Sciortino, S. 2003, A&A, 407, L63
Preibisch, T., Kim, Y.-C., Favata, F., *et al.* 2005, ApJS, 160, 401
Rebull, L. M., Stauffer, J. R., Ramirez, S. V., *et al.* 2006, AJ, 131, 2934
Romanova, M. M. & Lovelace, R. V. E. 2006, Ap. Lett., 645, L73
Romanova, M. M., Ustyugova, G. V., Koldoba, A. V., Wick, J. V., & Lovelace, R. V. E. 2003, ApJ, 595, 1009
Shu, F., Najita, J., Ostriker, E., *et al.* 1994, ApJ, 429, 781
Stassun, K. G., van den Berg, M., & Feigelson, E. 2007, ApJ, 660, 704
Stelzer, B. & Schmitt, J. H. M. M. 2004, A&A, 418, 687
Symington, N. H., Harries, T. J., Kurosawa, R., & Naylor, T. 2005, MNRAS, 358, 977
Valenti, J. A. & Johns-Krull, C. M. 2004, Ap&SS, 292, 619
von Rekowski, B. & Brandenburg, A. 2004, A&A, 420, 17
Yang, H., Johns-Krull, C. M., & Valenti, J. A. 2005, ApJ, 635, 466
Yang, H., Johns-Krull, C. M., & Valenti, J. A. 2007, AJ, 133, 73

Discussion

REGEV: I don't think potential field models are appropriate for modelling T Tauri stars. Can you comment?

JARDINE: The only way to test a theory is to compare its prediction with observations. We are currently collaborating with Tim Harries to calculate the rotationally-modulated line profiles that would be predicted by our models and to compare these with the observed line profiles. The nature of this comparison will tell us if our field structures are largely consistent with the accretion indicators.

Star-Disk Interaction in Young Stars
Proceedings IAU Symposium No. 243, 2007
J. Bouvier & I. Appenzeller, eds.

doi:10.1017/S1743921307009428

Submillimetre polarimetric observations of magnetic fields in star-forming regions

Rachel L. Curran[1], **Antonio Chrysostomou**[2] **and Brenda C. Matthews** [3]

[1]School of Cosmic Physics, Dublin Institute for Advanced Studies,
5 Merrion Square, Dublin 2, Ireland
email: rcurran@cp.dias.ie

[2]Joint Astronomy Centre, 880 N. A'ohoku Place, University Park, Hilo, Hawaii USA
email: Antonio.Chrysostomou@jach.hawaii.edu

[3]Herzberg Institute of Astrophysics, 5071 West Saanich Road, Victoria, Canada
email: brenda.matthews@nrc-cnrc.gc.ca

Abstract. Submillimetre imaging polarimetry is one of the most powerful tools at present for studying magnetic fields in star-forming regions, and the only way to gain significant information on the structure of these fields. We present analysis of the largest sample (to date) of both high- and low-mass star-forming regions observed using this technique. A variety of magnetic field morphologies are observed, with no single field morphology favoured. Both the continuum emission morphologies and the field morphologies are generally more complex for the high-mass sample than the low-mass sample. The large scale magnetic field (observed with the JCMT; 14″ resolution) of NGC1333 IRAS2 is interpreted to be weak (compared to the energetic contributions due to turbulence) from the random field pattern observed. On smaller scales (observed with the BIMA array; 3″ resolution) the field is observed to be almost radial, consistent with the polarisation nulls in the JCMT data – suggesting that on smaller scales, the field may be more important to the star formation process. An analysis of the magnetic field direction and the jet/outflow axis is also discussed. Cumulative distribution functions of the difference between the mean position angle of the magnetic field vectors and the jet/outflow axis reveal no correlation. However, visual inspection of the maps reveal alignment of the magnetic field and jet/outflow axis in 7 out of 15 high-mass regions and 3 out of 8 low-mass regions.

Keywords. Techniques: polarimetric, submillimetre, stars: formation, stars: magnetic fields.

---

1. Magnetic fields in star formation

Two main aspects of star formation which are still not fully understood are the initial support of the cloud (cloud lifetimes are much longer than originally predicted by theory), and the launching and collimation of the observed jets and outflows emanating from protostars. Whilst immense progress has been made through theoretical work and simulations – which suggest magnetic fields could be important in both of these aspects – observational evidence of these required fields is needed before these models can be corroborated.

There are currently two models for cloud support in star formation, one which relies on magnetic fields – in which strong fields regulate the collapse via ambipolar diffusion, and one which relies on turbulence, where the pressure from turbulent eddies provides the required support to the cloud. Observations of magnetic fields (especially at early stages of star formation) may prove important in ascertaining the cloud support mechanism, whether from magnetic fields or turbulence alone, or a combination of the two. It has been shown that strong fields (compared to the energetic contributions of the turbulence)

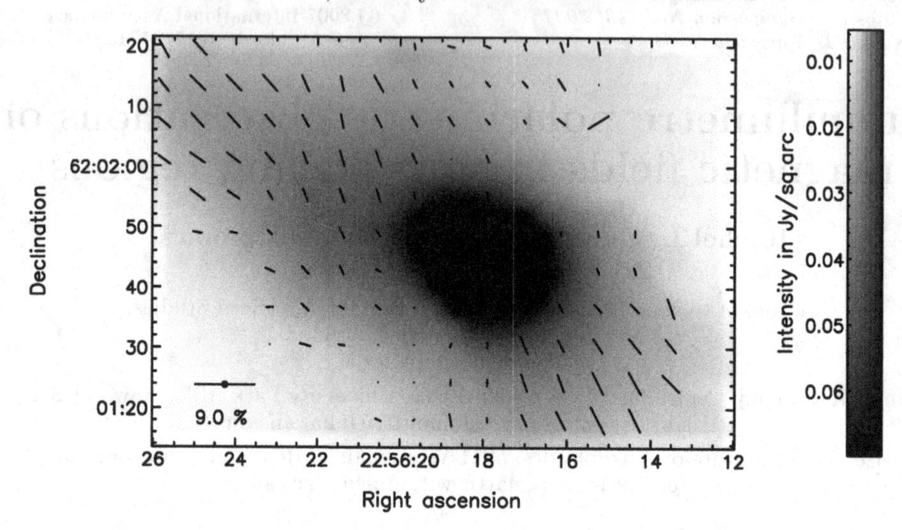

Figure 1. Cepheus A. The greyscale is the 850μm continuum emission, with the B-vectors overlaid to represent the plane-of-the-sky magnetic field component. The field is uniform away from the central core, with a northwest-southeast direction with typical polarisation percentages of $\sim$ 3–4%. Across the core the polarisation percentage drops to almost zero. There is a largescale east-west outflow in this region, as well as a smaller scale northwest-southeast jet, which aligns well with the magnetic field direction. Epoch J2000.

produce uniform field patterns, whereas weak fields produce more random, disorganised field patterns.

Jet and outflow launching and collimation – according to theory – relies on the magnetic field being $\sim 60°$ to the disk plane (for the launching, see the article by J. Ferreira in this volume), and then toroidal in order to collimate the jet. Observations of the direction of the field, both through the disk and along the jet, will help to place constraints on such models. Measuring the magnetic field on different scales may illuminate whether the gas in jets/outflows is dragged along by the magnetic field, or if the magnetic field is dragged along by the gas..

2. Observing magnetic fields

There are currently two main approaches to studying magnetic fields in star-forming regions observationally. Zeeman measurements provide the line-of-sight magnetic field strength, although in star-forming regions these estimates come primarily from maser emission and so only reveal the magnetic field for the position of the maser (i.e. from a point source). The second method is polarised thermal emission in the far infrared and submillimetre, which reveals the plane-of-the-sky magnetic field direction throughout the cloud.

Submillimetre polarisation occurs when cool spinning elongated dust grains become aligned to the magnetic field such that their minor axis is parallel to the magnetic field (Davis & Greenstein 1951). The thermal emission from a volume of such aligned grains is thus polarised, and polarimetry therefore allows for a unique determination of the direction and magnitude of the **E**-vector from which the **B**-vector may be inferred. Previous submillimetre polarisation measurements of such star-forming regions have revealed typical polarisation percentages of <3% (Matthews & Wilson 2002, Chrysostomou *et al.* 2002) although percentages of <8 or 9% have been observed in some star-forming regions (Curran *et al.* 2004). The millimetre/submillimetre wavelength regime is an ideal

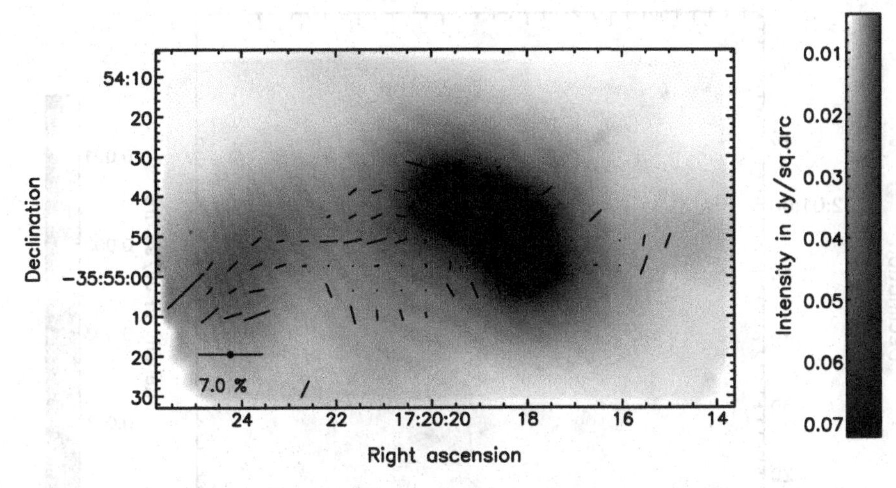

Figure 2. NGC6334A. Greyscale and vectors as for Fig. 1.

probe for star formation as the radiation is optically thin, and therefore is dominated by the densest structures where protostars form. The polarimetry data provide a column-averaged (and intensity weighted) measurement of the magnetic field direction through the cloud and projected onto the plane-of-the-sky. We have used the Submillimetre Common User Bolometer Array (SCUBA; Holland *et al.* 1999), in conjunction with the polarimeter (Greaves *et al.* 2003), on the James Clerk Maxwell Telescope (JCMT) in Hawaii to observe 18 high-mass and 8 low-mass star-forming regions.

3. Submillimetre emission & field morphology

The morphology of the continuum emission is much more varied in the high-mass sample (Curran *et al.* 2004, Curran & Chrysostomou 2007a) than in the low-mass sample (Curran & Chrysostomou, *in prep.*), as is the morphology of the magnetic fields. The high-mass sample is composed of several (11 out of 18) sources which appear almost spherical (marginally resolved by the JCMT beam), of which 5 of these are associated with fainter, more extended emission. The magnetic field is well traced throughout the extended emission, as well as across the cores. For example, Cepheus A (Fig. 1) consists of a marginally resolved main core and fainter, extended emission detected to its northwest and southeast. The polarimetry reveal that the plane-of-the-sky magnetic field is parallel to this extended emission, in a northwest-southeast direction. Interestingly, the field seems to 'pinch-in', and the polarisation percentage decreases sharply, across the core. This field pattern is consistent with pinched or 'hourglass' field models (Aitken *et al.* 2002). In this case, the decrease in polarisation percentage across the core would be due to the direction of the magnetic field changing significantly on scales smaller than the JCMT beam, resulting in a lower net polarisation along those lines of sight. Other explanations for the decrease in polarisation percentage with increasing total intensity may be that the grains become more spherical in regions of high density, or that the grains suffer from more collisions, and therefore become un-aligned to the field.

There are also four cores in the high-mass sample that appear elongated, although these could be multiple cores unresolved in the JCMT beam, for example NGC6334A (Fig. 2). The magnetic field of this core is complex, with the polarimetry revealing abrupt changes in direction of the magnetic field. The field appears to be parallel to the extension of gas and dust connecting the main source with the Eastern core, but it then turns

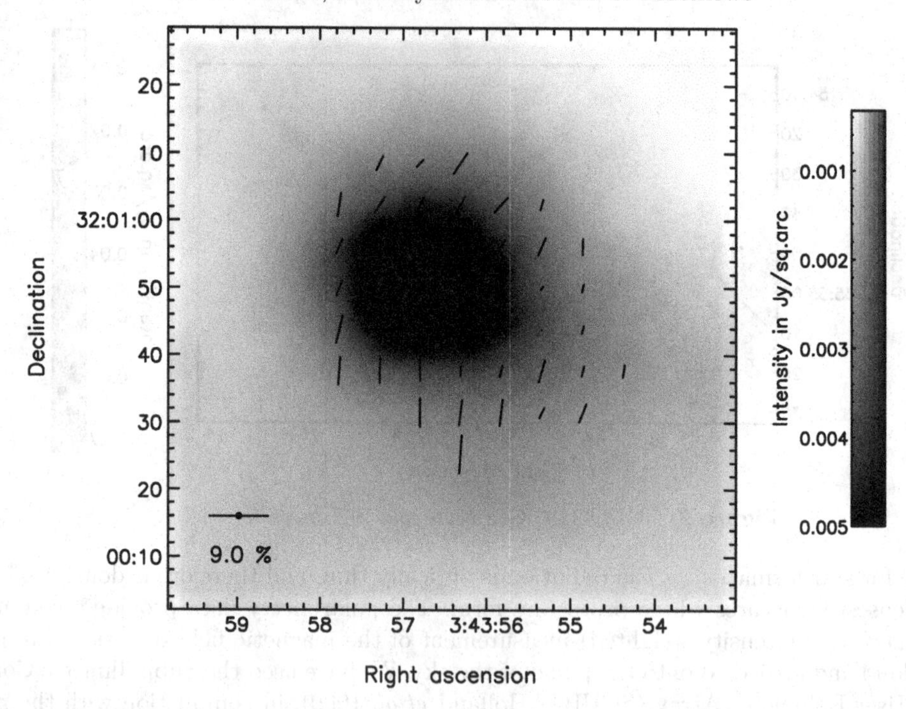

Figure 3. HH211. Greyscale and vectors as for Fig. 1.

perpendicular to the major axis of emission across the southwestern part of the core. There are also several polarisation nulls, some of which are not coincident with regions of high intensity.

The low-mass sample is mainly composed of spherical cores. The magnetic field is generally only traced across the peaks and not in associated extended emission, as these sources are much fainter than the high-mass sources. Even with submillimetre emission similar between sources, we observe a variety of field morphologies. The polarimetry of HH211 (Fig. 3) reveal an ordered field pattern, suggesting a strong magnetic field. The field appears to curve slightly, as it has a north-south direction in the south of the source, curving to become northwest-southeast in the north of the source. The polarisation percentage also remains consistent across the source. In contrast to this ordered field, NGC1333 IRAS2 (Fig. 4) was observed to have an almost random field pattern. There are several polarisation nulls observed across the source. To the north, the magnetic field is roughly north-south in direction, but to the south, the vectors have a large scatter in position angle. This field morphology may indicate a weak magnetic field.

4. Interferometry

It has been found in some cases that the magnetic field becomes more important on smaller scales (for example W51, see Chrysostomou *et al.* 2002 and the higher resolution data of Lai *et al.* 2001). In order to fully understand the magnetic field of NGC1333 IRAS2, we observed this source with the Berkeley Illinois Maryland Association (BIMA) array, with a resolution of $\sim 3''$. The BIMA data are presented in Fig. 5. The polarimetry trace the magnetic field close in to the protostar, as interferometry 'filters out' the more diffuse emission. The BIMA data agree with the JCMT data –towards the northeast of the peak, the magnetic field is roughly north-south in direction. Moving westwards the

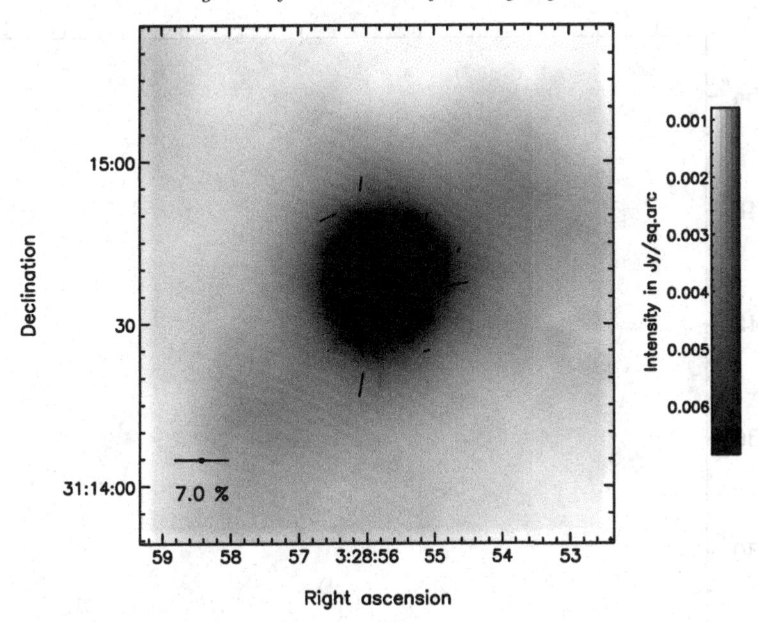

Figure 4. NGC1333 IRAS2. Greyscale and vectors as for Fig. 1.

field then changes to become almost east-west orientated. To the south of the peak, the field is orientated northeast-southwest in direction. This almost radial field at this scale explains the polarisation nulls in the JCMT data – the field direction is significantly different as to cause cancellation of the polarisation within the JCMT beam. At even smaller scales, the magnetic field may have a similar morphology, as once again, the polarisation percentage decreases with increasing intensity.

5. Magnetic fields and outflows

Magnetic fields are thought to be important in the launching and collimation of jets and outflows, and so it may be important to compare the observed magnetic field directions to the jet/outflow direction to check for any relationship on these large scales. Previous studies of T Tauri stars (Ménard & Duchêne 2004) found no relation between the magnetic field and the sample of T Tauri stars as a whole (both with and without jets), although there was a relation between the magnetic field direction and the T Tauri stars with jets. A relation between the magnetic field on these scales and the jet/outflow direction may indicate the large scale magnetic field influencing the outflow direction.

The weighted mean of the **B**-vector position angles were calculated and compared to the position angle of the jet/outflow axis. The magnetic field vectors are not true (i.e. undirectional) vectors. They have a 180° ambiguity and so range from 0° through to 180°. The smallest difference between the magnetic field direction and the outflow direction is taken (i.e. if the outflow position angle for the blueshifted lobe is 160° and the redshifted is 20°, with the magnetic field position angle of 0°, the difference is taken as 20° as the position angle of the redshifted lobe is used). Fig. 6 shows the cumulative distribution functions for the high-mass sample, low-mass sample, and both samples together. It also shows the function expected for a randomly orientated sample. All three distribution functions are close to the randomly orientated function, and the Kolmogorov-Smirnov test reveals that the high-mass sample has a 84.9% chance of being randomly orientated, the low-mass 98.8% and both samples together 99.9%.

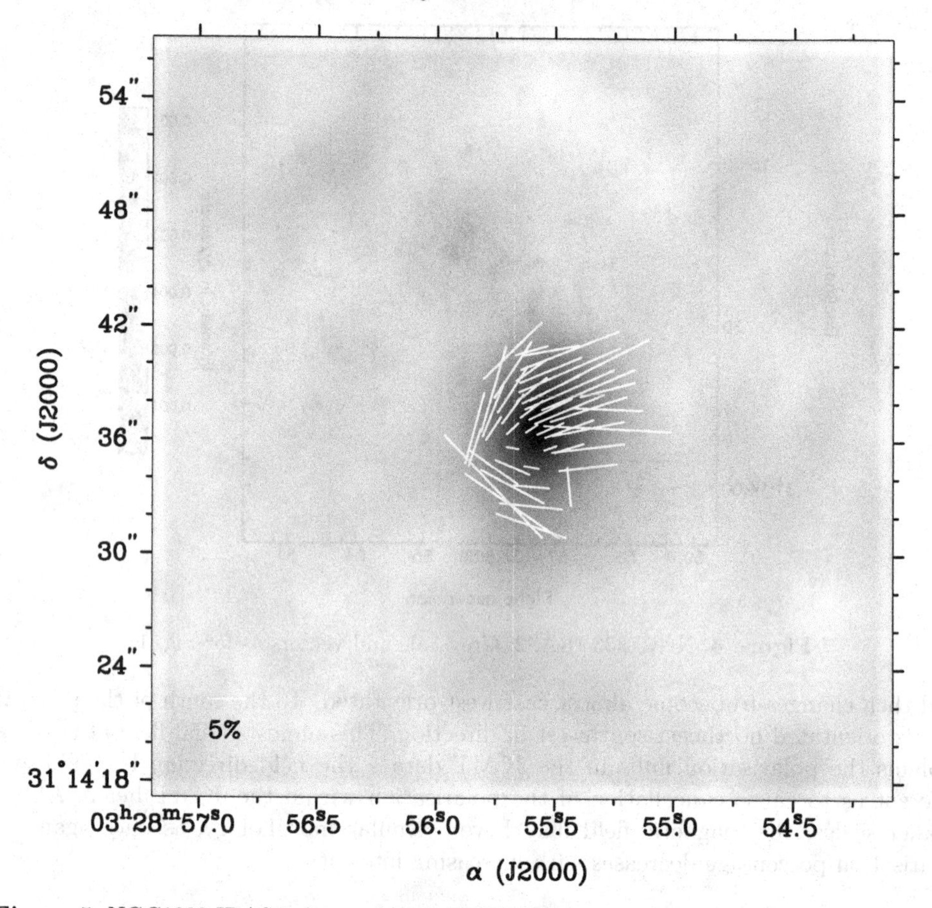

Figure 5. NGC1333 IRAS2 observed with BIMA. The greyscale is the 1.3 mm continuum emission, with the B-vectors overlaid.

This alignment analysis needs to be treated with caution. For some regions, for example Monoceros R2 (Fig. 7), the mean magnetic field direction may not accurately represent the magnetic field direction. Monoceros R2 has a complex field, which changes direction abruptly. The field is north-south in the north of the region, east-west in the east of the region and northeast-southwest in the southwest of the region. The mean magnetic field direction for this region is 63.5°, or northeast-southwest. Interestingly, the outflow system in this region is also complex, with a large scale north-southwest outflow and a smaller easterly outflow. The above analysis implies there is alignment between the mean magnetic field direction and the southwest part of the outflow, however, visual inspection of the field pattern reveals that there may be alignment between the magnetic field and all three observed directions of the outflows. Visual inspection of the field maps may be a better way of determining outflow alignment.

6. Conclusions

We have mapped the plane-of-the-sky magnetic field morphology using submillimetre imaging polarimetry for a sample of 18 high-mass and 8 low-mass star-forming regions. To date, these are the largest samples of this kind. We observe varied field morphologies, with those of the high-mass sample being more complex than those found in the low-mass

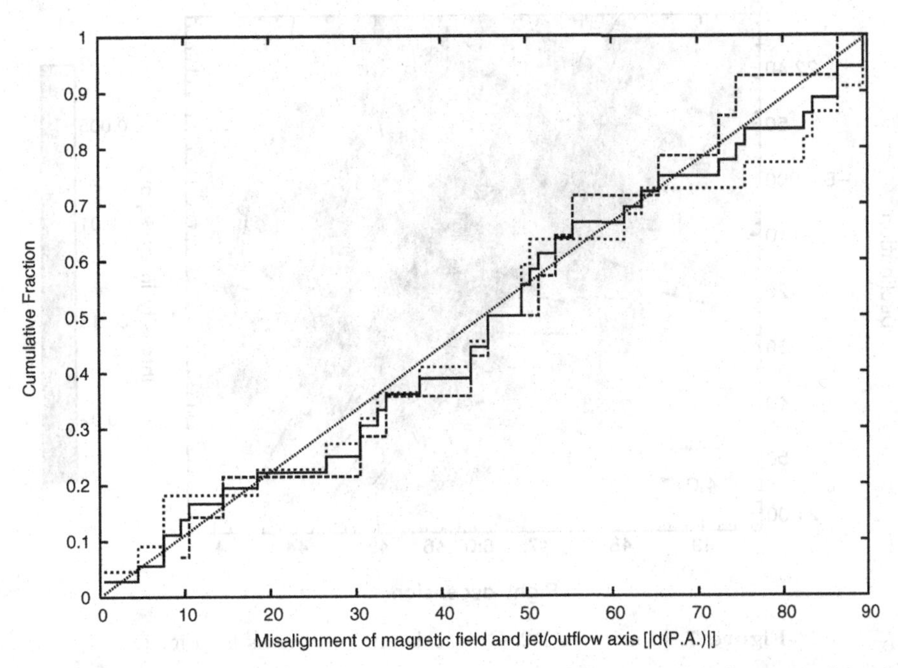

Figure 6. The cumulative distribution function of the difference in PAs between the weighted mean of the magnetic field vectors and the jet/outflow axis. The short-dashed histogram is the high-mass sample, the long-dashed histogram is the low-mass sample. The solid histogram shows the distribution for both samples. The dotted line is the function expected for an infinite randomly orientated sample.

sample. Whilst some of the magnetic field patterns can be interpreted by existing models (e.g. Cepheus A can be compared to the 'hourglass' model from Aitken *et al.* 2002), others cannot (e.g. Monoceros R2).

Half of the observed sources exhibit decreasing polarisation percentage with increasing total intensity. These may have one of three explanations: either the magnetic field is twisted/changes direction rapidly within the beam of the telescope, the dust grains are more spherical in regions of high density, or the dust grains suffer more collisions in regions of high density and so become unaligned to the magnetic field. All three of these explanations result in a lower net polarisation being measured across the intensity peak.

The JCMT data of NGC1333 IRAS2 has a random field pattern with several polarisation nulls across the source, which may indicate a weak field on these scales. Higher resolution data reveal a field pattern that may be consistent with a radial field morphology. Even at 3″ resolution, there is depolarisation across the core, suggesting the same field morphology may be present at even smaller scales.

Analysis of the mean position angle of the magnetic field vectors and the jet/outflow axis reveal that neither the high- or low-mass samples show significant alignment. Kolmogorov-Smirnov tests reveal that there is a 98.8% chance that the low-mass sample is randomly orientated, 84.9% for the high-mass sample and 99.9% chance that both samples are randomly orientated. The mean magnetic field vectors' position angle may not be the best way of summarising the magnetic field direction within such regions, as we observe several sources in which there are abrupt changes in direction of the magnetic field (see for example Monoceros R2, Fig. 7). For regions where the magnetic field direction changes abruptly, modal averages (in reasonable sized bins) may be more

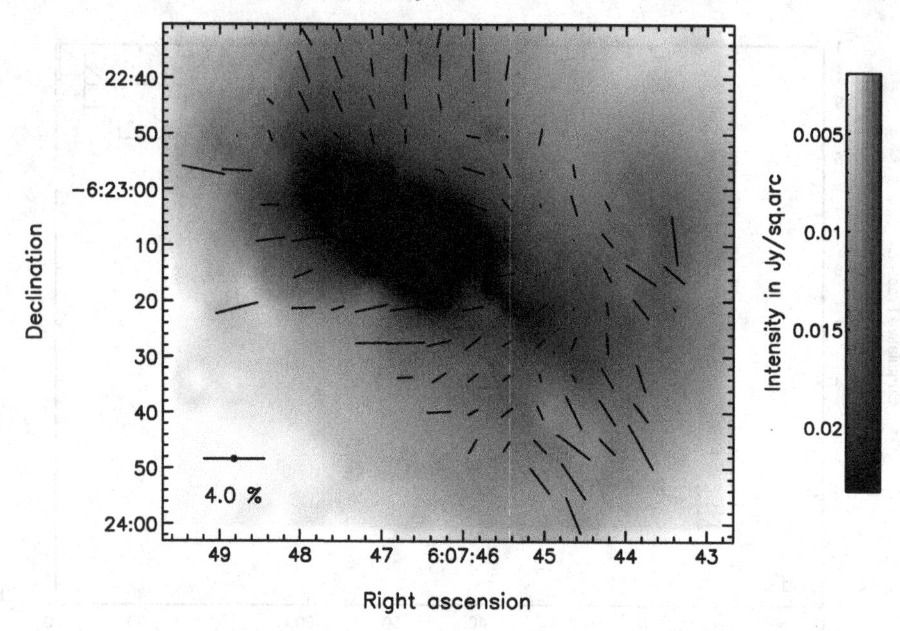

Figure 7. Monoceros R2. Greyscale and vectors as for Fig. 1.

representative. Also, visual inspection of the polarimetry maps is a good way of analysing the alignment.

Acknowledgements

The James Clerk Maxwell Telescope is operated by The Joint Astronomy Centre on behalf of the Science and Technology Facilities Council of the United Kingdom, the Netherlands Organisation for Scientific Research, and the National Research Council of Canada. The BIMA array was operated with support from the National Science Foundation under grants AST-02-28963 to UC Berkeley, AST-02-28953 to U. Illinois, and AST-02-28974 to U. Maryland. R. L. C. is supported by SFI under grant number 04/BRG/P02741.

References

Aitken, D.K., Estathiou, A., Hough, J.H. & McCall, A. 2002, *MNRAS* 329, 647
Chrysostomou, A., *et al.* 2002 *A&A* 385, 1014
Curran, R.L., Chrysostomou, A., Collett, J.L. *et al.* 2004, *A&A* 421, 195
Curran, R.L. & Chrysostomou, A. 2007a, *MNRAS* Submitted.
Curran, R.L. & Chrysostomou, A. 2007b, in prep.
Davis, L. & Greenstein, J.L. 1951, *ApJ* 114, 206
Greaves, J.S., *et al.* 2003 *MNRAS* 340, 353
Holland, W.S., *et al.* 1999 *MNRAS* 303, 659
Lai, S.-P., *et al.* 2001, *ApJ* 561, 864
Matthews, B.C. & Wilson, C.D. 2002 *ApJ* 571, 356
Ménard, F. & Duchêne, G. 2004, *A&A*, 425, 973

Star-Disk Interaction in Young Stars
Proceedings IAU Symposium No. 243, 2007
J. Bouvier & I. Appenzeller, eds.

doi:10.1017/S174392130700943X

Time variable funnel flows

Silvia H. P. Alencar[1]

[1]Departamento de Física – ICEx – UFMG, CP 702, 31270-901, Belo Horizonte, MG, Brazil
email: silvia@fisica.ufmg.br

Abstract. Magnetospheric accretion models are the current consensus to explain the main observed characteristics of classical T Tauri stars. In recent years the concept of a static magnetosphere has been challenged by synoptic studies of classical T Tauri stars that show strong evidence for the accretion process to be dynamic on several timescales and governed by changes in the magnetic field configuration. At the same time numerical simulation results predict evolving funnel flows due to the interaction between the stellar magnetosphere and the inner disk region. In this contribution we will focus on the main recent observational evidences for time variable funnel flows and compare them with model predictions.

Keywords. Stars: pre–main-sequence, stars: magnetic fields, circumstellar matter, accretion.

---

1. From static to dynamical models

Young solar-type stars (T Tauri stars, TTS) typically accrete material from their circumstellar disks for a few million years. Accretion is thought to be the source of the powerful jets observed in young stars and has a deep impact on the evolution of low mass stars by providing mass and angular momentum and by influencing disk evolution at a time when planets are forming. Understanding the accretion process in TTS is therefore an important issue for star and planet formation theories.

Magnetospheric accretion models (Hartmann, Hewett & Calvet 1994; Shu, Najita, Ostriker, *et al.* 1994; Muzerolle, Hartmann & Calvet 2001; Kurosawa, Harries & Symington 2006) are the current consensus to describe the main observational characteristics of accreting TTSs (classical T Tauri stars, CTTSs). In these models, accretion from the disk to the star is mediated by the stellar magnetic field that interacts with the inner circumstellar disk. The inner disk is truncated at a point where the magnetic pressure overcomes the ram pressure due to accretion and the accreting material is channeled to the star following magnetic field lines at near free-fall velocities. As the accreting material hits the star, a shock is formed at the stellar surface and a hot continuum, known as veiling, is emitted. Stellar and disk winds may be present in the models, and help drive away a magnetically driven outflow. The general characteristics of magnetospheric accretion models are supported by observational results. Magnetic fields of 1 to 3 kG have been measured at the surface of T Tauri stars (Johns-Krull, Valenti, Hatzes, *et al.* 1999; Valenti & Johns-Krull 2004; Symington, Harries, Kurosawa, *et al.* 2005), which are strong enough to disrupt the inner disk and channel magnetospheric accretion in funnel flows. Accretion columns are inferred through the common observation of inverse P Cygni profiles with redshifted absorptions reaching several hundred km s^{-1}, which indicates that gas is accreted onto the star from a distance of a few stellar radii (Edwards, Hartigan, Ghandour, *et al.* 1994). Accretion column evidence also comes from the successful modeling of strong optical and IR emission lines that are formed, at least in part, in the magnetospheric accretion flow (Hartmann *et al.* 1994; Muzerolle *et al.* 2001). Evidence of accretion shocks at the stellar surface comes from the successful fitting of the UV and optical excesses observed in CTTSs by accretion shock models (Calvet &

Gullbring 1998; Gullbring, Calvet, Muzerolle, *et al.* 2000) and from the analysis of light curve variability due to hot spots at the stellar surface (Bouvier, Covino, Kovo, *et al.* 1995). Numerical simulations of the magnetic star-disk interaction have also given support to the general magnetospheric accretion scenario. Starting from a large-scale dipolar magnetosphere configuration, simulation results show that an inner magnetospheric cavity develops and material is accreted to the star through funnel flows. As the accreting material hits the star, a hot spot or ring is formed, as predicted. The structure of funnel flows and hot spots depends on the accretion rate, on the topology of the magnetosphere and its inclination (Romanova, Ustyugova, Koldoba, *et al.* 2002; Romanova, Ustyugova, Koldoba, *et al.* 2003).

Although both observations and numerical simulations give a good support to the general predictions of magnetospheric accretion models, some problems with the standard models still remain. Most magnetospheric accretion models assume an axisymmetric configuration for the stellar magnetosphere. Observational results have shown however that the stellar magnetic axis can be inclined with respect to the rotation axis, causing periodical modulation of accretion and outflow diagnostics. Johns & Basri (1995) have shown that the CTTS SU Aur presents inflow and outflow signatures that vary periodically at the stellar rotation period but are enhanced at opposite phases. They interpreted their results as due to an inclined dipole field with respect to the rotation axis that favors accretion and outflow 180° out of phase. These results were later confirmed by Petrov, Gullbring, Ilyin, *et al.* (1996) with observations separated by several years from those analysed by Johns & Basri (1995). Rotational modulation by hot spots (Bouvier *et al.* 1995) is also expected to arise due to an inclined magnetosphere, since an axisymmetric magnetospheric configuration would produce a uniform hot ring at the stellar surface whose emission would not vary as the star rotates. Rotational modulation by hot spots will also cause periodic veiling variations that are observed in CTTS systems (Bouvier, Grankin, Alencar, *et al.* 2003; Bouvier, Alencar, Boutelier, *et al.* 2007). Evidence for inclined magnetospheres also comes from the rotational modulation of the circular polarization signal measured in the HeI 5876Å line of several CTTSs (Valenti & Johns-Krull 2004). Very recently, the first emission line profiles were calculated based on 3D MHD simulation results of Romanova *et al.* (2003) and Romanova, Ustyugova, Koldoba, *et al.* (2004b), who considered accretion onto a CTTS with a misaligned dipole with respect to the rotation axis. Kurosawa, Romanova & Harries (2007) took the density, velocity and temperature structures of the MHD simulations and used them to calculate the rotationally induced line variability of several emission lines (Hβ, Paβ and Brγ) that are commonly observed in CTTSs. This will allow for the first time a direct comparison between the emission line profiles originating from a complex magnetospheric structure and the time variable observed line profiles of several CTTSs.

Another common assumption of standard magnetospheric accretion models is steady-state accretion. The variability observed on a timescale of hours in emission lines of CTTSs such as TW Hya (Alencar & Batalha 2002), which is seen almost pole-on and is therefore not subjected to rotational modulation effects, seems to rule out steady-state accretion as a straightforward assumption. Overall the accretion process appears to be variable on several timescales, in hours due to non-steady accretion, like the observed veiling changes of RU Lup that varies during a single night (Stempels & Piskunov 2002), in days due to rotational modulation effects caused by an inclined magnetosphere, like the periodical variability of the redshifted absorption component of Hβ in SU Aur (Johns & Basri 1995), in weeks due to global instabilities of the stellar magnetosphere caused by differential rotation between the star and the disk, as discussed in the following

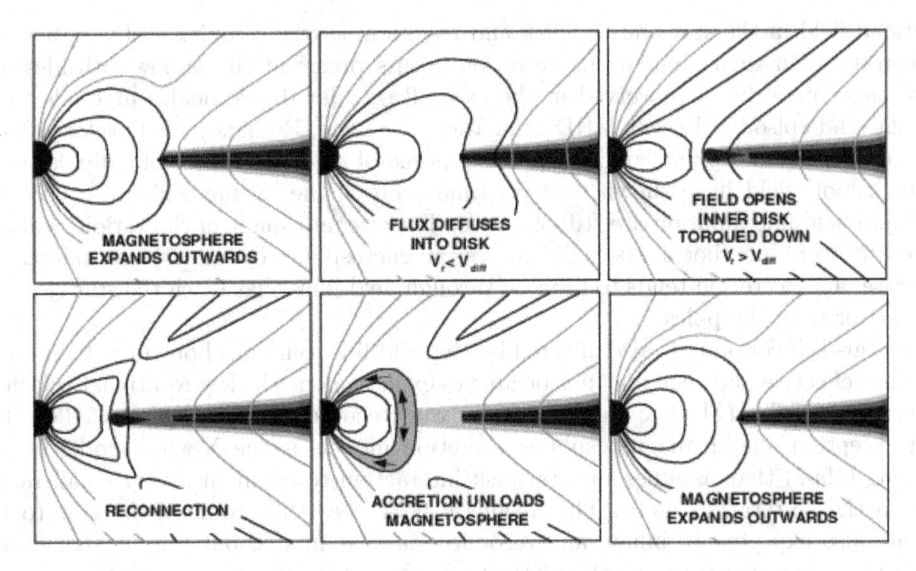

Figure 1. Differential rotation along the field lines leads to their expansion, opening and reconnection. From Goodson & Winglee (1999).

paragraphs, and in years in EXor and FUor like events (Herbig 2007; Reipurth & Aspin 2004).

Most magnetospheric accretion models assume a static magnetospheric configuration, where the stellar magnetic field may interact with the disk over a significant region. However, it is then expected that many stellar field lines will thread the disk at points that rotate differentially from the star. As the system rotates, this will give rise to magnetic field lines distorted by differential rotation, as commonly observed in numerical simulations. Some numerical simulations of the star-disk interaction predict that differential rotation between the star and the inner disk regions where the stellar magnetosphere is anchored, leads to the field lines expansion, opening and reconnection, which eventually restores the initial magnetospheric configuration (Fig. 1; Goodson, Winglee & Böhm 1997; Goodson & Winglee 1999; Matt, Goodson, Winglee, *et al.* 2002; Uzdensky, Königl & Litwin 2002; Romanova, Ustyugova, Koldoba, *et al.* 2004a; von Rekowski & Brandenburg 2004). Magnetospheric inflation cycles are expected by most numerical models to develop in a few Keplerian periods at the inner disk, and be accompanied by strong periodic outflow events, as reconnection takes place, and by time dependent accretion onto the star. The timescale of field line inflation is actually determined by the diffusion of the magnetic flux through the inner disk regions, which depends on the magnetic diffusivity in the disk that is unfortunately an uncertain parameter to calculate from first principles.

Most magnetospheric accretion models also assume a magnetosphere with a dipolar field geometry. Recently several observational results have indicated that the surface magnetic fields of young low-mass stars may be very complex and include high order multipoles (Valenti & Johns-Krull 2004; Jardine, Cameron, Donati, *et al.* 2006; Daou, Johns-Krull & Valenti 2006; Donati, Jardine, Petit, *et al.* 2007). Accretion may then eventually proceed through multipole filed lines if the interaction region between star and disk is not very far away from the stellar surface for only the dipole component to be important. This is more likely to happen in stars that present high accretion rates, since in that case the inner gas disk will extend closer to the star than in the low mass accretion rate systems. Multipolar axisymmetric fields were recently modeled by von Rekowski & Brandenburg (2006). Their time-dependent simulations include a dynamo generated field in the star and the disk and results in a very complex and variable field configuration. In their simulations accretion occurs preferentially at low latitudes and is no longer periodic, as observed in the case of a stellar dipole field, but tends to be irregular and episodic. Recent MHD simulations by Long, Romanova & Lovelace (2007) of an initial magnetosphere configuration composed of a quadrupole or of a dipole added to quadrupole field have shown that, as time evolves, the magnetospheric cavity, hot spots and field line inflation are still observed. The size and shape of the various features related to accretion (hot spots, light curves, magnetosphere dynamics) are different in each case, and accretion tends to proceed through field lines that reach the star closer to the equator than the poles.

The star-disk coupling is also affected by the field line dynamical opening. It becomes much less effective and then may not be able to explain alone the low rotational velocities observed in CTTSs (Matt & Pudritz 2005a; von Rekowski & Brandenburg 2006). The main exception among magnetospheric accretion models is the X-wind model of Shu & Shang (1997) that assumes the star-disk interaction to occur in a very small region close to the co-rotation point. The magnetic field lines that connect the star to the disk are also expected to inflate and reconnect in that model, but the departure from corotation is on average very small and the long-term interaction between the star and the disk can be analysed as a steady state of corotation. Therefore the angular momentum transport between the star and the disk is not much affected by the opening of the field lines in the X-wind model. Most models however do have an angular momentum problem and stellar winds powered by accretion were suggested to help regulate the stellar angular momentum of accreting T Tauri stars (Matt & Pudritz 2005b; Kwan, Edwards & Fischer 2007; see also the contributions of Suzan Edwards and Sean Matt in this volume). To better understand the star-disk interaction it is therefore important to look for observational evidence of magnetic field line opening in CTTS systems.

Observational results have in recent years started to confirm the time-dependent nature and the many predictions of numerical simulations of the star-disk interaction. Episodic high velocity outbursts, possibly related to magnetospheric reconnection, have been reported for a few systems based on the velocity variation of blueshifted absorption components in emission line profiles on a timescale of hours to days (Alencar, Johns-Krull & Basri 2001; Ardila, Basri, Walter, *et al.* 2002; Oliveira, Foing, van Loon, *et al.* 2000). Possible evidence for twisted magnetic field lines due to differential rotation between the star and the disk was proposed for SU Aur by Oliveira *et al.* (2000). A time delay of a few hours was measured between the appearance of high velocity redshifted absorption components in line profiles formed at different altitudes in the accretion column. This was interpreted as the crossing of an azimuthally twisted accretion column on the line of sight. Another possible evidence for magnetic field lines being twisted by differential rotation between the star and the disk and leading to quasi-periodic reconnection events was reported for the embedded protostellar source YLW 15, based on the observations of quasi-periodic X-ray energetic flares coming from large plasma-confining loops (loop length of $\sim 14\ R_\odot \sim 3\ R_\star$, Montmerle, Grosso, Tsuboi, *et al.* (2000) and Fig. 2). Evidence for very large flaring structures (much larger than the stellar radius), likely coming from reconnection events of magnetic loops that link the young stars to their disks, has also been reported by Favata, Flaccomio, Reale, *et al.* (2005) for about 15 stars in Orion based on the analysis of some of the most energetic X-ray flares observed in the Chandra Orion Ultradeep Project.

There is nowadays observational evidence that accretion and wind activity are non-steady processes acting in the star-disk systems throughout the lifetime of accretion disks, which in certain cases may extend up to ~ 10 Myr (Jayawardhana, Coffey, Scholz

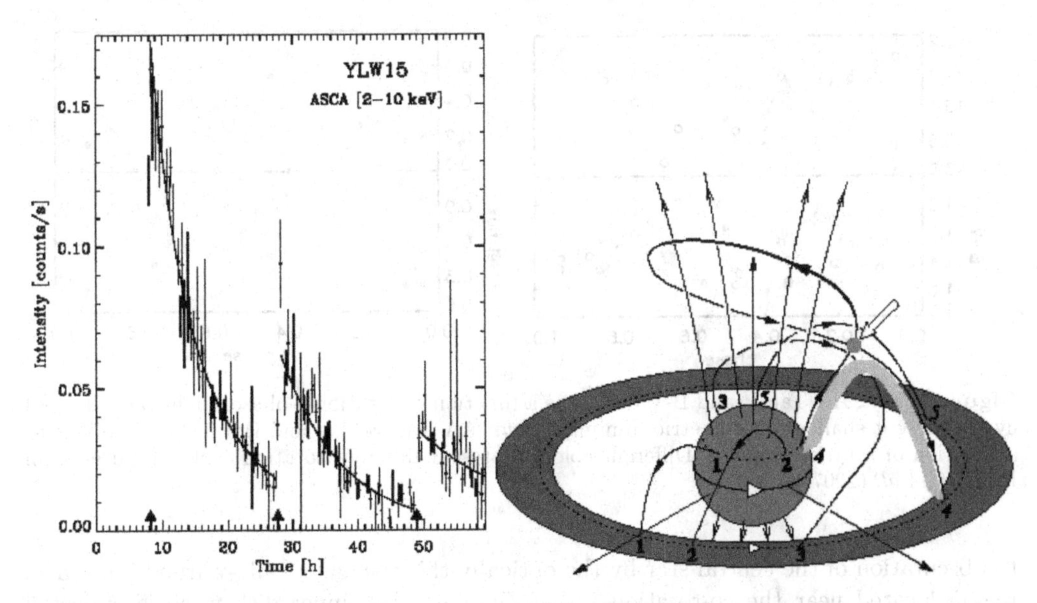

Figure 2. *Left*: light curve of YLW 15 obtained with ASCA (Tsuboi, Imanishi, Koyama, *et al.* 2000) showing three energetic flare events. *Right*: sketch of a plausible evolution of magnetic field configuration leading to reconnection and flaring as observed in the light curve of YLW 15. Figures from Montmerle *et al.* (2000).

et al. 2006). There is also recent observational evidence for time variable magnetospheric accretion onto brown dwarfs (Scholz & Jayawardhana 2006). These results suggest that the dynamic star-disk interaction probably controls the accretion process in very different mass domains and over a large time span, as long as an accretion disk is present. As discussed in this volume by Robert Mathieu, close binary systems, where circumstellar disks cannot survive, but only a circumbinary disk is present are also seen to accrete with the same characteristics as the stars we nowadays believe to be single (Mathieu, Stassun, Basri, *et al.* 1997; Alencar, Melo, Dullemond, *et al.* 2003; Jensen, Dhital, Stassun, *et al.* 2007). Accretion, as we know it, then manages to proceed in environments with very different physical characteristics too.

The ideal way to study the dynamical star-disk interaction is to follow young star-disk systems for several rotation periods and to look for evidences of variable accretion and outflow events that can be related to time variable funnel flows and other predictions of numerical models. This was done for the CTTS AA Tau at different epochs and the main results are discussed in the following section.

2. AA Tau: a test case for the dynamical evolution of magnetospheric accretion

AA Tau is one of the few CTTSs that was studied synoptically with simultaneous photometry and high-resolution spectroscopy covering several stellar rotation periods at different epochs. The 1995, 1999 and 2004 observing campaigns yielded a huge amount of detailed information about the accretion and outflow processes and their connection (Bouvier, Chelli, Allain, *et al.* 1999; Bouvier *et al.* 2003; Bouvier *et al.* 2007).

The AA Tau system presents modulation of photometric and spectroscopic diagnostics with a period of 8.2 days, which is the rotational period of the star. The light curves show quasi-cyclic recurrent minima with very little color changes that were attributed

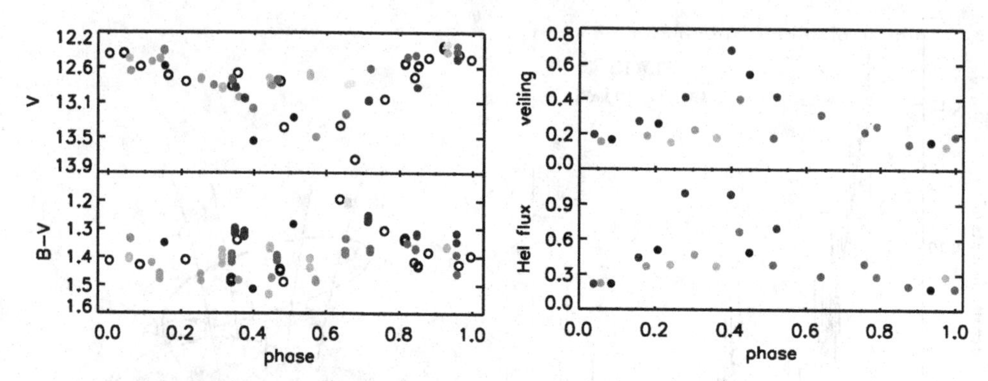

Figure 3. *Left*: AA Tau V and B-V curves as a function of rotational phase. Light blue and red cycles present shallow photometric minima. *Right*: AA Tau veiling and HeI 5876Å line flux as a function of rotational phase. Different colors represent different rotation cycles. Figures from Bouvier *et al.* (2007).

to obscuration of the central star by the optically thick, magnetically-warped inner disk region located near the co-rotation radius (Fig. 3). The inner disk warp is expected by magnetospheric models to develop due to the interaction of the disk with an inclined (non-axisymmetric) magnetosphere (Terquem & Papaloizou 2000; Romanova *et al.* 2003). A blueshifted absorption component is seen all the time in the Balmer lines and is supposed to arise from a cold inner disk wind. The redshifted absorption components of the Balmer line profiles exhibit a periodic modulation at the stellar rotation period with a maximum strength when the main accretion funnel flow passes through our line of sight, which also corresponds to the photometric minima. Redshifted absorption is thought to appear when the hot spot region is seen through the accretion funnel and is therefore a strong accretion diagnostic. The veiling is also modulated by the appearance of the accretion shock and the photospheric and HeI radial velocities present periodic variations attributed to spots (cold and hot) at the stellar surface. All the above characteristics are present in the two observational campaigns that had simultaneous photometric and spectroscopic measurements (1999 and 2004) and give support to the general predictions of magnetospheric accretion models.

The AA Tau system also presented evidence for large scale changes occurring in the magnetospheric structure of the star on timescales of weeks and years. These cannot be explained by static magnetospheric models and require time-variable accretion funnels.

The AA Tau light curve presents cycles of deep and shallow eclipses, where the obscuring screen characteristics are very distinct. Furthermore, the cycles that only show a shallow photometric minimum do not exhibit a very deep redshifted absorption component in the Balmer lines and they present low veiling and HeI 5876Å line flux values (light blue and red dots in Fig. 3). The HeI 5876Å line in AA Tau is composed most of the time of only a narrow component that is thought to come from a region close to the accretion hot spot (Beristain, Edwards & Kwan 2001), so like the veiling, it should also trace the physical conditions near the accretion spot. The spectroscopic characteristics described above indicate that accretion is at a very low level when most of the occulting circumstellar material is absent. There is therefore a clear correlation between the accretion level, as measured by spectroscopic features such as veiling and line components related to the accretion shock region, and the formation of the inner disk warp, as measured by the photometric minimum depth, both supposedly depending on the time variable magnetic configuration at the disk truncation radius.

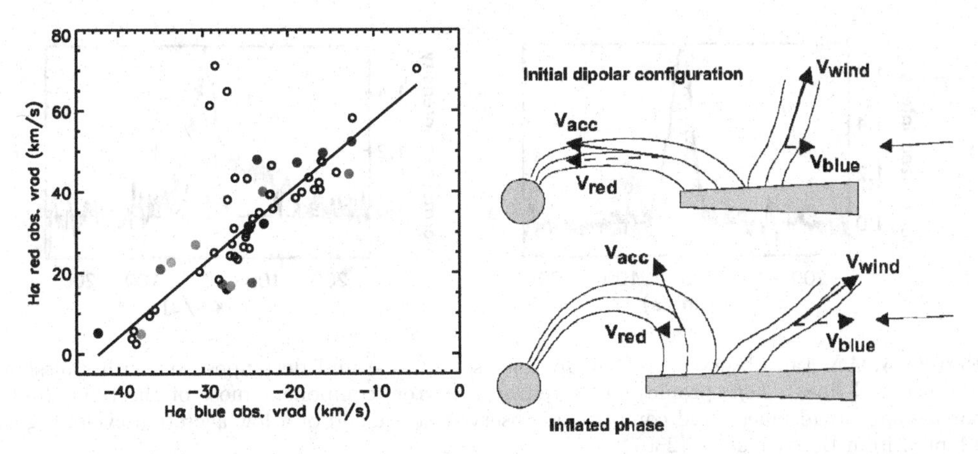

Figure 4. *Left*: measured relation between the radial velocities of Hα redshifted and blueshifted absorption components. Open symbols correspond to the 1999 campaign and filled symbols to the 2004 campaign. Different colors refer to different rotation cycles. *Right*: sketch of magnetospheric inflation showing the radial velocities of the blue (wind) and red (accretion) absorption components of Hα. The arrow on the right side indicates the line of sight to the AA Tau system (Bouvier *et al.* 2003; Bouvier *et al.* 2007).

There is also a very tight correlation between the radial velocities of the redshifted (accretion) and blueshifted (wind) absorption components of the Hα line that was observed in both the 1999 and 2004 campaigns. This provides evidence for a physical connection between time dependent accretion and outflow in CTTSs. It was interpreted by Bouvier *et al.* (2003) as a first observational evidence of inflation of the stellar magnetosphere caused by differential rotation between the stellar magnetic field lines and the inner disk, which are predicted by numerical simulations of star-disk interaction (see § 1). The idea is illustrated in Figs. 1 and 4. Starting from a initial dipolar configuration, differential rotation between the inner disk edge and the stellar surface inflates the magnetic field lines that thread the inner disk region, reducing the accretion flow onto the star. At a certain critical point, the field lines open and reconnect leading to enhanced wind outflow. After reconnection, the initial dipolar configuration is restored. It was proposed by Bouvier *et al.* (2003) that the projected radial velocity of the redshifted absorption component of Hα measures the curvature of the accretion funnel, as shown in Fig. 4. If the wind traced by the blueshifted absorption component of the Balmer lines originates close to the inner disk edge, the projected radial velocity of this component is expected to decrease as that of the redshifted absorption component increases, thus providing the observed correlation. This is exactly what is observed in AA Tau in both campaigns. Just before and during the shallow eclipses and low accretion cycles the field lines are inflated and opening, as measured by the projected radial velocities of the wind and accretion components. In this phase a hot wind outflow develops, traced by the appearance of a broad blueshifted component in the HeI 5876Å line (Fig. 5), which is attributed to hot winds according to Beristain, Edwards & Kwan (2001). After this, the deep eclipses and normal veiling values are restored, as is the initial configuration of the accretion funnel. These results tend to support the idea of magnetospheric accretion cycles on a timescale of several rotation periods in accreting T Tauri stars, although the periodicity of such episodes as predicted by dynamical models still remains to be tested with longer time series.

AA Tau also presented long term photometric variations. Comparing four observational photometric campaigns from 1995, 1999, 2003 and 2004, Bouvier *et al.* (2007) showed

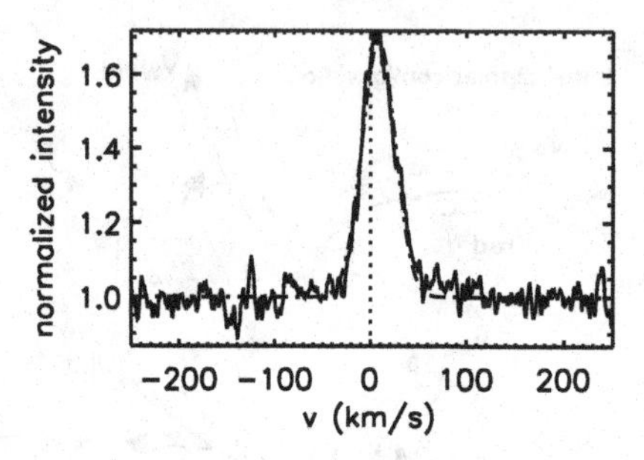

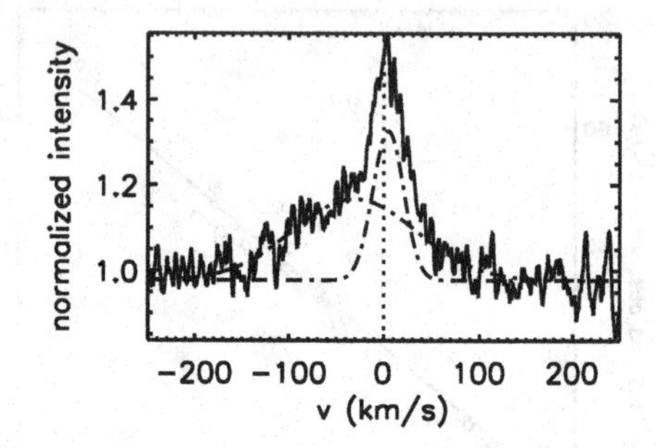

Figure 5. AA Tau observed HeI 5876Å line (solid lines) and decomposition with gaussians (dash-dotted lines). *Left*: profile dominated by a narrow component most of the time. *Right*: narrow and broad blueshifted components observed at the end of a low accretion cycle. Figure adapted from Bouvier *et al.* (2007).

that the shape and depth of the eclipses can change dramatically in years, although they vary only slightly on a timescale of weeks within each observing campaign. The general characteristics of the light curve is the same in the several observing seasons, all presenting a constant maximum interrupted by episodes of recurring minima with little color change. However, the long term changes of the light curve shape in a timescale of years can be quite important. The light curves can change from a broad very deep eclipse per cycle that does not show much variability amplitude (1995 campaign), to two shallower and variable eclipses per cycle (1999 campaign), back to a major minimum per cycle with variable amplitude (2003 and 2004 campaigns). This indicates that the shape of the obscuring screen can change drastically over the years, which suggests that the stellar magnetic field interaction with the inner disk, that is thought to produce the inner disk warp, varies significantly over the years too.

The AA Tau system showed that the accretion process is variable on several timescales, all of them related to the asymmetric, non-steady magnetospheric configuration. It varies in days, as shown by the veiling changes due to an inclined magnetosphere with respect to the rotation axis, it varies in weeks due to inflation cycles caused by differential rotation between the stellar magnetosphere and inner disk regions and it varies in years, as picture by the variable overall configuration of the inner disk warp that is thought to arise due to the interaction of an inclined magnetic field and the inner circumstellar disk.

3. Conclusion

We have shown that magnetospheric accretion models can account for the general observed characteristics of CTTSs. However, axisymmetric steady-state magnetospheric models cannot explain the time variable characteristics of such accreting young systems.

The star-disk interaction was shown to be very dynamical on several timescales (hours, weeks, months and years) and is thought to be mediated by the stellar magnetic field. This dynamical aspect of the star-disk interaction may have implications to the short and long term variability of inflows and outflows and to the angular momentum evolution of the star-disk system.

Finally, in CTTSs, accretion and outflow are strongly related processes and models that try to explain these systems should always take both into account.

Acknowledgements

S.H.P.A. would like to acknowledge support from IAU, CNPq and Fapemig.

References

Alencar, S. H. P., & Batalha, C. 2002, *ApJ* 571, 378
Alencar, S.H.P., Johns-Krull, C.M. & Basri, G. 2001, *AJ* 122, 3335
Alencar, S.H.P., Melo, C.H.F., Dullemond, C.P., Andersen, J., Batalha, C., Vaz, L.P.R., & Mathieu, R.D. 2003, *A&A* 409, 1037
Ardila, D.R., Basri, G., Walter, F.M., Valenti, J.A., & Johns-Krull, C.M. 2002, *ApJ* 566, 1100
Beristain, G., Edwards, S., & Kwan, J. 2001, *ApJ* 551, 1037
Bouvier, J., Covino, E., Kovo, O. *et al.* 1995, *A&A* 299, 89
Bouvier, J., Chelli, A., Allain, S., Carrasco, L., Costero, R., Cruz-Gonzalez, I., Dougados, C., Fernandez, M., Martin, E.L., Menard, F., Mennessier, C., Mujica, R., Recillas, E., Salas, L., Schmidt, G., & Wichmann, R. 1999, *A&A* 349, 619
Bouvier, J., Grankin, K.N., Alencar, S.H.P., Dougados, C., Fernandez, M., Basri, G., Batalha, C., Guenther, E., Ibrahimov, M.A., Magakian, T.Y., Melnikov, S.Y., Petrov, P.P., Rud, M.V., & Zapatero Osorio, M.R. 2003, *A&A* 409, 169
Bouvier, J., Alencar, S.H.P., Boutelier, T., Dougados, C., Balog, Z., Grankin, K., Hodgkin, S.T., Ibrahimov, M.A., Kun, M., Magakian, T.Yu., & Pinte, C. 2007, *A&A* 463, 1017
Calvet, N. & Gullbring, E. 1998, *ApJ* 509, 802
Daou, A.G., Johns-Krull, C.M. & Valenti, J.A. 2006, *AJ* 131, 520
Donati, J.-F., Jardine, M.M., Petit, P., Morin, J., Bouvier, J., Cameron, A.C., Delfosse, X., Dintrans, B., Dobler, W., Dougados, C., Ferreira, J., Forveille, T., Gregory, S.G., Harries, T., Hussain, G.A.J., Menard, F., & Paletou, F. 2007, in: G. van Belle (ed.), *14th Cambridge Workshop on Cool Stars, Stellar Systems, and the Sun*, ASP Conf. Series, in press
Edwards, S., Hartigan, P., Ghandour, L., & Andrulis, C. 1994, *AJ* 108, 1056
Favata, F., Flaccomio, E., Reale, F., Micela, G., Sciortino, S., Shang, H., Stassun, K.G., & Feigelson, E.D. 2005, *ApJS* 160, 469
Goodson, A.P., Winglee, R.M. & Böhm, K.-H. 1997, *ApJ* 489, 199
Goodson, A.P. & Winglee, R.M. 1999, *ApJ* 524, 159
Grankin, K. N., Melnikov, S. Y., Bouvier, J., Herbst, W., & Shevchenko, V. S. 2007, *A&A* 461, 183
Gullbring, E., Calvet, N., Muzerolle, J., & Hartmann, L. 2000, *ApJ* 544, 927
Hartmann, L., Hewett, R. & Calvet, N. 1994, *ApJ* 426, 669
Herbig, G.H. 2007, *AJ* 133, 2679
Jardine, M., Cameron, A.C., Donati, J.-F., Gregory, S.G., & Wood, K. 2006, *MNRAS* 367, 917
Jayawardhana, R., Coffey, J., Scholz, A., Brandeker, A., & van Kerkwijk, M.H. 2006, *ApJ* 648, 1206
Jensen, E.L.N., Dhital, S., Stassun, K.G., Patience, J., Herbst, W., Walter, F.M., Simon, M., & Basri, G. 2007, *AJ* 134, 241
Johns, C.M., & Basri, G. 1995, *ApJ* 449, 341
Johns-Krull, C.M., Valenti, J.A., Hatzes, A.P., & Kanaan, A. 1999, *ApJ* (Letters) 510, L41
Kurosawa, R., Harries, T.J. & Symington, N.H. 2006, *MNRAS* 370, 580
Kurosawa, R., Romanova, M.M. & Harries, T.J. 2007, *Star-disk interaction in young stars*, IAU Symposium No. 243, poster available at http://www.iaus243.org/
Kwan, J., Edwards, S. & Fischer, W. 2007, *ApJ* 657, 897
Long, M., Romanova, M.M. & Lovelace, R.V.E. 2007, *MNRAS* 374, 436
Mathieu, R.D., Stassun, K., Basri, G., Jensen, E.L.N., Johns-Krull, C.M., Valenti, J.A., Hartmann, L.W. 1997, *AJ* 113, 1841
Matt, S., Goodson, A.P., Winglee, R.M., Böhm, K.-H. 2002, *ApJ* 574, 232
Matt, S. & Pudritz, R.E. 2005a, *MNRAS* 356, 167
Matt, S. & Pudritz, R.E. 2005b, *ApJ* (Letters) 647, L45
Montmerle, T., Grosso, N., Tsuboi, Y., Koyama, K. 2000, *ApJ* 532, 1097
Muzerolle J., Hartmann L. & Calvet N. 2001, *ApJ* 550, 944

Oliveira, J.M., Foing, B.H., van Loon, J.T., & Unruh, Y.C. 2000, *A&A* 362, 615
Petrov, P.P., Gullbring, E., Ilyin, I., Gahm, G.F., Tuominen, I., Hackman, T., & Loden, K. 1996, *A&A* 314, 821
Reipurth, B. & Aspin, C. 2004, *ApJ* (Letters) 608, L65
Romanova, M.M., Ustyugova, G.V., Koldoba, A.V., & Lovelace, R.V.E. 2002, *ApJ* 578, 420
Romanova, M.M, Ustyugova, G.V., Koldoba, A.V., Wick, J.V., & Lovelace, R.V.E. 2003, *ApJ* 595, 1009
Romanova, M.M., Ustyugova, G.V., Koldoba, A.V., & Lovelace, R.V.E. 2004a, *ApJ* (Letters) 616, L151
Romanova, M.M., Ustyugova, G.V., Koldoba, A.V., & Lovelace, R.V.E. 2004b, *ApJ* 610, 920
Scholz, A., & Jayawardhana, R. 2006, *ApJ* 638, 1056
Shu, F., Najita, J., Ostriker, E., Wilkin, F., Ruden, S., & Lizano, S. 1994, *ApJ* 429, 781
Shu, F.H., & Shang, H. 1997, in: B. Reipurth & C. Bertout (eds.), *Herbig-Haro Flows and the Birth of Stars*, IAU Symposium No. 182 (Kluwer Academic Publishers), vol. 182, p. 225
Stempels, H.C. & Piskunov, N. 2002, *A&A* 391, 595
Symington, N.H., Harries, T.J., Kurosawa, R., & Naylor, T. 2005, *MNRAS* 358, 977
Terquem, C. & Papaloizou, J.C.B. 2000, *A&A* 360, 1031
Tsuboi, Y., Imanishi, K., Koyama, K., Grosso, N., & Montmerle, T. 2000, *ApJ* 532, 1089
Uzdensky, D.A., Königl, A. & Litwin, C. 2002, *ApJ* 565, 1191
Valenti, J.A. & Johns-Krull, C.M. 2004, *Ap&SS* 292, 619
von Rekowski, B. & Brandenburg, A. 2004, *A&A* 420, 17
von Rekowski, B. & Brandenburg, A. 2006, *Astron. Nachr.* 327, 53

Discussion

EDWARDS: We don't normally see CTTS light curves with deep recurrent minima like the ones observed in AA Tau. Is there anything special about this system that makes it more likely to have an inner disk warp that eclipses the stellar photosphere ? Is the inclination of the magnetic axis with respect to the rotation axis particularly large ?

ALENCAR: In a recent paper by Grankin *et al.* (2007), they show that only about 15% of the CTTSs in their sample exhibit light curves similar to AA Tau. This is indeed a small number of objects and the authors suggest that this could be related to an inclination effect. Only systems seen at high inclination with respect to the line of sight, which is the case of AA Tau, would exhibit a light curve pattern related to the occultation of the photosphere by circumstellar material from the disk. Regarding AA Tau specifically, Kurosawa and collaborators also had to use a large inclination ($> 60°$) of the magnetic axis with respect to the rotation axis to try to qualitatively reproduce the observed time variable Hβ line with the magnetospheric configuration and physical parameters of the Romanova *et al.* (2003) 3D MHD simulations. However, Bouvier *et al.* (2007) estimated the tilt between the axis of the large scale magnetosphere and the rotation axis to be around 20° based on the modeling of radial velocity variations of the HeI 5876Å line, supposed to be due to a major hot spot at the stellar surface. Valenti & Johns-Krull (2004) derived an inclination of 12° between the two axis from spectropolarimetric measurements. So it is not clear if the inclination of the magnetic to the rotation axis is very large in AA Tau and I would say that so far there is no conclusive indication that AA Tau is a special system among CTTSs. It is apparently just a system seen at high inclination with respect to the line of sight.

SOMEONE: How do the observed line profiles compare with those calculated from the magnetospheric configuration of the Romanova MHD simulations ?

ALENCAR: The first time I saw emission line profiles calculated using the density, temperature and geometric structure of the 3D MHD simulations was in this conference in the poster by Ryuchi Kurosawa. I talked to him and they have just started to compare general features of theoretical and observed line profiles for AA Tau. Some characteristics are well reproduced by the model profiles, like the overall scale of variability across the Hβ line, but the theoretical profiles do not reproduce other observed fearures, like the absorption near the Hβ line center. They plan to do a direct comparison between observed and calculated line profiles soon.

JOHNS-KRULL: I would like to add a comment. We published time series spectroscopic observations of T Tau, which is a CTTS viewed pole-on, as is TW Hya which was discussed in the talk. T Tau also displays strong emission line variability, indicating that the mass accretion rate is variable on this star on short timescales as well.

SHU: We must interpret with care the importance of dynamical predictions of numerical simulations, since the simulations still run over a small timescale compared to the PMS evolution.

ALENCAR: I agree. The simulations typically run only over a few hundred (thousand ?) Keplerian periods at the inner disk edge.

ROMANOVA: I disagree. We have reached a steady state in our simulations. So our predictions are robust.

Star-disk interaction in young stars
Proceedings IAU Symposium No. 243, 2007
J. Bouvier & I. Appenzeller, eds.

doi:10.1017/S1743921307009441

Radiative-transfer modelling of funnel flows

Tim J. Harries

School of Physics, University of Exeter, Stocker Road, Exeter, EX4 4QL, UK
email : th@astro.ex.ac.uk

Abstract. Emission line profiles from pre-main-sequence objects accreting via magnetically-controlled funnel flows encode information on the geometry and kinematics of the material on stellar radius scales. In order to extract this information it is necessary to perform radiative-transfer modelling of the gas to produce synthetic line profiles. In this review I discuss the physics that needs to be included in such models, and the numerical methods and assumptions that are used to render the problem tractable. I review the progress made in the field over the last decade, and summarize the main successes and failures of the modelling work.

Keywords. Radiative transfer, line: formation, stars: formation, stars: magnetic fields.

---

1. Introduction

The permitted emission lines from Classical T Tauri stars (CTTS) are thought to arise from the hot funnel flows and winds that occur within a few stellar radii of the stellar photosphere. These lines display a wide variety of morphologies, including double- and single-peaked profiles along with classical and inverse P Cygni absorption dips (see for example Reipurth *et al.* 1996). The lines often show high levels of variability on a time-scale of hours to days, with the former likely to be characterstic of the infall timescale and the latter representing rotational modulation (see contribution by Alencar in these proceedings).

Although the accretion rates are best determined from the blue continuum excess, this is only possible for the highest accretion rates. However the strength and shape of the emission lines are also an accretion rate diagnostic, albeit a more model-dependent one. The broad aim of this field is to find a self-consistent model for the geometry and kinematics of the material as it falls onto, or is ejected from, the protostellar system. In order to construct such a model it is first necessary to have a representation of the magnetosphere, perhaps from a canonical analytical description, or from an MHD model, or even from a magnetic field description derived from observation. This density and velocity structure may then be used in combination with a radiative-transfer code to compute line profiles, and comparison with observations then allows one to refine the magnetospheric description.

We are, frankly, a long way from such a self-consistent model, but the last decade has seen significant advances which mean that the idea of developing such a model is not quite as fanciful as it was 10 years ago. In the following section I detail the physical ingredients that go in to the radiative-transfer model, and then I summarize the main numerical techniques that are used to solve the equation of statistical equilibrium and to compute the profiles. I give a review of the increasingly sophisticated models in subsequent sections, and finish with a critical review of the state-of-the-art.

2. The physical model

The material in the magnetosphere is certainly not distributed in spherical symmetry, and any calculations need to be two, or possibly three, dimensional in nature. The gas is also accelerating, and the velocity pattern is non-monotonic. These properties combine to make the solution of radiative-transfer equations for the magnetospheric a difficult numerical problem, requiring sophisticated modelling codes. Here I detail some of the equations that apply, and the methods and approximations that are used to solve them.

2.1. *Radiative equilibrium*

How hot is the gas in the magnetosphere? How does its temperature vary as a function of height above the stellar surface? These are important questions, as the temperature distribution has a strong effect on the line source function, and therefore the line's shape and morphology. The solution of the radiative equilibrium is, on the face of it, simply a question of comparing the local heating rate with the local cooling rate and iterating on the temperature to balance the two. The problem comes with calculating the rates. The cooling is relatively straightfoward, with processes such as bremstrahlung, collisional recombination, and emission from ionic species all contributing. The heating terms that we can easily identify include adiabatic heating due to the convergence of the flow towards the protostar, and photoionization by UV radiation from the hot spots. However, as with the solar corona, there are almost certainly magnetic heating effects that are very difficult to quantify and are in fact probably dominant.

The only paper I am aware of that explicitly addresses the question of radiative-equilibrium in magnetospheric accretion in CTTS is that of Martin (1996), who considered a simple dipolar field with accretion occuring in free-fall along convergent field lines. He did not include any magnetic heating, and discovered that cooling by Mg II and Ca II was important near the photosphere, and that adiabatic compression was the dominant heating source. However, Hartmann, Hewett & Calvet (1994) found that line profiles from their radiative-transfer calculations based on the Martin temperature structure do not match observations.

Hartmann, Hewett & Calvet (1994) proposed temperature structure in which a schematic cooling rate (from Hartmann 1982) is balanced with a volumetric heating rate that scales as r^{-3} (taken to represent magnetic heating from a dipolar field). It was found that this temperature structure resulted in line profiles that better-matched observations, and hence it has been widely-used in subsequent calculations. However it should be noted that its schematic nature is one of the main weaknesses of the current models.

2.2. *Statistical equilibrium*

Formally the equation of statistical equilibrium should be solved simultaneously with that of radiative equilibrium, but as pointed out above the majority of contemporary calculations assume a temperature structure. The equation of statistical equilibrium boils down to

$$\frac{dN_i(\mathbf{r})}{dr} = R_i + C_i = 0 \tag{2.1}$$

where $N_i(\mathbf{r})$ are the level populations (including the continuum) of a particular species, and R_i and C_i are the radiative and collisional rates into and out of level i. The key problem here is that the radiative rates depend on the radiation field, which in turn depends on the level populations: an iterative solution is required. The intensity I of the radiation field may be found by integrating the radiation-transport equation, given by

$$\mathbf{n} \cdot \nabla I_\nu(\mathbf{r}, \mathbf{n}) = -k_\nu(\mathbf{r})\phi\left(\nu - \frac{\nu_0}{c}\mathbf{n} \cdot \mathbf{v}(\mathbf{r})\right)(I_\nu - S_\nu) \tag{2.2}$$

where $\mathbf{n}$ is a given direction, k_ν is the line opacity, S_ν is the source function, and ϕ is the normalized line profile function.

The integral of the above equation is much simplified if the velocity gradient is large enough, since the line profile term then becomes effectively a delta-function and the integral becomes a summation over the resonance zones along $\mathbf{n}$. This is the Sobolev, or large velocity gradient, approximation which underpins the majority of magnetospheric accretion radiative transfer models in the literature to date. The Sobolev approximation is only valid if the conditions of the gas do not change over the Sobolev length l_s, given by

$$l_s = \frac{v_{\rm therm}}{\frac{dv}{dr}} \tag{2.3}$$

where $v_{\rm therm}$ is the thermal line width and dv/dr is the velocity gradient. The approximation is poor at the start of the accretion funnels where the velocity gradient is lowest, and improves near the base of the flows. Furthermore, the Sobolev approximation does not allow for non-thermal line broadening. Finally the Sobolev approximation does not treat density discontinuities properly (such as the surfaces of the accretion streams), since the line optical depth depends solely on the velocity gradient, and it may lead to an overestimate of the line optical depth at such boundaries. The alternative approach is to use comoving frame line transfer, in which Equation 2.2 is integrated formally along many rays in order to adequately estimate the mean intensity J_ν. The major disadvantage of this method is that it is much more computationally expensive, although we have developed an algorithm employing co-moving frame transfer (see § 7).

3. Development of the radiative-transfer model

Hartmann, Hewett & Calvet (1994) used a simple axisymmetric dipole model for the magnetosphere, with the field lines connecting to the disk at a range of radii around corotation and converging to rings in the northern and southern hemispheres of the protostar. A two-level approximation was used to compute hydrogen line profiles, with the radiation field computed under the Sobolev approximation. It was found that the profiles displayed some of the characteristics of the observations, including asymmetries and inverse P Cygni morphologies.

The same geometrical model was employed by Muzerolle, Calvet & Hartmann (1998), but the two-level approximation was dropped in favour of a full solution of the statistical equilibrium equation a multi-level hydrogen atom. The line profiles computed using this more complete physical model did not differ significantly from those of Hartmann, Hewett & Calvet (1994), indicating that the two-level approximation was a good one. It was demonstrated that the profile strengths provided a constraint on the accretion funnel temperatures as a function of mass accretion rate. For a given accretion rate there is an upper-limit to the temperature above which the continuum becomes optically thick and the lines go into absorption, and a lower-limit below which the model line flux does not match observations.

The profiles detailed above were calculated under the Sobolev approximation, which does not allow for broadening of the profiles via line damping. Muzerolle, Calvet & Hartmann (2001) published models in which the Sobolev approximation was used for the statistical equilibrium equation but the profiles themselves were computed using an exact integration of the transport equation in the co-moving frame – it had previously been shown that the main source of error in the profiles arises from the use of the Sobolev approximation in the computation of the line profiles rather than in the statistical

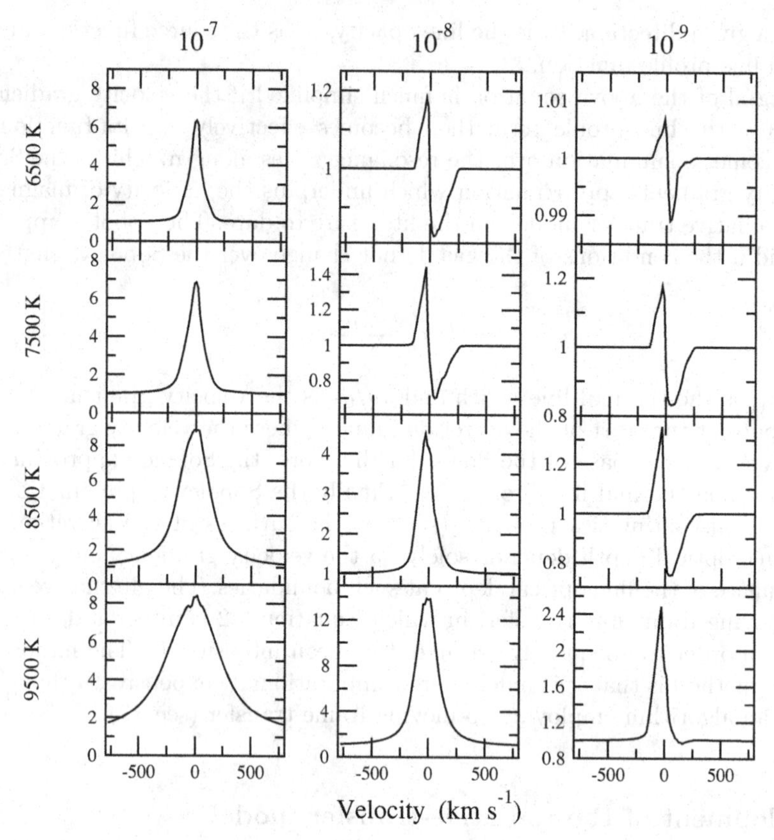

Figure 1. Hα line profiles computed by the TORUS radiative-transfer code and detailed in Kurosawa, Harries & Symington (2006). The profiles are shown (in velocity space and normalized to the local continuum) as a function of the mass accretion rate and the maximum temperature in the magnetosphere. The models with lower $T_{\rm max}$ and $\dot{M}$ show classic inverse P Cygni morphology, while the strongest accretors show no evidence for absorption and are broadened significantly by line damping. Similar grids of models have been published by Muzerolle, Calvet & Hartmann (1998) and Muzerolle, Calvet & Hartmann (2001).

equilibrium calculation e.g. Bastian *et al.* (1980). The inclusion of line damping meant that the lower-order Balmer line profiles extended to velocities significantly greater than the terminal velocity of the material, providing much better agreement with observation. This paper was also the first to show a direct comparison (or fit) with the observations of a T Tauri star (specifically BP Tau). We illustrate the kinds of profiles produced using this method in Figure 1.

The developments reviewed here represent a step-change in physical realism over previous models. Nonetheless there are significant problems with these models when they are compared with the observations. Firstly the areal hot spot coverage of the dipolar geometry used in the models is $\sim 8\%$, which is significantly higher from estimates based on photometric rotational modulation, which indicate about 1% area coverage. This has implications for both the geometry of the flow and the radiation field from the star. Furthermore the models are two-dimensional and do not produce the variability that seems to be a ubiquitous feature of magnetospheric accretion. Finally since the models only include an inflowing component they cannot reproduce the classical P Cygni profiles that are frequently observed. Recent models have attempted to address some of these shortfalls, and I review these efforts in the following sections.

4. Rotational modulation of the emission line profiles

Rotational modulation of the line profile shapes indicates a departure from axisymmetry in the magnetospheres. Symington, Harries & Kurosawa (2005) used the canonical description of a dipolar magnetosphere, and broke the symmetry by simply only allowing accretion over a range of azimuthal angles (see Figure 2). This 'accretion curtains' geometry approximated to the situation proposed by observers to explain the rotational variability, and also resembles the geometries predicted by MHD models. From the time-series synthetic spectra Symington, Harries & Kurosawa (2005) were able to compute observables, such as temporal variance spectra and cross-correlation matrices, which could be compared with data.

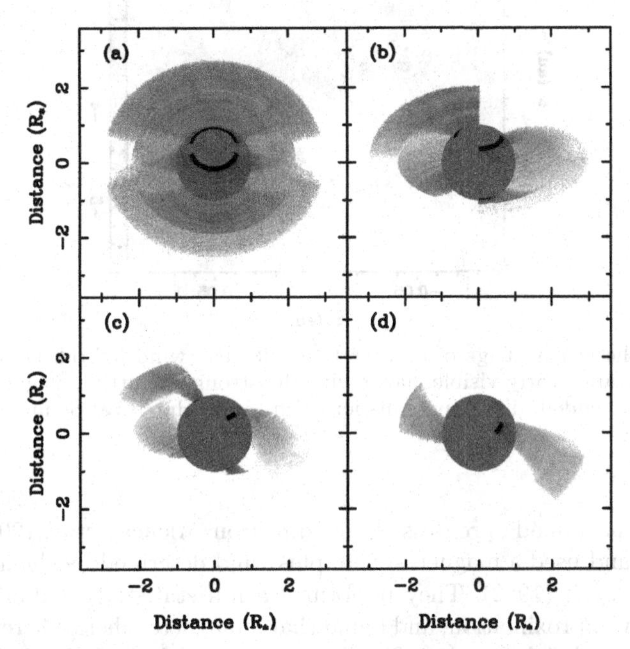

Figure 2. Synthetic Paβ images of the accretion curtains models considered by Symington, Harries & Kurosawa (2005). The models demonstrate increasing departures from axisymmetry, with curtains of (a) 150°, (b) 90° (c) 30° extent. Model (d) has two 30° curtains in opposite hemispheres, and includes a dipole offet.

It was found that the line variability predicted by most models exceeded the observed variability, excepting those that had high degrees of rotational symmetry. Models with a dipole offset, and narrow accretion curtains (which are similar to the MHD models) displayed line variability far in excess of that observed. It was also observed that although the line variability was generally too high, the continuum variability caused by the hotspots was typically too low. In order to reconcile these differences it is necessary to have a magnetosphere that displays a high degree of axisymmetry at large radii, and then becomes more structured, perhaps by converging into narrow funnels, towards the photosphere.

5. Models with inflow and outflow

Magnetospheric accretion is often (always?) accompanied by mass ejection thought to arise via MHD effects e.g. disk winds, stellar winds, jets, X-winds etc. These outflows influence the line profiles principally by adding blue-shifted absorption, but in some objects (which show strong double-peaked profiles) there must be significant line emission as well. Distinguishing between the different outflow mechanisms necessitates determining the launching region, and it is possible that a modelling analysis of line profiles will allow this.

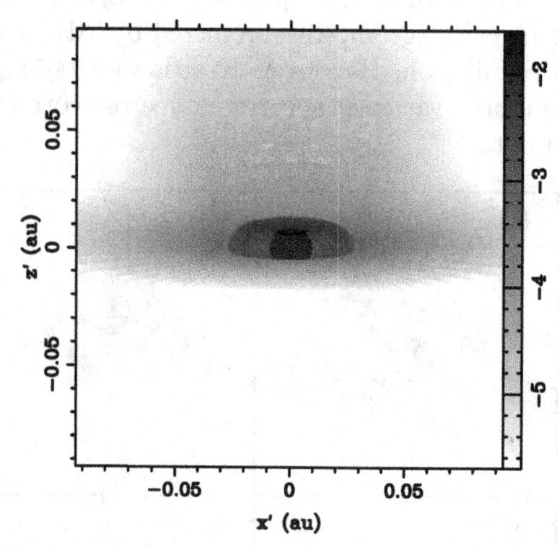

Figure 3. A synthetic Hα image of an accretion plus disk wind hybrid model. The hot rings in the photosphere are clearly visible, along with the strongly emitting magnetosphere and the fainter, but more extended, disk wind emission (Kurosawa, Harries & Symington 2006).

Recent papers have made progress in this direction. Alencar *et al.* (2005) studied the CTTS RW Aur, and used a magnetosphere plus wind described by the analytical model of Blandford & Payne (1982). They performed a full statistial equilibrium calculation under the Sobolev approximation, and found that a magnetospheric-accretion-only model could not reproduce the observed profiles, but that a model in which the outflow emission dominated showed much better agreement. It was found collimated disk winds starting from a small region near the disk inner radius produced the best match, whereas winds with large opening angles were precluded.

A promising direction for research into wind launching is the He I λ10830 line, which has a high line opacity and may show classical *and* inverse P Cygni profiles simulatenously. Kwan, Edwards & Fischer (2007) published outflow profiles based on a parameterized description of the line emissivity and opacity and using Monte Carlo methods. In summary, these models demonstrate the power of the λ10830 line in diagnosing the wind geometry: this work is covered in detail by Edwards' contribution to these proceedings.

Kurosawa, Harries & Symington (2006) computed grids of Hα profiles based on a canonical magnetosphere and a parameterization of a disk wind (see Figures 3 and 4). It was found that this hybrid model was capable of reproducing the wide variety of observed profile shapes, even with the mass accretion rate to mass-loss rate ratio fixed at 0.1. It is found that the most frequently observed profiles occupied the widest volume of parameter space, whereas the rarest profiles required very specific sets of parameters. We attempted to place the Reipurth *et al.* (1996) spectral classification system on a more physical basis, by determining what range of parameters are necessary to produce each profile sub-type. We summarize these findings in Figure 5 and Table 1.

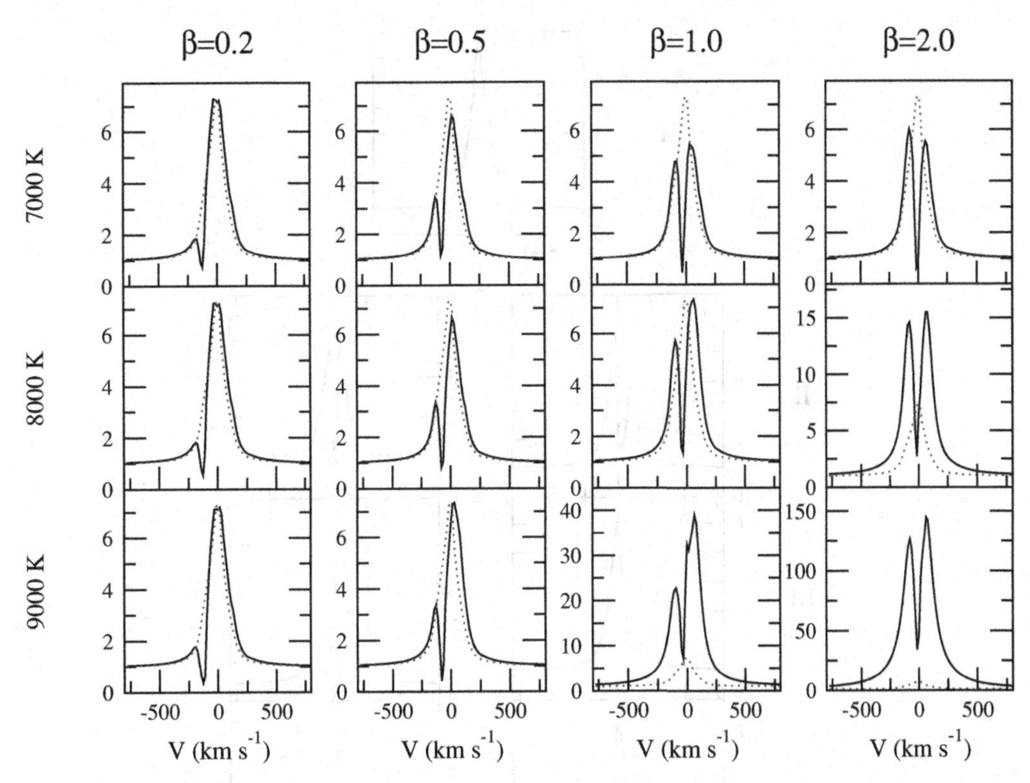

Figure 4. Synthetic Hα line profiles as a function of maximum accretion funnel temperature and wind acceleration parameter, β. Models with a larger β accelerate more slowly, leading to larger densities near the wind-launching region and more line emission from the wind. The models include both the wind and the magnetosphere, and the dotted lines show the equivalent magnetospheric-accretion-only profile. Profiles at the top left of the diagram are dominated by emission from the accretion funnels, those at the bottom right are dominated by the wind.

Table 1. The wind, accretion, and viewing angle requirements for different line profile classes in the Reipurth *et al.* (1996) classification scheme (from Kurosawa, Harries & Symington 2006).

Class	Comment
I	Accretion dominated. Wide range of inclinations
II-B	Wide range of wind acceleration. Mid-to-high inclination.
II-R	Slow wind acceleration rate. High inclination.
III-B	Fast wind acceleration. Mid inclination.
III-R	Mid wind acceleration. Very high inclination.
IV-B	Fast wind acceleration. Mid-to-high inclination.
IV-R	Accretion dominated. Low mass-accretion rate. Mid inclination.

6. Models for low- and high-mass protostars

Attempts have been made to apply the magnetospheric accretion paradigm across the mass spectrum from brown dwarfs (BDs) to Herbig AeBe stars. It is now clear that young BDs share many observational properties with CTTS, including near-to-mid IR excesses indicating the presence of circumstellar disks (e.g. Scholtz *et al.* 2007) and broad, asymmetric, variable Hα emission that may be a signature of accretion (e.g. Muzerolle *et al.* 2005; Mohanty *et al.* 2005). The accretion rates for many of VLM protostars and young BDs seem to be too low to produce a significant blue excess, and the line emission

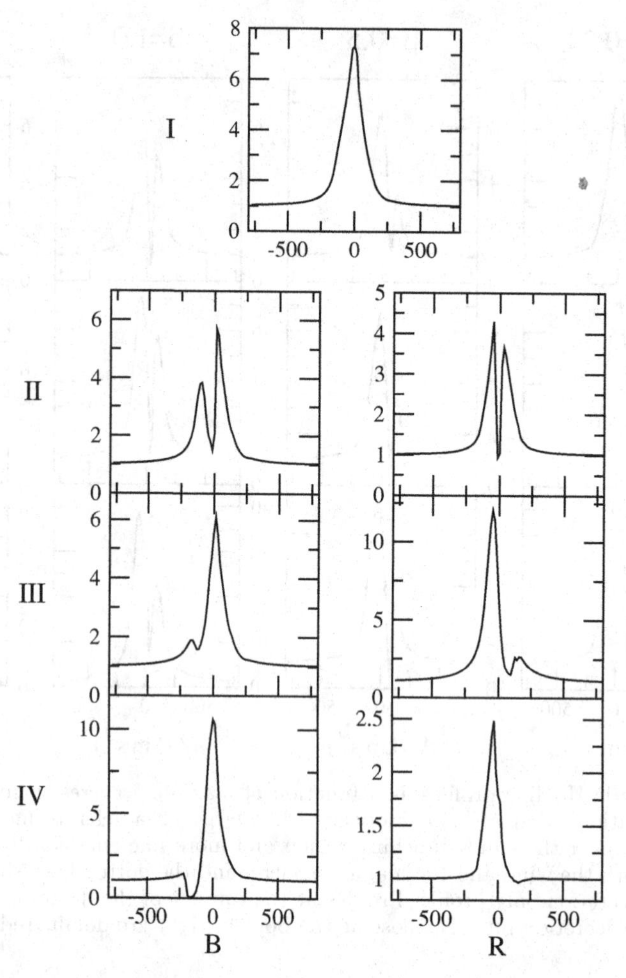

Figure 5. Sample Hα line profiles from the grid computed by Kurosawa, Harries & Symington (2006), labelled by their Reipurth sub-types. A description of the physical parameters used to compute each of the sub-types is given in Table 1.

is the sole route to measuring the accretion rates. This can be done by extrapolating the line flux/mass accretion rate relationship that is found for higher mass stars (e.g. Natta, Testi & Randich 2006), or by applying the magnetospheric accretion model e.g. Muzerolle *et al.* (2005). These studies indicate that the accretion rate drops rapidly with stellar mass (approximately $\dot{M} \propto M^2$, see Mohanty's contribution to these proceedings).

It now appears that a simple application of the magnetospheric accretion model to BDs is probably not correct. In order to get sufficient line emission at these low accretion rates Muzerolle *et al.* (2005) had to increase the temperature of the accretion funnels to $\sim$10 kK, at which point the material is completely ionized and optically thin in the line. The justification for this is that the cooling rates, which are density-squared processes, being much less efficient in these rarified streams. However, measurements of the Paβ to Brγ line ratios in BDs (Gatti *et al.* 2006) suggest that they form in optically thick gas, perhaps in the area around the shocks at the BD surface. Further work is needed here, but the fact that the line fluxes appear to follow (albeit with significant scatter) the same flux/accretion rate relationship as the CTTS suggests that the empirical calibration may be sufficient to obtain accretion rates without reference to the radiation-transfer modelling.

Muzerolle *et al.* (2004) applied the magnetospheric accretion model (including a gas disk) to Herbig AeBe stars, and obtained reasonable matches to the line profiles for UX Ori. However the evidence for magnetic fields in Herbig AeBe stars is somewhat weaker than that for CTTS, and it is yet to be seen whether the magnetospheric accretion model can be successfully and routinely applied at these higher masses.

7. Magnetic field extrapolation models

The models reviewed so far are based on canonical descriptions of the magnetosphere, but new methods are emerging which may allow us to generate a more realistic picture of the accretion funnel pattern. The technique of Zeeman Doppler Imaging (ZDI) uses the rotational modulation of the circular polarization signature within photospheric absorption lines to map the strength and polarity of the surface field. The ZDI method has been successfully applied to main sequence objects, and attention is now being focussed on pre-main sequence objects.

The field pattern around the star may be determined, under the assumption of a potential field, by extrapolating field lines from the surface. One may then determine which field lines intersect the disk and have an inward pointing effective gravity–these are the accreting field lines (see Figure 6). The density and velocity along the field lines can be found by assuming a mass-accretion rate and free-fall kinematics e.g. Gregory *et al.* (2006), Gregory, Wood & Jardine (2007), and Jardine's and Gregory's contributions to these proceedings.

We have created synthetic spectra based on the extrapolated field geometry, via the following process: First, the stellar surface is divided into a large number of area elements. The density and velocity of the accretion flow immediately above each area element is used to find the kinetic power dumped into that element. This power is assumed to be completely reprocessed at the photosphere and emitted as blackbody radiation, enabling us to assign a hot spot temperature to each element. After solving the statistical equilibrium for 15 levels of hydrogen, line profiles can be computed for any transition and viewing angle using a simple ray-tracing integration of the transfer equation. The line profiles (an example is shown in Figure 6) show some of the characteristic features one would expect of an accreting T Tauri star, including an asymmetric, broad Hα profile and a strong inverse P Cygni (IPC) at Hβ. The strength of the IPC profile is due to a combination of the temperature structure we have adopted, and line-damping.

The eventual aim of this research is to perform this analysis on data obtained for CTTS, and to compare the resulting synthetic time-series data with the echelle spectroscopy that was used to create the magnetic field maps. This will provide a strong test of the magnetic field extropolation technique, and may provide a route to a self-consistent model of the magnetosphere.

8. The successes and failures of the model

Radiation-transfer models are broadly successful in predicting the gross characteristics of the line profiles in terms of shapes and strengths. Furthermore the models have been applied to individual objects in order to find mass-accretion rates, and the rates determined this way show good agreement with the rates derived using the less model-dependent technique of blue-continuum excess. The few objects that have made a

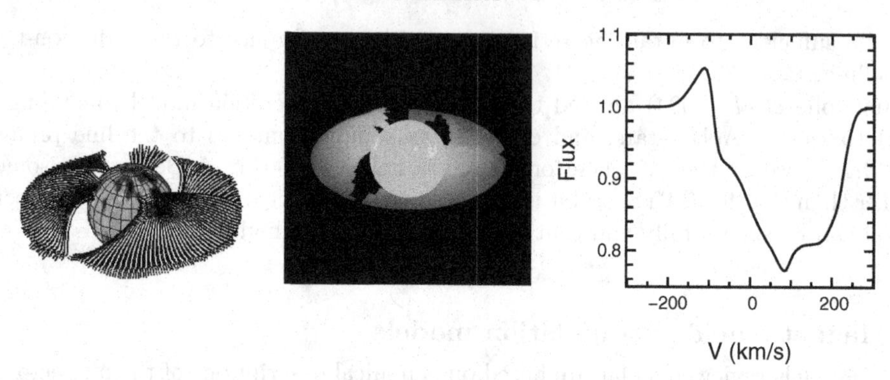

Figure 6. The leftmost figures shows the structure of the accreting magnetic field for a model CTTS with a mass of $0.5\,M_\odot$, radius $2\,R_\odot$, rotation period 6 d, with an isothermal corona at 20 MK extrapolated from a surface magnetogram of AB Dor. Accretion is assumed to take place over a range of radii within corotation (Gregory, Wood & Jardine 2007). The figures on the right show an Hβ intensity image and spectrum respectively, computed in the co-moving frame and based on the geometry of the field extrapolation.

multi-line analysis show that the model works simultaneously over a variety of spectral diagnostics, which is very encouraging.

Progress has also been made in understanding the origin of line profile variability in CTTS, by generating simple accretion curtains models. Rudimentary comparisons have been made between the characteristic variability patterns observed and model grids, and attempts have been made in tailoring such models to individual objects (Kurosawa, Harries & Symington 2005). It seems that the simple description of the geometry provided by the curtains model is too crude to explain the complex variability seen in the observed profiles. It may be that the magnetosphere is composed of many streams that average to something that approximates to an even azimuthal distribution at large radii, and that then converge as they approach the surface in order to produce the small hotspot areal coverage that is observed.

Several groups are looking at including both inflow and outflow in their radiation transfer models. Although this inevitably leads to yet more free parameters in the models, the inclusion of the outflowing material is vital if the wide-range of line profiles that is observed are to be adequately modelled. Furthermore, analysis of spectral diagnostics that are sensitive to the outflow (e.g. λ10830) have the potential to strongly constrain the wind-launching mechanism and thus the MHD models. Conversely attempts are being made to produce synthetic profiles based on the MHD modelling (see Romanova's contribution to these proceedings). Once again comparison with the strength, shape, and variability of observed profiles should allow the refinement of the hydrodynamical models.

Despite these successes there are some serious deficiencies in the models that need to be addressed. In particular there needs to be a concerted effort to address the problem of radiative equilibrium in the funnel flows. This is a formidable problem, that has parallels with research into the solar corona. But perhaps if the cooling pathways are properly understood it may be possible to at least constrain the magnetice heating terms? It is also important to try and produce more tailored models for individual targets, using multiple spectral diagnostics to provide the strongest constraints on the array of free parameters in the magnetospheric accretion models. The difficulty here is obtaining observations at sufficient resolution across a sufficiently wide spectral range (preferably the entire visible and near-IR regimes) simultaneously, which is necessary to overcome uncertainties due to variability that hamper current studies.

Acknowledgements

Much of research conducted at Exeter and cited here was performed by Ryuichi Kurosawa and Neil Symington. Scott Gregory and Moira Jardine provided the extrapolated field maps for Section 7, and Christophe Pinte helped with generating the profiles and figures for that section. I thank the SOC for the invitation and the LOC for their hospitality.

References

Alencar, S.H.P., Basri, G., Hartmann, L., & Calvet, N. 2005, *A&A* 440, 595
Bastian, U., Bertout, C., Stenholm, L., & Wehrse, R. 1980, *A&A* 86, 105
Blandford, R.D., & Payne, D.G. 1982, *MNRAS* 199, 883
Gatti, T., Testi, L., Natta, A., Randich, S., & Muzerolle, J. 2006, *A&A* 460, 547
Gregory, S.G., Wood, K., & Jardine, M. 2007, astro-ph-07042958
Gregory, S.G., Jardine, M., Collier Cameron, A., & Donati, J.-F. 2006, *MNRAS* 373, 827
Hartmann, L. 1982, *ApJS* 48, 109
Hartmann, L., Hewett, R., & Calvet, N. 1994, *ApJ* 426, 669
Kwan, J., Edwards, S., & Fischer, W. 2007, *ApJ* 657, 897
Kurosawa, R., Harries, T.J., & Symington, N.H. 2005, *MNRAS* 358, 671
Kurosawa, R., Harries, T.J., & Symington, N.H. 2006, *MNRAS* 370, 580
Martin, S. 1996, *ApJ* 470, 537
Mohanty, S., Jayawardhana, R., & Basri, G., 2005, *ApJ* 626, 498
Muzerolle, J., Calvet, N., & Hartmann, L. 1998, *ApJ* 492, 743
Muzerolle, J., Calvet, N., & Hartmann, L. 2001, *ApJ* 550, 944
Muzerolle, J., D'Alessio, P., Calvet, N., & Hartmann, L. 2004, *ApJ* 617, 406
Muzerolle, J., Luhman, K.L., Briceño, C., Hartmann, L., & Calvet, N. 2005, *ApJ* 625, 906
Natta, A., Testi, L., & Randich, S. 2006, *A&A* 452, 245
Reipurth, B., Pedrosa, A., & Lago, M.T.V.T. 1996, *A&AS* 120, 229
Scholz, S., Jayawardhana, R., Wood, K., Meeus, G., Stelzer, B., Walker, C., & O'Sullivan, M. 2007, *ApJ* 660, 1517
Symington, N.H., Harries, T.J., & Kurosawa, R. 2005, *MNRAS* 356, 1489

Star-disk Interaction in Young Stars
Proceedings IAU Symposium No. 243, 2007
J. Bouvier & I. Appenzeller, eds.

doi:10.1017/S1743921307009453

Measuring the physical conditions of accreting gas in T Tauri systems

Jeffrey S. Bary[1] and Sean P. Matt[1]

[1]Department of Astronomy, University of Virginia, Charlottesville, VA 22902, USA
email: jbary@virginia.edu, spm5x@virginia.edu

Abstract. Hydrogen emission lines observed from T Tauri stars (TTS) are associated with the accretion/outflow of gas in these young star forming systems. Magnetospheric accretion models have been moderately successful at reproducing the shapes of several HI emission line profiles, suggesting that the emission arises in the accretion funnels. Despite considerable effort to model and observe these emission features, the physical conditions of the gas confined to the funnel flows remain poorly constrained by observation. We conducted a mutli-epoch near-infrared spectroscopic survey of 16 actively accreting classical TTS in the Taurus-Auriga star forming region. We present an analysis of these *simultaneously* acquired line flux ratios of many Paschen and Brackett series emission lines, in which we compare the observed ratios to those predicted by the Case B approximation of hydrogen recombination line theory. We find that the line flux ratios for the Paschen and Brackett decrements as well as a comparison between Brγ and Paschen transitions agree well with the Case B models with $T < 5000$ K and $n_e \approx 10^{10}$ cm^{-3}.

Keywords. Accretion, stars: low-mass, formation, variables: T Tauri stars, infrared: stars

1. HI forms in accretion flows?

Where and how the neutral hydrogen emission lines are formed in the environment of accreting T Tauri stars remains an outstanding problem. Initially, the hydrogen lines were associated with stellar winds and mass loss. However, in the last ten to fifteen years we have witnessed a sea-change in this interpretation and the HI lines have come to *signify* accretion, as the concept of magnetospherically guided accretion (Ghosh & Lamb 1979; Königl 1991; Shu *et al.* 1994) has gained wide acceptance, partly due to its success at explaining the line shapes of some of the Balmer series, Paβ, and Brγ emission lines (Hartmann *et al.* 1994; Muzerolle *et al.* 1998a, 2001).

Balmer series lines appear quite complicated, displaying a variety of structures which vary from star-to-star and from epoch-to-epoch (Basri & Batalha 1990). On the other hand, the infrared Paschen and Brackett series features, with their lower opacity may not experience the same self-absorption or other complicating optical depth effects. Folha & Emerson (2001) collected high resolution near-infrared spectra of 50 TTS to study the Paβ and Brγ emission lines and found what they described as a 'conspicuous' lack of blueshifted absorption features. The absence of blueshifted absorption in the infrared lines may imply that they are not produced in an outflowing wind. However, this work and others (e.g., Johns & Basri 1995; Alencar & Basri 2000), that have conducted large scale high resolution surveys of HI conclude that much work needs to be done to determine if the observed HI line shapes can be accurately modeled by the magnetospheric accretion model. For the purpose of discussion, in these proceedings we make the initial assumption that the HI gas emission is produced in the accretion column.

2. The need for simultaneous observations

Previous attempts to determine the temperature of the emitting HI gas using line ratios have been inconclusive with the measured values not agreeing with those expected for optically thick or thin gas in local thermodynamic equilibrium. Additionally, comparisons of these ratios to those predicted by the Case B approximation of Baker *et al.* (1938) recombination line theory, likewise have proven unsuccessful. However, in most cases the line ratio, usually only Paβ/Brγ, was determined from spectra collected at different times. Given the short timescale variability (on the order of hours) that has been observed for the fluxes of these individual lines, a line ratio calculated from two observations separated by a day or more will easily vary significantly from a value determined from spectra taken simultaneously. Therefore, it is not surprising that the line ratio comparisons have failed to contribute anything meaningful to our understanding of the physical conditions of the accreting gas.

The lack of agreement between the measured and predicted line ratios have lead astronomers to conclude that the emitting region is quite complicated. We suggest that this interpretation of the extant line flux data may be non-sensical given the variability of the emission features and the lack of simultaneous observations. In order to address the shortcomings of the previous line ratio comparisons, we present an analysis of multiple near-infrared HI line ratios measured from 104 multi-epoch spectral observations of 16 actively accreting cTTS in the Taurus-Aurga star forming region. The excellent wavelength coverage of CorMASS, a cross-dispersed low resolution (R$\sim$300) spectrometer with continuous coverage from 0.8 to 2.5 μm, allows us to *simultaneously* measure the strength of any detectable HI emission feature in both the Paschen and Brackett series. The data improve upon previous analyses, not only by acquiring the line data at the same time, but by increasing the number of line ratios and range of n-states included in the analysis.

3. Simultaneous spectral observations

The survey spectra were collected using CorMASS on the 1.8m Vatican Advanced Technology Telescope (VATT) atop Mt. Graham, AZ over a two year period beginning in December 2003 and ending January 2005. On VATT, CorMASS slit dimensions are $1''.6$ in width and $12''.2$ in length. Observations of both the target cTTS and telluric calibration sources were made using a standard ABBA nod procedure to allow for efficient removal of sky background, sky emission lines, and dark current. Standard data reduction was performed using the IDL program CORMASSTOOL, which was adapted from SpexTOOL (Cushing *et al.* 2004). As an illustration of the wavelength coverage and the information included in the CorMASS spectra, we include Figure 1, which contains ten epochs of observations of DR Tau.

4. Measuring HI line ratios

We measured line fluxes for nine Paschen (Paβ, γ, δ, 8, 9, 10, 11, 12, and 14) and eight Brackett series (Brγ, 10, 11, 12, 13, 14,15, and 16) lines. Single-epoch observations of all sources show a broad range of HI line fluxes for the entire sample, while multi-epoch observations of the individual sources find significant variability in these same features. Since it has been shown that a correlation likely exists between the strength of the infrared HI line fluxes and the mass accretion rates of TTS (Skrutskie *et al.* 1996; Muzerolle *et al.* 1998b), our spectral observations sample a set of actively accreting TTS with a variety of mass accretion rates.

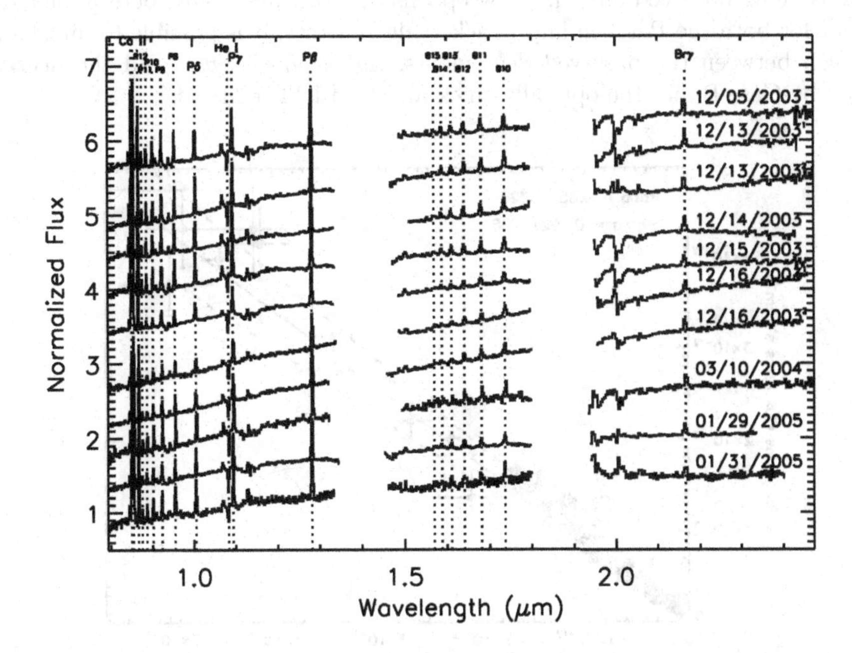

Figure 1. As an example of the spectra taken for the multi-epoch survey, we present all of the spectra obtained for the source, DR Tau. Each spectrum contains multiple HI recombination emission features, the Ca II infrared triplet in emission, and the He I (1.083 μm) feature, seen here in absorption. Individual spectra are labeled with the UT date on which it was observed. Vertical dotted lines locate the spectral features. Note that telluric water features have been removed near 1.4 and 1.9 μm.

In Figure 2, we plot all of the dereddened line fluxes measured for Paγ against those measured for Paβ, including errors, and find there to be a strong correlation between the two. Surprisingly, we find little variation in the line ratios from source-to-source and for different epochs of observations of the same source where even the most substantial fluctuations in line strengths were observed. If these line fluxes are, indeed, correlated with mass accretion rate, this result suggests that the line ratios are insensitive to the mass accretion rate and one 'global' Paγ/Paβ ratio may be determined for all the stars in our survey.

In order to find a representative 'global' line ratio, we determined the average line ratio, weighted by the uncertainty in the line flux measurements, for all of the stars in our sample. In addition, we calculated the weighted standard deviation from the mean to estimate the scatter in this measurement. In Figure 2, we plot a solid line whose slope is the 'global' line ratio for Paγ/Paβ, while the dashed line represent the scatter measurement.

Although not shown here, similar linear relationships were found for the rest of the HI emission features observed and 'global' line ratios were determined for those using the same method. Each ratio uses the Paschen or Brackett series transition with the lowest excited state as the reference transition (Paβ with $n_u = 5$ and Brγ with $n_u = 7$). In the case of the Paschen series, eight emission lines were ratioed with the reference transition. For the Brackett series, which was weaker and detectable only in the most active TTS, we measured seven emission features. The number of stars possessing measurable emission at the higher transitions in the Brackett series diminished quickly with the highest three transitions being observed only in a few epochs of DR Tau. Having determined a set of line ratios for both the Paschen and Brackett decrements, it is possible to make a direct comparison between the observed decrements, *not a single ratio*, and the decrements predicted by Case B and the optically thick and thin LTE regimes.

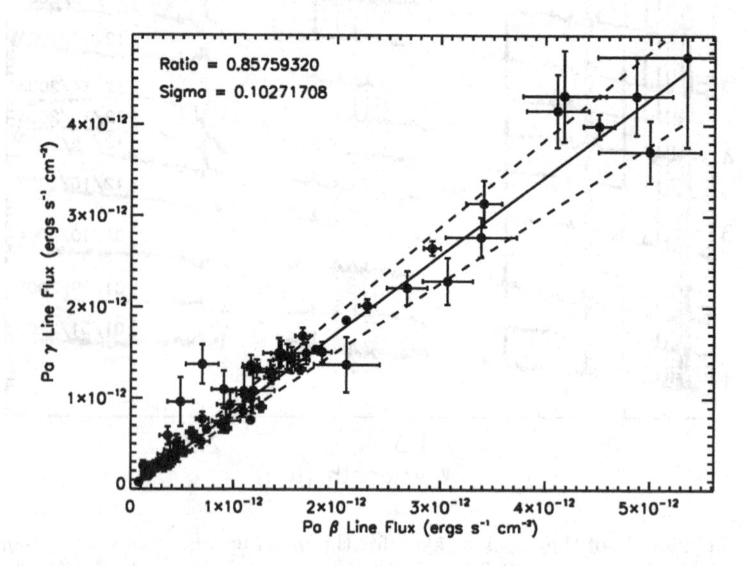

Figure 2. For each spectra in our sample, we plot the measured Paβ line flux versus that measured for Paγ with associated errors and find a strong linear correlation between these emission features. The solid line plot through the data has a slope equal to the weighted average of the data points. The dashed lines represent the weighted standard deviation and provide a useful measure of the scatter about the weighted mean.

5. Case B comparison

We used the fortran program provided by Storey & Hummer (1995) (on-line at http:-//vizier.u-strasbg.fr/viz-bin/VizieR?-source=VI/64 and accompanied by the necessary data files) to produce models of the Paschen and Brackett decrements as well as the line ratios of Brγ to Pan for 130 different temperature and electron density combinations for $500 \leqslant T \leqslant 30000$ K and $100 \leqslant \mathrm{n_e} \leqslant 10^{14}$ cm^{-3}. We calculated the reduced Chi-squares to determine which, if any, models were good fits to our data.

In Figure 3, we plot a line ratio curve for the observed Paschen decrement using Paβ as the reference transition and overplot the four Case B models with the smallest χ^2 values. The 'scatter' bars on each of the data points represent the weighted standard deviation discussed in §4. The closest model (the one with the smallest χ^2) has a temperature of 1000 K and an electron density of 10^{10} cm^{-3} and a reduced χ^2 of 0.89. A similar comparison was made for the Brackett decrement and the closest model was determined to have T = 500 K, $\mathrm{n_e}$ = 10^{10} cm^{-3}, and χ^2 = 0.43. The fact that the closest Case B models to fit the Paschen and Brackett series independently have the similar T and the same $\mathrm{n_e}$ is strong evidence that the assumptions of Case B theory approximate the conditions in the emitting gas.

We proceed with a comparison of line ratios between the Paschen series and the strongest Brackett feature, Brγ. The large separation in wavelength between the Paschen series and Brγ allows us to search for any dependence in our Case B comparison on the

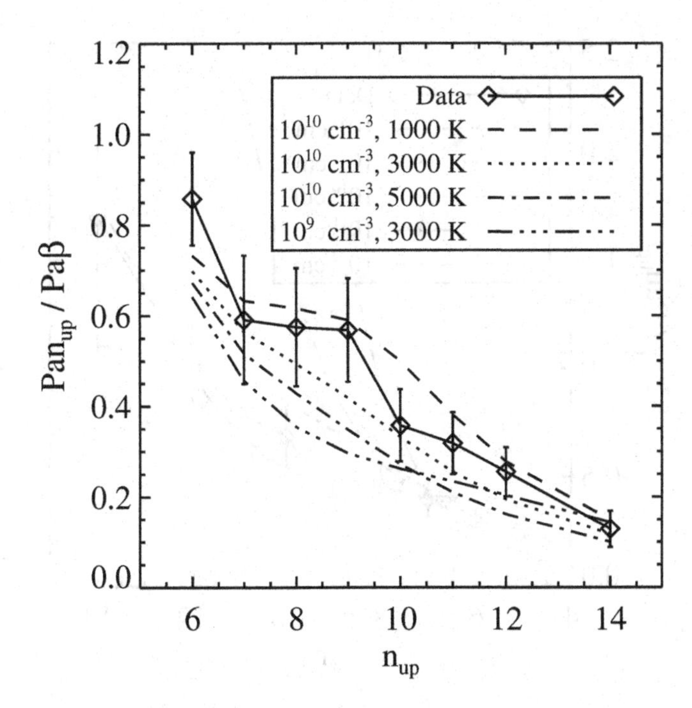

Figure 3. We plot the weighted mean Pan/Paβ ratio versus the principal quantum number, n, of the upper level as a solid line connecting the diamond-shaped data points. 'Scatter' bars, determined from the weighted standard deviation from the mean line ratio value, are plotted on top of the data points. Of the possible 130 models, we overplot the four models with the smallest reduced χ^2 values. The temperatures and electron densities for these models are given in the figure legend.

reddening correction. In Figure 4, we present the line ratio curve for Brγ to Pan including the closest model with T = 10^3 K and $n_e = 10^{10}$ cm^{-3}. We overplot four additional Case B line ratio curves holding the temperature constant (T = 1000 K) and using four densities in the Storey & Hummer (1995) grid of models adjacent to the density of the best fit. At low n-values, the spread in predicted line ratios is quite small for this temperature. However, for the line ratios of the $n_{\rm upper}$ states of the Paschen transitions, the models separate quite well, clearly distinguishing the model that most closely matches data. As a result, the Brγ to Pan line ratios place a strong constraint on n_e of the emitting gas. Again, the temperature and density of the closest Case B model for Brγ/Pan agree well with those determined for both the Paschen and Brackett decrements. This result suggests that the spectra were properly corrected for reddening.

6. LTE optically thick and thin

In Figure 5, we compare the observed Brγ to Pan line ratio curves to those predicted by both an optically thick ($\tau \gg 1$) and thin ($\tau \ll 1$) region of HI gas. We find no correlation between the LTE line ratios and the measured line ratios for any temperature, with the exception of the Brγ to Paδ line ratio. Since both transitions begin with the $n = 7$ principal quantum state, the line ratio in the optically thin limit is temperature independent. For the densities considered here, the Case B ratio approaches the same value as in the LTE optically thin case. The fact that our observed value of Brγ/Paβ

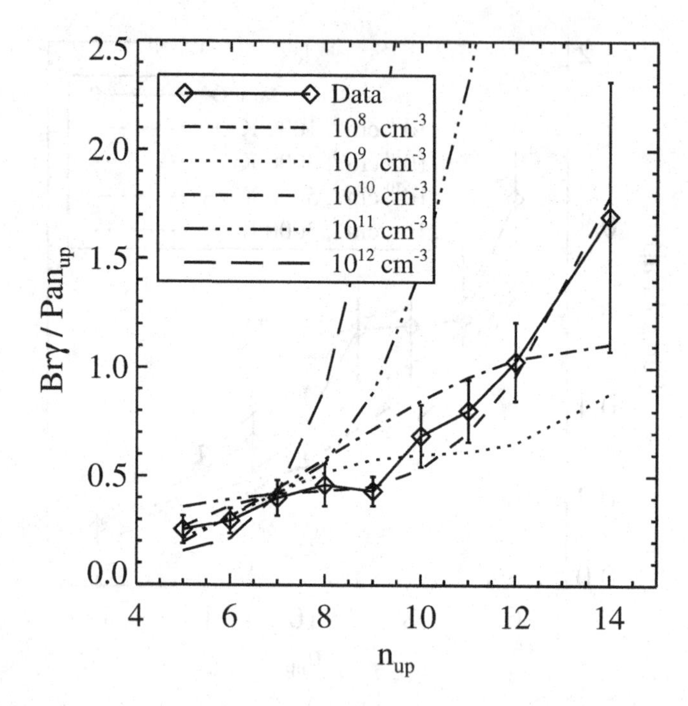

Figure 4. We plot the values of the line ratios of Brγ to the Paschen series. The broad wavelength coverage between the Brackett and Paschen lines provides a stronger constraint on density as the predicted ratios diverge significantly at high n. We demonstrate this sensitivity by plotting four models of constant temperature (T = 1000 K) and separated by one order of magnitude in density along with the data. Once again, we find the model with the smallest Chi-square is $n_e = 10^{10}$ cm^{-3}.

matches both the LTE optically thin and the Case B value is independent evidence that these two lines are optically thin.

As the temperature increases for the optically thin case, the curves quickly approach an asymptotic limit and will never match the observed line ratio curve. As for the optically thick cases, two curves are plotted that bracket the previously predicted temperatures for the accreting gas ($\approx$ 6000–12000 K: Martin 1996; Muzerolle *et al.* 1998a) and clearly do not fit the shape of the observed values. Figure 5 clearly rules out both the LTE optically thin and thick cases for the emitting HI lines.

7. Conclusions

In contrast to previous comparisons of observed HI line ratios to those predicted by Case B approximation in line recombination theory, we find good agreement to our observed values for a tightly constrained range of T and n_e. We do not find any agreement between our observed values and those predicted for the optically thick and thin LTE cases suggesting that the level populations are *not* in LTE. Therefore, we conclude that the emitting gas is optically thin to the detected infrared HI lines and that the bulk of this line emission is produced in a non-LTE recombining gas with $T < 5000$ K and $n_e \approx 10^{10}$ cm^{-3} (see Figure 6).

Does the emission originate in the accretion flow? The measured temperature range is far lower than the T = 10^4 K previously predicted for gas confined to an accretion

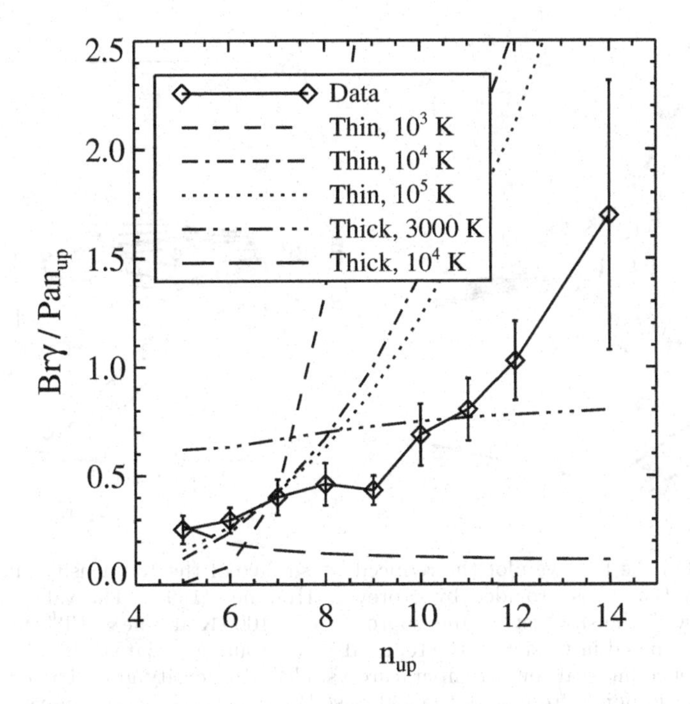

Figure 5. We plot the values of the line ratios of Br γ to the Paschen series as a solid line connecting the measured line ratios. 'Scatter' bars , determined from the weighted standard deviation from the mean line ratio value, are plotted on top of the data points. Overplotted are models for the LTE optically thick and thin cases for a variety of temperatures consistent with those expected for gas in an accreting column. In the optically thin limit, one model with a temperature of 10^5 K is plotted to show the asymptote approached by the LTE models at high T.

flow and would suggest a problem with the incident radiation field or heating vs. cooling rates currently considered in the calculations of the gas temperature. On the other hand, the electron density agrees with the value expected for the neutral hydrogen density of accreting gas (Muzerolle *et al.* 2001) for plausible ionization fractions of 10^{-2}–10^{-3}. Regardless of its origin, we argue that this Case B analysis makes a strong case that the bulk of the emitting gas is spatially coincident and shares similar emission characteristics. Whether the infrared line emission arises from gas in the accretion flows, at the inner edge of the disk, or in a dense wind flowing away from the star will remain a 'hotly' debated issue.

Acknowledgements

J. Bary acknowledges financial support from the NSF Astronomy and Astrophysics Postdoctoral Fellowship program. The research of S. Matt was supported by the University of Virginia through a Levinson/VITA Fellowship, partially funded by The Frank Levinson Family Foundation through the Peninsula Community Foundation. Also, thanks to Phil Arras, Mike Skrutskie, Craig Sarazin, John Wilson, Matt Nelson, & Dawn Peterson for help with the data collection, reduction, and understanding atomic physics.

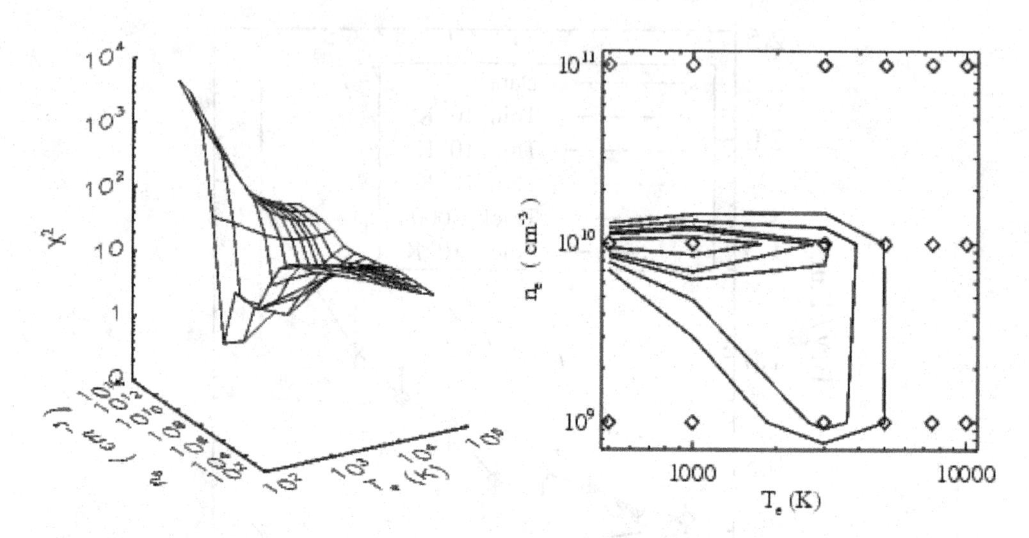

Figure 6. (a) On the left, we plot the reduced χ^2 surface of the 130 density and temperature dependent Case B models provided by Storey & Hummer (1995). The valley of the surface locates models with the smallest χ^2 and approach $T = 1000\,\mathrm{K}$ and ne $= 10^{10}\ \mathrm{cm}^{-3}$. (b) On the right, we plot a zoomed in version of the reduced χ^2 contours associated with the 60, 90, 95, 99, & 99.9% confidence intervals on a temperature vs. electron density grid. The empty diamonds designate T, n_e grid points from 18 of the 130 Case B models used in our analysis. The contours show the tight constraint placed on densities higher than $10^{10}\ \mathrm{cm}^{-3}$. Temperatures greater than 5000 K are ruled out at the 99.9% confidence level.

References

Alencar, S. H. P., & Basri, G. 2000, *AJ*, 119, 1881
Baker, J. G., Menzel, D. H., & Aller, L. H. 1938, *ApJ*, 88, 422
Basri, G., & Batalha, C. 1990, *ApJ*, 363, 654
Cushing, M. C., Vacca, W. D., & Rayner, J. T. 2004, *PASP*, 116, 362
Folha, D. F. M., & Emerson, J. P. 2001, *A&A*, 365, 90
Ghosh, P., & Lamb, F. K. 1979, *ApJ*, 234, 296
Hartmann, L., Hewett, R., & Calvet, N. 1994, *ApJ*, 426, 669
Johns, C. M., & Basri, G. 1995, *AJ*, 109, 2800
Königl, A. 1991, *ApJ*, 370, L39
Martin, S. C. 1996, *ApJ*, 470, 537
Muzerolle, J., Calvet, N., & Hartmann, L. 1998a, *ApJ*, 492, 743
Muzerolle, J., Calvet, N., & Hartmann, L. 2001, *ApJ*, 550, 944
Muzerolle, J., Hartmann, L., & Calvet, N. 1998b, *AJ*, 116, 455
Shu, F., Najita, J., Ostriker, E., Wilkin, F., Ruden, S., & Lizano, S. 1994, *ApJ*, 429, 781
Skrutskie, M. F., Meyer, M. R., Whalen, D., & Hamilton, C. 1996, *AJ*, 112, 2168
Storey, P. J., & Hummer, D. G. 1995, *MNRAS*, 272, 41

Star-disk Interaction in Young Stars
Proceedings IAU Symposium No. 243, 2007
J. Bouvier & I. Appenzeller, eds.

doi:10.1017/S1743921307009465

Observations of accretion shocks

David R. Ardila[1]

[1]Spitzer Science Center, Infrared Processing and Analysis Center, MS 220-6, California Institute of Technology, Pasadena, CA 91125, USA
email: ardila@ipac.caltech.edu

Abstract. I review our current understanding of accretion shocks in classical T Tauri stars (CTTs), from a UV and X-ray perspective. The region of the accretion shock is a good candidate as a source of UV transition region lines from Li/Na-like ions, which are stronger in CTTs than in naked atmospheres. Disk gas captured by the stellar magnetic field produces a strong radiative shock upon falling on the stellar surface. Radiation from the shock creates a radiative precursor and heats the stellar surface resulting in a hot spot. Stellar and shock models indicate that unless the post-shock column is very large, it will be buried on the stellar photosphere. Models of the continuum emission produced by this configuration can roughly reproduce the observed excess spectra down to 1650 Å. Transition region lines in CTTs are broad, very variable, and present blueshifted, centered, and redshifted centroids. Detailed models of the line emission have so far failed to reproduce the fluxes, line shapes, and line ratios. High resolution X-ray line observations indicate the presence of larger amounts of cool plasma in CTTs with respect to WTTs. Observations of density sensitive line ratios of He-like ions suggest high plasma densities, as expected from lines originating in the accretion shock. For most stars, the interpretation of these ratios in terms of density remains equivocal due to the presence of the strong UV continuum.

Keywords. Stars: atmospheres, coronae, winds, outflows; line: formation; magnetic fields; ultraviolet: stars; X-rays: stars

1. Introduction

The presentations in this volume testify to the substantial progress made in our understanding of T Tauri stars since Alfred Joy first described them (Joy 1945). We now believe that the strong excesses seen in CTTs are the result of the presence of an accretion disk and its interaction with the stellar magnetic field. In particular, the excess line emission at optical wavelengths has been successfully modeled (e.g. Muzerolle *et al.* 1998) as being due to the presence of gas captured in the extended magnetosphere. The continuum excess is believed to be primarily due to heating of the stellar photosphere by the same infall gas (Calvet & Gullbring 1998).

Strong continuum and line excesses are also observed at ultraviolet wavelengths, and in this sense the observational problem is conceptually similar to the optical one (Fig. 1), although the shorter wavelengths imply that the lines trace higher energy processes. As in the optical, surface fluxes are much larger than expected from a naked atmosphere. For example, using *International Ultraviolet Explorer* data, Johns-Krull *et al.* (2000) showed that the surface flux in the C IV resonant lines can be as much as an order of magnitude larger than the largest flux observed in Weak T Tauri stars (WTTs), main sequence dwarfs, or RS CVn stars. They also showed that the excess flux in the lines is strongly correlated with accretion rate, suggesting that the lines are powered by the accretion process.

In recent years we have also seen an explosion of new observational results at X-ray wavelengths thanks to the Chandra Orion Ultradeep Project (Feigelson *et al.* 2002)

and, more recently, to the XMM-Newton Extended Survey of the Taurus molecular cloud (Güdel *et al.* 2007). Those measurements have confirmed that CTTs have lower L_X/L_{Bol} values than WTTs, by a factor of ~2, but harder emission. This would suggest the presence of an additional source of extinction, not accounted for in the calculation of the occulting column. However, CTTs are deficient in L_X/L_{Bol} with respect to WTTs even at the highest energies, perhaps suggesting that a fraction CTTs emission measure has been cooled to non-detectable temperatures (Telleschi *et al.* 2007a). Indeed, temperature sensitive X-ray spectroscopic diagnostics reveal the presence of soft X-ray excesses in CTTs (Telleschi *et al.* 2007b). In addition, anomalously low line ratios of He-like ions are measured in CTTs, perhaps indicating the presence of high plasma densities (see Kastner's poster in this conference).

These observational constraints fit, at least qualitatively, with our understanding of the behavior of the accretion stream. The gas captured by the magnetosphere falls along the magnetic field lines, reaching speeds comparable to the free-fall velocity (~300 km/s). The density of the accretion stream depends on the accretion rate (see § 2) but it is typically of the order of $\sim 10^{12}$ cm^{-3}. The supersonic flow, confined by the magnetic field, produces a strong J-shock upon reaching the star and converts most of the kinetic energy of the gas into thermal energy. The gas heats to temperatures of the order of a million degrees immediately after the shock surface and it cools radiatively until it reaches the stellar photosphere. Therefore, one expects temperature diagnostics to reveal a range of temperatures ranging from a few times 10^6 K down to the hot spot temperature, and density diagnostics to reveal higher densities than those found even in active stellar atmospheres (10^9 cm^{-3} to 10^{11} cm^{-3}, see Ness *et al.* 2004).

In this contribution, I will review the current understanding of UV and X-ray line and continuum observations in the context of this accretion shock model. Fundamentally, the questions are the following: what region of the T Tauri system are the UV and X-ray observations describing? Is the emission from the accretion shock region enough to explain those observations, or are other regions of the system contributing to the emission? An understanding of the radiation coming from the accretion shock and its interaction with the rest of the T Tauri system is crucial to map the disk photoevaporation (Alexander *et al.* 2004), model the heating of jets in their launching regions (Gómez de Castro & Verdugo 2007), and interpret broad-band X-ray observations of young stars (Preibisch *et al.* 2005).

2. The shock region

The density of the gas immediately before the accretion shock is given by (Calvet & Gullbring 1998):

$$n_{pre} = \frac{\dot{M}}{\mu m_H f 4\pi R_*^2} = 5.8\times 10^{12} cm^{-3}\left(\frac{\dot{M}}{10^{-8}M_\odot/yr}\right)\left(\frac{M}{0.5M_\odot}\right)^{1/2}\left(\frac{R}{2R_\odot}\right)^{-3/2}\left(\frac{f}{0.01}\right)^{-1} \tag{2.1}$$

where $\dot{M}$ is the accretion rate, μ is the mean molecular mass, m_H is the mass of a hydrogen atom, R_* is the stellar radius and f is the shock filling factor. This last parameter specifies size of the accretion shock surface on the star. After the shock, the gas density and velocity are increased and decreased by a factor of four, respectively. The kinetic energy is transferred to thermal energy and the gas temperature is given by:

$$T_s = \frac{3}{16}\frac{\mu m_H}{k}v_{ps}^2 = 8.6\times 10^5 K\left(\frac{M}{0.5M_\odot}\right)\left(\frac{R}{2R_\odot}\right)^{-1} \tag{2.2}$$

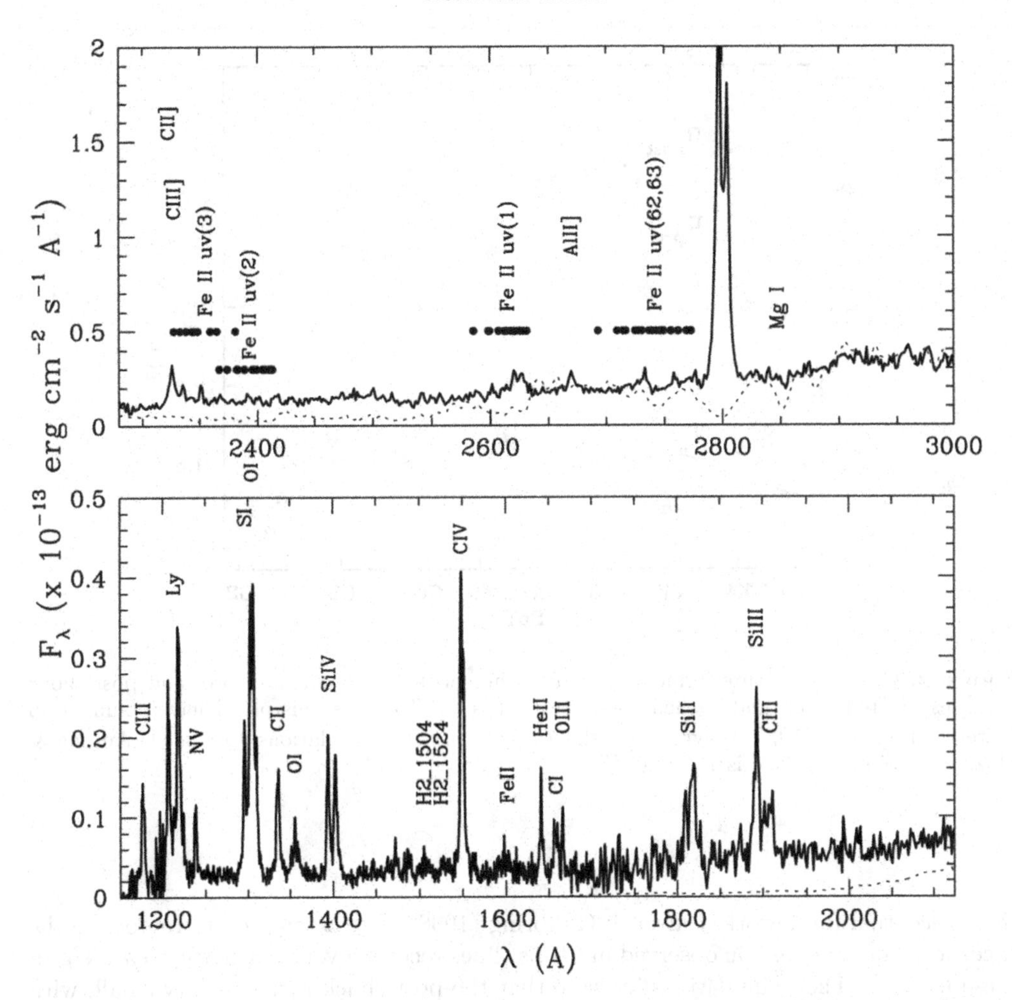

Figure 1. UV spectrum of EZ Ori, taken from Calvet *et al.* (2004). Top: 2270 to 3000 Å. Bottom: 1150 to 2120 Å. The dotted line is the expected emission from the naked G0 photosphere.

where k is the Boltzmann constant and v_{ps} is the pre-shock velocity of infall gas. The typical densities and temperatures of the gas are given in Figure 2. The post-shock gas immediately after the shock surface is hot enough to produce soft X rays (§ 5) and it will cool over a distance given by the cooling length:

$$l_{post} = \frac{3kTv_{post}}{n\Lambda} \tag{2.3}$$

where Λ is the plasma cooling function. Half of the cooling radiation from the post-shock gas will be emitted toward the star, creating a hot spot with temperature of

$$T_{eff} = 5421K\left(\frac{\dot{M}}{10^{-8}M_\odot/yr}\right)^{1/4}\left(\frac{M}{0.5M_\odot}\right)^{1/4}\left(\frac{R}{2R_\odot}\right)^{-3/4}\left(\frac{f}{0.01}\right)^{-1/4} \tag{2.4}$$

For some stars, the hot spot temperature can be as high 10,000 K. The other half will be emitted toward the incoming accretion flow, and it will create a radiative precursor to the shock (the "pre-shock") with typical temperatures of the order of 10,000 K. The emergent emission from the shock region is the sum of radiation emitted by optically thin gas from the pre- and post-shock, plus radiation from the heated photosphere. Within

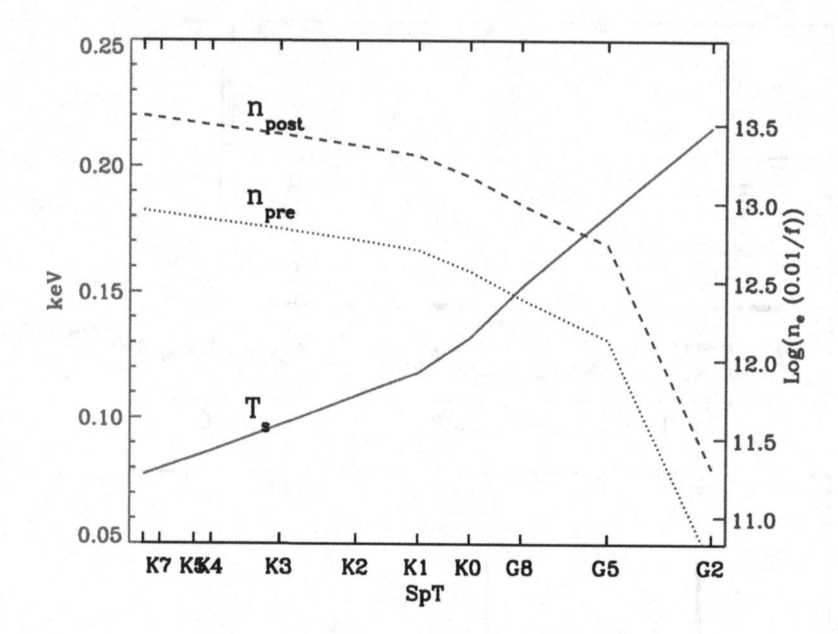

Figure 2. Gas temperature immediately after the shock (solid red) and pre- and post-shock densities (dotted green and dashed green, respectively). To make this plot I have assumed an accretion rate $\dot{M} = 10^{-8}$ $M_{\odot}yr^{-1}$ and the 1 Myr mass-radius relationship from D'Antona & Mazzitelli (1997). 0.1 keV is $1.2 \times 10^6 K$.

this conceptual framework, Calvet & Gullbring (1998) have attempted to reproduce the excess continuum emission observed in CTTs (lines receive a very schematic treatment in these models). Their simulations indicate that the post-shock region is very small, with depths (from the shock surface to the stellar surface) ranging from 1 to 1,000 km, for typical parameters. The size of the pre-shock is of the order of 1,000 km, from the shock surface to the point in which the gas becomes neutral. The models assume that all the radiation is observed with the shock surface perpendicular to the observer (face-on). Pre- and post-shock regions are mostly transparent to the hot spot radiation at wavelengths longer than Ly_{α}, and this hot spot radiation dominates at optical wavelengths. The contribution from the pre-shock, appearing mainly wavelengths < 3000 Å, becomes more important as the gravitational potential well of the star becomes larger. The continuum contribution of the post-shock emission is small in all cases.

Based on comparisons between the observed excesses and the models, Calvet & Gullbring (1998) derive accretion rates of the order of $\dot{M} = 10^{-8}$ $M_{\odot}yr^{-1}$, as well as filling factors $f \sim 0.001 - 0.01$. The models are successful reproducing the empirical L_U vs. L_{acc} relationship (Gullbring *et al.* 1998), and the gross shape of the excess spectra (Figure 3). Interestingly, the models underpredict the continuum below 1650 Å. Bergin *et al.* (2004) argue that collisionaly excited (by hot electrons) molecular hydrogen is responsible for the observed excess continuum below 1650 Å. One limitation of the Calvet & Gullbring (1998) models is their assumption of coronal equilibrium in calculating the cooling function (Figure 4). The correct ionization distribution will alter the geometry of the post-shock and will result in a different emitted spectrum.

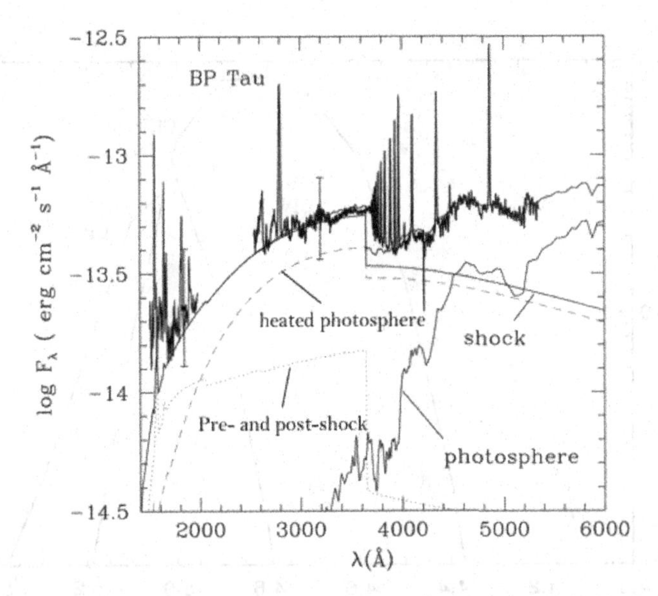

Figure 3. Results of continuum fitting to BP Tau, based on the models by Calvet & Gullbring (1998). The observations are shown with a dark trace. The solid line ("shock") is the sum of the heated photosphere, the pre-shock, and the post-shock. Adapted from Gullbring *et al.* (2000).

3. Observations of transition-region lines

Line emission in the UV, due to atomic species, has been described in detail (e.g. Ardila *et al.* 2002), although the exact characteristics of the material they are tracing remains very poorly understood. Here I will concentrate primarily in the well-studied family of Lithium and Sodium-like ions (filled shells with an unpaired electron in a S orbital) whose resonant transitions are commonly seen in the UV spectra of CTTs. Examples include C IV (λλ 1548, 1550 Å), Si IV (λλ 1394, 1403 Å), N V (λλ 1238, 1243 Å), and O VI (λλ 1032, 1038 Å). These are the "transition region" lines, so called because in stellar atmospheres they are produced in the region in which the temperature profile transitions from chromospheric to coronal temperatures. In ionization equilibrium, the maximum population of these ions occur at $\sim 10^5$ K. In all these ions, if the emission comes from an optically thin, static plasma, the flux in the doublets should be in a 2:1 ratio.

As Figure 5 shows, the lines of TW Hya are fairly broad (~200 to ~300 km/s, FWHM) and the bulk of their emission is redshifted. In this, they are reminiscent of H_α lines. However, as it is often the case in CTTs, single observations do not do justice to the diversity of lines observed in other stars. Figure 6 shows a set of observations of the C IV doublet (Ardila *et al.* 2002). For comparison the lines of the WTTs HBC 388 are also shown. The lines in the WTTs are narrow (~ 60 km/s) and well centered on the star, and the doublet fluxes are in a 2:1 ratio. For the CTTs the lines are broad (200 to 300 km/s) with line ratios not always 2:1 (for example in DG Tau, DR Tau, RU Lup, and RW Aur). The line centroids vary from blue (DG Tau, RY Tau), to zero (RU Lup, BP Tau, DF Tau, T Tau), to red (upper right panel of DR Tau, TW Hya in Fig. 5). The two doublet lines from RW Aur do not look anything like each other, suggesting a steeper density or temperature profile in the emitting region, with respect to other stars.

As in other wavelength ranges, variability is a common feature of the UV lines. Multiple observations of the same target show that the lines may be strongly variable. The two observations of DR Tau shown in the top row of Figure 6 are separated by two years. Strong blueshifted emission (at -250 km/s) is observed in both members of the doublet in

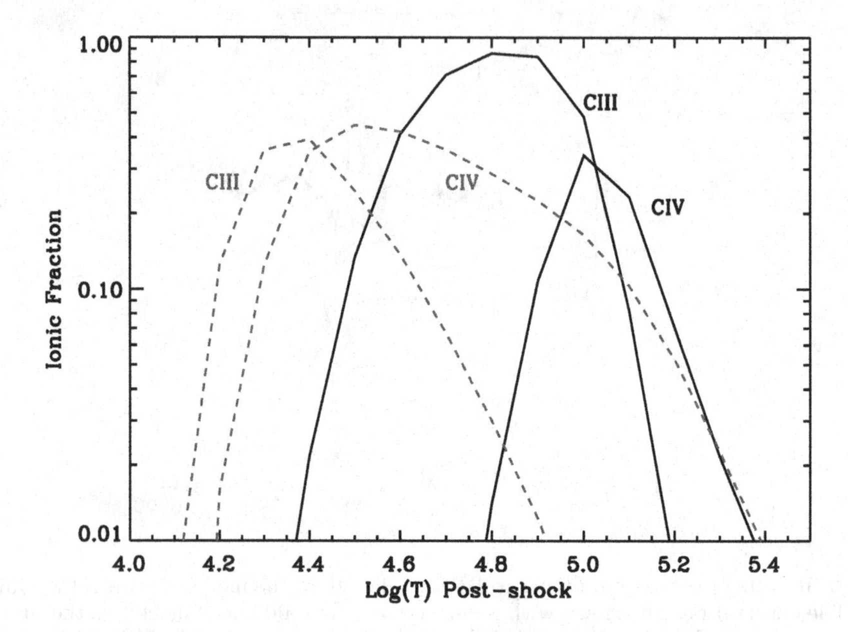

Figure 4. Given the large temperature gradient of the post-shock, the gas will not be in ionization equilibrium. The figure shows the ionization state of C III and C IV both in ionization equilibrium (solid black, from Mazzotta *et al.* 1998) and in the accretion shock (dashed red, from Ardila & Johns-Krull 2007).

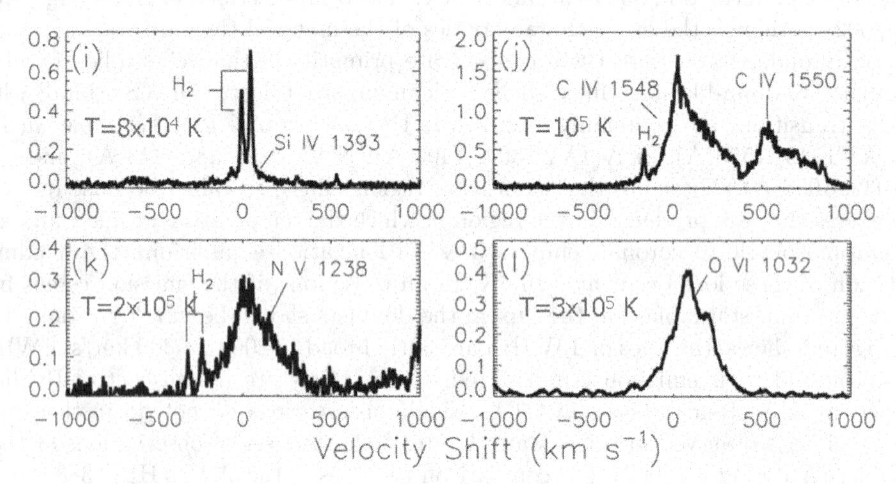

Figure 5. Examples of transition region lines in the CTTs TW Hya. For all panels, the ordinate is the flux density in units of 10^{-14} ergs s^{-1} cm^{-2} Å^{-1}, while the abscissa is the wavelength scale in km/s centered in the stellar rest velocity. In each panel, the temperature at which the maximum population of the ion occurs under conditions of ionization equilibrium has been indicated. As is common in CTTs (Ardila *et al.* 2002; Herczeg *et al.* 2006), Si IV is buried beneath the fluorescent molecular hydrogen emission. Adapted from Herczeg *et al.* (2002).

the 1993 (4th top panel, from the left). This emission is absent from the 1995 observations (5th top panel, from the left), for which the DR Tau lines look similar to those of TW Hya. Figure 7 shows another example of variability, this time for RU Lup. Given the similarities in atomic parameters, one would expect all the resonant lines from Li/Na-like ions to look similar. Indeed, the STIS 2000 observations show similarly broad, well

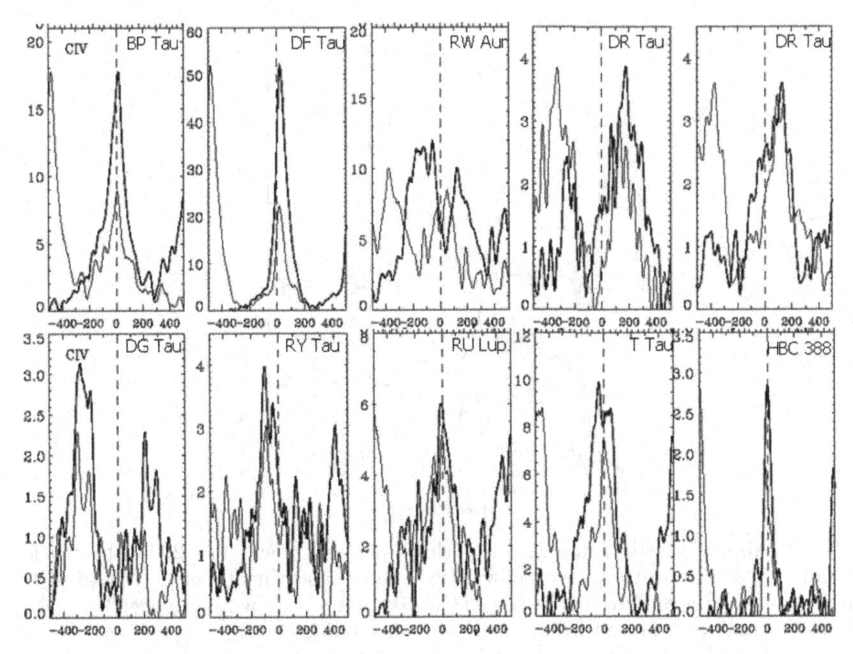

Figure 6. Examples of C IV lines in a variety of CTTs and one WTTs. The units are the same as in Figure 5. Both C IV lines are plotted in each panel, with the optically thicker member of the doublet indicated by the thicker line. The WTTs is HBC 388 (lower right) and it serves as a benchmark for the emission from a naked atmosphere. Two epochs are shown for DR Tau: 1993, 4th panel from the left on top; 1995, 5th panel from the left on top. Adapted from Ardila *et al.* (2002).

centered emission for C IV, Si IV, and N V. However, the FUSE 2001 observations show a very blueshifted O VI line. Herczeg *et al.* (2005) also show that the C III lines at λ 1175 Å and λ 1177 Å, which trace the accretion flow, are strong in the first epoch, but absent in the second. They attribute this behavior to the disappearance of the accretion stream in the second epoch, perhaps leaving behind only the signature of a wind.

The accretion shock region provides a natural source of energy for these emission lines. If produced in the pre-shock, the lines should be centered at the large infall velocities (ignoring projection effects), while in the post-shock their velocities will depend on where exactly they are created, but will always be redshifted. Producing blueshifted lines in the accretion shock requires that the near accretion stream (to the observer) be absent or occulted. Alternatively, an additional line source, like an outflow, may be responsible for the blueshifted emission.

4. Predictions of models

The most detailed published models of line emission from the accretion shock regions of CTTs have been developed by Lamzin and collaborators (see Lamzin 2003 and references therein). These semi-analytic models consider the role of the post-shock in heating the radiative precursor and the stellar surface, and trace the optical depth of the lines in all the regions. From these models they conclude that the pre-shock should be the primary source of the transition region lines for infall velocities larger than 300 km/s. If so, one would expect to see very redshifted lines in a large fraction of the sample, but this is not what is observed. As a matter of fact, realistic attempts to explain the line shapes and centroids have so far failed (Lamzin 2003), perhaps indicating that the field lines that

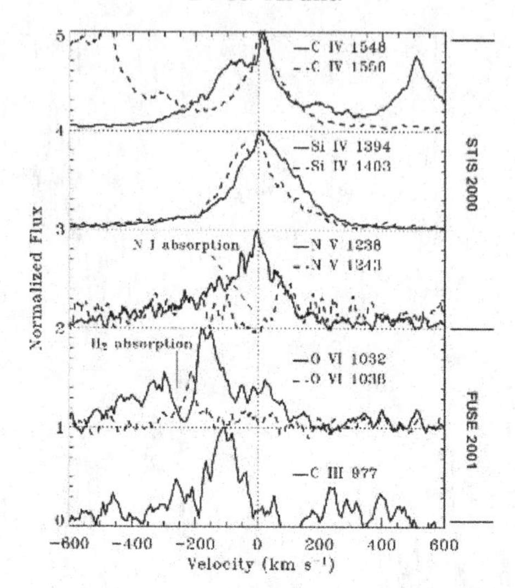

Figure 7. Variability in RU Lup. The top three panels show STIS 2000 observations of C IV, Si IV, and N V. The bottom two show FUSE 2001 observations of O VI and C III. Given the similarity in atomic parameters of the Li/Na-like iones, one would expect their shapes and centroids to be similar. Adapted from Herczeg *et al.* (2005).

confine the flux depart from a parallel configuration very close to the star (Lamzin 2000, 2003). The models also predict line fluxes that are 10 to 100 times larger than observed, and transition line ratios 10 times smaller than observed. At this point it is not clear if the models need to be developed further or if the lack of agreement with observations is pointing toward novel physical conditions. The contribution by S. Lamzin in this volume addresses some of these issues.

In addition to the accretion shock, other regions have been invoked as a source of the emission lines. Alexander *et al.* (2005) argues that photoionizing radiation with luminosities between 10^{41} to 10^{44} photons/sec is necessary to dissipate the disk and to explain the fraction of classical and weak T Tauri stars in star formation regions. Differential emission analysis on the UV spectra indicates that these ionizing luminosities are available in the region producing the UV lines. However, their models suggest that the flux cannot be produced in the accretion shock because it would be absorbed by the incoming accretion column (Alexander *et al.* 2004). Therefore, they conclude that the lines observed in UV spectra of CTTs have to come from the stellar atmosphere: they are real transition region lines. Their models assume a 1 $R_\odot$ long pre-shock column, longer than any predicted by models, and no radiation escaping from the column sides (which may be negligible anyway, see § 5). Therefore, they overestimate the importance of the accretion column as a source of extinction. If one accepts that the UV lines are primarily atmospheric in origin this implies that the presence of accretion radically alters the physical characteristics of the stellar atmosphere, as these strong UV lines are not seen in WTTs.

Gómez de Castro & Verdugo (2007) analyze the spectrum of RY Tau, and argue that the species with the smallest critical densities are more blueshifted than those with large critical densities, as one would expect from a wind. They do not address how can such wind be heated to the large measured temperatures (60,000 K, based on their C IV/Si III] line ratios). In this context, DG Tau A provides an interesting example. The star is well known to have a jet that has been seen in a variety of wavelengths, perhaps even in X-ray images (Güdel *et al.* 2005). Its C IV line is blueshifted by 250 km/s, which may indicate

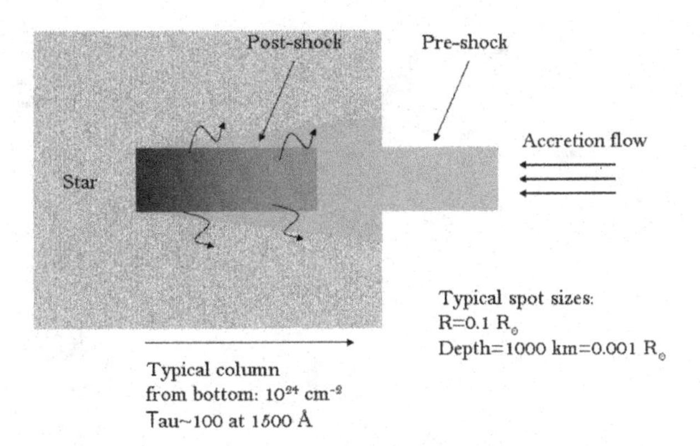

Figure 8. A cartoon of the shock region. In this cartoon, the aspect ratio of the pre- and post-shock is incorrect because for typical filling factors the shock is wider than its long.

a wind origin or a very asymmetric accretion configuration. If heated by a shock in the wind, the line should come from the radiative precursor of the shock. Wind velocities as large as 600 km/s have been measured for DG Tau A in He I (Beristain *et al.* 2001).

Multiple processes may be at work here. It is possible that weak blueshifted wind emission is present in all cases but masked by redshifted accretion shock emission. Observationally, they way to attack this problem is with multi-epoch UV observations, with high enough resolution (R$\sim$ 20,000, like that provided by the *Cosmic Origins Spectrograph* - COS) to trace the line shape. As the accretion spot disappears from sight, one should see only the wind signature, if present.

5. X-rays from shocks

The gas immediately after the shock surface has enough thermal energy to produce soft X-ray emission, but a large fraction of this emission will be absorbed by the pre-shock. If observed, this soft X-ray emission should primarily come from the edges of the post-shock column. As argued by Drake (2005), the position of the bottom of the post-shock column is determined by the equilibrium between the ram pressure of the incoming flow and the stellar gas pressure. For example, based on a Kurucz model of a star with T_{eff}=4250 K, the bottom of the post-shock column in BP Tau ($p_{ram} = 7 \times 10^3$ dyn/cm^2, $l_c \sim 1$ km, Calvet & Gullbring 1998) should be buried under a hydrogen column of a few times 10^{23}cm^{-2}. At this column, all soft X-ray post-shock radiation will be absorbed by the stellar photosphere. For TW Hya ($p_{ram} = 3 \times 10^3$ dyn/cm^2, $l_c \sim 1000$ km, based on an accretion rate of $\dot{M} = 10^{-9}$ M$_\odot$yr^{-1} and a post-shock length from Ardila & Johns-Krull 2007), the column density at the bottom of the post-shock is also of the order of 10^{23}cm^{-2}, but the post-shock is large enough that the stellar column density at the shock surface is two orders of magnitude less. Therefore, radiation from the post-shock may escape from the sides of the column and be observed (Figure 8). In other words, X-ray emission should only be observed in cases when the flow has low ram pressures (low accretion rate, low infall velocity, or very large spot sizes).

X-ray observations are roughly consistent with this picture. As indicated in Figure 9, XMM-Newton/EPIC observations reveal that TW Hya has a strong soft X-ray excess, compared to other CTTs. Günther *et al.* (2007) has modeled EPIC observations of TW Hya, by predicting the soft X-ray emission from the post-shock column. Their models

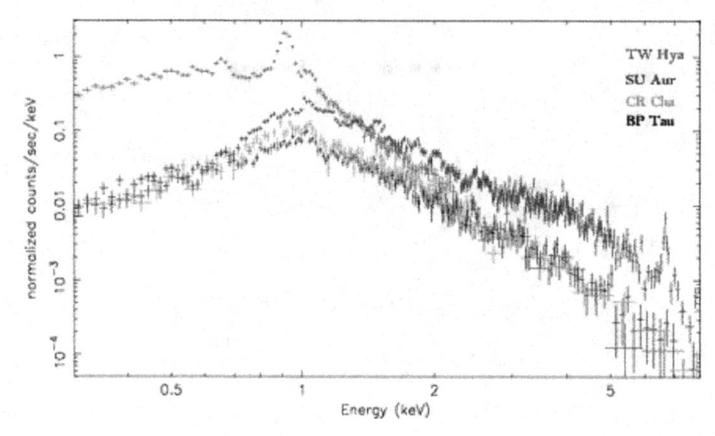

Figure 9. XMM-Newton/EPIC observations of four CTTS, adapted from Robrade & Schmitt (2006).

consider only emission from the post-shock: no absorption due to pre-shock or stellar atmosphere. They can reproduce the observed spectral shape with a three-component spectrum that includes, in addition to the shock emission, plasmas with temperatures of 3×10^6 K and 16×10^6 K. From their models, they derive $\dot{M} = 10^{-10}$ M$_\odot$yr^{-1} and $f = 0.2 - 0.4\%$. The fit is marginally better than the one obtained in a simple three-plasma component model (Stelzer & Schmitt 2004). The values of the accretion rate are one order of magnitude lower than those derived from optical and UV observations (Alencar & Batalha 2002; Herczeg *et al.* 2004), which may point to the fact that the shock column is indeed partially buried. The model has also been applied to observations of V4046 Sgr (Günther & Schmitt 2007).

As Figure 9 shows, strong soft excess is not as noticeable in other CTTs. However, these kind of low-resolution observations are not very sensitive to <0.2 keV emission. Measurements of the temperature sensitive O VII/O VIII line ratio (tracing gas between 2×10^6 K and 4×10^6 K) suggest that as a group CTTs have larger amounts of cool gas than WTTs (Telleschi *et al.* 2007a) as expected from the accretion shock. If the emission is due to post-shock gas, the density of this cool plasma should be larger in CTTs, compared to the same emission in naked atmospheres ($\sim 10^9$ cm^{-3} to $\sim 10^{11}$ cm^{-3}, see Ness *et al.* 2004). Observations of density-sensitive forbidden (for) and intercombination (int) lines of He-like ions (like Ne IXand O VII) reveal for/int values smaller than ~ 0.5 in CTTs, suggesting plasma densities larger than $\sim 10^{11}$cm^{-3} (see for example Kastner *et al.* in this volume, and references therein).

However, questions remain as to the interpretation of these values. For example, for for/int in O VII for T Tau is >4, indicating low plasma densities, unlike what is expected for CTTs (Güdel *et al.* 2007). In addition, the forbidden line of He-like ions is depopulated by ultraviolet radiation, which will lead to overestimating the plasma density. To circumvent this problem, Ness & Schmitt (2005) have used observations of Fe XVII line ratios in TW Hya. They find very low ratios, indicative of higher than atmospheric densities. Unfortunately, the constraints on density are ambiguous because the theoretical calculations do not cover the full range of interactions between the levels involved. One additional problem is that all the density predictions are based on ionization equilibrium calculations, which will not be valid for the post-shock. This effect remains to be fully explored.

6. Summary

Disk gas captured by the stellar magnetic field produces a strong radiative shock upon falling on the stellar surface. The gas reaches temperatures $\sim 10^6$ K immediately after the shock surface and it cools radiatively until it reaches the star. Radiation from the shock creates a radiative precursor (~10,000 K) and it heats the stellar surface resulting in a hot spot (5,000 to 10,000 K). Stellar and shock models indicate that unless the post-shock column is very large, it will be buried on the stellar photosphere.

Models of the continuum emission produced by this configuration reveal that most of the excess veiling in a CTTs is due to hot spot radiation, with the emission from the rest of the column (pre- and post-shock) becoming important as the gravitational potential well of the star increases. Such models can reproduce the observed excess spectra down to 1650 Å. For shorter wavelengths, continuum emission due to molecular hydrogen in the disk has been proposed as a continuum source.

Given the range of temperatures present, the region of the accretion shock is an excellent candidate to explain the observations of UV transition region lines from Li/Na-like ions such as C IV, Si IV, O VI, and N V. Transition region lines in CTTs tend to be broad (200 to 300 km/s) and present blueshifted, centered, and redshifted centroids. Repeated observations of the same target reveal that the centroids may shift substantially in a timescale of years.

Detailed models of the line emission have so far failed to reproduce the fluxes, line shapes, and line ratios. It is not clear if the models need further development or if these failures are pointing toward a different shock configuration. In particular, the blueshifted emission from RY Tau and DG Tau may come, not from the accretion shock, but from a wind.

Low resolution X-ray observations indicates strong soft emission from TW Hya, and detailed shock models have been successful modeling the spectra. High resolution X-ray line observations indicate the presence of larger amounts of cool plasma in CTTs with respect to WTTs. Observations of density sensitive line ratios of He-like ions suggest high plasma densities, as expected from lines originating in the accretion shock. However, the interpretation of the commonly used line ratios is rendered equivocal by the strong UV continuum of the hot spot. Fe XVII line ratios may alleviate this problem and at least in the case of TW Hya, they confirm the high plasma densities derived from other diagnostics.

The COS instrument, to be installed on-board *HST* in 2008, will provide us with new UV spectroscopic data, which will help clarify the exact contribution of the accretion shock to the UV lines. The synergy of the X-ray facilities with *HST* will provide new insights and new mysteries, insuring, for years to come, the presence of a strong community of researchers interested in these complex, amazing objects, the T Tauri stars.

Acknowledgements

I would like to thank the organizers of the IAU Symposium 243 for the invitation to Grenoble and for a wonderful conference. I would also like acknowledge Christopher Johns-Krull and Gregory Herczeg, who provided clarity and encouragement for this work.

References

Alencar, S. H. P. & Batalha, C. 2002, ApJ, 571, 378
Alexander, R. D., Clarke, C. J., & Pringle, J. E. 2004, MNRAS, 354, 71
Alexander, R. D., Clarke, C. J., & Pringle, J. E. 2005, MNRAS, 358, 283
Ardila, D. & Johns-Krull, C. M. 2007, in preparation.

Ardila, D. R., Basri, G., Walter, F. M., Valenti, J. A., & Johns-Krull, C. M. 2002, ApJ, 566, 1100
Bergin, E., *et al.* 2004, ApJL, 614, L133
Beristain, G., Edwards, S., & Kwan, J. 2001, ApJ, 551, 1037
Calvet, N. & Gullbring, E. 1998, ApJ, 509, 802
Calvet, N., Muzerolle, J., Briceño, C., Hernández, J., Hartmann, L., Saucedo, J. L., & Gordon, K. D. 2004, AJ, 128, 1294
D'Antona, F. & Mazzitelli, I. 1997, MemSAI, 68, 807
Drake, J. J. 2005, in ESA Special Publication, Vol. 560, ESA Special Publication, ed. F. Favata & *et al.*, 519
Feigelson, E. D., Broos, P., Gaffney, III, J. A., Garmire, G., Hillenbrand, L. A., Pravdo, S. H., Townsley, L., & Tsuboi, Y. 2002, ApJ, 574, 258
Gómez de Castro, A. I. & Verdugo, E. 2007, ApJL, 654, L91
Güdel, M., Skinner, S. L., Briggs, K. R., Audard, M., Arzner, K., & Telleschi, A. 2005, ApJL, 626, L53
Güdel, M., Skinner, S. L., Mel'Nikov, S. Y., Audard, M., Telleschi, A., & Briggs, K. R. 2007, A&A, 468, 529
Gullbring, E., Calvet, N., Muzerolle, J., & Hartmann, L. 2000, ApJ, 544, 927
Gullbring, E., Hartmann, L., Briceno, C., & Calvet, N. 1998, ApJ, 492, 323
Günther, H. M. & Schmitt, J. H. M. M. 2007, MemSAI, in press
Günther, H. M., Schmitt, J. H. M. M., Robrade, J., & Liefke, C. 2007, A&A, 466, 1111
Herczeg, G. J., Linsky, J. L., Valenti, J. A., Johns-Krull, C. M., & Wood, B. E. 2002, ApJ, 572, 310
Herczeg, G. J., Linsky, J. L., Walter, F. M., Gahm, G. F., & Johns-Krull, C. M. 2006, ApJS, 165, 256
Herczeg, G. J., *et al.* 2005, AJ, 129, 2777
Herczeg, G. J., Wood, B. E., Linsky, J. L., Valenti, J. A., & Johns-Krull, C. M. 2004, ApJ, 607, 369
Johns-Krull, C. M., Valenti, J. A., & Linsky, J. L. 2000, ApJ, 539, 815
Joy, A. H. 1945, Contributions from the Mount Wilson Observatory / Carnegie Institution of Washington, 709, 1
Lamzin, S. A. 2000, Astron. Rep., 44, 323
Lamzin, S. A. 2003, Astron. Rep., 47, 498
Mazzotta, P., Mazzitelli, G., Colafrancesco, S., & Vittorio, N. 1998, A&AS, 133, 403
Muzerolle, J., Calvet, N., & Hartmann, L. 1998, ApJ, 492, 743
Ness, J.-U., Güdel, M., Schmitt, J. H. M. M., Audard, M., & Telleschi, A. 2004, A&A, 427, 667
Ness, J.-U. & Schmitt, J. H. M. M. 2005, A&A, 444, L41
Preibisch, T., *et al.* 2005, ApJS, 160, 401
Robrade, J. & Schmitt, J. H. M. M. 2006, A&A, 449, 737
Stelzer, B. & Schmitt, J. H. M. M. 2004, A&A, 418, 687
Telleschi, A., Güdel, M., Briggs, K. R., Audard, M., & Palla, F. 2007a, A&A, 468, 425
Telleschi, A., Güdel, M., Briggs, K. R., Audard, M., & Scelsi, L. 2007b, A&A, 468, 443

Star-Disk Interaction in Young Stars
Proceedings IAU Symposium No. 243, 2007
J. Bouvier & I. Appenzeller, eds.

doi:10.1017/S1743921307009477

On the origin of continuum and line emission in CTTSs

S. A. Lamzin[1], M. M. Romanova[2] and A. S. Kravtsova[1]

[1]Sternberg Astronomical Institute, Universitetskij prospect 13, Moscow, 119991, Russia
email: lamzin@sai.msu.ru, kravts@sai.msu.ru

[2]Department of Astronomy, Cornell University, Ithaca, NY 14853-6801, USA
email: romanova@astro.cornell.edu

Abstract. We calculated profiles of CIV 1550Å, Si IV 1400Å, NV 1240Å and OVI 1035Å doublet lines using results of 3D MHD simulations of disc accretion onto young stars with a dipole magnetic field. It appeared that our calculations cannot reproduce the profiles of these lines observed (HST/GHRS-STIS and FUSE) in CTTSs spectra. We also found that the theory predicts much larger C IV 1550Å line flux than observed (up to two orders of magnitude in some cases) and argue that the main portion of accretion energy in CTTSs is liberated outside the accretion shock. We conclude that the reason of disagreement between the theory and observation is the strongly non-dipolar character of CTTS magnetic field near its surface.

Keywords. Stars: pre–main-sequence, shock waves, ultraviolet: stars, magnetic fields.

1. Introduction

Since the beginning of the 1990s, there has been a consensus that the line and continuum emission observed in the spectra of classical T Tauri stars (CTTSs) results from the magnetospheric accretion of circumstellar material. More precisely, the magnetic field of the star is believed to stop the accretion disk from reaching the stellar surface. In some way the disk material becomes frozen in the magnetospheric field lines and slides along them toward the stellar surface, eventually being accelerated to velocities $V_0 \sim 300$ km/s. The gas is then decelerated in an accretion shock, whose radiation presumably gives rise to the observed line and continuum emission.

The radial extention of pre- and post-shock radiating regions of CTTS accretion shock is much smaller than the stellar radius – $Z_{pre}, Z_{pst} \ll R_*$ in Figure 1, – making it possible to calculate the structure and spectrum of the accretion shock in the 1-D approximation (Lamzin 1995). Calculations of Lamzin (1998) and Calvet & Gullbring (1998) indicated that the structure of the flow can be specified nearly unambiguously by two parameters: the velocity V_0 and density ρ_0 (or particle number density N_0) of the gas far in front of the shock.

Calvet & Gullbring (1998) used results of their calculations to derive the parameters of the accretion shock via modeling of the continuum spectral energy distributions of classical T Tauri stars. However, they did not take into account limb darkening effects. Furthermore, the agreement between the calculated and observed spectra of the veiling continuum cannot be considered as a decisive support for the magnetospheric model, since boundary-layer models provide equally good agreement – see e.g. Basri & Bertout (1989). The line spectrum is far more informative, and comparisons of the calculated and observed intensities and profiles of emission lines enable detailed studies of the accretion processes.

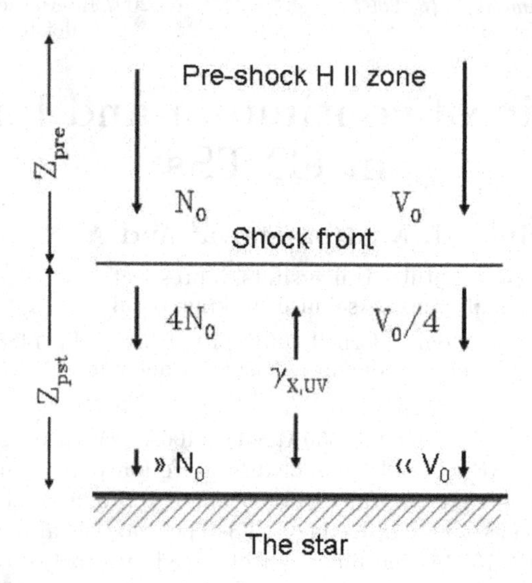

Figure 1. Schematic structure of CTTS accretion shock.

Optically thin lines are best suited for this purpose: their intensity ratios can be used to derive physical conditions independent of the geometry of the region where they are formed, and the line profiles provide information about both the velocity field and the geometry of this region. Calculations by Lamzin & Gomez de Castro (1999) demonstrated that the O III] 1663Å, Si III] 1892Å and C III] 1909Å lines should display the highest intensities among the optically thin lines, and these lines were used to determine the accretion-shock parameters for several young stars in that paper. However, Gomez de Castro & Verdugo (2001, 2007) questioned whether these lines form in the accretion shock. Spectral lines of neutral or singly ionized atoms apparently form not only in the accretion shock but in the magnetospheric flow and wind as well – see e.g. Edwards, this volume – and thus likewise cannot be used as diagnostics of the accretion shock.

Resonant UV lines of the C IV, Si IV, N V and O VI uv1 doublets look as the most suitable diagnostics of CTTSs accretion shock, especially lines of the C IV 1550Å doublet: they are strong in CTTS spectra and the calculation of their intensities is relatively simple.

2. Intensities of the C IV 1550Å doublet lines: theory vs. observations

Generally speaking, lines of C^{+3}, Si^{+3}, N^{+4} and O^{+5} ions form before and behind the shock front. Gas temperature in the pre-shock (precursor) zone of CTTSs does not exceed 20,000 K, but ions up to O^{+5} (at $200 < V_0 < 400$ km/s) exist here due to the photoionization of accreted matter by X-ray and UV quanta from the post-shock cooling zone. When infalling gas crosses the shock front its temperature raises up to 1-3 MK and for example C^{+3} ions almost completely transform to C^{+6} ions. Then the gas cools down and ions of interest appear again but at a gas velocity close to zero – see Lamzin (1998) for details.

Lamzin (2003a) carried out non-LTE calculations of profiles of C IV 1550Å, Si IV 1400Å, N V 1240Å and O VI 1035Å doublet lines for a plane-parallel shock viewed at various angles. Calculations were performed for a range of preshock gas parameters V_0, N_0 appropriate for CTTSs. Intensities of C IV 1548+1551Å lines, normalized to $\mathcal{F} = \rho_0 V_0^4/4$, as a function of cosine μ of the angle between the normal to the shock front and the line

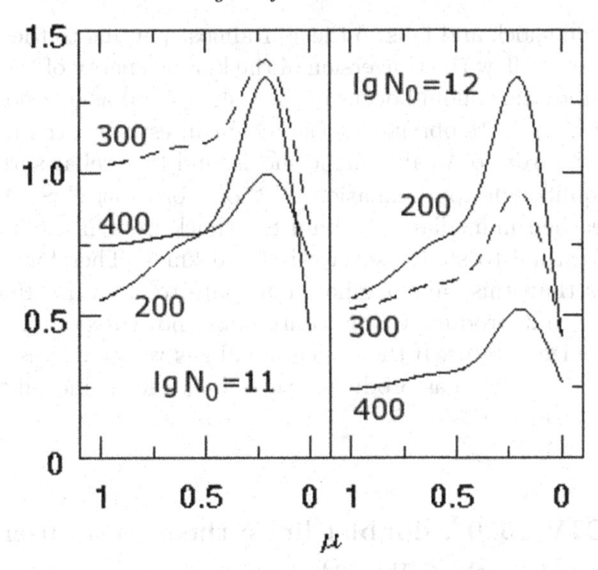

Figure 2. Relative intensities of the C IV 1550Å doublet lines expressed in percent. See text for details.

of sight are presented in Figure 2. The ratio is expressed in percent, such as different lines in the figure correspond to different infall gas velocities ($V_0 = 200$, 300 and 400 km/s). The results were calculated for a gas particle density of $N_0 = 10^{11}$ cm^{-3} (left panel) and 10^{12} cm^{-3} (right panel).

The value $\mathcal{F}$ was chosen for normalisation because this value is expected to be equal to the bolometric flux of the veiling continuum emission produced by the accretion shock. Indeed, according to the current paradigm half of the X-ray and UV quanta from the post-shock cooling zone reach the stellar surface. These quanta are absorbed in the upper layers of the stellar atmosphere and then should be predominantly reradiated in the continuum – see Calvet & Gullbring (1998) for details. Thus the ratio δ, depicted in Figure 2, is the theoretical prediction for the ratio of C IV 1550Å doublet line flux to the bolometric flux of the veiling continuum. Thus this ratio is expected to be $\simeq 1\%$ and almost independent of V_0 and N_0.

Meanwile it was found that the observed ratio is much smaller – see Table 1 in which we summarised the results of our analysis of CTTSs UV spectra observed by the Hubble Space Telescope (Kravtsova & Lamzin 2002a; Kravtsova & Lamzin 2002b; Kravtsova 2003; Lamzin *et al.* 2004). The descrepancy between theory and observation is significant – more than two orders of magnitude in the case of RY Tau and DR Tau, such as this conclusion does not depend on current uncertainty on the value and law of interstellar extinction in the direction of the investigated stars.

Table 1. Relative contribution of C IV lines to the emission of accretion shock

Star:	RY Tau	DR Tau	T Tau	DS Tau	BP Tau	DG Tau	Theory
δ, %	0.002	0.003	0.02	0.02	0.04	0.07	~ 1.0

We suppose that this discrepancy means that the main portion of the veiling continuum (up to 99 % in some cases juging from Table 1) originates outside the C IV 1550Å line formation region, i.e., outside the *strong* (Mach number $M_{sh} \gg 1$) accretion shock. In other words, we conclude that the main portion of the accreted matter does not pass through the accretion shock and falls to the star almost parallel to the stellar surface. In this part of the accretion flow the conversion of the kinetic energy of the infalling gas into heat and then into radiation should occur in the same way(s) as in a boundary layer, i.e., in a series of *weak* ($M_{sh} \simeq 1$) oblique shocks. One can estimate the maximum possible angle γ_{sh} between the front of weak oblique shocks and the stellar surface as follows.

In the coronal equlibrium approximation C^{+3} ions forms at $T \simeq 10^5$ K. Such a temperature can be reached immediately behind the shock front if the accreted gas velocity component V_r normal to stellar surface is $\simeq$ 70 km/s. Therefore oblique accretion shocks with V_r less than this value cannot contribute to the C IV 1550Å line emission of the accretion flow but produce veiling continuum and emission of neutral lines, and of singly or twice ionized atoms. If the typical infall gas velocity V_0 is $\simeq$ 300 km/s, then $\gamma_{sh} = \sin^{-1}(V_r/V_0) < 15^o$, i.e. gas producing such weak shocks indeed falls onto the star almost parallel to its surface.

3. Profiles of C IV 1550Å doublet lines: theory vs. observations

As was mentioned above the CIV 1550Å doublet lines form in two spatially distinct regions of strong accretion shock: in the radiative precursor and in the post-shock zone. Gas velocity in these regions are different: $V \simeq V_0 \sim 300$ km/s in the pre-shock zone and $\sim 5 - 10$ km/s in the post-shock line formation region. As a result the profile of e.g. the CIV 1548Å line in the spectrum of a plane-parallel shock, viewed from the direction perpendicular to the surface of the shock front, should have two components: the first one is at almost zero-velocity and the second is redshifted to V_0. As follows from our calculations (Lamzin 2003a), both components are optically thick, resulting in a FWHM for each component of $\sim 20 - 30$ km/s, with a relative strength depending on V_0 : at $V_0 < 300$ km/s the "zero-velocity" component is stronger than the "high-velocity" one and vice versa at $V_0 > 300$ km/s. If the shock is viewed from a direction that makes an angle θ with the perpendicular to the shock surface, then the "zero-velocity" component should be seen practically at the same position but the redshifted "high-velocity" peak should now be at $V_0\,\mu$, where $\mu = \cos\theta$. The same is true (in a qualitative way) for lines of Si IV 1400Å, NV 1240Å and OVI 1035Å doublets.

Consider now a part of CTTS surface occupied with strong accretion shock (accretion zone). The observed profile of e.g. the CIV 1548Å line emitted by the accretion shock is the sum (an integral) of the double-peaked profiles from all elementary areas ΔS of the accretion zone (multiplied by the $\mu\,\Delta S$ factor). All elementary areas are viewed at different angles due to the curvature of stellar surface and these angles vary with time due to stellar rotation. One can expect that the intensities of "zero-velocity" components from all parts of the accretion zone will sum up and the (weighted) sum of high-velocity components will result in a more or less wide red wing or separated redshifted component depending on the distribution of V_0 and N_0 in the accretion zone and on its geometry. Obviously the profile should vary with time due to stellar rotation and non-stationary accretion as well.

Lamzin (2003b) calculated the profiles of the C IV 1550Å doublet lines from a strong accretion shock assuming that: 1) matter falls to the star in the radial direction; 2) V_0 and N_0 are constant within the accretion zone; 3) the zone has the shape of a circular spot or a spherical belt. The results of the calculations were compared to the profiles of UV lines in the spectra of CTTSs observed with the Goddard High Resolution Spectrograph (GHRS) and Space Telescope Imaging Spectrograph (STIS). Observational data were extracted from the scientific archives of the HST. The calculated profiles differ significantly from the observed ones presumably because our assumptions about the character of the accretion flow near the stellar surface were not realistic enough.

One can expect better agreement by using parameters of an accretion flow derived from 3D MHD simulations of disc accretion to a slowly rotating magnetized young star with its dipole moment inclined at an angle α to the stellar rotation axis. Accretion rate $\dot{M}_{acc}$, polar magnetic field strength B as well as mass and radius of the central star are free parameters of these simulations in addition to the angle $0 \leqslant \alpha \leqslant 90^o$ – see Romanova *et al.* (2003) for details. The velocity field $\mathbf{V}_0$ and gas density ρ_0 at the stellar surface, adopted from the simulations, were used as input parameters to calculate profiles of CIV 1550Å, SiIV 1400Å, NV 1240Å and OVI 1035Å doublet lines. For all models we adopted $M_* = 0.8\ \mathrm{M}_\odot$, $R_* = 1.8\ \mathrm{R}_\odot$, $B = 1-3$ kG and varied α, $\dot{M}_{acc}$ in the $0^o - 90^o$ and $10^{-8} - 3 \cdot 10^{-7}\ \mathrm{M}_\odot$/yr intervals respectively. Profiles were calculated for each accretion zone model with different values of the angle between the stellar rotation axis and the line of sight ($0^o \leqslant i \leqslant 90^o$) as well as for a set of phases of stellar rotation periods, i.e., for different angles ψ (in 2π units) between the magnetic dipole axis and the plane which contains the rotation axis and the line of sight ($0 \leqslant \psi \leqslant 1$).

Observed profiles of the C IV 1550Å doublet lines in DS Tau, BP Tau, DF Tau and DR Tau spectra are shown in Figure 3. Solid and dashed lines depicts C IV 1548Å and C IV 1551Å componens of the doublet. Profiles of the C IV 1550Å doublet components in the spectrum of T Tau are shown in Figure 4 (left column). T Tau is the only star where there is more than one high resolution UV spectrum and one can observe variability of the C IV doublet lines profiles. Only in the case of TW Hya (right panel of the figure) is there a possibility to obtain information about lines of the O VI 1035Å doublet – see Ardila, this volume for references and details.

We plot in Figure 5 the results of our calculations for the model with $\alpha = 30^o$, $\dot{M}_{acc} \simeq 4 \cdot 10^{-8}\ \mathrm{M}_\odot$/yr, $i = 10^o$ (left panel) and $i = 70^o$ (right panel). The vertical row of profiles in each panel corresponds to the following set of rotational phases (from top to bottom): $\psi = 0$, 0.25, 0.5, 0.75. Profiles were normalized to the maximum intensity of the line at

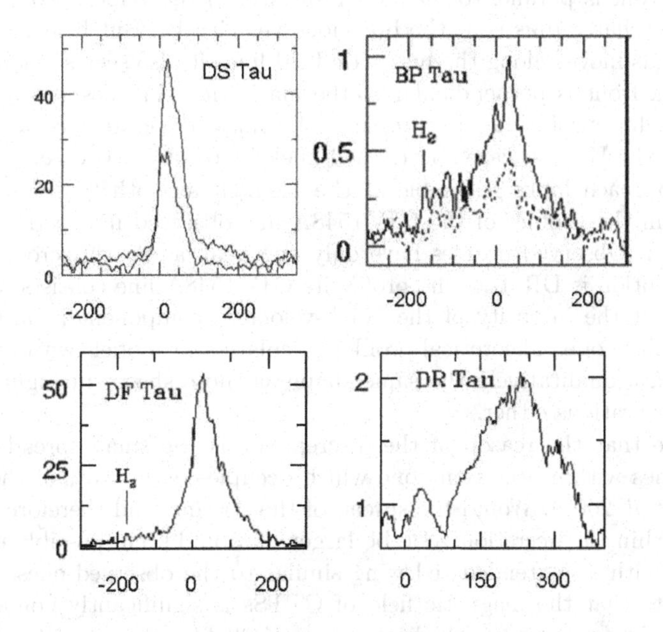

Figure 3. Profiles of the C IV 1550Å doublet lines in the spectra of some CTTSs.

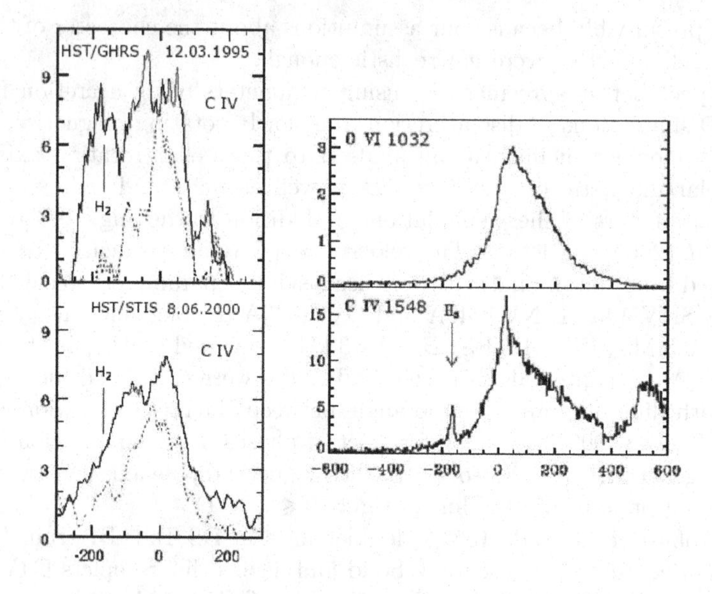

Figure 4. Profiles of the C IV 1550Å and O VI 1035Å doublet components in the spectra of T Tau (left panel) and TW Hya (right panel).

$\psi = 0$. It was assumed that the accretion disk does not prevent to observe the part of the star that is situated below the disk midplane – this is the reason why some profiles have an extended blue wing.

Matter falls to the star with a dipole magnetic field at an angle $\theta < 90^o$ relative to its surface. In the absence of a magnetic field oblique shocks would arise, which means that: 1) the shock front is parallel to the stellar surface; 2) the velocity component V_r which is parallel to stellar radius is at the pre-shock velocity V_0. But bear in mind that since the accreted gas moves along the magnetic field lines it also seems resonable to suppose that the shock front is perpendicular to the magnetic field lines and therefore $V_0 = V$. We calculated line profiles for both cases: solid lines in Figure 5 correspond to profiles calculated for the $V_0 = V_r$ case and dashed lines for the $V_0 = V$ case.

Such an approach looks resonable at the moment as both types of theoretical profiles differ from the profiles of the C IV 1548Å line observed in spectra of CTTS shown in Figures 3, 4. Observed profiles have only one peak at nearly zero velocity position. The only exception is DR Tau: the profile its C IV 1548Å line consists of two redshifted components but the intensity of the "high-velocity" component is larger than that of the "low-velocity" one. Theoretical profiles calculated for models with other values of α, $\dot{M}_{acc}$ and i have qualitatively the same shape as those shown in Figure 5, i.e., cannot reproduce observations either.

We suppose that the reason of the discrepancy is the small spread of the accreted gas stream lines within accretion zone which occupies only $\sim 5\%$ of the stellar surface (Romanova *et al.* 2003). Would the spread of the stream (and therefore of the magnetic field) lines within the accretion zone be larger, it would seem possible to obtain single-peak profiles with an extended red wing similar to the observed ones. In any case our results indicate that the magnetic field of CTTSs is significantly non-dipolar near the stellar surface in agreement with direct magnetic field measurements (see Johns-Krull, this volume).

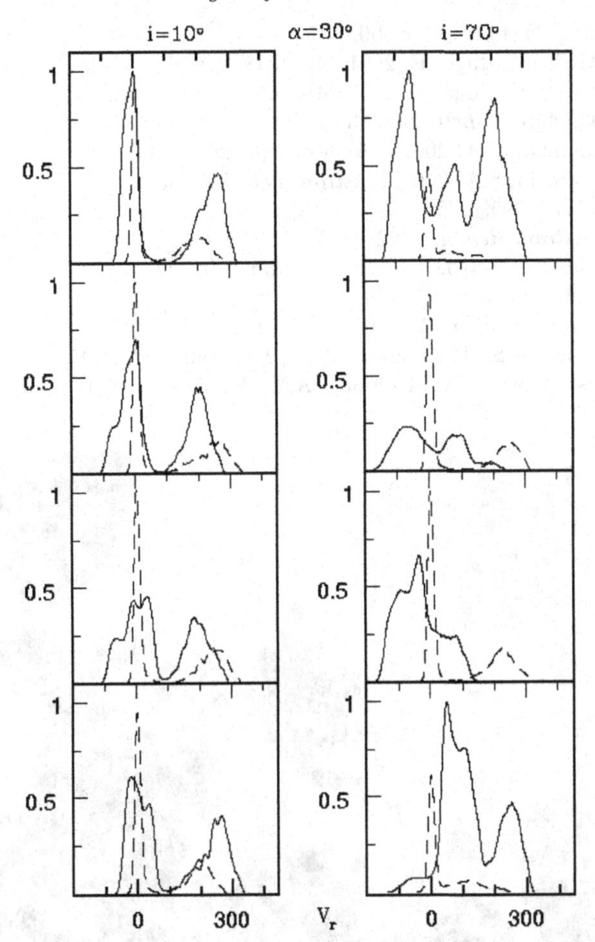

Figure 5. Theoretical profiles of the C IV 1548Å line calculated for a CTTS with a dipole magnetic field whose axis is inclined at $\alpha = 30^o$ to the rotation axis of the star. See text for details.

4. Conclusion

We demonstrated that the observed intensity and profiles of C IV 1550Å doublet lines significantly differ from theoretical predictions based on the assumption that the magnetic field of CTTSs near the stellar surface is close to a dipole. We conclude therefore that the geometry of CTTS magnetic field near the stellar surface is strongly non-dipolar. Multipole components of the global magnetic field of young star and/or small-scale magnetic fields in active regions probably produce a large divergence of the accreted gas stream lines within the accretion zone that presumably accounts for the disagreement between theory and observations.

Acknowledgements

We thank the SOC and the LOC of the Symposium for the invitation, financial support and hospitality.

References

Basri, G. & Bertout, C. 1989, *ApJ* 341, 340

Calvet, N. & Gullbring, E. 1998, *ApJ* 509, 802
Gómez de Castro, A.I. & Verdugo, E. 2001, *ApJ* 548, 976
Gómez de Castro, A.I. & Verdugo, E. 2007, *ApJ* 654, 91
Kravtsova, A.S. 2003, *Astron. Lett.* 29, 463
Kravtsova, A.S., & Lamzin, S.A. 2002a, *Astron. Lett.* 28, 676
Kravtsova, A.S., & Lamzin, S.A. 2002b, *Astron. Lett.* 28, 835
Lamzin, S.A. 1995, *A&A* 295, L20
Lamzin, S.A. 1998, *Astron. Rep.* 42, 322
Lamzin, S.A. & Gomez de Castro, A.I. 1999, *Astron. Lett.* 24, 748
Lamzin, S.A. 2003a, *Astron. Rep.* 47, 498
Lamzin, S.A. 2003b, *Astron. Rep.* 47, 540
Lamzin, S.A., Kravtsova, A.S., Romanova, M.M., & Batalha, C. 2004, *Astron. Lett.* 30, 413
Romanova, M.M., Ustyugova, G.V., Koldoba, A.V., Wick, J.V., & Lovelace, R.V.E. 2003, *ApJ* 595, 1009

Star-Disk Interaction in Young Stars
Proceedings IAU Symposium No. 243, 2007
J. Bouvier & I. Appenzeller, eds.

doi:10.1017/S1743921307009489

Inner disk regions revealed by infrared interferometry

Fabien Malbet

Laboratoire d'Astrophysique de Grenoble,
UJF/CNRS, BP 53, F-38041 Grenoble cedex 9, France
Email: fabien.malbet@obs.ujf-grenoble.fr

Abstract. I review the results obtained by long-baseline interferometry at infrared wavelengths on the innermost regions around young stars. These observations directly probe the location of the dust and gas in the disks. The characteristic sizes of these regions are found larger than previously thought. These results have motivated in part a new class of models of the inner disk structure. However the precise understanding of the origin of these low visibilities is still in debate. Mid-infrared observations have probed disk emission over a larger range of scales revealing mineralogy gradients in the disk. Recent spectrally resolved observations allow the dust and gas to be studied separately. The few results show that the Brackett gamma emission can find its origin either in a wind or in a magnetosphere but there are no definitive answers yet. In a number of cases, the very high spatial resolution seems to reveal very close companions. In any case, these results provide crucial information on the structure and physical properties of disks surrounding young stars especially as initial conditions for planet formation.

Keywords. Accretion disks, stars: pre–main-sequence, stars: emission-line, stars: mass loss, stars: winds, outflows, planetary systems: protoplanetary disks, infrared: stars, techniques: interferometric, techniques: spectroscopic.

---

1. Introduction

Many physical phenomena occur in the inner regions of the disk which surrounds young stars. The matter which falls onto the stellar surface spirals in a more or less accreting circumstellar disk subject to turbulence, convection, external and internal irradiation. The disks which are rotating in a quasi Keplerian motion are probably the birth location of future planetary systems. Strong outflows, winds and even jets often find their origin in the innermost regions of many young stellar systems. The mechanisms of these ejection processes are not well understood but they are probably connected to accretion. Most of young stellar systems are born in multiple systems which can be very tight and therefore have a strong impact on the physics of the disk inner regions.

The details of all these physical processes are not yet well understood because of lack of data to constrain them. The range of physical parameters which best define the inner regions of disk in young stellar objects are:

- radius ranging from 0.1 AU to 10 AU
- temperature ranging from 150 K to 4000 K
- velocities ranging from 10 km/s to a few 100 km/s

The instrumental requirements to investigate the physical conditions in such regions are therefore driven by the spectral coverage which must encompass the near and mid infrared from 1 to 20 μm. Depending on the distance of the object (typically between 75 pc and 450 pc) the spatial resolution required to probe the inner parts of disks ranges between fractions and a few tens of milli-arcseconds. Since the angular resolution of astronomical instruments depends linearly on the wavelength and inversely on the telescope diameter, observing in the near and mid infrared wavelength domain points toward telescopes of sizes ranging from ten to several hundred meters. The only technique that allows such a spatial resolution is therefore infrared interferometry.

Millan-Gabet *et al.* (2007) review the main results obtained in infrared interferometry in the domain of young stars between 1998 and 2005. The purpose of the present review is to concentrate on the inner disk regions and to give the latest results in this field. § 2 briefly explains the principles of infrared interferometry and lists the literature on the observations carried out with this technique. § 3 focuses on the main results obtained on disk physics (sizes, structures, dust and gas components,...) and § 4 presents results on other phenomena constrained by interferometry (winds, magnetosphere, multiple systems,...). In § 5 I describe the type of results that can be expected in the future.

2. Infrared interferometry

2.1. *Principle and observations*

Long baseline optical interferometry consists in mixing the light received from an astronomical source and collected by several independent telescopes separated from each other by tens or even hundreds of meters. The light beams are then overlapped and form an interference pattern if the optical path difference between the different arms of the interferometer—taking into account paths from the source up to the detector—is smaller than the coherence length of the incident wave (typically of the order of several microns). This interference figure is composed of fringes, i.e., a succession of stripes of faint (destructive interferences) and bright (constructive interferences) intensity. By measuring the contrast of these fringes, i.e. the normalized flux difference between the darkest and brightest regions, information about the morphology of the observed astronomical source can be recovered. Figure 1 illustrates this principle.

2.2. *Instruments available for inner regions studies*

Interferometric observations of young stellar objects were and are still performed at six facilities on seven different instruments (see Table 1). We can classify these observations into three different categories:

- **Small-aperture interferometers**: PTI, IOTA and ISI were the first facilities to be operational for YSO observations in the late 1990's (see Fig. 2 & 3). They have provided mainly the capability of measuring visibility amplitudes and lately closure phases. The latest one, CHARA, has an aperture diameter of 1 m. The instruments are mainly accessible through team collaboration.
- **Large-aperture interferometers**: KI, VLTI and soon LBT are facilities with apertures larger than 8 m. The instruments are widely open to the astronomical community through general calls for proposals. Lately, these facilities have significantly increased the number of young objects observed.
- **Instruments with spectral resolution:** CHARA, MIDI and AMBER provide a spectral resolution from a few hundred up to 10,000 whereas other instruments mainly provided broadband observations. The spectral resolution allows the various phenomena occurring in the environment of young stars to be separated.

2.3. *Elements of bibliography*

Figure 2 displays the number of published results, and show that it is increasing with time and improved facilities. At the date of the conference there were 31 refereed articles published in the field of young stars corresponding to 66 young stellar objects observed (see by chronological order: Malbet *et al.* 1998; Millan-Gabet *et al.* 1999; Akeson *et al.*

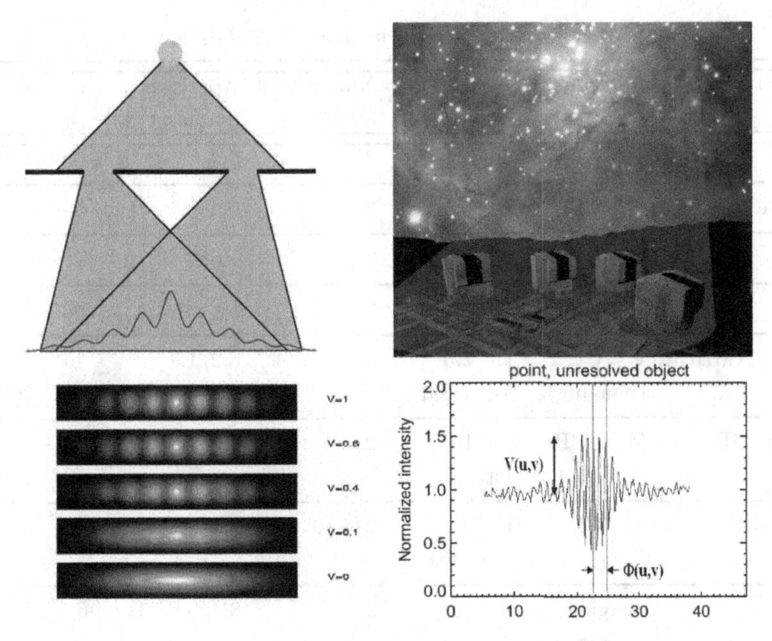

Figure 1. Principle of interferometry. Upper panels: the Young's slit experiment (left) compared to optical interferometry (right): in both cases the light travels from a source to a plane where the incoming wavefront is split. The telescope apertures play the same role as the Young's slits. The difference lies in the propagation of light after the plane. In the case of optical interferometry, the instrument controls the propagation of light down to the detectors. At the detector plane, the light beams coming from the two apertures are overlapped. Lower panels: interference fringes whose contrast changes with the morphology of the source. Left panel shows fringes whose contrast varies from 0 to 1. Right panel displays actual stellar fringes but scanned along the optical path. The measure of the complex visibilities corresponds to the amplitude of the fringes for the visibility amplitude and the position of the fringes in wavelength units for the visibility phase.

2000; Millan-Gabet *et al.* 2001; Tuthill *et al.* 2002; Eisner *et al.* 2003; Colavita *et al.* 2003; Wilkin & Akeson 2003; Leinert *et al.* 2004; Eisner *et al.* 2004; van Boekel *et al.* 2004; Malbet *et al.* 2005; Akeson *et al.* 2005b; Eisner *et al.* 2005; Monnier *et al.* 2005; Boden *et al.* 2005; Akeson *et al.* 2005a; Eisner *et al.* 2006; Millan-Gabet *et al.* 2006a; Preibisch *et al.* 2006; Ábrahám *et al.* 2006; Millan-Gabet *et al.* 2006b; Monnier *et al.* 2006; Quanz *et al.* 2006; Eisner *et al.* 2007; Malbet *et al.* 2007; Tatulli *et al.* 2007; Kraus *et al.* 2007; Lachaume *et al.* 2007; Eisner 2007; Ratzka *et al.* 2007).

Graphs in Fig. 3 show that the distribution of observed objects is rather well distributed among the various facilities. Several categories of young stellar systems have been observed at milli-arcsecond scales mainly in the near-infrared wavelength domain, but also in the mid-infrared one. They include the brightest Herbig Ae/Be stars, the fainter T Tauri stars and the few FU Orionis. Finally most observations were carried out in broad band but the advent of large aperture interferometers like the VLTI and KI allow higher spectral resolution to be obtained.

3. Inner disk physics

Most of the studies carried out on YSOs are focused on the physics of inner regions of disks. They started with the determination of rough sizes of emission then led to more constraints on the disk structure. Mid infrared spectrally resolved observations were able

Table 1. Interferometers involved in YSO science

Facility	Instrument	Wavelength (microns)	Numbers of apertures	Aperture diameter (m)	Baseline (m)
PTI	V^2	H, K	3	0.4	80 – 110
IOTA	V^2, CP	H, K	3	0.4	5 – 38
ISI	heterodyne	11	2 (3)	1.65	4 – 70
KI	V^2, nulling	K	2	10	80
VLTI/AMBER	V^2, CP (imaging)	1 – 2.5 /spectral	3 (8)	8.2/1.8	40 – 130 /8 – 200
VLTI/MIDI	V^2 (/CP)	8 – 13 /spectral	2 (4)	8.2/1.8	40 – 130 /8 – 200
CHARA	V^2, CP (imaging)	1 – 2.5 /spectral	2/4 (6)	1	50 – 350
LBT	imaging, nulling	1 – 10	2	8.4	6 – 23

V^2: visibility measurement; CP: closure phase.

Acronyms. PTI: *Palomar Testbed Interferometer*; IOTA: *Infrared and Optical Telescope Array* (closed since 2006); ISI: *Infrared Spatial Interferometer*; KI: *Keck Interferometer*; VLTI: *Very Large Telescope Interferometer*; CHARA: *Center for High Angular Resolution Array*; LBT: *Large Binocular Telescope* (not yet operational).

to identify different types of dust grains. Near infrared spectrally resolved observations are coming out and permits to spatially discriminate between gas and dust.

3.1. *Sizes of circumstellar structures*

About 10 years ago, the paradigm was that disks were present around a majority of young stars. These disks were believed to behave "normally" with a radial temperature distribution following a power-law $T \propto r^{-q}$ with q ranging between 0.5 and 0.75. The value of q depends on the relative effect of irradiation from the central star in comparison to heat dissipation due to accretion. This model was successful in reproducing ultraviolet and infrared excesses in spectral energy distributions (SEDs). Malbet & Bertout (1995) investigated the potential of optical long baseline interferometry to study the disks of T Tauri stars and FU Orionis stars. They found that the structure would be marginally resolved but observations would be possible with baselines of the order of 100 m with a visibility amplitude remaining high.

First observations of brighter Herbig Ae/Be stars showed that the observed visibilities were much smaller than the expected ones especially in these objects were the accretion plays a minor role. Monnier & Millan-Gabet (2002) pointed out that the interferometric sizes of these objects were much larger than expected from the standard disk model. They plotted the sizes obtained as a function of the stellar luminosity and found a strong correlation following a $L^{0.5}$ law over two decades. This behavior is consistent with the variation of radius of dust sublimation with respect to the central star luminosity: the

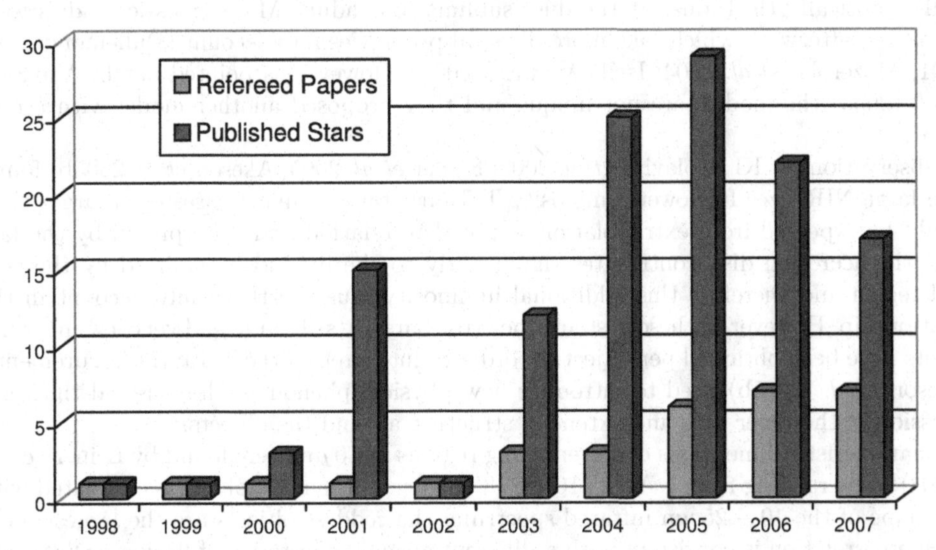

Figure 2. Young stellar objects observed by interferometry and number of refereed papers published in the period 1998-2007. The statistics of the year 2007 is not complete.

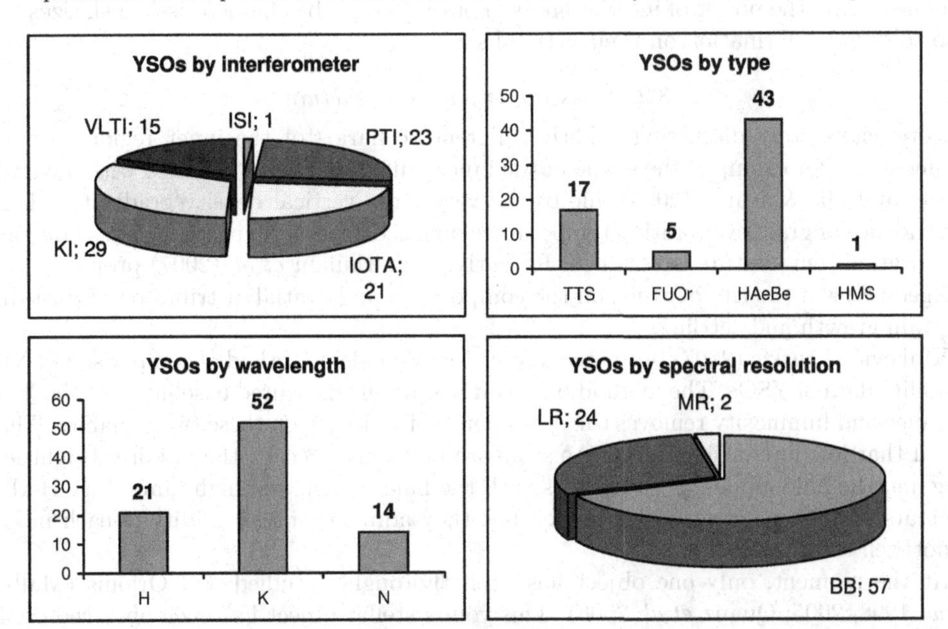

Figure 3. Young stellar objects observed by interferometry in the period 1998-2007. Upper left: distribution by interferometer. Upper right: distribution by YSO type. Lower left: distribution by wavelength of observation. Lower right: distribution by spectral resolution. The statistics is the same as the one of Fig. 2.

more luminous the object the further the dust can survive in solid form at a temperature lower than the sublimation limit ($\sim 1000 - 1500\,\mathrm{K}$). Only the most massive Herbig Be stars seem to be compliant with the standard accretion disk model.

In the meantime, in order to account for the near-infrared characteristics of SEDs and in particular a flux excess around $\lambda = 3\mu\mathrm{m}$, Natta *et al.* (2001) proposed that disks around Herbig Ae/Be stars have optically thin inner cavity and create a puffed-up inner wall of optically thick dust at the dust sublimation radius. More realistic models were developed afterward which take more physical properties into account (Dullemond *et al.* 2001; Muzerolle *et al.* 2004; Isella & Natta 2005). However as pointed out by Vinković *et al.* (2003) the models are not unique and they proposed another model with a disk halo.

Observations at KI (Colavita *et al.* 2003; Eisner *et al.* 2005; Akeson *et al.* 2005b) found also large NIR sizes for lower-luminosity T Tauri stars, in many cases even larger than would be expected from extrapolation of the HAe relation. It is interpreted by the fact that the accretion disk contributes significantly to the luminosity emitted by the central region and therefore this additional luminosity must be taken into account in the relationship. However in these systems the error bars are still large and very few measurements have been obtained per object. In order to interpret all the T Tauri measurements, Akeson *et al.* (2005b) need to introduce new physical phenomena like optical thick gas emission in the inner hole and extended structure around the objects.

Characteristic dimensions of the emitting regions at 10 μm were found by Leinert *et al.* (2004) to be ranging from 1 AU to 10 AU. The sizes in their sample stars correlated with the slope of the $10 - 25\,\mu$m infrared spectrum: the reddest objects are the largest ones. Such a correlation is consistent with a different geometry in terms of flaring or flat (self-shadowed) disks for sources with strong or moderate mid-infrared excess, respectively, demonstrating the power of interferometry not only to probe characteristic disk sizes but also to derive information on their vertical structure.

3.2. *Constraints on disk structure*

Theoreticians start discussing slightly different scenarios of the inner regions around young stars. For example, the shape of the inner puffed-up wall is modeled with a curved shape by Isella & Natta (2005) due to the very large vertical density gradient and the dependence of grain evaporation temperature on gas density as expected when a constant evaporation temperature is assumed. Recently, Tannirkulam *et al.* (2007) proposed that the geometry of the rim depends on the composition and spatial distribution of dust due to grain growth and settling.

Vinković & Jurkić (2007) presented a model-independent method of comparison of NIR visibility data of YSOs. The method based on scaling the measured baseline with the YSO distance and luminosity removes the dependence of visibility on these two variables. They found that low luminosity Herbig Ae stars are best explained by the uniform brightness ring and the halo model, T Tauri stars with the halo model, and high luminosity Herbig Be stars with the accretion disk model, but they admit that the validity of each model is not well established.

At the moment, only one object has been thoroughly studied: FU Orionis (Malbet *et al.* 1998, 2005; Quanz *et al.* 2006). This young stellar object has been observed on 42 nights over a period of 6 years from 1998 to 2003 with 287 independent measurements of the fringe visibility at 6 different baselines ranging from 20 to 110 m in length, in the H and K bands. The data not only resolves FU Ori at the AU scale, but also allows the accretion disk scenario to be tested. The most probable interpretation is that FU Ori hosts an active accretion disk whose temperature law is consistent with the standard model. In the mid infrared, Quanz *et al.* (2006) resolved structures that are also best explained with an optically thick accretion disk. A simple accretion disk model fits the observed SED and visibilities reasonably well and does not require the presence of any additional structure such as a dusty envelope. This is why one should remain careful with results coming from surveys having only few measurements per object.

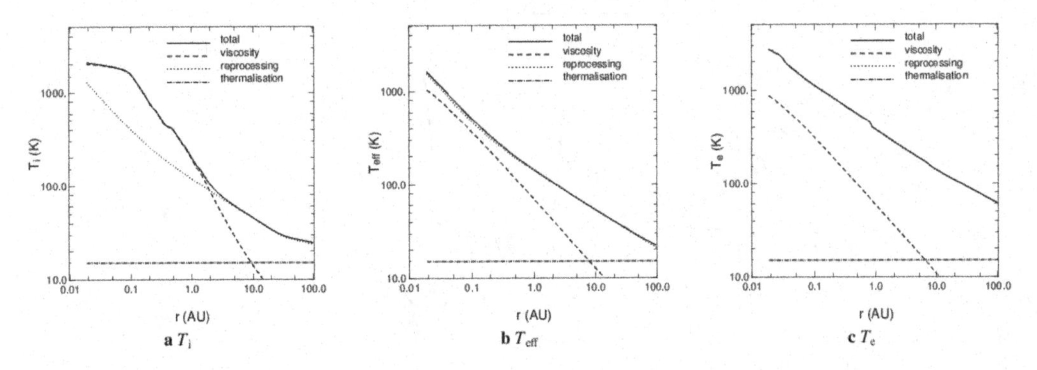

Figure 4. Radial distribution of temperature at different location in the disk computed with a two-layer model by Lachaume *et al.* (2003). Left: temperature in the equatorial plane. Center: effective temperature. Right: surface temperature. The different lines represent the various contribution to the heating: viscosity, reprocessing of the stellar light, thermalization with the ambient temperature.

Millan-Gabet *et al.* (2006a) obtained K-band observations of three other FU Orionis objects, V1057 Cyg, V151 Cyg, and Z CMa-SE and found that all three objects appear significantly more resolved than expected from simple models of accretion disks tuned to fit the SEDs. They believe that emission at the scale of tens of AU in the interferometer field of view is responsible for the low visibilities, originating in scattering by large envelopes surrounding these objects. In a not yet published study, Li Causi *et al.* have measured again interferometric visibilities of Z CMa with VLTI/AMBER and propose to interpret the data by the presence of a very close companion.

On the theoretical side, very few physical models achieved to fit interferometric data simultaneously with SEDs. Using a two-layer accretion disk model, Lachaume *et al.* (2003) found satisfactory fits for SU Aur, in solutions that are characterized by the midplane temperature being dominated by accretion, while the emerging flux is dominated by reprocessed stellar photons (see Fig. 4). Since the midplane temperature drives the vertical structure of the disk, there is a direct impact on the measured visibilities that are not necessarily taken into account by other models.

Very interesting results have been presented at this conference by Kraus *et al.* (this volume) showing that they are able to derive the temperature radial distribution of the disk around MWC 147 from interferometric measurements using the spectral variation of the visibilities at low resolution. A similar work has been attempted at PTI with larger error bars (Eisner *et al.* 2007).

3.3. *Dust mineralogy*

The mid-infrared wavelength region contains strong resonances of abundant dust species, both oxygen-rich (amorphous or crystalline silicates) and carbon-rich (polycyclic aromatic hydrocarbons, or PAHs). Therefore, spectroscopy of optically thick protoplanetary disks offers a diagnostic of the chemical composition and grain size of dust in disk atmosphere.

van Boekel *et al.* (2004) spatially resolved three protoplanetary disks surrounding Herbig Ae/Be stars across the N band. The correlated spectra measured by MIDI at VLTI correspond to disk regions ranging from 1 to 2 AU. By combining these measurements with unresolved spectra, the spectrum corresponding to outer disk regions at 220 AU can also be derived. These observations have revealed that the dust in the inner regions was highly crystallized (40 to 100%), more than any other dust observed in young

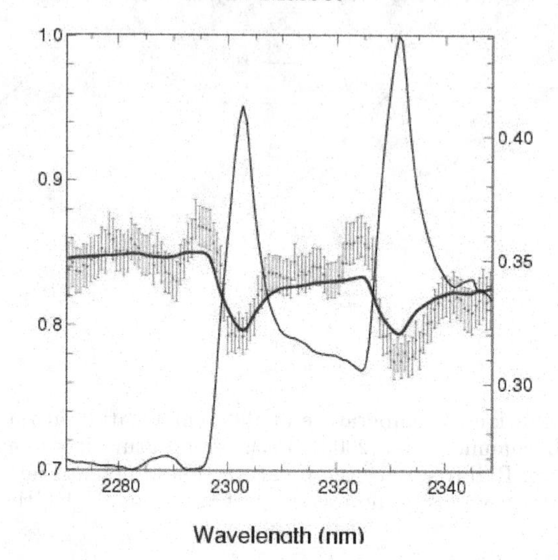

Figure 5. Spectrally dispersed visibility amplitudes of 51 OPh in the CO bandhead spectral region. Overimposed is the spectrum as measured by VLTI/AMBER (black line). The blue curve corresponds to the addition of simple uniform disk model for the excess emission in the line with a typical diameter of 0.2 AU. From Tatulli et al. (priv. comm.)

stars until now. The spectral shape of the inner-disk spectra shows surprising similarity with Solar System comets. Their observations imply that silicates crystallize before terrestrial planets are formed, consistent with the composition of meteorites in the Solar System. Similar measurements were also carried out by Ratzka *et al.* (2007) on the T Tauri system, TW Hya. According to the correlated flux measured with MIDI, most of the crystalline material is located in the inner, unresolved part of the disk, about 1 AU in radius.

3.4. *Gas/dust connection*

Gil *et al.* (2005) observed the young stellar system 51 Oph confirming the interpretation of Thi *et al.* (2005) and more recently Berthoud *et al.* (2007) of a disk seen edge-on: the radial distribution of excitation temperatures for the vibrational levels of CO overtone ($\Delta v = 2$) emission from hot gas is consistent with the gas being in radiative thermal equilibrium except at the inner edge, where low vibrational bands have higher excitation temperatures. In yet unpublished results, Tatulli *et al.* (priv. comm.) confirm the high inclination of the disk but also detect the CO bandheads allowing the dust responsible for the continuum to be separated from the gas emitting these CO bands. As a matter of fact, the visibilities in the CO bands is lower than the ones measured in the continuum implying that the region responsible for this gas emission is larger than the region responsible for the dust emission. Figure 5 illustrates this result.

This result shows that the combination of very high spatial information with spectral resolution opens brand new perspectives in the studies of the inner disk properties by discriminating between species.

4. Other AU-scale phenomena

Several other physical phenomena have been investigated in the innermost region of disks: wind, magnetosphere and close companions.

4.1. *Outflows and winds*

The power of spectrally resolved interferometric measurements provides detailed wavelength dependence of inner disk continuum emission (see end of § 3.2). These new capabilities enable also detailed studies of hot winds and outflows, and therefore the physical conditions and kinematics of the gaseous components in which emission and absorption lines arise like Brγ and H_2 lines. With VLTI/AMBER, Malbet *et al.* (2007) spatially resolved the luminous Herbig Be object MWC 297, measuring visibility amplitudes as a function of wavelength at intermediate spectral resolution (R = 1500) across the 2.0 − 2.2 μm band, and in particular the Brγ emission line. The interferometer visibilities in the Brγ line are about 30% lower than those of the nearby continuum, showing that the Brγ emitting region is significantly larger than the NIR continuum region. Known to be an outflow source, a preliminary model has been constructed in which a gas envelope, responsible for the Brγ emission, surrounds an optically thick circumstellar disk. The characteristic size of the line-emitting region is 40% larger than that of the NIR disk. This model is successful at reproducing the VLTI/AMBER measurements as well as previous continuum interferometric measurements at shorter and longer baselines (Millan-Gabet *et al.* 2001; Eisner *et al.* 2004), the SED, and the shapes of the Hα, Hβ, and Brγ emission lines. The precise nature of the MWC 297 wind, however, remains unclear; the limited amount of data obtained in these first observations cannot, for example, discriminate between a stellar or disk origin for the wind, or between competing models of disk winds (see e.g. Ferreira's and Shu's contributions in this volume).

4.2. *Magnetosphere*

The origin of the hydrogen line emission in Herbig Ae/Be stars is still unclear. The lines may originate either in the gas which accretes onto the star from the disk, as in magnetospheric accretion models (Hartmann *et al.* 1994), or in winds and jets, driven by the interaction of the accreting disk with a stellar (Shu *et al.* 1994) or disk (Casse & Ferreira 2000) magnetic field. For all models, emission in the hydrogen lines is predicted to occur over very small spatial scales, a few AUs at most. To understand the physical processes that happen at these scales, one needs to combine very high spatial resolution with enough spectral resolution to resolve the line profile.

One one hand, Tatulli *et al.* (2007) performed interferometric observations of the Herbig Ae star HD 104237, obtained with the VLTI/AMBER instrument with $R = 1500$ spectral resolution. The observed visibility was identical in the Brγ line and in the continuum, even though the line represents 35% of the continuum flux. This immediately implies that the line and continuum emission regions have the same apparent size. Using simple toy models to describe the Brγ emission, they showed that the line emission is unlikely to originate in either magnetospheric accreting columns of gas or in the gaseous disk but more likely in a compact outflowing disk wind launched in the vicinity of the rim, about 0.5 AU from the star. The main part of the Brγ emission in HD 104237 is unlikely to originate in magnetospheric accreting matter.

On the other hand, Eisner (2007) measured an increase of the Brγ visibility in MWC 480 implying that the region of emission of the hydrogen line is very compact, less than 0.1 mas in radius which could be interpreted as an emission originated in the magnetosphere of the system.

At the present time, given the limited number of samples, it is difficult to derive a general tendency but it seems that all possible scenari can be found.

4.3. *Binaries and multiple systems*

Boden *et al.* (2005) performed the first direct measurement of pre-main sequence stellar masses using interferometry, for the double-lined system HD 98800-B. These authors established a preliminary orbit that allowed determination of the (subsolar) masses of the individual components with 8% accuracy. Comparison with stellar models indicates the need for subsolar abundances for both components, although stringent tests of competing models will only become possible when more observations improve the orbital phase coverage and thus the accuracy of the stellar masses derived.

In another instance, based on a low-level oscillation in the visibility amplitude signature in the PTI data of FU Ori, Malbet *et al.* (2005) claim the detection of an off-centered spot embedded in the disk that could be physically interpreted as a young stellar or protoplanetary companion located at ~ 10 AU, and could possibly be at the origin of the FU Ori outburst itself. Using another technique, Millan-Gabet *et al.* (2006b) reported on the detection of localized off-center emission at 1-4 AU in the circumstellar environment of AB Aurigae. They used closure-phase measurements in the near-infrared. When probing sub-AU scales, all closure phases are close to zero degrees, as expected given the previously determined size of the AB Aurigae inner-dust disk. However, a clear closure-phase signal of $-3.5° \pm 0.5°$ is detected on one triangle containing relatively short baselines, requiring a high degree of asymmetry from emission at larger AU scales in the disk. They interpret such detected asymmetric near-infrared emission as a result of localized viscous heating due to a gravitational instability in the AB Aurigae disk, or to the presence of a close stellar companion or accreting substellar object.

5. Future prospects and conclusion

As emphasized in this review, more interferometric data is required with better accuracy and also wider coverage of the baselines in order to better constrain the models that have been proposed. Like for radio astronomy, these supplementary data will allow image reconstruction without any prior knowledge of the observed structure. Several projects are ready to obtain interferometric images although with few pixels across the field: MIRC at CHARA and AMBER at the VLTI in the near-infrared. However at the moment MIRC is limited in sensitivity and AMBER in number of telescopes (3) which makes it difficult to routinely achieve imaging. In the mid-infrared the MATISSE instrument is being proposed to ESO to provide imaging with 4 telecopes at the VLTI. VSI is also a proposed VLTI instrument of second generation which can combine from 4 to 8 beams at the same time so that imaging becomes easier. LBT will also provide imaging capability.

All these instruments provide spectral resolution that make them indeed spectro-imagers. Therefore in the future, one should be able to obtain a wealth of information from the innermost regions of disks around young stars. However in the meantime, observations are already mature enough to allow detailed modeling of the phenomena occuring in these inner regions.

References

Ábrahám, P., Mosoni, L., Henning, T., *et al.* 2006, *A&A*, 449, L13

Akeson, R. L., Boden, A. F., Monnier, J. D., *et al.* 2005a, *ApJ*, 635, 1173

Akeson, R. L., Ciardi, D. R., van Belle, G. T., Creech-Eakman, M. J., & Lada, E. A. 2000, *ApJ*, 543, 313

Akeson, R. L., Walker, C. H., Wood, K., *et al.* 2005b, *ApJ*, 622, 440

Berthoud, M. G., Keller, L. D., Herter, T. L., Richter, M. J., & Whelan, D. G. 2007, *ApJ*, 660, 461
Boden, A. F., Sargent, A. I., Akeson, R. L., *et al.* 2005, *ApJ*, 635, 442
Casse, F. & Ferreira, J. 2000, *A&A*, 353, 1115
Colavita, M., Akeson, R., Wizinowich, P., *et al.* 2003, *ApJ*, 592, L83
Dullemond, C. P., Dominik, C., & Natta, A. 2001, *ApJ*, 560, 957
Eisner, J. A. 2007, *Nature*, 447, 562
Eisner, J. A., Chiang, E. I., & Hillenbrand, L. A. 2006, *ApJ*, 637, L133
Eisner, J. A., Chiang, E. I., Lane, B. F., & Akeson, R. L. 2007, *ApJ*, 657, 347
Eisner, J. A., Hillenbrand, L. A., White, R. J., Akeson, R. L., & Sargent, A. I. 2005, *ApJ*, 623, 952
Eisner, J. A., Lane, B. F., Akeson, R. L., Hillenbrand, L. A., & Sargent, A. I. 2003, *ApJ*, 588, 360
Eisner, J. A., Lane, B. F., Hillenbrand, L. A., Akeson, R. L., & Sargent, A. I. 2004, *ApJ*, 613, 1049
Gil, C., Malbet, F., Schoeller, M., Chesneau, O., & Leinert, C. 2005, in "The Power of Optical / IR Interferometry: Recent Scientific Results and 2nd Generation VLTI Instrumentation", Garching, April 4-8, 2005, ed. C. A. Richichi A., Delplancke F. & P. F., Vol. in press, (ArXiv preprint: astro–ph/0508052)
Hartmann, L., Hewett, R., & Calvet, N. 1994, *ApJ*, 426, 669
Isella, A. & Natta, A. 2005, *A&A*, 438, 899
Kraus, S., Balega, Y. Y., Berger, J.-P., *et al.* 2007, *A&A*, 466, 649
Lachaume, R., Malbet, F., & Monin, J.-L. 2003, *A&A*, 400, 185
Lachaume, R., Preibisch, T., Driebe, T., & Weigelt, G. 2007, *A&A*, 469, 587
Leinert, C., van Boekel, R., Waters, L. B. F. M., *et al.* 2004, *A&A*, 423, 537
Malbet, F., Benisty, M., de Wit, W.-J., *et al.* 2007, *A&A*, 464, 43
Malbet, F., Berger, J.-P., Colavita, M. M., *et al.* 1998, *ApJ*, 507, L149
Malbet, F. & Bertout, C. 1995, *A&AS*, 113, 369
Malbet, F., Lachaume, R., Berger, J.-P., *et al.* 2005, *A&A*, 437, 627
Millan-Gabet, R., Malbet, F., Akeson, R., *et al.* 2007, in Protostars and Planets V, ed. B. Reipurth, D. Jewitt, & K. Keil, 539–554
Millan-Gabet, R., Monnier, J. D., Akeson, R. L., *et al.* 2006a, *ApJ*, 641, 547
Millan-Gabet, R., Monnier, J. D., Berger, J.-P., *et al.* 2006b, *ApJ*, 645, L77
Millan-Gabet, R., Schloerb, F. P., & Traub, W. A. 2001, *ApJ*, 546, 358
Millan-Gabet, R., Schloerb, F. P., Traub, W. A., *et al.* 1999, *ApJ*, 513, L131
Monnier, J. D., Berger, J.-P., Millan-Gabet, R., *et al.* 2006, *ApJ*, 647, 444
Monnier, J. D. & Millan-Gabet, R. 2002, *ApJ*, 579, 694
Monnier, J. D., Millan-Gabet, R., Billmeier, R., *et al.* 2005, *ApJ*, 624, 832
Muzerolle, J., D'Alessio, P., Calvet, N., & Hartmann, L. 2004, *ApJ*, 617, 406
Natta, A., Prusti, T., Neri, R., *et al.* 2001, *A&A*, 371, 186
Preibisch, T., Kraus, S., Driebe, T., van Boekel, R., & Weigelt, G. 2006, *A&A*, 458, 235
Quanz, S. P., Henning, T., Bouwman, J., Ratzka, T., & Leinert, C. 2006, *ApJ*, 648, 472
Ratzka, T., Leinert, C., Henning, T., *et al.* 2007, *A&A*, 471, 173
Shu, F., Najita, J., Ostriker, E., *et al.* 1994, *ApJ*, 429, 781
Tannirkulam, A., Harries, T. J., & Monnier, J. D. 2007, *ApJ*, 661, 374
Tatulli, E., Isella, A., Natta, A., *et al.* 2007, *A&A*, 464, 55
Thi, W.-F., van Dalen, B., Bik, A., & Waters, L. B. F. M. 2005, *A&A*, 430, L61
Tuthill, P. G., Monnier, J. D., Danchi, W. C., Hale, D. D. S., & Townes, C. H. 2002, *ApJ*, 577, 826
van Boekel, R., Min, M., Leinert, C., *et al.* 2004, *Nature*, 432, 479
Vinković, D., Ivezić, Ž., Miroshnichenko, A. S., & Elitzur, M. 2003, *MNRAS*, 346, 1151
Vinković, D. & Jurkić, T. 2007, *ApJ*, 658, 462
Wilkin, F. P. & Akeson, R. L. 2003, *Ap&SS*, 286, 145

Star-Disk Interaction in Young Stars
Proceedings IAU Symposium No. 243, 2007
J. Bouvier & I. Appenzeller, eds.

doi:10.1017/S1743921307009490

Gas at the inner disk edge

John S. Carr[1]

[1]Naval Research Laboratory, Remote Sensing Division, Washington, DC 20375, USA
email: carr@nrl.navy.mil

Abstract. Infrared molecular spectroscopy is a key tool for the observation of gas in the innermost region of disks around T Tauri stars. In this contribution, we examine how infrared spectroscopy of CO can be used to study the inner truncation region of disks around T Tauri stars. The inferred inner gas radii for T Tauri star disks are compared to the inner dust radii of disks, to the expectations of models for disk truncation, and to the orbital distribution of short-period extra-solar planets.

Keywords. Accretion disks, protoplanetary disks, stars: formation.

---

1. Introduction

The star-disk interaction takes place within a region at several stellar radii from the star. In the magnetospheric accretion paradigm, the inner accretion disk is truncated by a strong stellar magnetic field, and material accretes onto the star along magnetic field lines (Königl 1991; Cameron & Campbell 1993; Shu *et al.* 1994). The disk is thought to be truncated near the corotation radius, where the Keplerian angular velocity in the disk equals the angular velocity of the star. The magnetic coupling between the star and disk, often referred to as disk locking, has been proposed as a mechanism to regulate the angular momentum of the star and explain the slow rotation of accreting T Tauri stars (Königl 1991). Shu *et al.* (1994) combine this picture of accretion and stellar angular momentum regulation with a magnetocentrifugal wind that originates at the X-point (the corotation radius) and carries away angular momentum.

Hence, the interaction of the star and disk via magnetospheric accretion is central to current ideas of accretion, mass loss, and angular momentum evolution. The observational signatures of magnetospheric accretion are plentiful (see papers in these proceedings), but the majority of these phenomena are tied to accretion onto the star, i.e., the hot gas in the accretion columns and the shocks at the stellar surface. We have little information on the interaction of the magnetosphere with the disk. In fact, something as fundamental as the inner truncation radius of the disk, which must be near the corotation radius in disk locking models, has been poorly constrained.

In this contribution, we discuss measurements of disk gas in the inner truncation region of the disk. We focus on the classical T Tauri stars (CTTS), accreting solar-mass pre-main-sequence stars in which the star-disk interaction is historically best studied. After presenting measurements of the inner gas radius based on velocity-resolved, but spatially unresolved, spectra, we compare these results to recent interferometric measurements of the dust, to the corotation radii for individual stars, and to orbital radii of extrasolar planets.

2. Gas probes of the inner disk

The useful probes of gas in the inner disk are determined by the physical conditions of the gas and dust. Within about 1 AU of the star, temperatures are expected to range from a few 100 K to few 1000 K. Under these conditions, molecules will be abundant in the gas phase, though molecular hydrogen may be dissociated near the inner disk edge or in a hot upper disk atmosphere. The temperatures and densities are sufficient to excite many rotational and ro-vibrational levels of molecules, whose transitions occur in the near- to mid-infrared. Hence, infrared molecular spectroscopy is a prime tool for studying the inner disk gas. High spectral resolution is particularly advantageous in the ability to provide kinematic information on the gas, which can be used to probe the gas structure and conditions as a function of Keplerian velocity. Some of the key molecules that have been observed and attributed to the inner disk are CO, H_2O, OH and both infrared and UV transitions of H_2 (see Najita *et al.* 2007 for an overview).

The infrared transitions of these molecules are normally observed in emission in CTTS. Emission lines can arise from disks under two broad scenarios. In the common case for CTTS, when the disk is optically thick, emission lines can be produced in a temperature inversion in the disk atmosphere. In this situation, the spectral features only probe the atmosphere and not the entire vertical column density of the disk. Emission lines can also be formed in regions of the disk that are optically thin in the continuum. This could be the situation at particular radii due to lower opacity as a result of dust sublimation or grain growth, or the entire column density could be greatly reduced due to dynamical clearing or disk dissipation.

3. CO emission in T Tauri stars

If one wishes to study the gas at the inner edge of T Tauri star disks, then the infrared transitions of CO are the logical choice. CO is an abundant molecule in many astrophysical contexts. The high dissociation energy of CO means that it can survive to temperatures of 4000-5000 K. Hence, CO is highly likely to be present in the disk gas around CTTS down to the inner disk edge. This situation is different from that for the higher temperature and more luminous Herbig Ae/Be stars, in which the inner gaseous disk will be hot enough, at least in the innermost regions, for CO to be dissociated. The gas in the innermost part of CTTS disks will be both dense and warm ($>$ 1000 K) enough that the near-infrared ro-vibrational transitions will be an ideal probe of the gas.

There are two different bands of CO that are observed in emission from CTTS. The first is the overtone band, $\Delta v = 2$ transitions, that are found as a series of bandheads near 2.3μm. CO overtone emission has been shown to originate from rotating disk gas in young stellar objects over a wide range in stellar mass (Najita *et al.* 2007). The emission traces gas in the 2000-4000 K range, and in low-mass young stellar objects the emission originates very close to the central star (Carr *et al.* 1993; Carr, Tokunaga & Najita 2004). However, CO overtone emission is only observed in a small fraction of CTTS, those with the highest mass accretion rates. This limits its usefulness in studying the inner disks in the general CTTS population.

In contrast, the CO fundamental ($\Delta v = 1$) transitions near 4.7μm are observed in emission in nearly all CTTS (Najita, Carr & Mathieu 2003). The larger transition probabilities (by two order of magnitude) and the lower energy of the $v = 1$ level, means that the CO fundamental can probe lower temperature and lower column density gas than the overtone. Evidently, while the conditions for CO overtone emission are rarely met in CTTS, the conditions required for CO fundamental emission are common. Hence, the CO fundamental lines are a good general tracer of gas in the inner disk of CTTS.

Examples of the CO fundamental spectrum observed in CTTS can be seen in Najita *et al.* (2003). The rotational transitions of the $v = 1-0$ band are the strongest, and the $v = 2-1$ lines are seen at lower strength when the S/N is sufficient. Higher vibrational

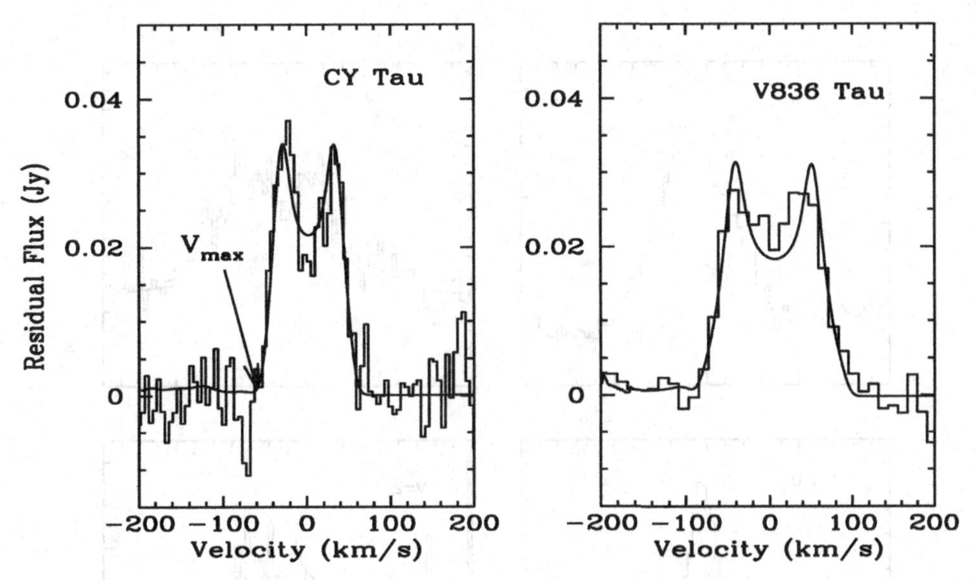

Figure 1. CO $v = 1 - 0$ line profiles for two CTTS that show a classic double-peaked disk profile (Najita, Carr & Mathieu 2003; Najita *et al.* 2007, in prep.). The smooth lines are model disk profiles. By measuring the maximum CO velocity in the profile, and knowing the stellar mass and inclination, the minimum CO gas radius follows for Keplerian rotation.

transitions (e.g., $v = 3 - 2$) and ^{13}CO $v = 1 - 0$ transitions may also be observed, depending on the gas temperature and CO optical depth. Excitation temperatures are in the general range of 500 to 1500 K. There exists a trend of stronger CO emission strength in objects with higher accretion rates (Najita, Carr & Mathieu 2003).

4. Probing gas at the inner disk edge

The disks in CTTS are expected to be truncated at distances of several stellar radii from the star, placing the region of interest inside of about 0.1 AU. This is an angular resolution of less than 1 mas at 140 pc, the distance of the Taurus star formation region. Hence, this region has not been spatially resolved until the recent advent of near-infrared interferometry (see Malbet, these proceedings). Recent work has provided the first spatially resolved information on the 2μm contiunuum in the innermost disk region (Eisner *et al.* 2005; Akeson *et al.* 2005). For study of the gas, spectro-interferometry that would allow one to spatially resolve the line emission separately from the continuum is desirable. Recent spectro-interferometry of the CO overtone bandhead emission in the Herbig Ae/Be star 15 Oph (see Malbet) illustrates the type of observations that one like to carry out for the inner gas in CTTS. However, CTTS are more challenging, given that they are much fainter and the CO emission region is expected to smaller in size. In the long term, spectro-interferometry of CO at 5μm is desirable, ideally at high spectral resolution.

Until this can be achieved, we can make progress on exploring the inner gas radii of disks by utilizing the kinematic information from line velocity profiles. For the ideal case of an emission line from a Keplerian disk, the line profile contains information on the radial variation of the gas emission. When line profiles for multiple transitions from the same molecule are included, the radial variation of gas temperature and column density can be constrained. Relative gas-phase abundances could also be derived from observations of different molecules.

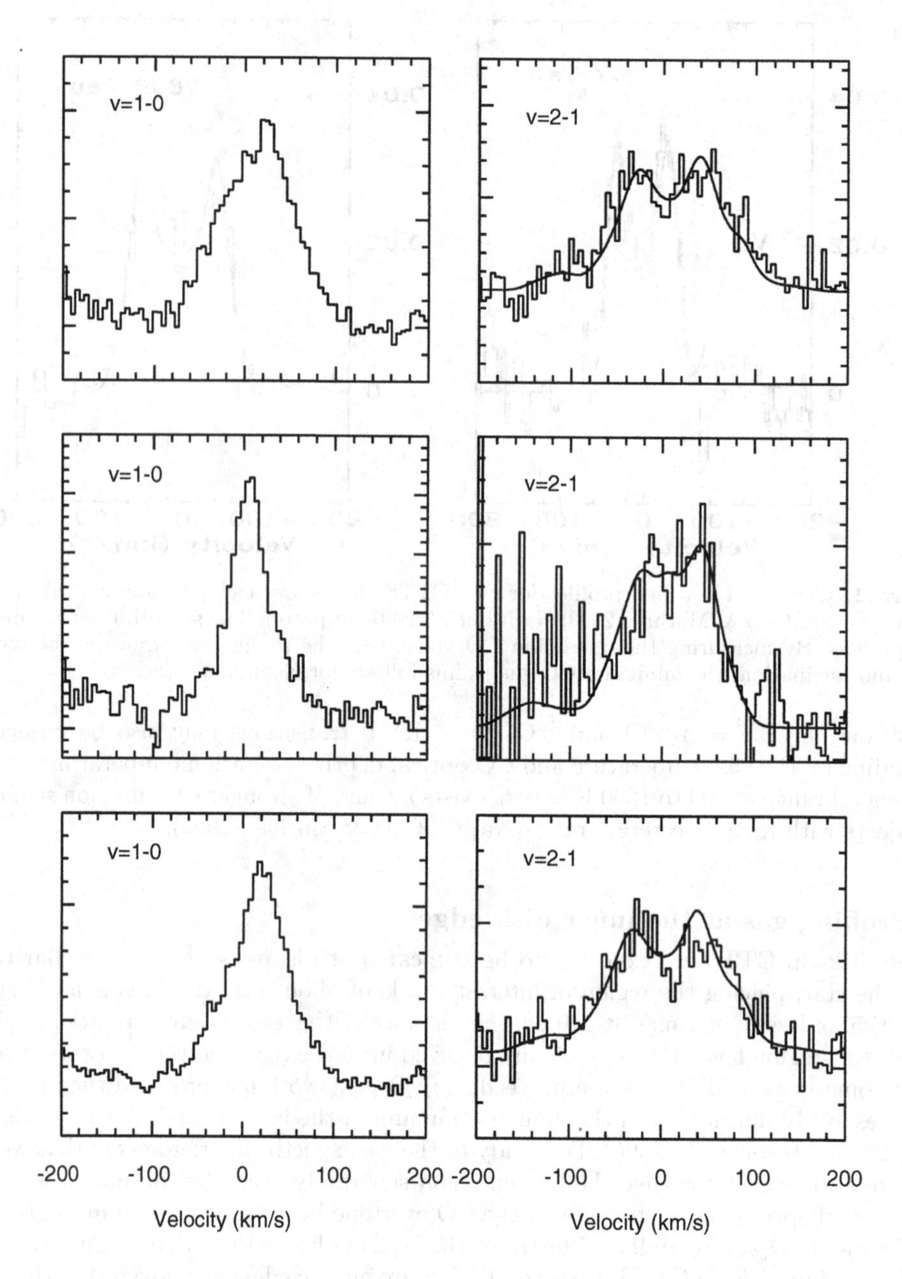

Figure 2. CO $v = 1-0$ and the $v = 2-1$ line profiles are compared for three CTTS (Carr *et al.* 2007, in prep.). The $v = 1-0$ profiles for these stars are centrally peaked, but at the same time the $v = 2-1$ profiles have a disk profile, as shown by the model disk profiles (smooth lines).

Figure 1 shows a CO $v = 1-0$ profile for two CTTS which show a classic double-peaked profile for emission from a disk. The smooth line is a model disk profile. For a Keplerian disk, the maximum observed velocity in the profile corresponds to the minimum radius at which the emission occurs. The velocity of the emission peak corresponds, roughly, to the maximum radius of the emission. Assuming that the gas is in Keplerian rotation, a measurement of the maximum velocity gives the minimum gas emission radius, provided we know the stellar mass and the inclination. This method was first applied to CTTS profiles by Najita *et al.* (2003).

5. CO fundamental profiles

The above interpretation of the CO emission depends upon the assumption that the gas is in Keplerian rotation. While a small number of CTTS have CO $v = 1 - 0$ emission lines with the expected double-peaked profile (Fig. 1), the more common $v = 1 - 0$ profile is one that is centrally peaked, sometimes with asymmetries in the line profile (Figure 2; also see Najita *et al.* 2003 for more examples of profiles). However, when the signal-to-noise is sufficient to measure the profiles of the higher excitation $v = 2 - 1$ transitions, these tend to show a double-peaked or flat-topped profile that is consistent with emission from a disk (Figure 2). Qualitatively, more centrally peaked $v = 1-0$ lines would be expected if the $v = 1 - 0$ emission originates over a greater range of disk radii out to fairly large radii. The FWHM velocity of the $v = 2 - 1$ profiles is normally larger than the FWHM of the $v = 1 - 0$ profiles. This fits a disk scenario where the $v = 2 - 1$ lines form in hotter gas at smaller radii, and hence larger Keplerian velocities, than the bulk of the $v = 1 - 0$ emission flux. At the same time, the full-width at zero intensity of the $v = 1 - 0$ and $v = 2 - 1$ lines are similar, showing that the lines share a common maximum velocity and minimum inner disk emission radius.

The observations support the idea that the bulk of the CO fundamental emission originates from the disk. Nevertheless, some details of the profiles suggest that the picture is not the ideal case. Moderately hot gas at large radii might explain the more sharply peaked lines, and asymmetric disk emission might account for line asymmetries. Potentially, CO emission (or absorption) from either the base of the funnel flow or from the base of a molecular wind could make contributions to the profiles.

6. The inner gas radius

Measurements of the inner gas radius are presented here based on a sample of CTTS that have measurements or estimates of stellar mass and inclination and 5μm echelle spectra measured at the Keck Observatory with NIRSPEC. The data are from Najita *et al.* (2003), additional data collected by Carr, Najita and Mathieu (in prep), and one star (GM Aur) published by Salyk *et al.* (2007). The stellar mass and inclination come from one of two methods. The more robust method uses the results of Simon *et al.* (2000), who used millimeter line interferometry to image the outer disks (100 AU) around a sample of CTTS. From the disk images and the measured Keplerian velocity, they determine the central stellar mass and the disk inclination. In practice, we do not require their actual values for mass and $\sin i$, but rather their measurement of the projected velocity at 100 AU. Essentially, by knowing the projected Keplerian velocity at a known radius, the unknown radius for a different measured Keplerian velocity directly follows. In the second method, the inclination is derived from a measured stellar rotational period, $v \sin i$ for the star, and the star's radius (as determined from the stellar luminosity and temperature). The stellar mass comes from evolutionary tracks, given the stellar temperature and luminosity. This method is judged to have larger uncertainties.

For each star in the sample, the maximum CO velocity was measured from the $v = 1-0$ emission lines and used to determine the inner CO radius. Figure 3 shows the distribution of the inner gas radius as measured by CO. The distribution shows a strong peak at about 0.04 AU. In the figure, the two stars indicated by unshaded bars are systems with

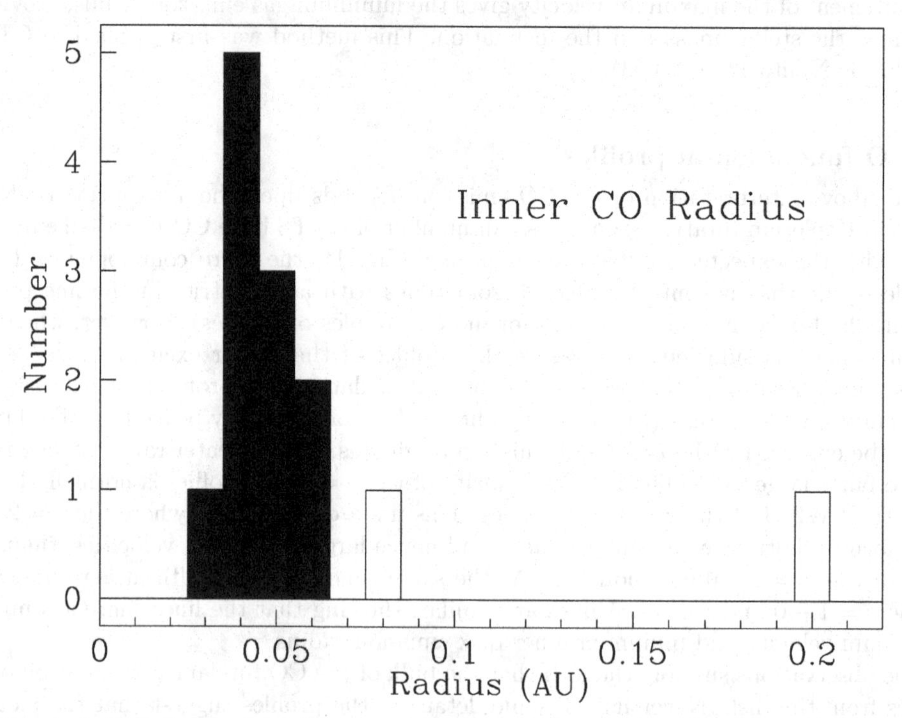

Figure 3. The distribution of inner gas radius as determined from the CO fundamental emission profiles in a sample of CTTS. The two unshaded bars are stars with transitional disks.

transitional disks, disks with large inner holes as determined from their spectral energy distributions. The inner gas radii for these systems fall at larger radii than their normal CTTS counterparts.

The inner gas radius can be compared to the co-rotation radius for each star as determined from the stellar rotational period and mass. In Figure 4, the inner gas radius is plotted vs. the co-rotation radius for the sample. The solid squares are for stars where the derived gas radii use of the results from Simon *et al.* (2000) for the stellar mass and inclination, while the open squares are for stars where the second method outlined above was used. The figure shows that the inner gas radius is typically smaller than the co-rotation radius. The average ratio of the inner gas to the co-rotation radii is about 0.7, excluding the two stars with transition disks. It is reassuring that the typical gas radius in the two group of stars, which use different methods for the mass and inclination, does not differ significantly. The error bars show the estimates of the uncertainties in the radii; the scatter in the inner gas radii appears to be consistent with the errors.

It is of major interest to compare the inner gas radius to determinations of the inner dust radius. This is shown in Figure 5. Unfortunately, the overlap in the two samples of stars is small, but the four stars in common agree with the overall comparison that dust radii are equal to or greater than the gas radii. The dust radii come from one of two techniques. The first group are inner dust radii determined from near-infrared interferometry (Eisner *et al.* 2005; Akeson *et al.* 2005). The interferometric dust radii depend on the model adopted for the geometry of the emission, usually either a flat disk or a ring, the latter being more equivalent to the inner dust rim scenario. The results for the ring or rim models are used in Figure 5. The second group of dust radii (unshaded

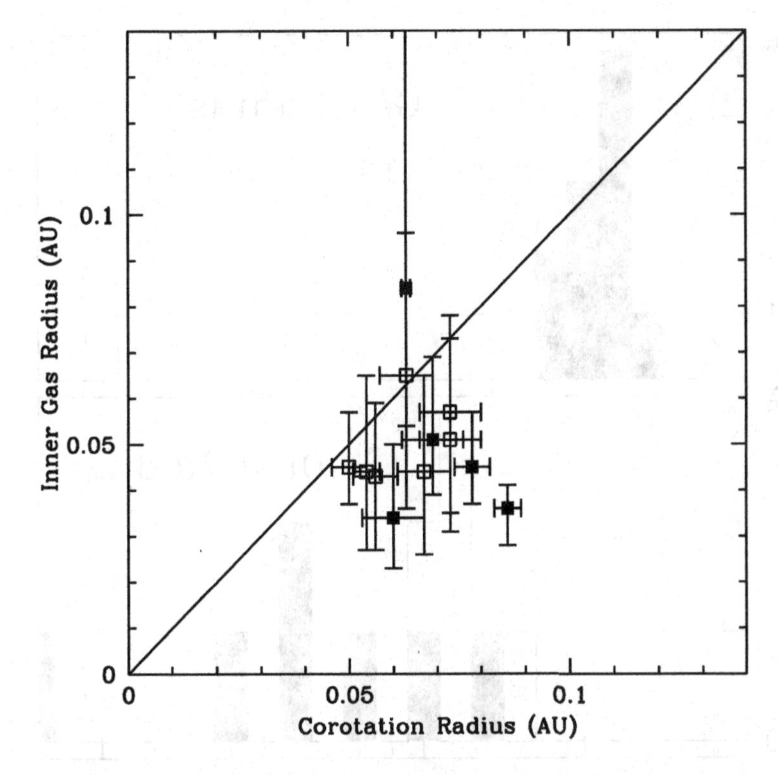

Figure 4. The inner CO gas radius plotted against the corotation radius. The solid points use the results of Simon *et al.* (2000) for determining the CO radius, while the open points use the stellar period and $v \sin i$ (see text).

histogram) are determinations of the dust inner rim radii based on modeling of the shape of the near-infrared excess continuum emission (Muzerolle *et al.* 2003).

Another way of examining the gas and dust radii is shown in Figure 6, in which the respective radii are ratioed to the co-rotation radius for each star (not all stars in the dust radius sample have a determined co-rotation radius). In contrast to the inner gas radius, the inner dust rim falls at or outside the co-rotation radius, a result that was previously noted by Muzerolle *et al.* (2003) and Eisner *et al.* (2005). Because the disk is theorized to be truncated near the corotation radius, these authors also anticipated that a gaseous disk must extend inside of the dust sublimation radius. Figures 5 and 6 show that this to indeed be the case. While this result should not be surprising, what is interesting is that the inner gas edge typically falls inside of the co-rotation radius.

Figure 7 compares the inner gas radius distribution to the distribution of orbital radii of known extrasolar planets. As has been known for some time, short-period extrasolar planets show a peak in their orbital period distribution at about 3 days. This piling up in period corresponds to an orbital radius of about 0.04 AU. As Figure 7 shows, there is an amazing coincidence in the peak of the inner gas radius distribution to the peak in planetary orbital radii. This provides strong support for the role of disk truncation in halting the inward migration of giant extrasolar planets, an idea that has been discussed in a number of papers (Lin *et al.* 1996; Kuchner & Lecar 2002; Romanova & Lovelace 2006). Note, however, that the peak in the gas inner radius is not at the outermost

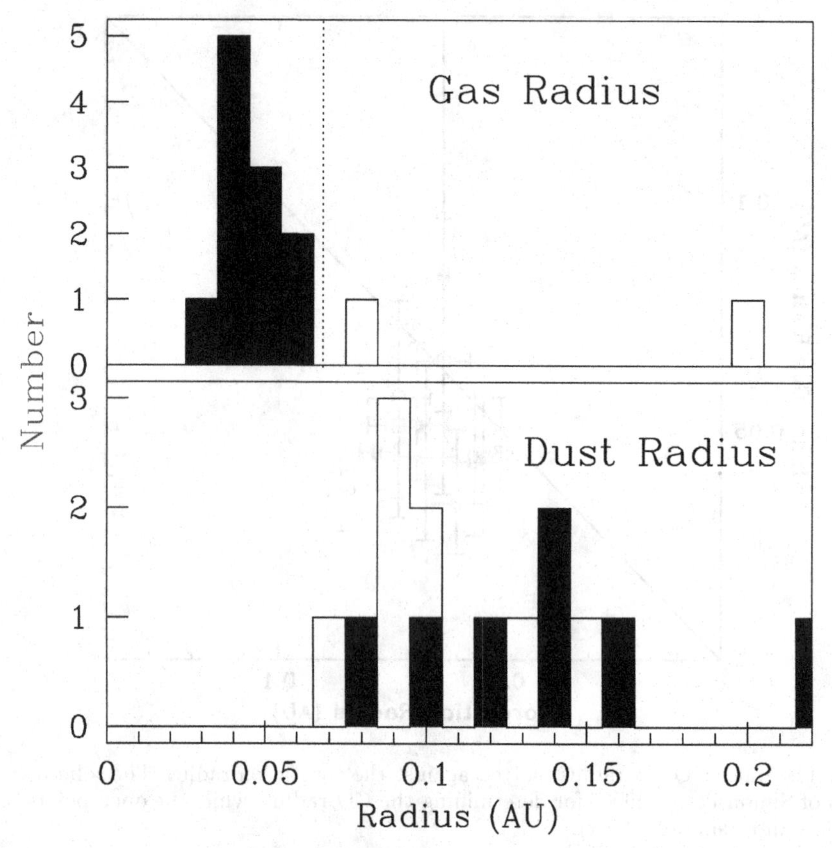

Figure 5. The distribution of inner gas radius is compared to the distribution of inner dust radius, for two different samples of CTTS. In the upper panel, symbols are the same as in Figure 4, and the vertical dotted line is the average corotation radius for the sample. In the lower panel, the shaded bars are dust radii determined from near-infrared interferometry, while the unshaded bars are from modeling the spectral energy distributions of the near-infrared excess (see text).

Lindblad resonance (1.59 times the radius) of the peak in the planetary orbit distribution, as might be expected if the CO gas radius is the physical truncation radius of the disk.

7. Discussion and summary

The results presented here for the inner gas radius as derived from spectroscopy of the CO fundamental emission in CTTS, when combined with the results for the inner dust radius, suggest a picture in which the gaseous disk extends inward of the dust sublimation radius to radii that are typically smaller than the corotation radius. In disk locking models, the magnetosphere truncates the disk near the corotation radius. To the extent that either the CO emission or near-infrared continuum emission trace the inner disk edge, it is reassuring to first order that the derived gas and dust radii are within a factor of two of the corotation radii.

For the dust, the innermost radius should be set by the dust sublimation temperature, and there is no *a priori* reason that the disk should terminate at this point. Hence, we would expect that the gas rather than the dust should be a better measure of the minimum radius of the disk.

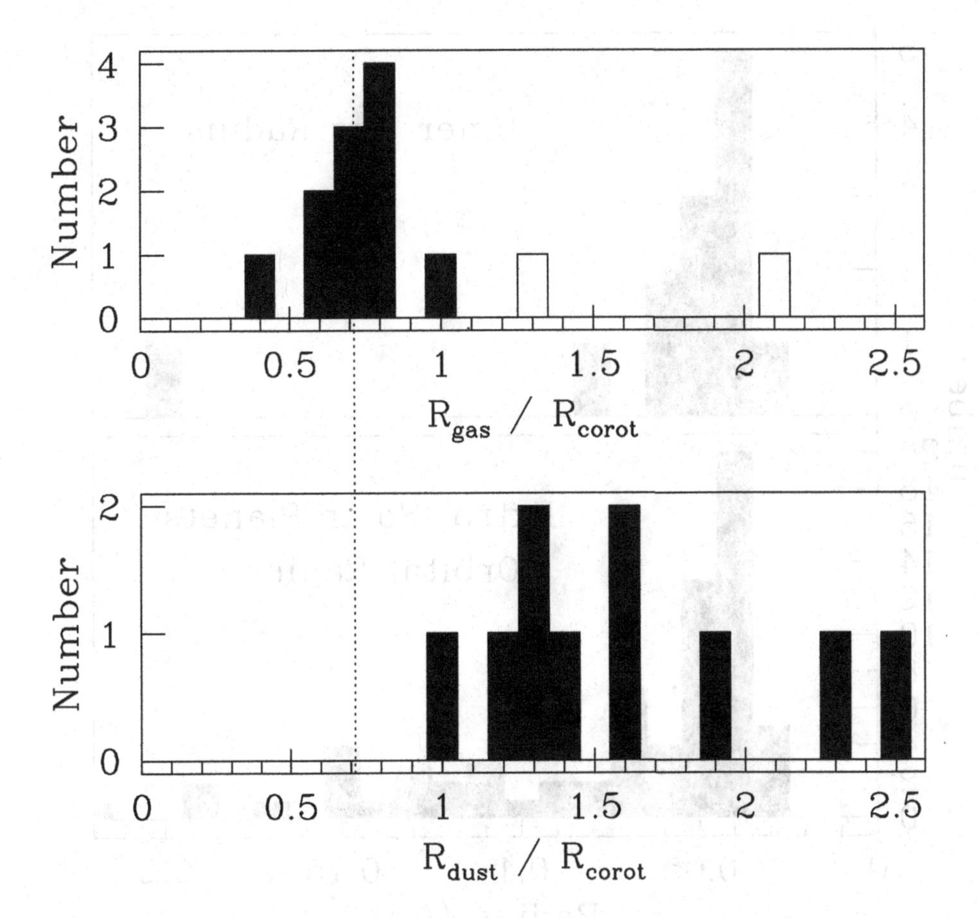

Figure 6. The distribution of inner gas and dust radii ratioed to the corotation radii are compared. The vertical dotted line is the average ratio of the gas radius to the corotation radius, equal to about 0.7. Symbols in the upper panel are the same as in Figure 4. Except for the two transitional disks, the gas radii are less than or equal to the corotation radii. The dust radii are greater or equal to the corotation radii.

How well do we know either the inner gas or dust radius? The inferred radii from near-infrared interferometry depend on the geometric model that is adopted, and results can differ by up to a factor of two in radius for a given star (Eisner *et al.* 2005; Akeson *et al.* 2005). In general, puffed-up inner rim or ring models (used in Figures 5 and 6) give better fits than a uniform flat disk, but adopting the flat or uniform disk models in Akeson *et al.* (2005) and Eisner *et al.* (2005) does not change the qualitative result that the dust radii are larger than the corotation radius.

The uncertainties for the inner CO gas radii, as shown by the error bars in Figure 4, include our best estimates of the uncertainties in the stellar parameters and the maximum CO velocity. While the errors are large enough for many stars to be consistent with the corotation radius, on average the CO emission gives a radius about 0.7 times the corotation radius. The radii that use the results of Simon *et al.* (2000) are independent of the values for mass and inclination and are not significantly different from the radii that do use mass and inclination estimates. In order to force the CO radii to agree with

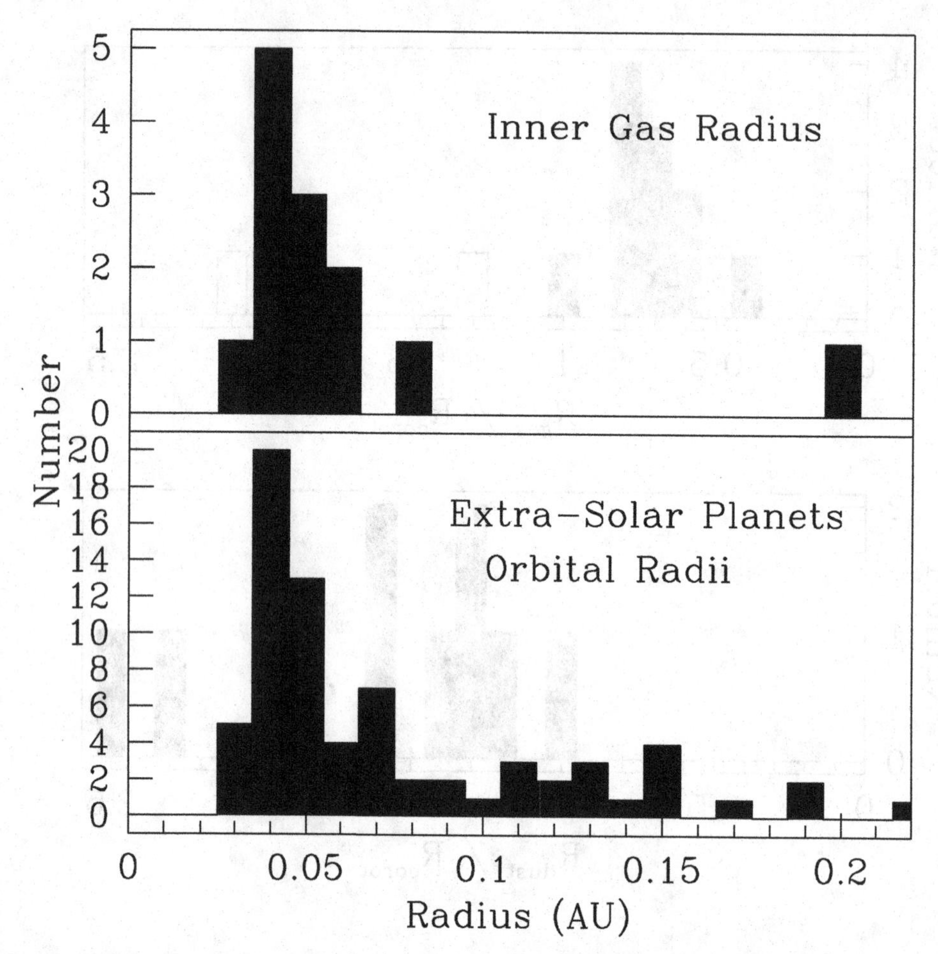

Figure 7. The distribution of inner gas radii are compared to the distribution of orbital radii for short-period extrasolar planets ("hot Jupiters"). The well known peak in the orbital radii of short-period planets coincides with the inner disk gas radius as determined from the near-infrared CO profiles.

corotation, the maximum CO velocities would have to systematically lower by a factor of 0.85, which on average is about 16 km s^{-1}. While this is excluded by the data, it is more difficult to exclude the possibility that the highest velocity in the wings of the profiles might be produced by weak emission from some kinematic component other than the disk.

An important caveat about the CO emission is that very little column density of gas is required to produce the fundamental emission. The emission observed in CTTS require total columns of 0.001 – 0.1 g cm^{-2}, assuming all of the C is in CO. These values are order of magnitudes lower than the expected disk column densities. For example, in the disk model of D'Alessio *et al.* (1998), the column density at 0.1 AU is about 500 g cm^{-2}, for a mass accretion rate of 10^{-8} and the viscosity parameter $\alpha = 0.01$. In the general CTTS case, the CO emission is believed to probe a small upper surface layer of the disk. However, it could also be the situation that the CO is essentially measuring the entire column of a very small column density of gas inside of the some radius. If this is the case, the profile wings could be measuring gas inside of the truncation radius, and the amount of mass would not be dynamically important, for example, in terms of effecting the migration of protoplanets. This scenario has its own caveat, because if disk locking is effective, then gas interior to corotation is expected to be sub-Keplerian, in near-rigid rotation (Shu *et al.* 1994; Ostriker & Shu 1995). Therefore, the maximum observed gas velocity will still correspond to the corotation radius.

High spectral resolution spectroscopy of the infrared transitions of CO provide a glimpse into disk gas at the inner edge of CTTS disks. At the same time, infrared interferometry has given us the first spatially resolved information on the hot dust continuum. We have begun to probe the disk truncation region of the star-disk interaction. Future refinement of the observations and analysis, and of the physical models for the interface of the disk and the magnetosphere, should enable us to better constrain and understand the physical processes centered in this region that are believed to be critical for star and planet formation.

References

Akeson, R. L. *et al.* 2005, *ApJ* 635, 1173

Cameron, A. C. & Campbell, C. G. 1993, *A&A* 274, 309

Carr, J. S., Tokunaga, A. T., Najita, J., Shu, F. H., & Glassgold, A. E. 1993, *ApJ* 411, L37

Carr, J. S., Tokunaga, A. T., & Najita, J. 2004, *ApJ* 603, 213

D'Alessio, P., Canto, J., Calvet, N., & Lizano, S. 1998, *ApJ* 500, 411

Eisner, J. A., Hillenbrand, L. A., White, R. J., Akeson, R. L., & Sargent, A. I. 2005, *ApJ* 623, 952

Königl, A. 1991, *ApJ* 370, L39

Kuchner, M. J. & Lecar, M. 2002, *ApJ* 574, L87

Lin, D. N. C., Bodenheimer, P., & Richardson, D. C. 1996, *Nature* 380, 606

Muzerolle, J., Calvet, N., Hartmann, L., & D'Alessio, P. 2003, *ApJ* 597, 149

Najita, J. R., Carr, J. S., & Mathieu, R. D. 2003, *ApJ* 589, 931

Najita, J. R., Carr, J. S., Glassgold, A. E., & Valenti, J. A. 2007, in: B. Reipurth, D. Jweitt & K. Keil (eds.), *Protoplanets and Planets V*, (Tucson: Univ. of Arizona), p. 507

Ostriker, E. & Shu, F. 1995, *ApJ* 447, 813

Romanova, M. M. & Lovelace, R. V. E. 2006, *ApJ* 645, L73

Salyk, C., Blake, G. A., Boogert, A. C. A., & Brown, J. M. 2007, *ApJ* 655, 105

Simon, M., Dutrey, A., & Guilloteau, S. 2000, *ApJ* 545, 1034

Shu, F., Najita, J., Ostriker, E., Wilkin, F., Ruden, S., & Lizano, S. 1994, *ApJ* 429, 781

Star-Disk Interaction in Young Stars
Proceedings IAU Symposium No. 243, 2007
J. Bouvier & I. Appenzeller, eds.

doi:10.1017/S1743921307009507

Observational constraints on disk photoevaporation by the central star

Gregory J. Herczeg

Caltech MC105-24, 1200 E. California Blvd., Pasadena, CA 9125

Abstract. We apply results from FUV and X-ray spectroscopy to evaluate the role of photoevaporation in dispersing the disk around TW Hya. Accretion produces bright EUV emission that may be smothered by the accretion column. Solar-like magnetic activity produces fewer ionizing photons, which may be absorbed by an accretion-powered neutral wind. We estimate a photoevaporation rate of $\sim 5 \times 10^{-11}$ $M_{\odot}$ yr^{-1} for the disk around TW Hya. These models can be tested by detecting gas in the ionized disk surface, including emission in the [Ne II] 12.8μm line. Photoevaporation is likely a minor process in disk dispersal during the accretion phase, but could remove ~ 1 M_J of remnant gas around a solar-mass star after accretion ceases.

Keywords. Stars: planetary systems: protoplanetary disks, stars: pre–main-sequence, accretion, stars: coronae

1. Introduction

Photoevaporation by EUV (< 912 Å) emission from the central star has been proposed as a dominant mechanism in disk dispersal (see review by Dullemond *et al.* 2007). In the detailed photoevaporation models of Alexander *et al.* (2006ab), ionizing radiation from the central star can heat the disk surface to $\sim 10^4$ K, leading to evaporation between 1–10 AU. If strong enough, solar-like magnetic activity could supply sufficient ionizing radiation to sustain photoevaporation rates of 10^{-10} to 10^{-9} M$_{\odot}$yr^{-1}, even as the accretion ceases.

Photoevaporation models require estimates for the ionizing flux incident upon the disk. However, the EUV emission from the central star is undetectable because of H absorption in our line of sight. Instead, the EUV radiation field at the disk surface can be estimated by calculating the flux from the star and correcting for any attenuation by neutral gas between the star and the disk. Alexander *et al.* (2005) estimated a ionizing photon flux of $10^{41} - 10^{44}$ phot s^{-1} from CTTSs based on the C IV λ1549 line luminosity. They assumed that the C IV emission is chromospheric and that all of the EUV emission from the star reaches the disk. Alternately, observational probes of the ionized disk surface could constrain the ionizing radiation incident upon the disk.

In this proceedings, we apply X-ray and FUV observations of the nearby classical T Tauri star (CTTS) TW Hya to evaluate photoevaporation models. TW Hya is an older CTTS with a mass accretion rate about an order of magnitude lower than that of younger Taurus CTTSs. The IR spectral energy distribution and imaging at 7mm indicates that the inner disk within 4 AU of the star is cleared of optically-thick dust (Calvet *et al.* 2002; Hughes *et al.* 2007), although gas and optically-thin micron-sized dust grains remain (Herczeg *et al.* 2002; Eisner *et al.* 2006). Such disk clearing may result from photoevaporation, the presence of a giant planet, or grain growth (Dullemond & Dominik 2005; Alexander *et al.* 2006b; Najita *et al.* 2007). TW Hya is therefore an ideal CTTS to determine whether photoevaporation is significant in disk dispersal.

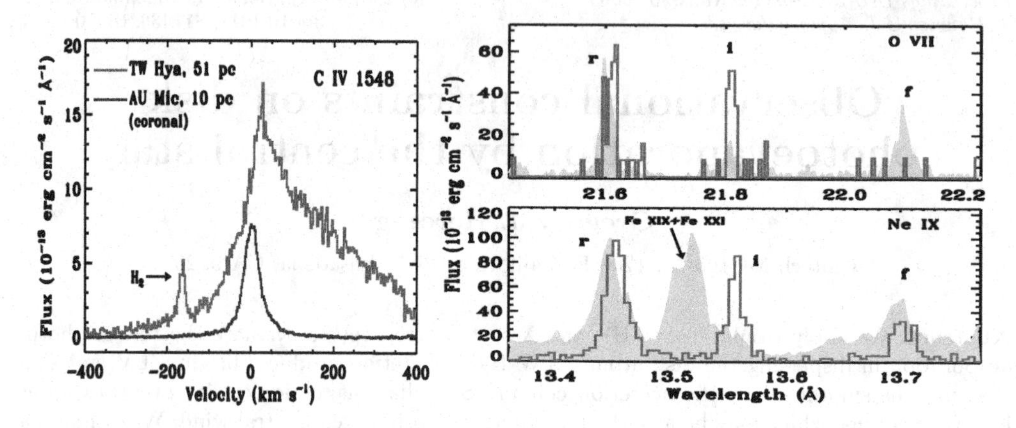

Figure 1. Left: Adapted from Johns-Krull & Herczeg (2007). The C IV emission from TW Hya (d=51 pc) is dominated by a broad red wing. In contrast, C IV emission from AU Mic (d=10 pc) is produced in the transition region, is centered at the stellocentric velocity, and is 200 times less luminous than that from TW Hya. **Right:** Adapted from Kastner *et al.* (2002). The forbidden line of He-like triplets from TW Hya is suppressed relative to coronal sources (Capella shown here in yellow), either due to densities of $\sim 10^{12.5}$ cm^{-3} or a strong FUV radiation field. Either explanation requires that the O VII and Ne IX lines are produced at or near the accretion shock.

Since hot lines in FUV and X-ray spectra of TW Hya are attributed to accretion, we suggest that the large estimate for EUV emission from TW Hya also applies to accretion. These photons may be smothered by the accretion column. Any photoevaporation of the disk around TW Hya may instead be dominated by the smaller amount of ionizing radiation produced by solar-like magnetic activity. We find that the amount of ionizing radiation is 1-3 orders of magnitude lower than that estimated by Alexander *et al.* 2005, and suggest that photoevaporation may not be significant until accretion ceases.

2. EUV irradiation of the disk

The amount of ionizing radiation that reaches the disk depends on the intrinsic ionizing luminosity from the star and attenuation of those photons by any neutral or molecular gas between the emission region and the disk. Ionizing photons can be produced by the accretion shock, which reaches temperatures of $\sim 10^6$ K (Calvet & Gullbring 1998), the 10^4 K accretion-heated stellar photosphere (e.g., Valenti *et al.* 1993), the transition region between a solar-like corona and chromosphere, or accretion-powered stellar outflows.

Alexander *et al.* (2004) found that the accretion column will attenuate any emission produced in the 10^4 K accretion continuum, since such emission occurs very close to the Lyman limit. In contrast, ionizing radiation produced by the accretion shock will be dominated by line emission and dispersed across a broad wavelength region. Such emission may be able to escape the accretion column. In the following subsections, we calculate the EUV luminosity from the accretion flow and from solar-like magnetic activity.

2.1. *The EUV luminosity of the accretion column*

Alexander *et al.* (2005) used line emissivities and the luminosity of C IV λ1549 doublet to estimate the ionizing radiation from CTTSs. Johns-Krull & Herczeg (2007) found that the asymmetric C III, C IV, N V, and O VI lines from TW Hya have a redshifted centroid and a broad wing that extends to ~ 400 km s^{-1}, consistent with formation in the accretion flow (Fig. 1a). Likewise, high-resolution X-ray spectra of TW Hya (Kastner *et al.* 2002;

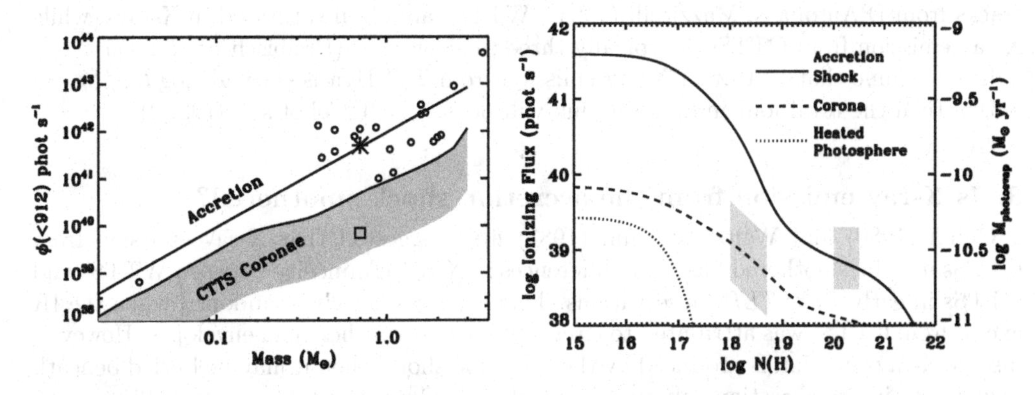

Figure 2. Left: The approximate ionizing photon fluxes from CTTSs, based on C IV luminosity in *HST*/STIS FUV spectra (circles, with TW Hya as the asterisk), correlates with stellar mass and mass accretion rate. The ionizing radiation from magnetic activity (shaded region with TW Hya, square) is much less, even if the X-ray emission is saturated (solid line). **Right:** The number of ionizing photons from the star, adjusted for any N(H I) between the emission region and the disk. The shaded regions show estimates for the ionizing radiation that reaches the disk.

Stelzer & Schmitt 2004) show that the forbidden-to-intercombination line ratio of the O VII and Ne IX He-like triplets is suppressed (Fig. 1b), which requires that the soft X-ray emission is produced at or near the accretion shock. Therefore, calculating EUV line emission from FUV or X-ray line fluxes from TW Hya provides an estimate for the ionizing radiation produced by the accreting gas.

We calculate a differential emission measure (DEM) and abundances for TW Hya from X-ray and FUV line fluxes (Kastner *et al.* 2002; Stelzer & Schmitt 2004; Herczeg *et al.* 2004) and *CHIANTI* line emissivities (Dere *et al.* 1997), assuming collisional ionization equilibrium. We then use the DEM to calculate line fluxes for EUV lines.

We estimate an ionizing photon flux from the accretion shock on TW Hya of $\sim 5\times10^{41}$ phot s^{-1} (Fig. 2a). Ardila (this volume) describes that collisional ionization may be a poor assumption for the accretion flow. Similarly, some uncertainty in relative abundances of C, N, O, Ne, and He, including a large relative Ne abundance, also adds to the uncertainty. However, since our estimates for EUV fluxes are tied to observed FUV and X-ray fluxes, we estimate an uncertainty of ~ 5 for this estimate for the ionizing photon flux.

2.2. *Estimating the EUV luminosity from the transition region*

Figure 1a shows that the C IV luminosity produced by the transition region of the 12 Myr old AU Mic is 200 times less than the total C IV luminosity from TW Hya. C IV emission from the transition region of TW Hya and most other CTTSs is masked by accretion onto the star. The coronal X-ray luminosity from TW Hya is also somewhat uncertain because of confusion with X-ray emission produced by accretion. Therefore, the total ionizing emission produced by solar-like magnetic activity on TW Hya is poorly constrained.

The differential emission measure for AU Mic was calculated by del Zanna *et al.* (2002) and is used here as a rough template for CTTSs. The emissivity at 10^5 K varies as $L_X^{0.5}$, based on an analysis of older coronal source by Ayres (1997). The shaded region of Figure 2a shows the estimated range of EUV flux from solar-like magnetic activity on TTSs as a function of mass. We calculate the approximate upper limit for EUV photon flux by assuming saturated X-ray emission ($L_X/L_{bol} = 10^{-3}$) for 1 Myr-old pre-main sequence tracks from D'Antona & Mazzitelli (1994). WTTSs are often saturated in X-rays, while X-ray emission from CTTSs is typically three times weaker (Preibisch *et al.* 2005).

If we assume that $\sim 10\%$ of X-ray emission from TW Hya is coronal ($\log L_X/L_{bol} = -3.7$), then the total ionizing photon flux will be $\sim 7 \times 10^{39}$ phot s^{-1} (Fig. 2).

3. Is X-ray emission from the accretion shock smothered?

Gahm (1980) and Walter & Kuhi (1981) first suggested that X-ray emission from CTTSs may be smothered based on differences in X-ray luminosity between WTTSs and CTTSs in early *EINSTEIN* observations. This idea was largely abandoned when the IR excess from CTTSs was attributed to circumstellar disks rather than envelopes. However, if some X-ray emission is produced by the accretion shock, then it may be buried beneath neutral gas in the accretion column (e.g., Calvet & Gullbring 1998; Drake 2005).

The N(H I) to CTTSs can be measured from either Lyman continuum absorption in X-ray spectra or from line absorption in FUV spectra. Figure 3 shows H I absorption from *HST*/STIS FUV and *FUSE* FUV spectra of TW Hya (Herczeg *et al.* 2004; Johns-Krull & Herczeg 2007). Lyα and Lyβ absorption lines are seen against emission in the same lines. Several absorption lines, including Lyδ and other lines not shown here, are seen against weak continuum emission. H Lyman absorption lines near 923 Å are detected, presumably against emission in the N IV multiplet.

The lack of any polarimetric signature in Hα emission from TW Hya (Yang *et al.* 2007) suggests that Hα emission, and therefore perhaps other H lines, are produced in a heated photosphere distributed on the star. Any difference in N(H I) between the two methods may be attributed to N(H I) between X-ray and Lyman line emission.

The absorption includes an interstellar component centered at -13 km s^{-1} and a wind component centered at -130 km s^{-1}, based on spectrally-resolved absorption components in O I, C II, and Mg II lines. We fit the absorption lines with Voigt profiles to measure N(H I). The optically-thick absorption in the *FUSE* lines extends to -280 km s^{-1}. If the N(H I) is the same to all FUV emission, such absorption requires that the Doppler broadening parameter is ~ 50 km s^{-1}. Based on the Lyα line, we find that N(H I)$=10^{19}$ cm^{-2} in the interstellar medium (Herczeg *et al.* 2004) and 3×10^{19} cm^{-2} in the wind, for a total N(H I)$=4 \pm 1 \times 10^{19}$ cm^{-2}.

These estimates assume that the H I scatters the Lyα photons out of our line of sight. The blueshifted absorption is not consistent with a self-reversed profile. However, the wind may be local to the star, in which case the Lyα photons will not scatter out of our line of sight. By modelling local scattering in a pure-hydrogen slab, we find that the N(H I) measured from a simple Voigt profile may be underestimated by $\sim 1.5 \times 10^{19}$ cm^{-2}.

Therefore, we estimate that N(H I)$=6\times10^{19}$ cm^{-2} to the Lyα emission region. Herczeg *et al.* (2004) also measured $N(H_2) < 10^{18}$ cm^{-2} from the non-detection of H_2 absorption lines in a *FUSE* spectrum of TW Hya. The N(H I) to the X-ray emission from TW Hya was measured to be $5.2 \pm 3.7 \times 10^{20}$ cm^{-2} in a *ROSAT* spectrum (Kastner *et al.* 1999) and $3.5 \pm 0.5 \times 10^{20}$ cm^{-2} in an *XMM* spectrum (Robrade & Schmitt 2006). Figure 3a shows that the Lyα absorption profile for such a large N(H I) is inconsistent with the observed absorption.

X-ray and FUV measurements of N(H I) are discrepant by $\sim 3 \times 10^{20}$ cm^{-2}. These measurements may be explained if N(H I) lies between the stellar photosphere and to the X-ray-emitting gas, or by invoking variability. We note that N(H I) measured from *ROSAT* and *XMM* observations of TW Hya, obtained years apart, are similar.

The N(H I) of 1.5×10^{21} cm^{-2} and 3.1×10^{21} cm^{-2} measured from the *XMM* spectra of the CTTSs BP Tau and SU Aur by Robrade & Schmitt (2006) are also much larger

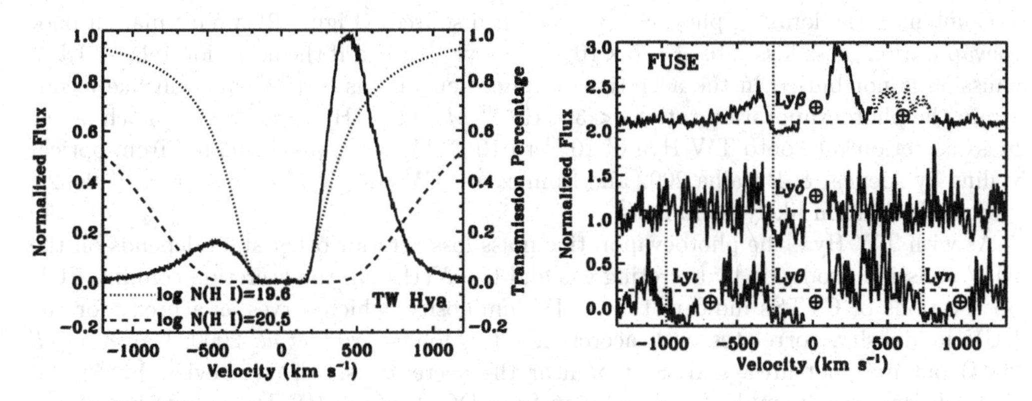

Figure 3. H I absorption in our line of sight to TW Hya can be measured using absorption lines in *HST*/STIS FUV spectra (**left**, adapted from Herczeg *et al.* 2004) and in *FUSE* spectra (**right**, top panel adapted from Johns-Krull & Herczeg 2007). The H I absorption includes a wind component at -130 km s^{-1} and an interstellar component at -13 km s^{-1}. The left panel shows the observed emission (solid line) compared with the transmission percentage for N(H I)$= 4 \times 10^{19}$ (dotted line) and 3×10^{20} cm^{-2} (dashed line). The N(H I) in our line of sight to the Lyα emission from TW Hya is less than the N(H I)$= 3.5 \times 10^{20}$ measured by Robrade & Schmitt (2006) from Lyman continuum absorption in an *XMM* spectrum of TW Hya.

than the N(H I) of 5×10^{19} cm^{-2} and 5×10^{20} cm^{-2}, respectively, measured from Lyα absorption by Lamzin (2006). These N(H I) measurements have a larger uncertainty than for TW Hya because they are measured from spectra with low resolution and lack complementary *FUSE* spectra, which means that $N(H_2)$ cannot be measured. However, the molecular fraction in our line of sight would have to be $\sim 95\%$ and 80%, respectively, to explain the discrepant N(H I) measurements from the FUV and X-ray spectra of BP Tau and SU Aur.

The difference in N(H I) measurements for BP Tau and SU Aur are much larger than that for TW Hya, possibly the result of larger mass accretion rates for the former two stars. At present we can only compare N(H I) measurements from FUV and X-ray spectra for those three stars, which have both STIS and *XMM* spectra. Strong outflows may corrupt similar N(H I) measurements from the Lyα line profile seen from RU Lup and T Tau.

4. Estimating photoevaporation rates

Hollenbach *et al.* (1994) estimate a photoevaporation mass loss rate of 3×10^{-10} $(\frac{\phi}{10^{41}\,\mathrm{phot}})^{0.5} M_\odot$ yr^{-1} for a 0.6 $M_\odot$ star, where ϕ is the flux of ionizing photons from the central star. Alexander *et al.* (2005) estimate an ionizing flux of $10^{41} - 10^{44}$ phot s^{-1} from CTTSs. However, if this emission from accretion is buried under $\sim 3 \times 10^{20}$ cm^{-2} of neutral gas, then the ionizing photons from the accretion shock that reach the disk are $\sim 10^{39}$ phot s^{-1}.

The neutral wind will also attenuate some EUV emission. Ardila *et al.* (2002) found that a low-temperature wind was present for all CTTSs, regardless of inclination. In an analysis of FUV H_2 emission, Herczeg *et al.* (2004) detected neutral gas between the Lyα emission region and the warm H_2 in the disk (Figure 4a). This gas is blueshifted in the frame of the H_2 with a N(H I)$\lesssim 5 \times 10^{18}$ cm^{-2}. Although only an upper limit because of uncertain Doppler broadening, such gas will attenuate any EUV emission from both the accretion and from the transition region.

Combining the ionizing photon flux from both sources (Figure 2), we estimate a photoevaporation mass loss rate of $\sim 5 \times 10^{-11}$ $M_\odot$ yr^{-1}. Even if the accretion-related EUV emission is not buried in the accretion column, neutral gas in the wind will likely constrain the photoevaporation rate to $< 3 \times 10^{-10}$ $M_\odot$ yr^{-1}. However, with a much larger mass accretion rate onto TW Hya of 10^{-9} to 10^{-8} $M_\odot$ yr^{-1} (as estimated from optical veiling by Alencar & Batalha 2002 and from excess UV emission by Herczeg *et al.* 2002), viscous dissipation likely dominates.

As with TW Hya, the photoevaporative mass loss rate for other stars depends on the EUV emission produced by accreting gas and the N(H I) in the accretion column. FUV observations of CTTSs indicate that C IV luminosity, which serves as a proxy for the EUV luminosity, correlates with accretion rate (Johns-Krull *et al.* 2000; Calvet *et al.* 2004) but may not always arise at or near the accretion shock (see review by Ardila, this volume). For example, C IV emission from DG Tau and RY Tau is blueshifted and therefore likely produced at or near the base of the outflow, while C IV emission from some other stars is seen at the radial velocity of the star.

For these stars with larger mass accretion rates, the ionizing flux may reach as high as 10^{43} phot s^{-1}, leading to a photoevaporation rate of 5×10^{-9} $M_\odot$ yr^{-1}. However, these stars also have stronger neutral outflows (Hartigan *et al.* 1995), which can attenuate much of the ionizing radiation before it irradiates the disk. The photoevaporation rate most likely stays below the rate of viscous dissipation for such stars.

Accreting low-mass stars and brown dwarfs have much smaller accretion rates than solar-mass CTTSs (e.g., Muzerolle *et al.* 2005). Based on the small C IV luminosity detected from the brown dwarf 2MASS1207-3932 (Gizis *et al.* 2005), the total EUV emission from the accretion shock will be much smaller for these stars. If the N(H I) in the accretion column is sufficiently small because the mass accretion rate is also small, then enough ionizing radiation may escape to lead to a photoevaporation rate of $\sim 10^{-11}$ $M_\odot$ yr^{-1}.

Once accretion ceases, any wind disappears and the photoevaporative rate due to solar-like magnetic activity may be as high as $\sim 3\times10^{-10}$ $M_\odot$ yr^{-1} for a solar-mass star. Thus, while photoevaporation might be negligible throughout the accretion process, as much as $\sim 3M_J$ of remnant gas could be dispersed over 10 Myr by photoevaporation after accretion ceases. This low photoevaporation rate requires one of the following possibilities: 1) long survival timescales for gas in disks after accretion ceases, which has not been seen (e.g., Roberge *et al.* 2005; Pascucci *et al.* 2006); 2) other disk dispersal mechanisms more efficient than photoevaporation, such as wind-disk interactions; or 3) accretion continues until < 1 M_J of gas is removed from the disk.

5. A direct probe of disk ionization

The results described above provide only indirect estimates of disk irradiation and suffer from substantial uncertainties. A probe of the ionized disk surface is needed test these estimates. Recently, Pascucci *et al.* (2007) and Lahuis *et al.* (2007) may have found such a probe by detecting emission in the [Ne II] 12.81μm fine-structure line in low-resolution *Spitzer* IRS spectra of CTTSs. [Ne II] was detected in four of six targets in the Pascucci *et al.* (2007) sample and in 15 of 76 targets in the Lahuis *et al.* (2007) sample. Lahuis *et al.* (2007) also detected [Ne III] emission from one source (Sz102).

We used MICHELLE on Gemini North to detect bright [Ne II] emission from TW Hya with $R \sim 30,000$ (Fig. 4b). This emission is located at the stellar radial velocity and has an intrinsic FWHM of 21 ± 4 km s^{-1}, which confirms speculation that the line is produced in the disk. The line width is broader than other narrow emission lines typically associated

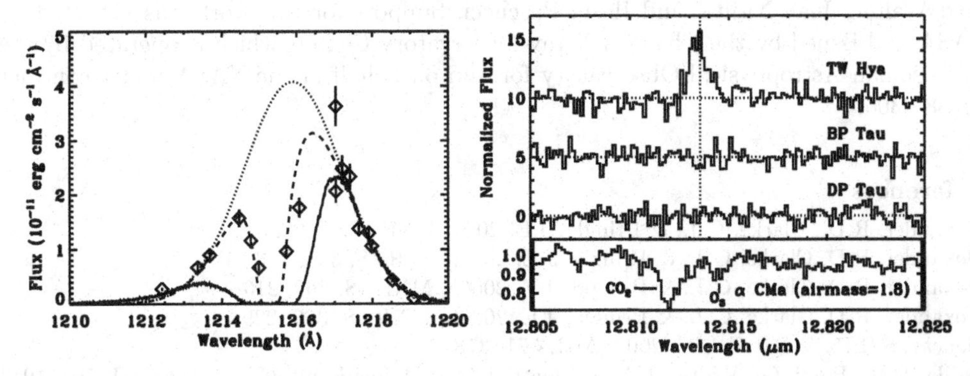

Figure 4. Left: Herczeg *et al.* (2004) reconstruct the Lyα emission incident upon H_2 gas in the disk (diamonds), compared to the observed Lyα profile (thick solid line). H I located between the star and disk, with a velocity of ~ -90 km s^{-1} from line center, absorbs some Lyα emission. **Right:** Strong [Ne II] emission in high-resolution Gemini North/MICHELLE spectra of TW Hya, with a dotted line indicating the stellocentric velocity for each source.

with the disk around TW Hya. If formed in a disk, the line broadening could result from turbulence in a warm disk atmosphere, Keplerian rotation at an average distance of 0.1 AU from the star, or a photoevaporative flow from the optically-thin region of the disk. Additional high-resolution spectra can discriminate between these scenarios.

Two ionization paths can ionize Ne and thereby produce the [Ne II] emission. Glassgold *et al.* (2007) propose that the line is formed because of Ne ionization by K-shell absorption of stellar X-rays at energies > 0.9 keV. The observed [Ne II] fluxes and the [Ne III]/[Ne II] flux ratio and lower limits are consistent with predictions from this X-ray ionization model. Alternately, EUV photons shortward of the Ne I ionization edge at 575 Å can produce [Ne II] emission if the EUV emission is able to penetrate through neutral gas in the accretion column and wind. Combining X-ray and EUV ionization models should provide estimates for the total EUV flux incident upon the disk surface.

6. Discussion

We find evidence that the EUV and soft X-ray emission from TW Hya may be smothered. The accretion column could attenuate most ionizing radiation produced by the accretion shock. We estimate photoevaporation rates of $\sim 5 \times 10^{-11}$ $M_\odot$ yr^{-1} from ionizing radiation produced by both the accretion shock and by solar-like magnetic activity.

The difference in N(H I) measurements need to be verified to ensure that the result is not attributable to variability. The results are also dependent on geometry, if the N(H I) of the accretion column is larger when seen face-on than when seen edge-on, as the disk views a dipolar accretion flow. Observations and modelling of [Ne II] emission will help to verify estimates of ionizing radiation incident upon the disk.

These results suggest that mass loss from disk photoevaporation is much smaller than viscous dissipation until accretion slows to undetectable rates. Alexander *et al.* (2006b) and Najita *et al.* (2007) both suggest that disk photoevaporation may be significant for CoKu Tau/4, which has a large inner disk hole and no previously identified accretion signatures. However, photoevaporation can only be a significant process if the timescale for gas dissipation is > 10 Myr or accretion leaves $\lesssim 1 M_J$ of gas in the disk.

Acknowledgements

I thank several collaborators who have contributed to this work, including Jeff Linsky, Chris Johns-Krull, Lynne Hillenbrand, Rachel Osten, Brian Wood, Jeff Valenti,

Fred Walter, Joan Najita, and Ilaria Pascucci. Support for this work was provided by NASA and issued by the Chandra X-ray Observatory Center, which is operated by the Smithsonian Astrophysical Observatory for and on behalf of the NASA under contract NAS8-03060.

References

Alexander, R.D., Clarke, C.J., & Pringle, J.E. 2004, MNRAS, 354, 71
Alexander, R.D, Clarke, C.J., & Pringle, J.E. 2005, MNRAS, 358, 283
Alexander, R.D, Clarke, C.J., & Pringle, J.E. 2006a, MNRAS, 369, 216
Alexander, R.D, Clarke, C.J., & Pringle, J.E. 2006b, MNRAS, 369, 229
Alencar, S.H.P., & Batalha, C. 2002, ApJ, 571, 378
Ardila, D.R., Basri, G., Walter, F.M., Valenti, J.A., & Johns-Krull, C.M. 2002, ApJ, 567, 1013
Ayres, T.R. 1997, JGR, 102, 1641
Calvet, N., & Gullbring, E. 1998, ApJ, 509, 802
Calvet, N., D'Alessio, P., Hartmann, L., *et al.* 2002, ApJ, 568, 1008
Calvet, N., Muzerolle, J., Briceno, C., *et al.* 2004, AJ, 128, 1294
D'Antona, F., & Mazzitelli, I. 1994, ApJS, 90, 467
del Zanna, G., Landini, M., & Mason, H.E. 2002, A&A, 385, 968
Dere, K.P., *et al.* 1997, A&AS, 125, 149
Drake, J.J. 2005, Proceedings of *13th Cool Stars Workshop*, eds. F. Favata, G. Hussein, & B. Battrick. 519
Dullemond, C.P., Hollenbach, D., Kamp, I., D'Alessio, P. 2007, proceedings of *P&PV*, eds. B. Reipurth, D. Jewitt, and K. Keil, 951, 555
Dullemond, C.P., & Dominik, C. 2005, A&A, 434, 971
Eisner, J.A., Chiang, E.I., & Hillenbrand, L.A. 2006, ApJ, 637, L133
Gahm, G.F. 1980, ApJ, 242, L163
Gizis, J.E., Shipman, H.L., Harvin, J.A. 2005, ApJ, 630, L89
Glassgold, A.E., Najita, J.R., & Igea, J. 2007, ApJ, 656, 515
Hartigan, P., Edwards, S., Ghandour, L. 1995, ApJ, 452, 736
Herczeg, G. J., Linsky, J. L., Valenti, J.A., Johns-Krull, C.M. 2002, ApJ, 572, 310
Herczeg, G.J., *et al.* 2004, ApJ, 607, 369
Hollenbach, D., Johnstone, D., Lizano, S., & Shu, F. 1994, ApJ, 428, 654
Hughes, A.M., *et al.* 2007, ApJ, accepted. astro-ph:0704.2422
Johns-Krull, C.M., Valenti, J.A., & Linksy, J.L. 2000, ApJ, 539, 815
Johns-Krull, C.M., & Herczeg, G.J. 2007, ApJ, 655, 345
Kastner, J.H., Huenemoerder, D.P., Schulz, N.S., & Weintraub, D.A. 1999, ApJ, 525, 837
Kastner, J. H. Huenemoerder, D. P., Schulz, N. S., *et al.* 2002, ApJ, 567, 434
Lahuis, F., van Dischoeck, E.F., Blake, G.A., *et al.* 2007, ApJ, accepted. astro-ph/07042305
Lamzin, S.A. 2006, AstL, 32, L176
Muzerolle, J., Luhman, K.L., Briceno, C., Hartmann, L., & Calvet, N. 2005, ApJ, 620, L107
Najita, J., Carr, J.S., Glassgold, A.E., & Valenti, J.A. 2007, PPV
Pascucci, I., *et al.* 2006, ApJ, 651, 1177
Pascucci, I., *et al.* 2007, ApJ, 663, 383
Preibisch, T., *et al.* 2005, ApJS, 160, 582
Roberge, A., Weinberger, A.J., & Malumuth, E.M. 2005, ApJ, 622, 1171
Robrade, J., Schmitt, J.H.M.M. 2006, A&A, 449, 737
Stelzer, B., Schmitt, J.H.H.M. 2004, A&A, 418, 687
Walter, F.M. & Kuhi, L.V. 1981, ApJ, 250, 254
Valenti, J.A., Basri, G., & Johns, C.M. 1993, ApJ, 106, 2024
Yang, H., Johns-Krull, C.M., & Valenti, J.A. 2007, AJ, 133, 73

Star-Disk Interactions in Young Stars
Proceedings IAU Symposium No. 243, 2007
J. Bouvier & I. Appenzeller, eds.

doi:10.1017/S1743921307009519

Accretion and outflow-related X-rays in T Tauri stars

Manuel Güdel[1,2], Kevin Briggs[1], Kaspar Arzner[1], Marc Audard[3], Jérôme Bouvier[4], Catherine Dougados[4], Eric Feigelson[5], Elena Franciosini[6], Adrian Glauser[1], Nicolas Grosso[4] †, Sylvain Guieu[4] ‡, François Ménard[4], Giusi Micela[6], Jean-Louis Monin[4], Thierry Montmerle[4], Deborah Padgett[7], Francesco Palla[8], Ignazio Pillitteri[6,9], Thomas Preibisch[10], Luisa Rebull[7], Luigi Scelsi[6,9], Bruno Silva[11], Stephen Skinner[12], Beate Stelzer[6] and Alessandra Telleschi[1]

[1]Paul Scherrer Institut, Würenlingen and Villigen, 5232 Villigen PSI, Switzerland
email: guedel@astro.phys.ethz.ch

[2]Max-Planck-Institute for Astronomy, Königstuhl 17, 69117 Heidelberg, Germany

[3]Integral Science Data Centre, Ch. d'Ecogia 16, 1290 Versoix, and Geneva Observatory, University of Geneva, Ch. des Maillettes 51, 1290 Sauverny, Switzerland

[4]Laboratoire d'Astrophysique de Grenoble, Université Joseph Fourier - CNRS, BP 53, 38041 Grenoble Cedex, France

[5]Department of Astronomy & Astrophysics, Penn State University, 525 Davey Lab, University Park, PA 16802, USA

[6]INAF – Osservatorio Astronomico di Palermo, Piazza del Parlamento 1, 90134 Palermo, Italy

[7]Spitzer Science Center, California Institute of Technology, Mail Code 220-6, Pasadena, CA 91125, USA

[8]INAF – Osservatorio Astrofisico di Arcetri, Largo Enrico Fermi, 5, 50125 Firenze, Italy

[9]Dipartimento di Scienze Fisiche ed Astronomiche, Università di Palermo, Piazza del Parlamento 1, 90134 Palermo, Italy

[10]Max-Planck-Institut für Radioastronomie, Auf dem Hügel 69, 53121 Bonn, Germany

[11]Centro de Astrofísica da Universidade do Porto, Rua das Estrelas, 4150 Porto, and Departamento de Matemática Aplicada, Faculdade de Ciêcias da Universidade do Porto, 4169 Porto, Portugal

[12]CASA, UCB 389, University of Colorado, Boulder, CO 80309-0389, USA

Abstract. We report on accretion- and outflow-related X-rays from T Tauri stars, based on results from the "XMM-Newton Extended Survey of the Taurus Molecular Cloud." X-rays potentially form in shocks of accretion streams near the stellar surface, although we hypothesize that direct interactions between the streams and magnetic coronae may occur as well. We report on the discovery of a "soft excess" in accreting T Tauri stars supporting these scenarios. We further discuss a new type of X-ray source in jet-driving T Tauri stars. It shows a strongly absorbed coronal component and a very soft, weakly absorbed component probably related to shocks in microjets. The excessive coronal absorption points to dust-depletion in the accretion streams.

Keywords. Accretion, stars: activity, stars: coronae, stars: formation, stars: magnetic fields, stars: pre–main-sequence, stars: winds, outflows.

---

† Present address: Observatoire Astronomique de Strasbourg, 11 rue de l'université, 67000 Strasbourg, France

‡ Present address: Spitzer Science Center, California Institute of Technology, Mail Code 220-6, Pasadena, CA 91125, USA

1. Introduction

Classical and weak-lined T Tauri stars (CTTS and WTTS) are pre-main sequence stars showing vigorous X-ray emission with X-ray luminosities (L_X) near the empirical saturation limit found for main-sequence (MS) stars, $L_X/L_{bol} \approx 10^{-3.5}$. Consequently, X-ray emission from T Tauri stars has been attributed to solar-like coronal activity. However, both accretion and outflows/jets may contribute to or alter X-ray production in the magnetic environment of strongly accreting young stars. X-rays thus provide an important diagnostic for the detection and study of accretion and outflow activity.

Support for X-ray *suppression* in CTTS has been reported from X-ray photometry. CTTS are, on average, less X-ray luminous than WTTS (e.g., Strom & Strom 1994; Neuhäuser *et al.* 1995). This finding has been partly supported by recent, deep surveys of the Orion Nebula Cluster (Getman *et al.* 2005; Preibisch *et al.* 2005).

On the other hand, X-rays may be *generated* by surface accretion shocks (Lamzin 1999). If gas collides with the stellar photosphere in free fall, shocks heat it to a few million K. Given the appreciable accretion rates, high shock densities of order $10^{12} - 10^{14}$ cm^{-3} are expected, as first reported for the CTTS TW Hya (Kastner *et al.* 2002). X-rays may similarly be produced in shocks forming in jets (Raga *et al.* 2002).

The *XMM-Newton Extended Survey of the Taurus Molecular Cloud (XEST)* (Güdel *et al.* 2007a) has provided new insights into these issues. XEST covers the most populated ≈5 sq. deg of the Taurus star-forming region. The average on-axis detection limit is $\approx 10^{28}$ erg s^{-1} for lightly absorbed objects, sufficient to detect about half of the observed brown dwarfs (Grosso *et al.* 2007a). We discuss X-ray results relevant for the star-disk interface in which accretion occurs and where (parts of) the jets may be accelerated.

2. Accretion and the "X-ray Soft Excess"

2.1. *Accretors in XEST*

Telleschi *et al.* (2007a) present statistical X-ray studies of the XEST CTTS and WTTS samples. The WTTS sample is essentially complete (all but one of the 50 surveyed objects detected), and the CTTS sample is nearly complete (85% of the 65 surveyed objects detected). The luminosity deficiency of CTTS is confirmed. More precisely, CTTS are statistically less luminous by a factor of ≈ 2 both in L_X and in L_X/L_{bol} (Fig. 1a) while the L_{bol} distributions of the two samples are drawn from the same parent population.

On the other hand, Telleschi *et al.* (2007a) report the average electron temperature, T, in the X-ray sources of CTTS to be higher than in WTTS, irrespective of the gas absorption column density (N_H). For WTTS, a trend also seen in MS stars is recovered, in the sense that the electron temperature increases with L_X. Such trends are expected in stochastic-flare heated coronae (Telleschi *et al.* 2005), but a similar trend is absent in CTTS in which the average temperature remains high for all activity levels.

In contrast to the apparent X-ray deficiency, a trend toward an ultraviolet excess is found in CTTS based on the Optical Monitor (OM) data (Audard *et al.* 2007). The UV excess supports an accretion scenario in which gas in accretion streams shock-heats near the surface to form hot spots (Calvet & Gullbring 1998). The OM has also recorded a slow U-band flux increase in a brown dwarf, most likely due to an "accretion event" covering a time span of several hours (Grosso *et al.* 2007b).

Accretion shocks may leave signatures in line-dominated high-resolution X-ray spectra. Given the typical mass accretion rates on T Tauri stars and their modest accretion hot spot filling factors of no more than a few percent (Calvet & Gullbring 1998), densities of order 10^{12} cm^{-3} or more are to be expected (Telleschi *et al.* 2007c; Güdel *et al.* 2007c),

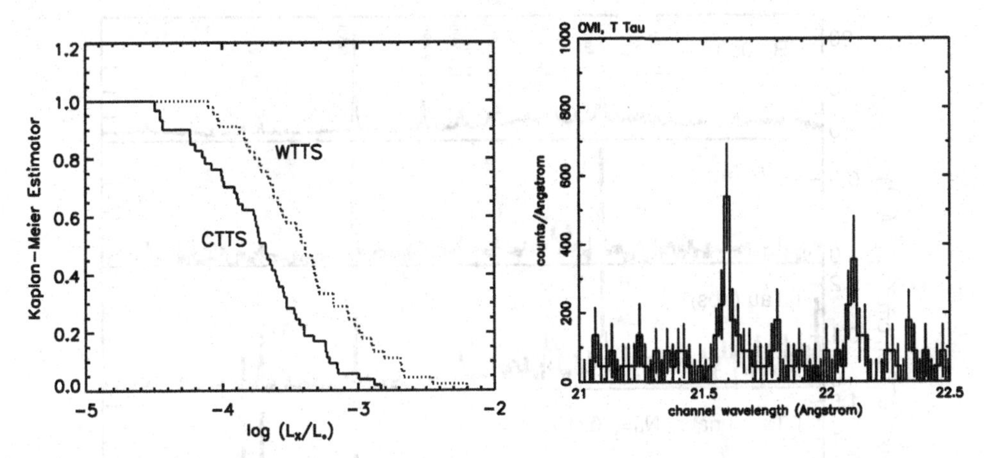

Figure 1. *Left (a)*: The cumulative distribution of the L_X/L_{bol} ratio for CTTS (solid) and WTTS (dotted) are different at the > 99.97% level (from Telleschi *et al.* 2007a). – *Right (b)*: The O VII triplet of T Tau, showing a strong forbidden line at 22.1 Å (after Güdel *et al.* 2007c).

and such densities are indeed indicated in the density-sensitive line ratios of O VII and Ne IX triplets of some CTTS (e.g., Kastner *et al.* 2002; Stelzer & Schmitt 2004). XEST has added relevant information on two further accreting pre-main sequence stars: T Tau N (Güdel *et al.* 2007c) and the Herbig star AB Aur (Telleschi *et al.* 2007b). However, in both *XMM-Newton* RGS spectra, the O VII triplet line ratios are compatible with density upper limits of a few times 10^{10} cm^{-3} (Fig. 1b), apparently not supporting the accretion-shock scenario. How important really is accretion for CTTS X-ray emission?

2.2. *The "X-Ray Soft Excess"*

Fig. 2 compares *XMM-Newton* RGS spectra of the active binary HR 1099 (X-rays mostly from a K-type subgiant; archival data) the weakly absorbed WTTS V410 Tau (Telleschi *et al.* 2007c), the CTTS T Tau (Güdel *et al.* 2007c), and the old F subgiant Procyon (archival data). HR 1099 and V410 Tau show the typical signatures of a hot, active corona such as a strong continuum, strong lines of Ne X and of highly-ionized Fe but little flux in the O VII line triplet. In contrast, lines of C, N, and O dominate the soft spectrum of Procyon, the O VII triplet exceeding the O VIII Lyα line in flux. T Tau reveals signatures of a very active corona shortward of 19 Å but also an unusually strong O VII triplet. Because its N_H is large (in contrast to N_H of V410 Tau), we have modeled the intrinsic, unabsorbed spectrum based on transmissions determined in XSPEC using N_H from EPIC spectral fits ($N_H \approx 3 \times 10^{21}$ cm^{-1}; Güdel *et al.* 2007a). *The* O VII *lines are the strongest lines in the intrinsic X-ray spectrum,* reminiscent of the situation in Procyon!

To generalize this finding, we plot in Fig. 3 the ratio between the intrinsic (unabsorbed) luminosities of the O VII r line and the O VIII Lyα line as a function of L_X, comparing CTTS and WTTS with a larger MS sample (Ness *et al.* 2004) and MS solar analogs (Telleschi *et al.* 2005). The TTS data are from Robrade & Schmitt (2006), Günther *et al.* (2006), Argiroffi *et al.* (2007), Telleschi *et al.* (2007c), and from our analysis of archival *XMM-Newton* data of RU Lup. For the TTS sample given by Telleschi *et al.* (2007c), we have approximated $L(\text{O VII } r) = 0.55L(\text{O VII})$ (Porquet *et al.* 2001). The trend for MS stars (black crosses and triangles, § 2.1) is evident: as the coronae get hotter toward higher L_X, the ratio of O VII r/O VIII Lyα line luminosities decreases. This trend is followed by the sample of WTTS, while CTTS again show a significant excess. This is the essence of the *X-ray soft excess* in CTTS first discussed by Telleschi *et al.* (2007c) and Güdel

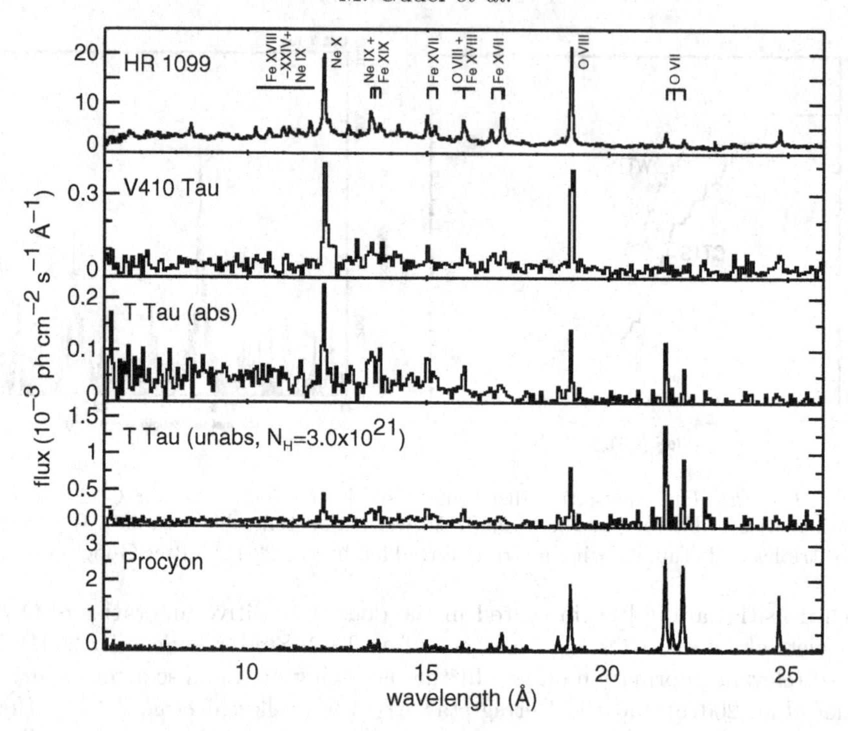

Figure 2. Comparison of fluxed *XMM-Newton* RGS photon spectra of (from top to bottom) the active binary HR 1099, the WTTS V410 Tau, the CTTS T Tau, T Tau modeled after removal of absorption, and the inactive MS star Procyon. The bins are equidistant in wavelength.

et al. (2007c): *accreting pre-main sequence stars reveal a strong excess of cool (1-2 MK) material, regardless of the overall X-ray deficiency at higher temperatures.*

2.3. *Summary and conclusions on accretion-related X-ray emission*

Crucial XEST results on X-ray production and accretion can be summarized as follows:

1. Accreting CTTS show a general deficiency in X-ray production when referring to the hot coronal gas recorded by X-ray CCD cameras (Telleschi *et al.* 2007a);
2. the average coronal T is higher in CTTS than in WTTS (Telleschi *et al.* 2007a);
3. All CTTS (except the two flaring sources SU Aur [also subject to high $N_{\rm H}$] and DH Tau) show a *soft excess* defined by an anomalously high ratio between the fluxes of the O VII He-like triplet and the O VIII Lyα line, when compared to WTTS and MS stars.

The origin of the additional cool plasma in CTTS is likely to be related to the accretion process. Accretion streams may shock-heat gas at the impact point to X-ray emitting temperatures. This model is supported by high electron densities inferred from the observed O VII or Ne IX triplets in some of the CTTS (e.g., Kastner *et al.* 2002; Stelzer & Schmitt 2004), although high densities were not seen in the two XEST accretors T Tau (Güdel *et al.* 2007c) and AB Aur (Telleschi *et al.* 2007b). Alternatively, the cool, infalling material may partly cool pre-existing heated coronal plasma, or reduce the efficiency of coronal heating in the regions of infall (Preibisch *et al.* 2005; Telleschi *et al.* 2007c; Güdel *et al.* 2007c). This model would at the same time explain why CTTS are X-ray weaker than WTTS (Preibisch *et al.* 2005; Telleschi *et al.* 2007a). We cannot assess what the relative importance of these processes is. It seems clear, however, that the soft excess described here argues in favor of a substantial influence of accretion on the X-ray production in pre-main sequence stars.

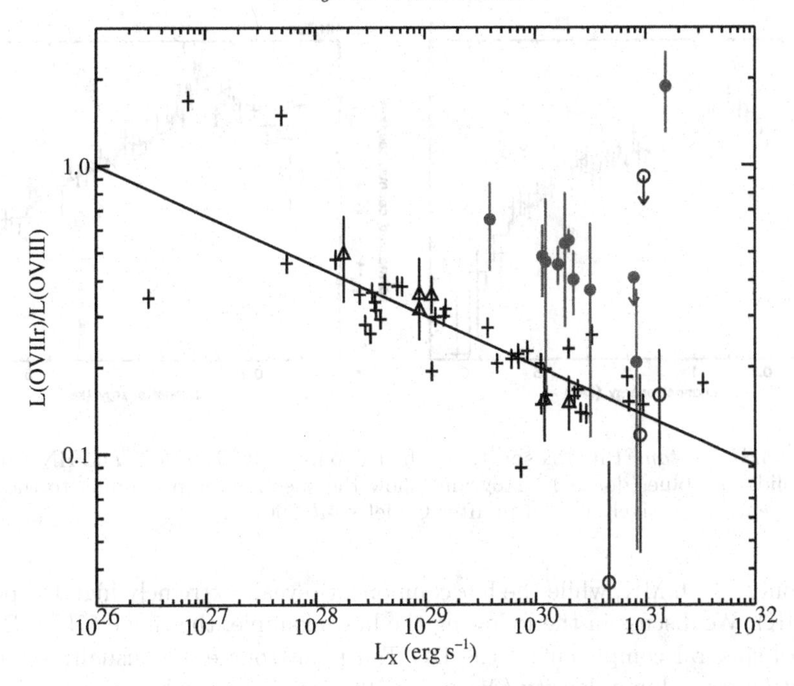

Figure 3. The ratio between O VII r and O VIII Lyα luminosities (each in erg s^{-1}) vs. the total L_X. Crosses mark MS stars, triangles solar analogs of different ages, filled (red) circles CTTS, and open (blue) circles WTTS. The solid line is a power-law fit to the MS stars (after Güdel & Telleschi 2007, submitted).

3. Outflow-related X-rays

The shock temperature in jets can be expressed as $T \approx 1.5 \times 10^5 v_{100}^2$ K (for fully ionized gas) where v_{100} is the shock speed in units of 100 km s^{-1} (e.g., Raga *et al.* 2002). Jet speeds are typically of order $v = 300 - 500$ km s^{-1} (Eislöffel & Mundt 1998; Anglada 1995; Bally *et al.* 2003), in principle allowing for shock speeds of similar magnitude.

Faint, soft X-ray emission has been detected from a few protostellar HH objects (Pravdo *et al.* 2001, Pravdo *et al.* 2004; Pravdo & Tsuboi 2005; Favata *et al.* 2002; Bally *et al.* 2003; Tsujimoto *et al.* 2004; Grosso *et al.* 2006). Bally *et al.* (2003) used a *Chandra* observation to show that X-rays form within an arcsecond of the protostar L1551 IRS-5 while the star itself is too heavily obscured to be detected. Strong absorption and extinction of protostars and their immediate environment make the launching region of their powerful jets generally inaccessible to optical, near-infrared, or X-ray studies. However, a class of strongly accreting, optically revealed CTTS also exhibit so-called micro-jets visible in optical lines (Hirth *et al.* 1997), with flow speeds similar to protostellar jets. CTTS micro-jets have the unique advantage that they can – in principle – be followed down to the acceleration region both in the optical and in X-rays.

3.1. *"Two-Absorber X-Ray" (TAX) Sources*

X-ray spectra of very strongly accreting, micro-jet driving CTTS exhibit an anomaly (Fig. 4, Güdel *et al.* 2005; Güdel *et al.* 2007b): the spectra of DG Tau, GV Tau, DP Tau, CW Tau, and HN Tau are composed of two components, a cool component subject to very low absorption and a hot component subject to photoelectric absorption about one order of magnitude higher. For similar phenomenology, see also Kastner *et al.* (2005) and Skinner *et al.* (2006). The cool component shows temperatures atypical for T Tau stars,

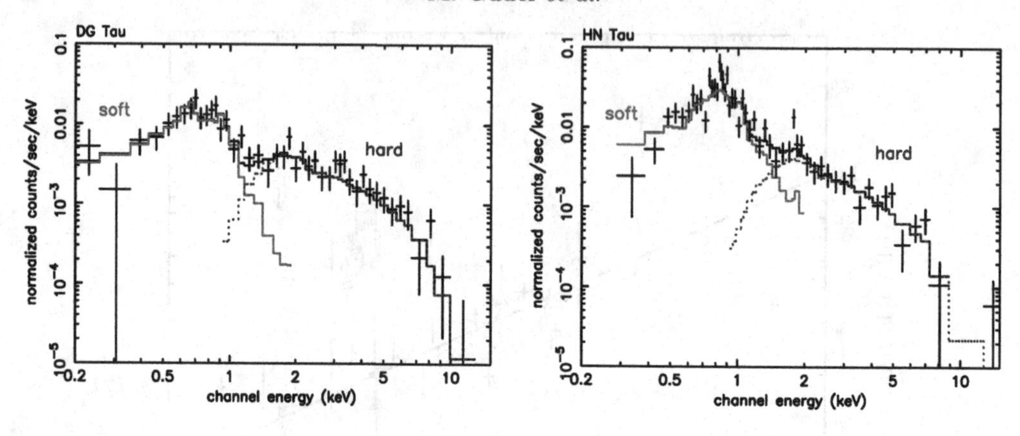

Figure 4. *XMM-Newton* EPIC PN spectra of the jet sources DG Tau (left) and HN Tau (right). The red (solid) and blue (dotted) histograms show the spectral fits pertaining to the soft and hard component, respectively (DG Tau after Güdel *et al.* 2007b).

ranging from $\approx 3-6$ MK, while the hot component reveals extremely high temperatures (10–100 MK). We discuss in the following the best example, the single CTTS DG Tau.

The hard spectral component of the DG Tau point source is unusually strongly absorbed, with a gas column density ($N_{\rm H} \approx 2 \times 10^{22}$ cm^{-2}) higher by a factor of ≈ 5 than predicted from the visual extinction $A_{\rm V}$ of 1.5–3 mag if standard gas-to-dust ratios are assumed (see Güdel *et al.* 2007b and references therein). Because the hard component occasionally flares, in one case being preceded by U band emission as in solar and stellar flares (Güdel *et al.* 2007b), it is most straightforwardly interpreted as coronal or "magnetospheric". The *excess absorption* is likely to be due to the heavy accretion streams falling down along the magnetic fields and absorbing the X-rays from the underlying corona/magnetosphere. The excess absorption-to-extinction ($N_{\rm H}/A_{\rm V}$) ratio then is an indicator of dust sublimation: *the accreting gas streams are dust-depleted.*

In contrast, $N_{\rm H}$ of the soft X-ray component, $N_{\rm H} = 1.3\ (0.7-2.4) \times 10^{21}$ cm^{-2} (90% error range) is *lower* than suggested from the stellar $A_{\rm V}$, $N_{\rm H}(A_{\rm V}) \approx (3-6) \times 10^{21}$ cm^{-2}. A likely origin of these X-rays is the base of the jet. Such an origin is suggested by i) the unusually soft emission not usually seen in T Tauri stars (Güdel *et al.* 2007a), ii) the low $N_{\rm H}$, and iii) the explicit evidence of jets in the *Chandra* image, as we will show below.

3.2. *X-rays and jets*

A *Chandra* X-ray image of the DG Tau environment is shown in Fig. 5b (pixel size 0.49″). This image was produced by combining counts from a total of 90 ks of *Chandra* exposure time. Also shown is a smoothed version. To suppress background and to emphasize the soft sources, only counts within the 0.6–1.7 keV range are plotted. There is clear evidence for a jet-like extension outside the stellar point spread function (PSF) to the SW along a position angle of ≈ 225 deg, but we also find a significant excess of counts in the NE direction (PA ≈ 45 deg). This is coincident with the jet optical axis, which for the SW jet has been given as 217-237 deg (Eislöffel & Mundt 1998). We verified, using raytrace simulations, that the jet sources are extended: a faint point source would occupy only a few pixels. We also find the counter jet to be harder, with photon energies mostly above 1 keV, while the forward jet shows a mixture of softer and harder counts. The spectral properties of the jet sources are reminiscent of the *soft* component in the "stellar" spectrum.

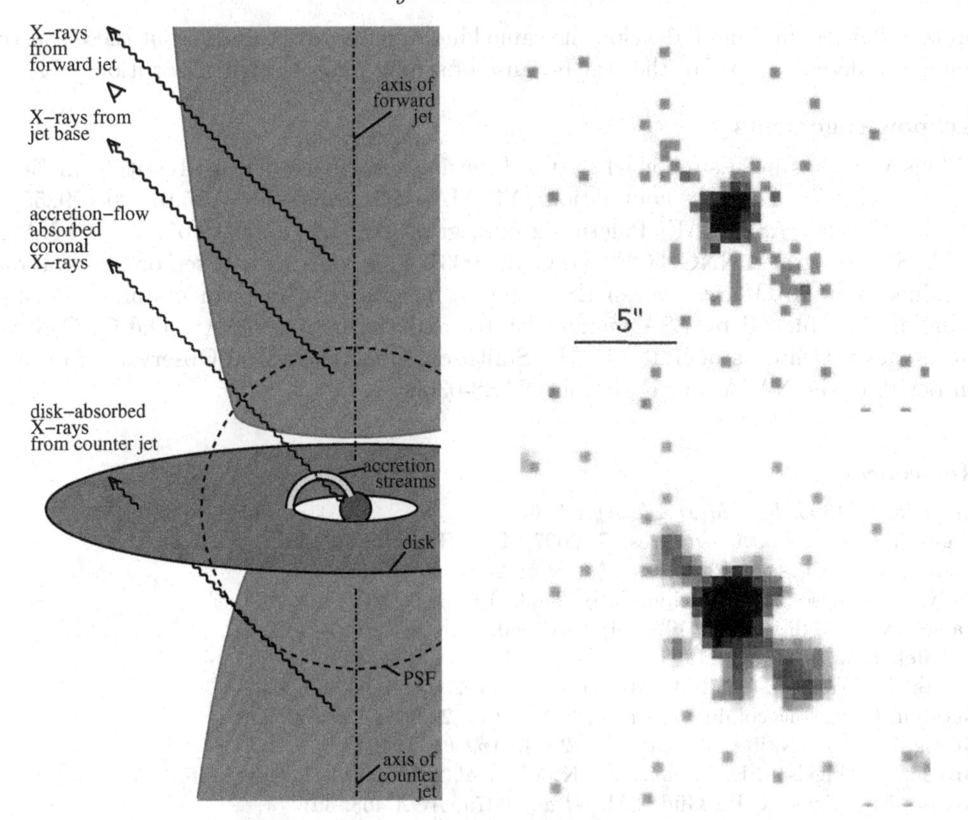

Figure 5. *Left (a):* Proposed model for the disk-jet interface with various X-ray sources, including: the spatially resolved forward jet; the spatially unresolved (within the stellar PSF) but spectrally resolved sources at the jet base; the accretion-stream absorbed corona/magnetosphere; and the disk-absorbed counter-jet sources. – *Right (b)*: Two *Chandra* ACIS-S 0.6–1.7 keV images of DG Tau and its jets; the lower figure has been smoothed (after Güdel *et al.* 2007, submitted).

3.3. *Summary on jets*

DG Tau is the prototype of a new class of jet-driving X-ray sources. It hosts at least four X-ray sources of different origin and subject to different $N_{\rm H}$ (see Fig. 5a), namely:

1. a weakly absorbed, diffuse, soft component along the forward-jet axis;
2. an intermediately absorbed, diffuse, soft component along the counter-jet axis;
3. a weakly absorbed, compact, non-variable, soft component (within the stellar PSF);
4. a strongly absorbed, compact, flaring, hard component (within the stellar PSF).

The low $N_{\rm H}$ of the *forward jet* and of the *soft stellar* component suggests that the soft spectral emission originates from a region "in front" of the star. We identify the soft component with X-ray emission from the base of the jets. In contrast, the harder *counter jet* suggests stronger absorption by the extended gas disk. A determination of the gas-to-dust ratio is in principle possible by measuring the differential absorption and extinction of the two jets. Finally, the *hard stellar* component is attributed to a (flaring) corona, with the excess photoelectric absorption due to dust-depleted accretion gas streams.

The combined power of the resolved jets and the unresolved soft spectral component is of order 10^{29} erg s^{-1}, similar to the X-ray output of a moderate T Tauri star. This emission, distributed above the accretion disk, may be an important contributor to X-ray heating and ionization of gaseous disk surfaces (Glassgold *et al.* 2004). We speculate that protostellar jets in general develop the same kind of jet X-ray emission, but these sources remain undetected close to the star because of strong photoelectric absorption.

Acknowledgements

This work has been supported by the International Space Science Institute in Bern, the Swiss National Science Foundation (AT, MA, MG: grants 20-66875.01, 20-109255/1, PP002–110504), ASI/INAF (Palermo group, grant ASI-INAF I/023/05/0), and NASA (MA, SS, DP: grants NNG05GF92G, GO6-7003). This research is based on observations obtained with *XMM-Newton*, an ESA science mission with instruments and contributions directly funded by ESA member states and the USA (NASA). The CXC X-ray Observatory Center is operated by the Smithsonian Astrophysical Observatory for and on behalf of the NASA under contract NAS8-03060.

References

Anglada, G. 1995, *Rev. Mexicana AyA* 1, 67
Argiroffi, C., Maggio, A., & Peres, G. 2007, *A&A* (Letters) 465, L5
Audard, M., Briggs, K. R., Grosso, N., *et al.* 2007, *A&A* 468, 379
Bally, J., Feigelson, E., & Reipurth, B. 2003, *ApJ* 584, 843
Calvet, N., & Gullbring, E. 1998, *ApJ* 509, 802
Eislöffel, J., & Mundt, R. 1998, *AJ* 115, 1554
Favata, F., Fridlund, C. V. M., Micela, G., *et al.* 2002, *A&A* 386, 204
Getman, K. V., Flaccomio, E., Broos, P. S., *et al.* 2005, *ApJS* 160, 319
Glassgold, A. E., Najita, J., & Igea, J. 2004, *ApJ* 615, 972
Grosso, N., Feigelson, E. D., Getman, K. V., *et al.* 2006, *A&A* (Letters) 448, L29
Grosso, N., Briggs, K. R., Güdel, M., *et al.* 2007a, *A&A* 468, 391
Grosso, N., Audard, M., Bouvier, J. *et al.* 2007b, *A&A* 468, 557
Güdel, M., Skinner, S. L., Briggs, K. R., *et al.* 2005, *ApJ* (Letters) 626, L53
Güdel, M., Briggs, K. R., Arzner, K., *et al.* 2007a, *A&A* 468, 353
Güdel, M., Telleschi, A., Audard, M., *et al.* 2007b, *A&A* 468, 515
Güdel, M., Skinner, S. L., Mel'nikov, S. Yu., *et al.* 2007c, *A&A* 468, 529
Günther, H. M., Liefke, C., & Schmitt, J. H. M. M., 2006, *A&A* (Letters) 459, L29
Hirth, G. A., Mundt, R., & Solf, J. 1997, *A&AS* 126, 437
Kastner, J. H., Huenemoerder, D. P., Schulz, N. S., *et al.* 2002, *ApJ* 567, 434
Kastner, J. H., Franz, G., Grosso, N., *et al.* 2005, *ApJS* 160, 511
Lamzin, S. A. 1999, *Astron. Lett.* 25, 430
Ness, J.-U., Güdel, M., Schmitt, J. H. M. M., *et al.* 2004, *A&A* 427, 667
Neuhäuser, R., Sterzik, M. F., Schmitt, J. H. M. M., *et al.* 1995, *A&A* 297, 391
Porquet, D., Mewe, R., Dubau, J., *et al.* 2001, *A&A* 376, 1113
Pravdo, S. H., Feigelson, E. D., Garmire, G., *et al.* 2001, *Nature* 413, 708
Pravdo, S. H., Tsuboi, Y., & Maeda, Y. 2004, *ApJ* 605, 259
Pravdo, S. H., & Tsuboi, Y. 2005, *ApJ* 626, 272
Preibisch, T., Kim, Y.-C., Favata, F., *et al.* 2005, *ApJS* 160, 401
Raga, A. C., Noriega-Crespo, A., & Velázquez, P. 2002, *ApJ* (Letters) 576, L149
Robrade, J., & Schmitt, J. H. M. M. 2006, *A&A* 449, 737
Skinner, S. L., Briggs, K. R., & Güdel, M. 2006, *ApJ* 643, 995
Stelzer, B., & Schmitt, J. H. M. M. 2004, *A&A* 418, 687
Strom, K. M., & Strom, S. E. 1994, *ApJ* 424, 237
Telleschi, A., Güdel, M., Briggs, K. R., *et al.* 2005, *A&A* 622, 653
Telleschi, A., Güdel, M., Briggs, K. R., *et al.* 2007a, *A&A* 468, 425
Telleschi, A., Güdel, M., Briggs, K. R., *et al.* 2007b, *A&A* 468, 541
Telleschi, A., Güdel, M., Briggs, K. R., *et al.* 2007c, *A&A* 468, 443
Tsujimoto, M., Koyama, K., Kobayashi, N., *et al.* 2004, *PASJ* 56, 341

Star-Disk Interaction in Young Stars
Proceedings IAU Symposium No. 243, 2007
J. Bouvier & I. Appenzeller, eds.

doi:10.1017/S1743921307009520

Why are accreting T Tauri stars less luminous in X-rays than non-accretors?

S. G. Gregory[1], K. Wood[1] and M. Jardine[1]

[1]SUPA, School of Physics and Astronomy, University of St Andrews,
St Andrews, KY16 9SS, UK
email: sg64@st-andrews.ac.uk

Abstract. Accreting T Tauri stars are observed to be less luminous in X-rays than non-accretors, an effect that has been detected in various star forming regions. To explain this we have combined, for the first time, a radiative transfer code with an accretion model that considers magnetic fields extrapolated from surface magnetograms obtained from Zeeman-Doppler imaging. Such fields consist of compact magnetic regions close to the stellar surface, with extended field lines interacting with the disk. We study the propagation of coronal X-rays through the magnetosphere and demonstrate that they are strongly absorbed by the dense gas in accretion columns.

Keywords. Radiative transfer, stars: coronae, stars: magnetic fields, stars: pre–main-sequence, stars: activity, stars: formation, X-rays: stars.

---

1. Introduction

Accreting T Tauri stars are observed to be less luminous in X-rays than non-accretors (Stelzer & Neuhäuser 2001; Flaccomio, Micela & Sciortino 2003a; Flaccomio, Damiani, Micela, *et al.* 2003c; Stassun, Ardila, Barsony, *et al.* 2004; Preibisch, Kim, Favata, *et al.* 2005; Flaccomio, Micela & Sciortino 2006; Telleschi, Güdel, Briggs, *et al.* 2007a). Accreting stars appear to be a factor of ~ 2 less luminous, and show a larger variation in their X-ray activity compared to non-accreting stars (Preibisch *et al.* 2005). However, it is only in recent years that this result has become clear, with previous studies showing conflicting results (e.g. Feigelson, Gaffney, Garmire, *et al.* 2003; Flaccomio, Damiani, Micela, *et al.* 2003b). The apparent discrepancy arose from whether stars were classified as accreting based on the detection of excess IR emission (a disk indicator) or the detection of accretion related emission lines. However, with careful re-analysis of archival data (Flaccomio *et al.* 2003a) and recent large X-ray surveys like the *Chandra* Orion Ultradeep Project (COUP; Getman, Flaccomio, Broos, *et al.* 2005) and the XMM-*Newton* Extended Survey of the Taurus Molecular Cloud (XEST; Güdel, Briggs, Arzner, *et al.* 2007a) the result is now clear, namely that accreting T Tauri stars are observed to be, on average, less luminous in X-rays than non-accreting stars. Although the difference is small it has been found consistently in various star forming regions: Taurus-Auriga (Stelzer & Neuhäuser 2001; Telleschi *et al.* 2007a), the Orion Nebula Cluster (Flaccomio *et al.* 2003c; Stassun *et al.* 2004; Preibisch *et al.* 2005), NGC 2264 (Flaccomio *et al.* 2003a, 2006) and Chamaeleon I (Flaccomio *et al.* 2003a).

It should be noted, however, that such observations from CCD detectors are not very sensitive to X-rays that are produced in accretion shocks. High resolution X-ray spectroscopic measurements have indicated emission from cool and high density plasma, most likely associated with accretion hot spots, in several (but not all) accreting stars (e.g.

Telleschi, Güdel, Briggs, *et al.* 2007b; Günther, Schmitt, Robrade, *et al.* 2007). In this work we only consider coronal X-ray emission such as is detected by CCD measurements.

It is not yet understood why accreting stars are under luminous in X-rays, although a few ideas have been put forward. It may be related to higher extinction due to X-ray absorption by circumstellar disks, however the COUP results do not support this suggestion (Preibisch *et al.* 2005). It may be related to magnetic braking, whereby the interaction between the magnetic field of an accreting star with its disk slows the stellar rotation rate leading to a weaker dynamo action and therefore less X-ray emission; although the lack of any rotation-activity relation for T Tauri stars has ruled out this idea (Flaccomio *et al.* 2003c; Preibisch *et al.* 2005; Briggs, Güdel, Telleschi, *et al.* 2007). A third suggestion is that accretion may alter the stellar structure affecting the magnetic field generation process and therefore X-ray emission (Preibisch *et al.* 2005). However, the most plausible suggestion is the attenuation of coronal X-rays by the dense gas in accretion columns (Flaccomio *et al.* 2003c; Stassun *et al.* 2004; Preibisch *et al.* 2005; Güdel, Telleschi, Audard, *et al.* 2007b). X-rays from the underlying corona may not be able to heat the material within accretion columns to a high enough temperature to emit in X-rays. Field lines which have been mass-loaded with dense disk material may obscure the line-of-sight to the star at some rotation phases, reducing the observed X-ray emission. Here we demonstrate this in a quantitative way by developing an accretion flow model and simulating the propagation of coronal X-rays through the stellar magnetosphere.

2. Realistic magnetic fields

In order to model the coronae of T Tauri stars we need to assume something about the form of the magnetic field. Observations suggest it is compact and inhomogeneous and may vary not only with time on each star, but also from one star to the next. To capture this behaviour, we use as examples the field structures of two different main sequence stars, LQ Hya and AB Dor determined from Zeeman-Doppler imaging (Donati, Cameron, Semel, *et al.* 2003). Although we cannot be certain whether or not the magnetic field structures extrapolated from surface magnetograms of young main sequence stars do represent the magnetically confined coronae of T Tauri stars, they do satisfy the currently available observational constraints. In future it will be possible to use real T Tauri magnetograms derived from Zeeman-Doppler images obtained using the ESPaDOnS instrument at the Canada-France-Hawaii telescope (Donati, Jardine, Gregory, *et al.* 2007). However, in the meantime, the example field geometries used in this work (see Fig. 1) capture the essential features of T Tauri coronae. They reproduce X-ray emission measures (EMs) and coronal densities which are typical of T Tauri stars (Jardine, Cameron, Donati, *et al.* 2006). The surface field structures are complex, consistent with polarisation measurements (Valenti & Johns-Krull 2004) and X-ray emitting plasma is confined within unevenly distributed magnetic structures close to the stellar surface, giving rise to significant rotational modulation of X-ray emission (Gregory, Jardine, Cameron, *et al.* 2006b).

2.1. *The coronal field*

For a given surface magnetogram we calculate the extent of the closed corona for a specified set of stellar parameters. We extrapolate from surface magnetograms by assuming that the magnetic field $\boldsymbol{B}$ is potential such that $\nabla \times \boldsymbol{B} = 0$. This process is described in detail by Jardine *et al.* (2006), Gregory, Jardine, Simpson, *et al.* (2006a) and Gregory *et al.* (2006b). We assume that the corona is isothermal and that plasma along field line loops is in hydrostatic equilibrium. The pressure is calculated along the path of field line loops and is set to zero for open field lines and for field lines where, at some point along the loop, the gas pressure exceeds the magnetic pressure. The pressure along a field line scales with the pressure at its foot point, and we assume that this scales with the magnetic pressure. This technique has been used successfully to calculate mean coronal densities and X-ray EMs for the Sun and other main sequence stars (Jardine, Wood, Cameron, *et al.* 2002) as well as T Tauri stars (Jardine *et al.* 2006). The AB Dor-like coronal field has an X-ray EM† of $\log \mathrm{EM} = 53.73\,\mathrm{cm}^{-3}$ (without considering accretion) and a mean EM-weighted coronal density of $\log \bar{n} = 10.57\,\mathrm{cm}^{-3}$, consistent with estimates from the modelling of individual flares (Favata, Flaccomio, Reale, *et al.* 2005). The LQ Hya-like field has a more extended corona and consequently a lower coronal density and EM, $\log \mathrm{EM} = 52.61\,\mathrm{cm}^{-3}, \log \bar{n} = 9.79\,\mathrm{cm}^{-3}$.

2.2. *The accreting field*

We assume that the structure of the magnetic field remains undistorted by the in-falling material and that the magnetosphere rotates as a solid body. The accreting field geometries shown in Fig. 1 are therefore only snap-shots in time, and in reality will evolve due to the interaction with the disk. The question of where the disk is truncated remains a major problem for accretion models. It is still unknown if the disk is truncated in the vicinity of the corotation radius, the assumption of traditional accretion models (e.g. Königl 1991), or whether it extends closer to the stellar surface (e.g. Matt & Pudritz 2005). In this work we assume that accretion occurs over a range of radii within the corotation radius. This is equivalent to the approach taken previously by e.g. Muzerolle, Calvet & Hartmann (2001) who have demonstrated that such an assumption reproduces observed spectral line profiles and variability. The accretion filling factors are of order 1%, consistent with observationally inferred values (e.g. Valenti & Johns-Krull 2004).

We assume that material is supplied by the disk and accretes onto the star at a constant rate. For a dipolar magnetic field accretion flows impact the stellar surface in two rings in opposite hemispheres centred on the poles. In this case, half of the mass supplied by the disk accretes into each hemisphere. For complex magnetic fields accretion occurs into discrete hot spots distributed in latitude and longitude (Gregory *et al.* 2006a). It is therefore not clear how much of the available mass from the disk accretes into each hot spot. We use a spherical grid and assume that each grid cell within the disk which is accreting supplies a mass accretion rate that is proportional to its surface area. If an accreting grid cell has a surface area that is 2% of the total area of all accreting grid cells, then this grid cell is assumed to carry 2% of the total mass that is supplied by the disk. Therefore, as an example, if grid cells which constitute half of the total area of all accreting cells in the disk carry material into a single hot spot, then half of the mass accretion rate is carried from the disk to this hot spot. In this way the accretion rate into each hot spot is different and depends on the structure of the magnetic field connecting the star to the disk.

2.3. *Accretion flow model*

We consider a star of mass $0.5\,\mathrm{M}_\odot$, radius $2\,\mathrm{R}_\odot$, rotation period 6 d, a coronal temperature of 20 MK and assume that the disk supplies a mass accretion rate of $10^{-7}\,\mathrm{M}_\odot\,\mathrm{yr}^{-1}$. In order to model the propagation of coronal X-rays through the magnetosphere we first need to determine the density of gas within accretion columns. Gregory, Wood & Jardine (2007) develop a steady state accretion flow model where material accretes

† The X-ray EM is given by $\mathrm{EM} = \int n^2 dV$ where n and V are the coronal density and volume. The EM-weighted density is $\bar{n} = \int n^3 dV / \int n^2 dV$.

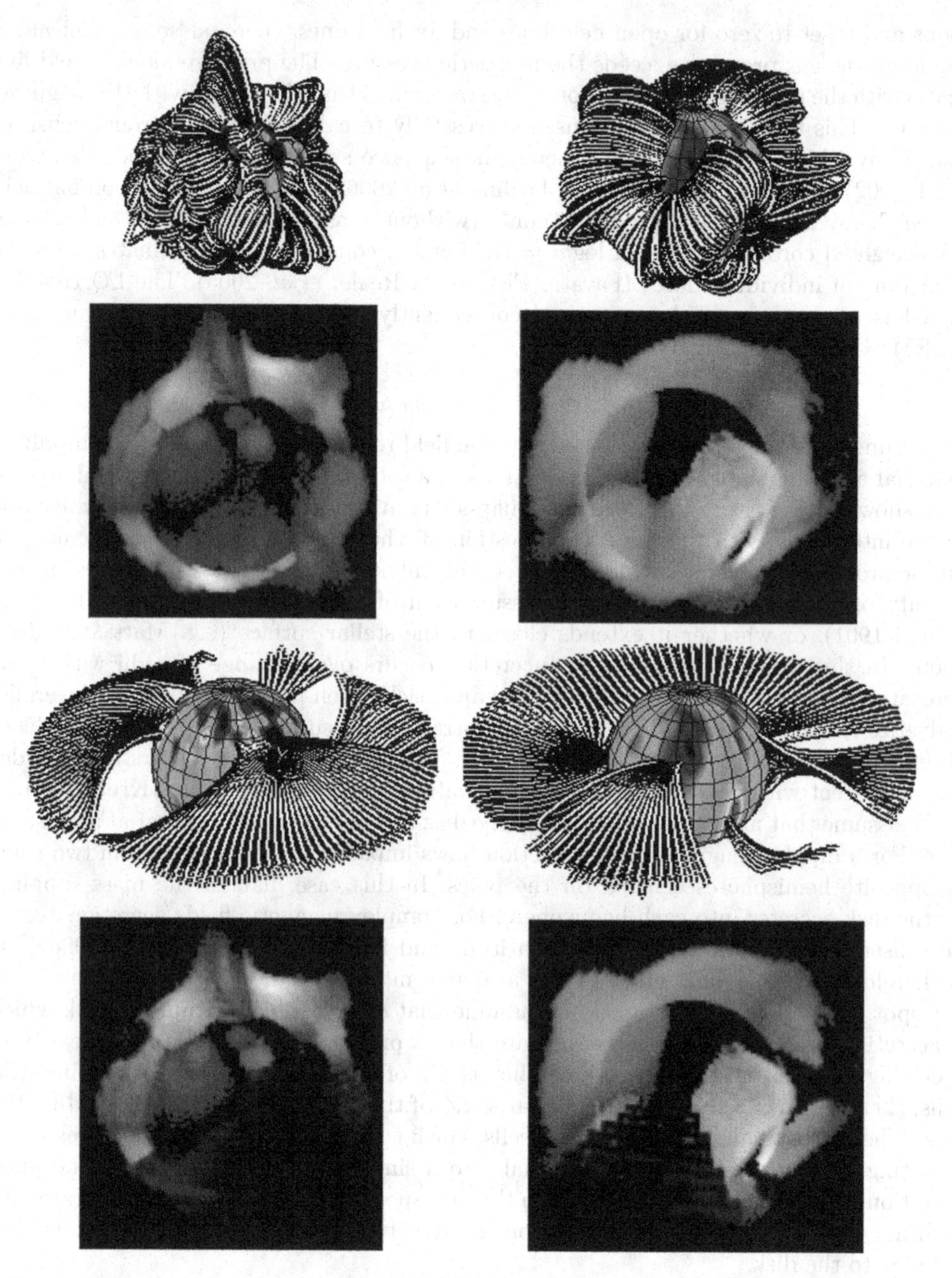

Figure 1. The model coronal (first row) T Tauri magnetic fields extrapolated from the AB Dor (left-hand column) and LQ Hya (right-hand column) surface magnetograms, with the corresponding X-ray corona (second row) assuming a stellar inclination of 60 °. Also shown is the structure of the accreting field (third row) and the X-ray emission images assuming that accretion is taking place (fourth row) - notice the occulted X-ray bright regions. For some lines-of-sight the X-ray bright regions are lightly obscured by the accretion columns, reducing the observed X-ray emission. For other lines-of-sight the coronal X-rays cannot penetrate the dense accreting gas. The average reduction in the observed X-ray EM across an entire rotation cycle is a factor of 1.4 (2.0) for the AB Dor-like (LQ Hya-like) field. The images are not to scale. Reproduced from Gregory *et al.* (2007).

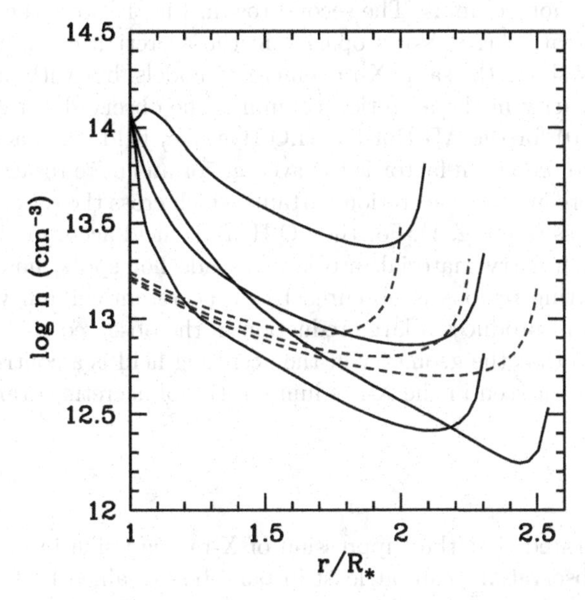

Figure 2. Some example density profiles (solid lines) for accretion along a small selection of the complex field lines shown in Fig. 1 (first column, third row) assuming a mass accretion rate of $10^{-7}\ \mathrm{M_\odot\,yr^{-1}}$. Also shown for comparison are the density profiles for accretion along dipolar field lines (dashed lines). r is the spherical radius. Reproduced from Gregory *et al.* (2007).

from a range of radii within corotation, free-falling along the field lines under gravity. The resulting density profiles do not depend on the absolute field strength, but instead on how the field strength varies with height above the star. The density profiles are typically steeper than those derived for accretion flows along dipolar field lines since the strength of a higher order field drops faster with height above the star. Fig. 2 shows the variation of the number density along the paths of a selection of accreting field lines, with those obtained for dipolar field lines shown for comparison (Gregory *et al.* 2007). For our assumed accretion rate of $10^{-7}\ \mathrm{M_\odot\,yr^{-1}}$ the flow densities range from $\log n \approx 12 - 14\,\mathrm{cm^{-3}}$, whilst for a lower accretion rate of $10^{-8}\ \mathrm{M_\odot\,yr^{-1}}$ the range is $\log n \approx 11 - 13\,\mathrm{cm^{-3}}$.

3. Simulated X-ray variability

We model the propagation of coronal X-rays through the magnetosphere by considering absorption by the dense gas within accretion columns. For the radiation transfer we use Monte Carlo techniques and discretise the emissivity and density onto a spherical polar grid (e.g. Jardine *et al.* 2002; Whitney, Wood, Bjorkman, *et al.* 2003). The stellar inclination is set to $i = 60^\circ$ and we assume the X-ray emission from the 20 MK corona is optically thin, but that the X-rays may be subsequently absorbed in the cool and hence optically thick accretion columns. For the X-ray absorptive opacity we adopt a value of $\sigma = 10^{-22}\ \mathrm{cm^2\,H^{-1}}$, typical of neutral gas at temperatures below 10^4 K at X-ray energies of a few keV (e.g. Krolik & Kallman 1984). At these energies the opacity of hot gas (above 10^7 K) is several orders of magnitude lower (e.g. Krolik & Kallman 1984, their Fig. 1) justifying our assumption that the coronal X-ray emission is optically thin.

In the Monte Carlo X-ray radiation transfer simulations we assume the scattering opacity is negligible, so our results in Fig. 1 show the effects of attenuation of the coronal emission by the accretion columns. The second row in Fig. 1 shows the X-ray images in the absence of attenuation (i.e. X-ray opacity in the accretion columns is set to zero) whilst the fourth row shows the same X-ray emission models, but with our adopted value for the soft X-ray opacity in the accretion columns. The observed X-ray EM is reduced by a factor of 1.4 (2.0) for the AB Dor-like (LQ Hya-like) field when accretion flows are considered, where the reduction factor is the average for an entire rotation cycle. For the AB Dor-like field there are large accretion curtains which cross the observers line-of-sight to the star as it rotates (see Fig. 1). For the LQ Hya-like field accretion is predominantly along field lines which carry material into low latitude hot spots, however, one of the brightest X-ray emitting regions is obscured by an accretion column which attenuates the coronal X-rays and produces a large reduction in the observed X-ray emission. This immediately suggests that the geometry of the accreting field is a contributory factor in causing the large scatter seen in the X-ray luminosities of accreting stars.

4. Summary

We have demonstrated that the suppression of X-ray emission in accreting stars apparent from CCD observations can, at least in part, be explained by the attenuation of coronal X-rays by the dense material in accretion columns. This suggests that both accreting and non-accreting stars have the same intrinsic X-ray luminosity, with accreting T Tauri stars being observed to be less luminous due to the effects of absorbing gas in accretion columns. The reduction in the observed X-ray emission depends on the structure of the accreting field. For stars where accretion columns rotate across the line-of-sight, X-rays from the underlying corona are strongly absorbed by the accreting gas which reduces the observed X-ray emission. A preliminary calculation indicates that the column densities from our simulations are large enough that the softer (cooler) coronal spectral components may be substantially, if not completely, absorbed by the accreting gas. The effect is greater the larger the accretion rate. Indeed Güdel *et al.* (2007b) have recently found that stars in XEST which have the largest accretion rates (and are driving jets) show a highly absorbed coronal spectral component, which is attributed to attenuation by accreting gas. This however does not rule out the fact that other mechanisms may also be responsible for reducing the X-ray emission in accreting stars. Jardine *et al.* (2006) have demonstrated that some stars (typically those of lower mass) have their outer coronae stripped away via the interaction with a disk. This also reduces the observed X-ray emission and this effect, combined with the radiative transfer calculations presented here, is likely to lead to a larger reduction in the observed X-ray emission. This would reduce the number of field lines which could be filled with coronal gas, such as is also suggested by Preibisch *et al.* (2005) and Telleschi *et al.* (2007a), with the observed X-ray emission being further reduced due to obscuration by the accreting gas.

References

Briggs, K.R., Güdel, M., Telleschi, A., Preibisch, T., Stelzer, B., Bouvier, J., Rebull, L., Audard, M., Scelsi, L., Micela, G., Grosso, N., & Palla, F. 2007, *A&A* 468, 413

Donati, J.-F., Jardine, M., Gregory, S.G., Petit, P., Bouvier, J., Dougados, C., Ménard, F., Cameron, A.C., Harries, T.J., Jeffers, S.V., & Paletou, F. 2007, *MNRAS* submitted

Donati, J.-F., Cameron, A.C., Semel, M., Hussain, G.A.J., Petit, P., Carter, B.D., Marsden, S.C., Mengel, M., López Ariste, A., Jeffers, S.V., & Rees, D.E. 2003, *MNRAS* 345, 1145

Favata, F., Flaccomio, E., Reale, F., Micela, G., Sciortino, S., Shang, H., Stassun, K.G., & Feigelson, E.D. 2005, *ApJS* 160, 469

Feigelson, E.D., Gaffney, J.A., III, Garmire, G., Hillenbrand, L.A., & Townsley, L. 2003, *ApJ* 584, 911

Flaccomio, E., Damiani, F., Micela, G., Sciortino, S., Harnden, F.R., Jr., Murray, S.S., & Wolk, S.J. 2003b, *ApJ* 582, 382

Flaccomio, E., Damiani, F., Micela, G., Sciortino, S., Harnden, F.R., Jr., Murray, S.S., & Wolk, S.J. 2003c, *ApJ* 582, 398

Flaccomio, E., Micela, G. & Sciortino, S. 2003a, *A&A* 397, 611

Flaccomio, E., Micela, G. & Sciortino, S. 2006, *A&A* 455, 903

Getman, K.V., Flaccomio, E., Broos, P.S., Grosso, N., Tsujimoto, M., Townsley, L., Garmire, G.P., Kastner, J., Li, J., Harnden, F.R., Jr., Wolk, S., Murray, S.S., Lada, C.J., Muench, A.A., McCaughrean, M.J., Meeus, G., Damiani, F., Micela, G., Sciortino, S., Bally, J., Hillenbrand, L.A., Herbst, W., Preibisch, T., & Feigelson, E.D. 2005, *ApJS* 160, 319

Gregory, S.G., Jardine, M., Cameron, A.C., & Donati, J.-F. 2006b, *MNRAS* 373, 827

Gregory, S.G., Jardine, M., Simpson, I., & Donati, J.-F. 2006a, *MNRAS* 371, 999

Gregory, S.G., Wood, K. & Jardine, M. 2007, *MNRAS* in press (astro-ph/0704.2958)

Güdel, M., Briggs, K.R., Arzner, K., Audard, M., Bouvier, J., Feigelson, E.D., Franciosini, E., Glauser, A., Grosso, N., Micela, G., Monin, J.-L., Montmerle, T., Padgett, D.L., Palla, F., Pillitteri, I., Rebull, L., Scelsi, L., Silva, B., Skinner, S.L., Stelzer, B., & Telleschi, A. 2007a, *A&A* 468, 353

Güdel, M., Telleschi, A., Audard, M., Skinner, S.L., Briggs, K.R., Palla, F., & Dougados, C. 2007b, *A&A* 468, 515

Günther, H.M., Schmitt, J.H.M.M., Robrade, J., & Liefke, C. 2007, *A&A* 466, 1111

Jardine, M., Cameron, A.C., Donati, J.-F., Gregory, S.G., & Wood, K. 2006, *MNRAS* 367, 917

Jardine, M., Wood, K., Cameron, A.C., Donati, J.-F., & Mackay, D.H. 2002, *MNRAS* 336, 1364

Königl, A. 1991, *ApJ* (Letters) 370, L39

Krolik, J. H. & Kallman, T.R. 1984, *ApJ* 286, 366

Matt, S. & Pudritz, R.E. 2005, *MNRAS* 356, 167

Muzerolle, J., Calvet, N. & Hartmann, L. 2001, *ApJ* 550, 944

Preibisch, T., Kim, Y.-C., Favata, F., Feigelson, E.D., Flaccomio, E., Getman, K., Micela, G., Sciortino, S., Stassun, K., Stelzer, B., & Zinnecker, H. 2005, *ApJS* 160, 401

Stassun, K.G., Ardila, D.R., Barsony, M., Basri, G., & Mathieu, R.D. 2004, *AJ* 127, 3537

Stelzer, B. & Neuhäuser, R. 2001, *A&A* 377, 538

Telleschi, A., Güdel, M., Briggs, K.R., Audard, M., & Palla, F. 2007a, *A&A* 468, 425

Telleschi, A., Güdel, M., Briggs, K.R., Audard, M., & Scelsi, L. 2007b, *A&A* 468, 443

Valenti, J.A. & Johns-Krull, C.M. 2004, *Ap&SS* 292, 619

Whitney, B.A., Wood, K., Bjorkman, J.E. & Wolff, M.J. 2003, *ApJ* 591, 1049

Discussion

ARDILA: Is absorption not corrected for when calculating X-ray luminosities?

GREGORY: The attenuation of X-rays is by gas in accretion columns, not by dust. Thus the absorption is larger than would be calculated from say the optical extinction. Indeed there is already evidence that for some stars the gas-to-dust ratio is larger than what is normally assumed, leading to a heavily absorbed coronal spectral component (for example, some stars in the XEST project). Although with the caveat that such stars have some of the largest inferred accretion rates, and therefore we may expect more X-ray attenuation by denser accretion columns. We are currently working on this with Ettore Flaccomio.

ARDILA: So do you think the difference in the observed X-ray luminosities would disappear if X-ray attenuation is accounted for properly?

GREGORY: Yes, I believe so.

KASTNER: I think the reason that you're getting disbelieving comments is that I'm unsure why absorption should modify the X-ray luminosity since it should be accounted for already.

FLACCOMIO: The derived column densities in the simulations can be high, which suggests that in accreting stars there may be a cool component that is completely or substantially absorbed and so is not detected in the spectrum.

JOHNS-KRULL: Would accretion columns rotating across the line-of-sight produce detectable sharp drops in X-ray light curves?

GREGORY: Modulation due to bright regions entering eclipse produces a much smoother variation with rotation phase than that due to accretion columns rotating across the line-of-sight. However, the problem with testing that is you require X-ray observations that span at least a couple of stellar rotation periods, which are difficult to get observing time for.

JOHNS-KRULL: But can't you use the COUP dataset for that?

GREGORY: Yes, although in the COUP paper on rotational modulation of X-ray emission they looked to see if the modulation occurred preferentially in accreting or non-accreting stars. However, they could not say anything conclusive as most stars are too heavily absorbed to have been studied spectroscopically from which their accretion status could have been determined.

BOUVIER: AA Tau may be an exception to your model. We find that the accretion hot spot, the accretion column and the disk warp exist at the same rotation phase, but if you look at the poster by Grosso, during an eclipse by the disk warp we saw an increase in X-ray emission.

GREGORY: Perhaps AA Tau is an exception, or perhaps the increase in X-ray emission is accretion related rather than coronal in origin. I need to think about AA Tau in more detail.

MATT: If half of the X-ray luminosity goes into heating the accretion columns then you may expect a correlation between X-ray luminosity and the flux in lines which form in the accretion columns. Has anyone looked for this?

GREGORY: I'm not aware that anyone has looked for that.

STELZER: The reduction factor was the average for a complete rotation cycle, so it can be higher over a smaller rotation phase?

GREGORY: Yes, it can be higher, or less, depending on the field geometry and the portion of the rotation cycle observed.

Star-Disk Interaction in Young Stars
Proceedings IAU Symposium No. 243, 2007
J. Bouvier & I. Appenzeller, eds.

doi:10.1017/S1743921307009532

Spectroscopic diagnostics of T Tauri inner winds

Suzan Edwards

Department of Astronomy, Smith College, Northampton, MA 01063, USA
email: sedwards@smith.edu

Abstract. The role of the star-disk interaction region in launching the high velocity component of accretion-driven outflows is examined. Spectroscopic indicators of high velocity inner winds have been recognized in T Tauri stars for decades, but identifying the wind launch site and the accompanying mass loss rates has remained elusive. A promising new diagnostic is He I λ10830, whose metastable lower level results in a powerful probe of the geometry of the outflowing gas in the interaction region. This, together with other atomic and molecular spectral diagnostics covering a wide range of excitation and ionization states, suggests that more than one launch site of the innermost wind is operational in most accreting stars.

Keywords. Stars:pre–main-sequence, mass loss, winds, outflows.

1. Historical perspective

The presence of high velocity winds from T Tauri stars, heralded by blueshifted absorption features cutting into strong and broad permitted emission lines, has been recognized for nearly half a century. A series of increasingly sophisticated attempts to determine mass loss rates from T Tauri stars over a period of several decades assumed that both the emission and absorption components of Hα were formed in spherically symmetric stellar winds (Kuhi 1964; Hartmann, Edwards & Avrett 1982; Natta, Giovanardi & Palla 1988; Hartmann *et al.* 1990). However, several issues were perplexing. One was the difficulty in powering energetic winds, since thermal coronal stellar winds were shown by de Campli (1981) to have a firm upper limit of 10^{-9} M$_\odot$ yr^{-1}, while mass loss rates from strong emission T Tauri stars could exceed $\dot{M}_w \sim 10^{-7}$ M$_\odot$ yr^{-1}. Another was that the line profiles predicted for stellar winds had a classic P Cygni character, unlike the majority of T Tauri Hα profiles where the blue absorption was typically at velocities considerably less than the broad emission wings and rarely penetrated the continuum.

The concept of a T Tauri stellar wind crumbled when it became clear that disk accretion was the energy source for T Tauri activity. Magnetospheric accretion, where funnel flows channel matter from the disk truncation radius to the star, became the favored source of line *emission* (Basri & Bertout 1989; Hartmann, Hewett & Calvet 1994). Blueshifted forbidden lines were recognized as superior diagnostics of mass outflows, leading to a decline of interest in the blueshifted absorption features in strong permitted lines. Forbidden lines became the basis for establishing what is now known as the accretion/outflow connection (Cabrit *et al.* 1990; Hartigan, Edwards & Ghandour 1995), and centrifugally powered disk winds were soon considered the likely source of T Tauri outflows. In recent years, however, the possibility that accretion-powered stellar winds may be significant contributors to mass ejection in accreting systems is now being re-examined. In this chapter we review the observational diagnostics of the inner regions of accretion-powered winds with the aim of clarifying where and how they originate.

2. Outflows from accreting protostars and T Tauri stars

A symbiotic relation between accretion and outflow persists from the youngest, most deeply embedded protostars through to the final stages of disk accretion when the central star is optically revealed as a classical T Tauri star. At spatial scales from tens of AU to several parsecs from the star, collimated jets of shocked gas and expanding lobes of swept up molecular gas provide the means of diagnosing outflow energies, momenta, and mass loss rates (Bally *et al.* 2007). The resultant correlations between diagnostics of the wind mass loss rate $\dot{M}_w$ and the disk accretion rate $\dot{M}_{acc}$ clarify that the outflows are accretion powered (see chapter by S. Cabrit), although how they are launched remains a mystery. Accretion powered outflows are suspected to play a major role in the angular momentum evolution of an accretion disk system, likely extracting angular momentum from the accreting star and/or the accretion disk (chapters by J. Bouvier, F. Shu, J. Ferreira). They also have the potential to disrupt infalling cores, thus limiting the mass of the forming star, and to affect cloud turbulence, especially on local scales (Nakamura & Li 2007).

The basic energy source for the outflows is thought to be magnetohydrodynamic in origin, with launch occurring in a region where open magnetic field lines anchor to a rotating object and collimation is provided by hoop stresses from the toroidal field, focusing the flow toward the rotation axis (Ferreira *et al.* 2006). Three basic steady state MHD ejection scenarios are under consideration, each with different implications for how angular momentum will be extracted from the star and disk. A widely explored option assumes that the inner disk has a sufficient magnetic field and ionization fraction to launch centrifugal winds over a range of disk radii from the inner truncation radius out to several AU (Pudritz *et al.* 2007). Another strong contender is a modified disk wind restricted to a narrow region near corotation where centrifugal launching is enhanced by a hijacked stellar field, providing strong magnetic channeling from an "X" point (Shu *et al.* 2000).

A third option is that some form of accretion-powered stellar wind is operating, where winds emerge along field lines anchored to the star (Kwan & Tademaru 1988; Hirose *et al.* 1997; Romanova *et al.* 2005). There are reasons to be skeptical of this option as an important source of mass loss, since X-ray fluxes are several orders of magnitude too small to provide launching via coronal thermal pressure (see chapter by S. Matt) and by the T Tauri phase the stars are spinning too slowly for effective centrifugal launching (see chapter by J. Bouvier). However, both theoretical and observational considerations have recently surfaced that suggest stellar winds need to remain as a contender. On the theoretical side, doubts have been raised as to whether disk winds can brake accreting stars to their observed slow spin rates (von Rekowski & Brandenburg 2006; Matt *et al.* 2005) and on the observational side the resonance profiles of a new spectroscopic diagnostic, He I λ10830, appear to require acceleration in a flow moving radially away from T Tauri stars with high disk accretion rates (Edwards *et al.* 2003; Kwan, Edwards & Fischer 2007). If so, then a robust means of accelerating an MHD stellar wind would need to be identified, possibly relying on Alfven waves or magnetic reconnection.

Ultimately the verdict on how accretion powered winds are launched will be established empirically. Observations of the collimation and kinematic structure of spatially resolved jets within 10-100 AU of the star hold valuable clues to wind origins (see chapters by T. Ray and S. Cabrit for a full discussion of this topic). For example, the poloidal velocity field inferred from channel maps reconstructed from multiple long-slit HST spectra of the jet from the high accretion rate TTS DG Tau suggests that the highest velocity gas (several hundred km s^{-1}) is confined to the jet axis, sheathed by concentric rings of slower moving gas (Bacciotti *et al.* 2002). Such kinematic structure is suggestive of extended MHD disk winds, where ejection velocities will scale with disk radii in proportion to the the associated Keplerian velocity, yielding higher velocity flows from the inner disk and slower flows from more distant regions. However, a growing number of spatial-velocity maps from long slit spectra acquired with adaptive optics on large ground based telescopes suggest that two separate ejection processes may be operational (Pyo *et al.* 2003). High and low velocity components (HVC and LVC, respectively) with distinctive spatial characteristics are often seen on the smallest spatial scales, but there is no consensus on what their separate origins might be. Possibilities include attributing the HVC to a disk wind and the LVC to entrained gas or attributing the HVC to a wind from the disk X-point (or a stellar wind), and the LVC to an extended disk wind (Takami *et al.* 2006; Pyo *et al.* 2006).

Clearly, definitive observational evidence for wind origins requires probes that reach closer to the launch sites. Interferometers working in the near IR are beginning to give some information on disk structure in the inner AU (see chapter by F. Malbet), but for now high resolution spectroscopy is the prime means of probing the inner disk and the star/disk interaction region where the winds are actually launched. From high resolution line profiles of molecular and atomic features we hope to elucidate the origin of the high and low velocity flows seen in the jets, using profile kinematics and physical conditions for line formation to provide clues to how/where accretion powered winds arise.

3. Winds from the inner disk?

A number of ro-vibrational molecular emission lines are attributed to formation in the inner disk, including C0, H_20, OH, H_2, HCN, C_2H_2 (Najita *et al.* 2007). The highest detection frequency in surveys of accreting stars is for the 4.6μ CO fundamental, which displays broad, centrally peaked, symmetric emission profiles centered on the stellar velocity. The lines are well modeled as simple Keplerian rotation over a range of radii in the inner disk, from the inner truncation radius ($\leqslant$ 0.1 AU) out to 1-2 AU (Najita *et al.* 2003). This range of radii is precisely the regime where centrifugally launched disk winds are expected to originate. Thus, if disk winds are launched between 0.1 to 2 AU, the acceleration region must lie higher above the disk plane than the temperature inversion layer where the CO fundamental arises, characterized by excitation temperatures $\sim 1000K$ and CO column densities $\sim 10^{18}\,\mathrm{cm}^{-2}$. Profiles from the other ro-vibrational species are also well modeled with Keplerian rotation rather than outflow, although a few exceptions are known (see chapters by J. Carr and S. Brittain).

In contrast to the paucity of outflow signatures from infrared ro-vibrational lines from the inner disk, evidence for winds is seen in the ultraviolet electronic transitions of H_2, pumped by broad Lyα emission. In a STIS study of a small number of accreting T Tauri stars Herczeg *et al.* (2006) find some stars with FUV H_2 profiles centered on the stellar velocity, consistent with origin from the warm inner disk surface, while others have centroids blueshifted by 10-30 $\mathrm{km\,s^{-1}}$. These profiles are shown in Figure 1, where it can be seen that several of the stars also have blue emission wings up to $\sim$ -100 $\mathrm{km\,s^{-1}}$. Similarly, in a GHRS study with lower quality data, Ardila *et al.* (2002a) found that the mean H_2 velocity for accreting T Tauri stars was negative, suggesting a wind contribution was present among stars in that sample as well.

The utility of the FUV H_2 transitions as tracers of winds is not fully understood. They are sensitive to far smaller quantities of hydrogen than the H_2 ro-vibrational lines formed in the outer disk and they require temperatures prior to fluorescence of T$\sim$ 2500K. Herczeg argues that at least some of the blueshifted gas must be close to the

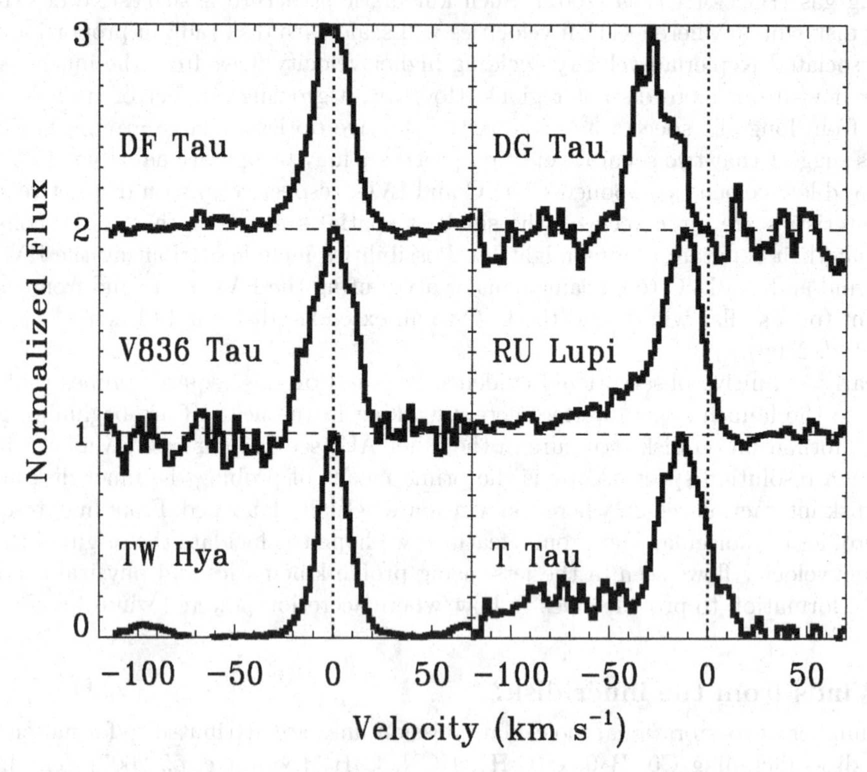

Figure 1. Summed FUV H_2 profiles for 6 accreting TTS, co-added from upper levels ($\nu = 0\ J = 1$), ($\nu = 0\ J = 2$), and ($\nu = 2\ J = 12$), adapted from Herczeg *et al.* (2006) by G. Herczeg. Stars in the left panel show profiles suggesting formation in the disk, stars in the right panel have blueshifted centroids and wings suggesting formation in a wind. Zero velocity corresponds to the rest velocity of the star.

accretion flow/inner wind in order for pumping to occur over the full range of the Lyα profile, as required by the observed transitions. While this cannot clarify precisely where the outflowing H_2 originates, the bulk of the emission is at velocities reminiscent of the low velocity components seen in spatially extended forbidden lines. Thus it is tempting to say that they trace the base of a disk wind, although why this would be seen in only some of the accreting stars is not clear.

4. High velocity winds from the star/disk interaction zone

The spectra of accreting T Tauri stars between 0.1-2μ are shaped by the flow of mass and the transfer of angular momentum in the complex region where the magnetic fields of the star and the disk encounter each other. Our uncertainty regarding the launching process for high velocity winds comes from the challenge of deciphering the bewildering array of profile morphologies displayed by lines formed with simultaneous contributions from the funnel flow, the accretion shock, the high velocity wind, and probably other activities as well. The strength and character of line profiles and the variety of observed excitation energies and ionization states are determined in large measure by the magnitude of the disk accretion rate. Temporal variations are also a factor, arising both from rotating non-axisymmetric magnetic structures and variations in the disk accretion rate, although the magnitude of the variations displayed by an individual star are typically small compared to those arising from factors of 100-1000 in accretion rate characterizing the T Tauri class (see chapter by S. Alencar). The empirical milestones in understanding T Tauri spectra have thus either been observational programs that characterize how stars behave over the full spectrum of accretion rates or those that devote intensive synoptic monitoring to one or a few stars, each yielding valuable but very different insight.

Blue absorption signifying high velocity winds has long been recognized in the strong emission lines of Hα, Na D, Ca II H&K, and MgII h&k (Mundt 1984), although it assumes a variety of shapes and it is not present in all stars. At least for Hα, both emission and absorption (when present) equivalent widths scale roughly with the mass accretion rate verifying these phenomena are accretion-powered (Alencar & Basri 2000). The higher Balmer lines rarely show absorption from the wind, but their emission breadths are comparable to Hα and they also often show redshifted absorption (Edwards *et al.* 1994). These profiles are usually attributed to formation in magnetospheric funnel flows (Muzerolle, Calvet & Hartmann 2001) with the exception of the blue absorption which is assumed to arise in a wind exterior to the accretion region. Simultaneous optical and ultraviolet spectra reveal that the general morphology of the NUV MgII h&k profiles is similar to Hα, both in the width of the emission and the velocity of the blueshifted absorption, although deeper absorptions in the Mg II lines indicate that this line samples larger volumes of the wind compared to Hα (Ardila *et al.* 2002b). Evidence that some of the hydrogen emission may arise in outflowing rather than infalling gas includes both spectrophotometry showing spatial extension in Paschen lines (Whelan *et al.* 2004) and large blue emission asymmetries in wings of various H lines (Folha & Emerson 2001; Beristain, Edwards & Kwan 2001, hereafter BEK).

The remaining metallic emission lines in the optical, which can be quite numerous in high accretion rate stars, almost never show absorption from a wind, but often have a two-component morphology to their emission, with a broad component that could arise in the funnel flow or a wind plus a narrow component that is likely from the accretion shock (Batalha *et al.* 1996; Beristain, Edwards & Kwan 1998). In contrast to the metallic lines in the optical, a STIS FUV study of the high accretion rate TTS RU Lup (Herczeg *et al.* 2005) revealed numerous neutral and singly ionized metallic lines with a clear P Cygni character, with blue absorption velocities from -70 to $-200\,\mathrm{km\,s^{-1}}$, in some cases penetrating the continuum. Ultraviolet lines with blue absorption from winds also include some with high excitation, up to at least CII, suggesting that some outflowing gas possesses temperatures of 10,000-30,000 K (Dupree *et al.* 2005; Johns-Krull & Herczeg 2007). In addition to P Cygni profiles, recent HST STIS+GHRS spectra of the accreting T Tauri star RY Tau shows blueshifted emission in semi-forbidden and forbidden lines of Si III], C III], and [O II] with small emitting volumes suggesting formation close to the star (Gomez de Castro & Verdugo 2007).

There is thus clear evidence for the presence of high velocity winds emerging from the star-disk interaction zone, although the less well studied ultraviolet appears to offer a richer variety of probes of the outflowing gas. The best studied optical lines can be complex and hybrid nature and difficult to interpret. Synoptic studies in a few low accretion rate stars suggest simple phased relations between infalling and outflowing gas (see chapter by S. Alencar), however among stars with high disk accretion rates synoptic data show chaotic behavior (Johns-Krull & Basri 1997) that provide no clue as to whether winds are launched from the inner disk, the X-point, or the star.

4.1. *Tracing the inner wind with helium lines*

A surprising new diagnostic of the inner wind region in accreting stars has the potential to break through the ambiguity in determining where winds are launched in the star/disk

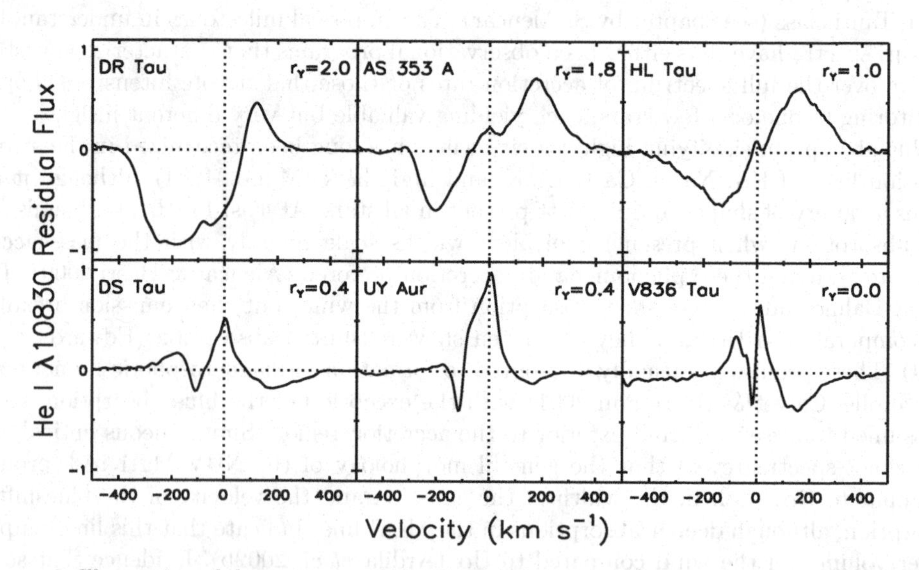

Figure 2. Illustration of the extremes of blueshifted absorption shown in He I λ10830 profiles of accreting TTS. The upper row shows P Cygni-like profiles with deep and broad blue absorption and the lower row shows examples of narrow blue absorption. The latter profiles also show red absorption from magnetospheric accretion columns. Simultaneous 1μ veiling, r_Y, is identified for each star.

interaction region. This is He I λ10830, falling in a wavelength regime only recently available to high dispersion spectrographs on large telescopes. A survey of 38 accreting T Tauri stars spanning a wide range of disk accretion rates shows that He I λ10830 has a far higher incidence of P Cygni profiles than any other line observed to date (Edwards *et al.* 2006), often with a velocity structure reminiscent of the spherical stellar wind models of Hartmann *et al.* (1990). Blueshifted absorption below the continuum is found in ~70% of the stars, in striking contrast to Hα where the fraction is ~10%. In some stars the blue absorptions display remarkable breadth and depth, where 90% of the 1μ continuum is absorbed over a velocity interval of 300-400 km s^{-1}, while others show narrow absorption with modest blueshifts, as illustrated in Figure 2. The extraordinary potential for He I λ10830 to appear in absorption derives from the high opacity of its metastable lower level ($2s^3S$), ~ 21 eV above the singlet ground state, which becomes significantly populated relative to other excited levels owing to its weak de-excitation rate via collisions to singlet states. There is no question that the inner wind probed by He I λ10830 derives from accretion, since it is not seen in non-accreting T Tauri stars and the strength of the combined absorption and emission correlates with the excess continuum veiling at 1μ.

The He I λ10830 profiles offer a unique probe of the geometry of the inner wind because this line is formed under conditions resembling resonance scattering, with only one allowed exit from the upper level and a metastable lower level. Thus it will form an absorption feature via simple scattering of the 1 μ continuum under most conditions. Any additional in-situ emission is modest in comparison to lines like $H\alpha$, which have net emission equivalent widths 10-100 times greater than He I λ10830 (Edwards *et al.* 2003). Moreover, the helium lines are restricted to form in a region of either high excitation or close proximity to a source of ionizing radiation, which is likely to be within the crucial 0.1 AU of the star where the inner high velocity wind is launched.

A recent study takes advantage of these sensitivities to model helium profile formation using Monte Carlo scattering calculations for two inner wind geometries: a disk wind emerging at a constant angle relative to the disk surface and a stellar wind emerging radially away from the star (Kwan, Edwards, & Fischer 2007). The specific configuration of the X-wind has not yet been explored as the flow geometry is currently being revised in order to take into account the interaction between the stellar and the disk field (see chapter by F. Shu). Both stellar and disk wind geometries were parameterized with a simple set of assumptions that allowed a variety of effects to be explored. For the disk wind, launching was confined between 2-6 R_* since helium excitation is unlikely to persist at great distances, and angular velocities along streamlines maintained either rigid rotation or conserved angular momentum. The stellar wind was assumed to originate between 2-4 R_*, both spherical and polar geometries were examined, and the effects of disk shadowing were explored. The latter ranged from extreme (disk truncation radius inside the wind origination radius) to negligible (disk truncation radius $\geqslant$ 5 times the wind origination radius). In each case, profiles were computed under the assumption of pure scattering, where all the 1μ continuum photons were assumed to come from the star, and also with the additional presence of in-situ emission.

The resulting suite of profiles show morphologies resembling the variety found in the data. One key difference between stellar and disk wind profiles is the breadth of velocities in the blue absorption feature. For a stellar wind, any line of sight to the star will intercept radial velocities corresponding to the full acceleration in the wind, giving rise to very broad blue absorptions, such as in the upper panel of Figure 2. For the disk wind however, a particular line of sight to the star will intercept line-of-sight velocities over a narrow interval, and the centroid of the narrow absorption will have a large or small blueshift, depending on the view angle to the star and the opening angle of the wind, as in the lower panel of Figure 2. Among the full sample of 38 accreting T Tauri stars there are roughly comparable numbers with blue absorptions resembling disk (30%) or stellar (40%) winds.

Another distinguishing characteristic between stellar and disk winds is the morphology of the emission component above the continuum. For stellar winds a range of emission morphologies are possible, depending on whether they are dominated by scattering or in-situ emission (which can fill in the blue absorption at low velocities), and whether disk shadowing is important. For the disk wind, however, emission from scattering would be spread out over such a range of velocity that it would appear very weak, while in-situ emission would either be entirely blueshifted or broad and double-peaked, depending on the view angle. Among the stars in the survey, none show He I λ10830 emission resembling the expectation for disk winds, implying that scattering is the only process forming helium lines in the disk winds. In contrast, most of the profiles show emission morphologies expected for stellar winds, including some of those with blue absorptions resembling disk winds. For the stars in Figure 2, the 3 in the top panel with blue absorptions resembling stellar winds all have net equivalent widths $\leqslant 0$, indicating that scattering is the dominant formation mechanism, and one of these, DR Tau, has quite a large net negative equivalent width, which is explained if the disk truncation radius is smaller than the wind origination radius. The 3 in the bottom panel with blue absorptions resembling disk winds all show helium emission as well as redshifted helium absorption from the funnel flow. Some of the emission will thus be from scattering in the funnel flow, while the rest may be from a polar stellar wind viewed at a large angle.

The restricted excitation conditions for He I λ10830 and its resonance scattering properties make it an unprecedented diagnostic of both winds and funnel flows in the star/disk interaction region, providing the first opportunity to diagnose wind geometries in stars

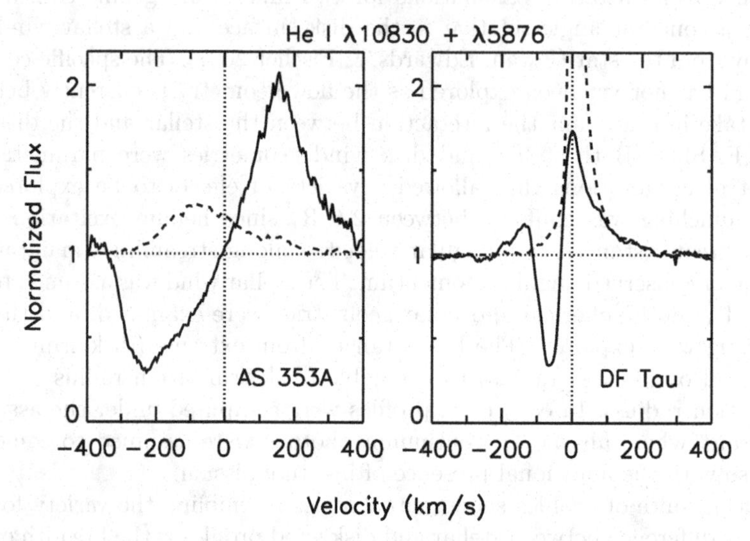

Figure 3. Simultaneous profiles for He I λ10830 (solid) and He I λ5876 (dotted) for two accreting T Tauri stars.

spanning nearly 3 orders of magnitude in disk accretion rates. In § 5 we suggest that while both disk and stellar winds are likely present in most accreting stars, the variety of observed He I λ10830 profiles probably results both from viewing star/disk systems over a range of angles and from an alteration in the geometry of the magnetospheric configuration between high and low accretion rate T Tauri stars.

4.2. *Wind energetics and mass loss rates*

Although there are a variety of atomic lines with spectroscopic signatures of high velocity winds formed in the inner 0.1 AU of accreting T Tauri stars, we have very limited quantitative information on the wind energetics. Recent attempts to model $H\alpha$ profiles suggest that a disk wind may be the source of the narrow blue absorptions often seen in this line (Alencar *et al.* 2005; Kurosawa *et al.* 2006). However, extracting mass loss rates from a hybrid line such as $H\alpha$ is difficult. From the frequent presence of blue absorption at Hα and its almost total absence in the higher Balmer lines Calvet (1997) estimated optical depths and temperatures of the wind, leading to $\dot{M}_w \sim 10^{-8}$ M$_\odot$ yr^{-1} for some of the high accretion rate stars. Ideally this would be compared with mass loss rates determined from forbidden lines to assess whether the bulk of the material in the more distant collimated jet (Hartigan, Edwards & Ghandour 1995) is carried by the high velocity wind seen at Hα. Unfortunately uncertainties in these values are currently too large for any meaningful comparison.

The He I λ10830 line seems to be penetrating the obscurity regarding the launch region, but by itself it is not a strong diagnostic of physical conditions in the wind. However, at least for the restricted region where He I λ10830 is excited and can scatter 1μ continuum photons, mass loss rates from both the stellar and disk winds will be able to be determined from comparison of He I λ10830 and its immediate precursor in a recombination/cascade sequence, He I λ5876. Although the kinematic properties of He I λ5876 are quite different from its near infrared sibling (appearing entirely in emission with kinematic properties that vary among stars, see BEK), this pair of lines have intimately coupled excitation conditions that will enable the relative contribution between continuum scattering and in-situ emission to be evaluated for He I λ10830 and the corresponding physical conditions, including line opacities, electron densities, kinetic temperatures, and limits on the ionizing flux, to be assessed.

In order to determine mass loss rates from this powerful pair of lines, simultaneous optical and near infrared spectra have been obtained with Keck's NIRSPEC + HIRES by L. Hillenbrand, S. Edwards and W. Fischer, and statistical equilibrium calculations for a 19-level helium atom and 6-level hydrogen atom have been made by J. Kwan. This work is in preparation, but examples of simultaneous profiles for two accreting T Tauri stars are shown in Figure 3. For AS 353 A, He I λ10830 shows a stellar wind profile that includes contributions from both scattering and in-situ emission. At the same moment, He I λ5876 shows blueshifted emission, arising from in-situ emission in the wind. In contrast, for DF Tau, He I λ10830 shows blue absorption formed by scattering in a disk wind, while both helium lines show in-situ emission from an accretion shock and redshifted infalling gas in the funnel flow. The absence of narrow He I λ5876 emission from an accretion shock is also found in other stars with strong stellar wind signatures and high accretion rates, suggesting that the size and/or geometry of the funnel flow is altered when disk accretion rates are high (BEK).

5. Multiple inner high velocity winds?

A mix of stellar and disk wind profiles in He I λ10830 would arise naturally if stellar winds emerge primarily from polar regions, inner disk winds are also present, and both are close enough to the star/accretion shock for helium to be excited either through ionization/recombination or (less likely) collisional excitation. Whether the blue absorption resembles a stellar or disk wind would depend on inclination, but emission from the stellar wind could be seen at any orientation. Inclination may thus be the explanation for the He I λ10830 profiles for two low accretion rate stars with well determined inclinations: TW Hya, nearly nearly pole-on (Dupree *et al.* 2005) has a stellar wind profile formed via scattering and in-situ emission, while AA Tau, nearly edge-on (Bouvier *et al.* 2007), shows blue absorption from a disk wind plus emission that could be from a polar stellar wind, along with red absorption from a funnel flow. Uncertainties in published inclinations can be large, but among the stars with blue absorptions resembling disk winds values range from 40°-80°, providing some support for the expectation that disk wind profiles in He I λ10830 are favored in more edge-on systems (Kwan, Edwards, & Fischer 2007). There is also a tendency to see redshifted absorption from funnel flows preferentially in the stars with disk wind profiles, which would also be favored in systems seen at larger inclinations.

In addition to inclination, the disk accretion rate also seems to be a factor in determining the observed morphology. Among the most heavily veiled stars in the Edwards *et al.* survey ($r_Y \geqslant 1$), neither disk wind or magnetospheric infall absorptions are seen at He I λ10830. Instead this group contains the clearest cases for stellar wind profiles (e.g. top panel of Figure 2) and also includes emission profiles with no subcontinuum absorption, as might result from a polar stellar wind seen edge on without any intervening absorption from the disk wind. The rarity of infall signatures for stars with high veiling suggests that the funnel flow may be reduced in size, as might result from a decrease in the disk truncation radius under the pressure of high disk accretion rates, or possibly there is a more drastic reconfiguration of the basic funnel flow geometry (see chapter by M. Romanova). Whatever the cause, the He I λ10830 profiles suggest that conditions for driving winds radially outward from the star are favored when disk accretion rates are high. Additional support for this line of reasoning is provided by $H\alpha$ profiles in some of the highest accretion rate systems such as AS353A and DR Tau, which show P Cygni structure resembling predictions for stellar winds (see Edwards *et al.* (2003)). In fact it was such $H\alpha$ profiles from a few of the brightest T Tauri stars that gave rise to the idea of stellar winds so many decades ago.

6. The future

The evidence to date suggests that it is oversimplified to imagine that accretion powered winds have a single origin. The question instead becomes "Which of the various means of launching winds from accreting systems is most influential in spinning down the star, in extracting angular momentum from the disk, and in feeding the extended jets and molecular outflows?" Ultimate understanding of the accretion/ejection connection will certainly require multiple lines of inquiry, but eventually the phenomenon Bertout (1989) called "The Twilight Zone" will be elucidated.

Acknowledgements

I gratefully acknowledge the work of my collaborators W. Fischer, L. Hillenbrand, and J. Kwan, without whom the results on He I $\lambda 10830$ would not exist, and NASA grant NAG5-12996 issued through the Office of Space Science. And thanks to J. Bouvier for taking the initiative to celebrate the scientific accomplishments and legacy of Claude Bertout!

References

Alencar, S. H. P. & Basri, G. 2000 *ApJ* 119, 1881
Alencar, S. H. P., Basri, G., Hartmann, L., & Calvet, N. 2005, *A&A* 440, 595
Ardila, D. R., Basri, G., Walter, F. M., Valenti, J. A., & Johns-Krull, C. M. 2002a *ApJ* 566, 1100
Ardila, D. R., Basri, G., Walter, F. M., Valenti, J. A., & Johns-Krull, C. M. 2002b *ApJ* 567, 1013
Bally, J., Reipurth, B. & Davis, C. 2007, in: B. Reipurth, D. Jewitt, & K. Keil (eds.) *Protostars and Planets V* (University of Arizona Press), p. 215
Basri, G. & Bertout, C. 1989, *ApJ* 341, 340
Batalha, C. C., Stout-Batalha, N. M., Basri, G., & Terra, M. A. O. 1996 *ApJS* 103, 211
Bacciotti, F., Ray, T.P., Mundt, R., Eisloffel, J., & Solf J. 2002 *ApJ* 576, 222
Bertout, C. 1989, *ARAA* 27, 351
Beristain, G., Edwards, S., & Kwan, J. 1998, *ApJ* 499, 828
Beristain, G., Edwards, S., & Kwan, J. 2001, *ApJ* 551, 1037 [BEK]
Bouvier, J. *et al.* 2007 *A&A* 463, 1017
Cabrit, S., Edwards, S., Strom, S.E., & Strom, K. 1990 *ApJ* 354, 687
Calvet, N. 1997 in: IAU Symp. 182, Herbig-Haro Flows and the Birth of Low Mass Stars, ed. B. Reipurth & C. Bertout (Dordrecht: Kluwer), p. 417
deCampli, W. M. 1891, *ApJ* 244, 124
Dupree, A. K., Brickhouse, N. S., Smith, G. H., & Strader, J. 2005 *ApJ* 625, L131
Edwards, S., Hartigan, P., Ghandour, L., & Andrulis, C. 1994 *AJ* 108, 1056
Edwards, S., Fischer, W., Kwan, J., Hillenbrand, L., & Dupree, A.K. 2003, *ApJ* 599, L41
Edwards, S., Fischer, W., Hillenbrand, L., & Kwan, J. 2006, *ApJ* 646, 319
Ferreira, J., Dougados, C, & Cabrit, S. 2006, *A&A* 453, 785
Folha, D. F. M. & Emerson, J. P. 2001 *A&A*, 365, 90
Gomez de Castro, A. & Verdugo, E 2007 *ApJ* 654, L91
Hartigan, P., Edwards, S., & Ghandour, L. 1995, *ApJ* 452, 736

Hartmann, L., Edwards, S., & Avrett, E. 1982, *ApJ* 261, 279
Hartmann, L.; Avrett, E, H., Loeser, R., & Calvet, N. 1990 *ApJ* 349,168
Hartmann, L., Hewett, R. & Calvet, N. 1994, *ApJ* 426, 669
Herczeg, G. *et al.* 2005 *AJ* 129, 2777
Herczeg, G., Linsky, J., Walter, F. M., Gahm, G., & Johns-Krull, C. 2006 *ApJS* 165, 256
Hirose, S., Uchida, Y., Shibata, K., & Matsumoto, R. 1997 *pasj*, 49, 193
Johns-Krull, C. & Basri, G. 1997 *ApJ* 474, 433
Johns-Krull, C. & Herczeg, G. 2007 *ApJ* 655, 345
Kuhi, L.V. 1964 *ApJ* 140, 1409
Kurosawa, R., Harries, T.J., Symington, N. H. 2006 *MNRAS* 370, 580
Kwan, J., & Tademaru, E. 1988 *ApJL*, 332, L41
Kwan, J., Edwards, S., & Fisher, W. 2007 *ApJ* 657, 897
Matt, Sean & Pudritz, R. 2005 *ApJ* 632,135
Mundt, R. 1984 *ApJ* 280, 749
Muzerolle, J., Calvet, N., & Hartmann, L. 2001 *ApJ* 550, 944
Nakamura, F. & Li, Z. 2007 *ApJ* 662, 395
Najita, J., Carr, J. S., & Mathieu, R. D. 2003 *ApJ* 589, 931
Najita, J., Carr, J., Glassgold, A., & Valenti, J. 2007 in: B. Reipurth, D. Jewitt, & K. Keil (eds.) *Protostars and Planets V* (University of Arizona Press), p. 507
Natta, A., Giovanardi, & Palla 1988 *ApJ* 332,921
Pudritz, R. E., Ouyed, R., Fendt, Ch., & Brandenburg, A, 2007, in: B. Reipurth, D. Jewitt, & K. Keil (eds.) *Protostars and Planets V* (University of Arizona Press), p. 277
Pyo, T-S. *et al.* 2003 *ApJ* 590, 340
Pyo, T-S. *et al.* 2006 *ApJ* 649, 836
Romanova, M. M., Ustyuogva, G. V., Koldoba, A. V., & Lovelace, R. V. 2005, *ApJ* 635, L165
Shu, F. H., Laughlin, G., Lizano, S., & Galli, D. 2000, in: V. Mannings, A.P. Boss, S. Russell (eds.) *Protostars and Planets IV* (University of Arizona Press), p. 789
Takami, M. *et al.* 2006 *ApJ* 641, 357
von Rekowski, B. & Brandenburg, A. 2006, *Astronomische Nachrichten*, 327, 53
Whelan, E. T., Ray, T. P., & Davis, C. J. 2004 *A&A* 417, 247

Star-Disk Interaction in Young Stars
Proceedings IAU Symposium No. 243, 2007
J. Bouvier & I. Appenzeller, eds.

doi:10.1017/S1743921307009544

The generation of jets from young stars: an observational perspective

T. P. Ray

School of Cosmic Physics, Dublin Institute for Advanced Studies, 31 Fitzwilliam Place, Dublin 2, Ireland
email: tr@cp.dias.ie

Abstract. The birth of a young star is accompanied not only by accretion but by the expulsion of matter as well in the form of a collimated outflow. These outflows are seen at various wavelengths from X-rays to the radio band, but ultimately the driving mechanism appears to be a highly collimated supersonic jet that contains not only atomic but molecular components as well. These jets may also play a key role in the star formation process itself since they could be one of the primary mechanism for removing angular momentum from the accretion disk thereby allowing accretion to occur. Whereas much is known about their propagation on large-scales (i.e., hundreds of AU to several parsecs) from both observations and simulations, we must explore the "central engine" in order to understand how they are generated. While this is particularly challenging, high spatial resolution studies are beginning to reveal interesting data from which we can confront the various models. In this review, I will summarise what these studies suggest and note how they already favour certain models over others. I will also describe some of the results from spectro-astrometry and interferometry that are revealing details of outflows on milliarcsecond scales from the source.

Keywords. Stars: pre–main-sequence, stars: low-mass, brown dwarfs, stars: winds, outflows, ISM: jets and outflows, ISM: Herbig-Haro objects.

1. Introduction

The phenomenon of jets from young stars is not only striking but may be fundamental to the star formation process itself. In particular they could be channels for removing angular momentum thereby allowing accretion to occur and thus regulating the final mass of the newborn star. Moreover they may add turbulence, energy and affect the chemical composition of the surrounding molecular cloud (Arce & Sargent 2004; Li & Nakamura 2006; Banerjee, Klessen & Fendt 2007) and therefore play a role in determining conditions in the circumstellar environment. Ultimately, in ways not currently understood, they could be a factor that helps shape the initial mass function (IMF).

Outflows from young stars were first discovered in molecular transitions in the early 1980s (e.g. Snell, Loren & Plambeck 1980) with the opening of the mm-band window for astronomy. Later it was realised that the optically visible nebulous patches known as Herbig-Haro (HH) objects (Herbig 1950; Haro 1952), uncovered many years before, were part of the same phenomenon. Deep optical images, made possible with the advent of CCD detectors, revealed that many well-known HH objects were just the brightest parts of a highly collimated jet, visible in low excitation emission lines (Mundt & Fried 1983; Ray 1987; Bührke, Mundt & Ray 1988), and that it was the jet that drove the slower gas seen, for example, in low order CO rotational transitions.

In this review, we will concentrate on what the most recent observations have to tell us about the driving mechanism for jets from young stars as well as sub-stellar objects, i.e., brown dwarfs. We will begin by summarising what is known about the outflow phenomenon with the emphasis on the highly collimated component be it atomic (ionised or neutral) or molecular (§2). The challenges in resolving the central engine from an observational perspective will be explored in §3 as well as the results of various high spatial resolution studies. Finally in §4, we will briefly discuss what the techniques of spectro-astrometry and interferometry are beginning to reveal about outflows within a few milliarcseconds from their cores.

2. The jet phenomenon in context

The combination of multi-epoch imaging (typically a few years apart) and spectroscopy quickly revealed many of the basic parameters of jets from young stellar objects (YSOs). As with HH objects, jets reveal themselves through their line emission which derives from radiative cooling in post-shock zones. This line emission is dominated by forbidden transitions, e.g. [OI]$\lambda\lambda$6300, 6363, [SII]$\lambda\lambda$6716, 6731, and [FeII]1.64μm (e.g. Nisini *et al.* 2005) although permitted, e.g. Hα, and semi-forbidden transitions, e.g. CIII]1909 (Böhm & Böhm-Vitense 1984) are also seen. Intermediate resolution spectroscopy can not only reveal the kinematic state of the jet, through radial velocity studies, but combination of lines can tell us other basic physical parameters such as electron density, n_e using the [SII] doublet. In addition, if we assume charge exchange, i.e., the ionised state of Oxygen and Nitrogen is determined via charge exchange with Hydrogen, and that Sulphur is singly ionised because of its low IP, other parameters such as electron temperature, T_e, and ionised fraction χ can be determined (Bacciotti & Eislöffel 1999). In turn this leads to total density estimates. Since we know the various line fluxes, and as the radiation is optically thin, such fundamental parameters as jet mass loss rates as well as mechanical luminosities can be derived.

Typical (hydrodynamic) Mach numbers M were found to be 10-30, with mass loss rates $\dot{M}_{jet} \approx 10^{-8} - 10^{-6} M_{\odot} yr^{-1}$ and velocities of around 100-300 $km\,s^{-1}$. In addition although initially it was thought that these collimated outflows were only a few thousand AU in length, it was soon established that many were comparable in size to the parent molecular cloud itself, i.e., parsec-scale (Ray 1987). Such lengths however are to be expected if, as seems to be the case, the outflow phenomenon lasts the same time as the accretion phase (i.e., $\approx 10^6$ years for solar mass YSOs) and giving their typical velocities. Moreover, in addition to their remarkable lengths, their degree of collimation, as indicated by the opening angle θ_{jet}, is striking. Typically this is only a few degrees and close to what is expected if such jets are freely expanding, i.e., $\theta_{jet} \approx 1/M$.

Close examination of YSO jets, such as HH 30 (Ray *et al.* 1996) invariably show they are not continuous but instead consist of a string of knots. On timescales of a few months to years new knots may appear close to the source and these either fade away or blend into existing knots. Shock velocities at the knots are usually low (a few tens of $km\,s^{-1}$) and considerably less than the jet velocity itself. This has led to the idea that the knots are due to temporal variations in the flow and are regions were faster jet material rams into slower moving gas ahead of it (e.g. Raga *et al.* 2007). Since the jet is moving supersonically this gives rise to a two shock structure: at the upstream shock the faster material is decelerated while at the downstream shock the slower material is accelerated. The region in between, known as the working surface, is over-pressured with respect to its environment and some gas is ejected sideways (although note that the gas is radiatively cooling). According to this model we would expect the knots to move outwards with velocities comparable to that of the jet. Such a result is found from proper motion studies (McGroarty, Ray & Froebrich 2007).

According to the picture painted above, we only see those regions of the jet that are currently being shocked. There must be other parts of the jet that are undisturbed and

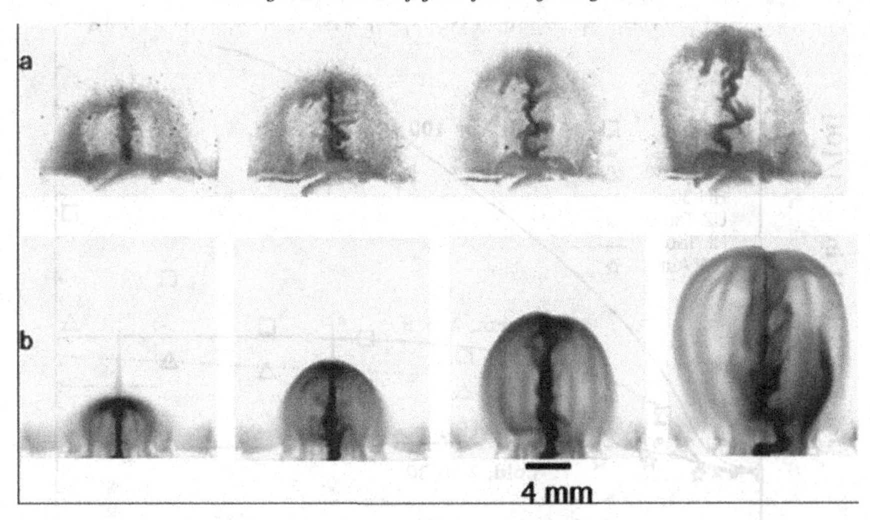

Figure 1. (a) Laboratory jets produced at Imperial College London's z-pinch facility. These jets are magnetically collimated by magnetic hoop stresses. Note the variability in emission, which in this case is probably due to the growth of instabilities, and the physically small scale of these jets. (b) Simulations of the magnetic tower jets seen in (a). Images courtesy of Andrea Ciardi and Sergey Lebedev.

hence invisible. Such regions are revealed however, through photo-ionisation, when the jet is sufficiently close to a UV source, e.g. an O-type star. These so-called irradiated jets can thus be found near HII regions, such as Orion and Carina, and give us more of an idea of what the true outflow history of a young star is like (Bally 2007).

Computer simulations of jets from young stars, at least on extended (0.01-1.0 pc) scales, have now reached a high degree of complexity. Not only do the codes allow for atomic and molecular cooling but also incorporate magnetic fields – the parameter that we perhaps know least in YSO jets (de Colle & Raga 2006). The problem is ideally done in a parallel processing environment and current codes can also incorporate adaptive mesh refinement (AMR) to track the physically most interesting parts of the flow (Cunningham *et al.* 2006). Such simulations are useful not only in revealing the expected morphology and properties of the working-surfaces but also in testing how the jet interacts with its environment. For example it has long been recognised that, in regions that are sufficiently dense, the atomic jet "pushes" ambient material to give rise to the slower (V$\sim$ a few tens of km s^{-1}) molecular outflow traced in CO rotational transitions. Simulations can address whether the molecular gas has the expected distribution of mass versus velocity $m(v)$ if it is jet-driven (Downes & Cabrit 2003).

In recent years however YSO jet simulations have come from an unexpected quarter, laboratories involved in fusion research. It is now possible, using either high power lasers or imploding z-pinch wire arrays, to produce high Mach number jets ($M \sim 5 - 20$) at a number of facilities (Lebedev *et al.* 2004; Stehlé *et al.* 2005; Sublett *et al.* 2007). Radiative cooling of the jet, at least for those produced through z-pinching is through X-ray emission, rather than UV, optical and near-infrared lines, but many fundamental numbers (e.g. the ratio of cooling length to dynamical length and the plasma β parameter) can be made to match. It is even possible, to introduce jet rotation (Ampleford *et al.* 2007)! While many of these simulations are at an early stage, the laboratory jets bear a remarkable resemblance to those seen from YSOs (see Figure 1). Perhaps just as importantly they can also be used to benchmark various jet codes, rigorously testing their limitations (Ciardi *et al.* 2007).

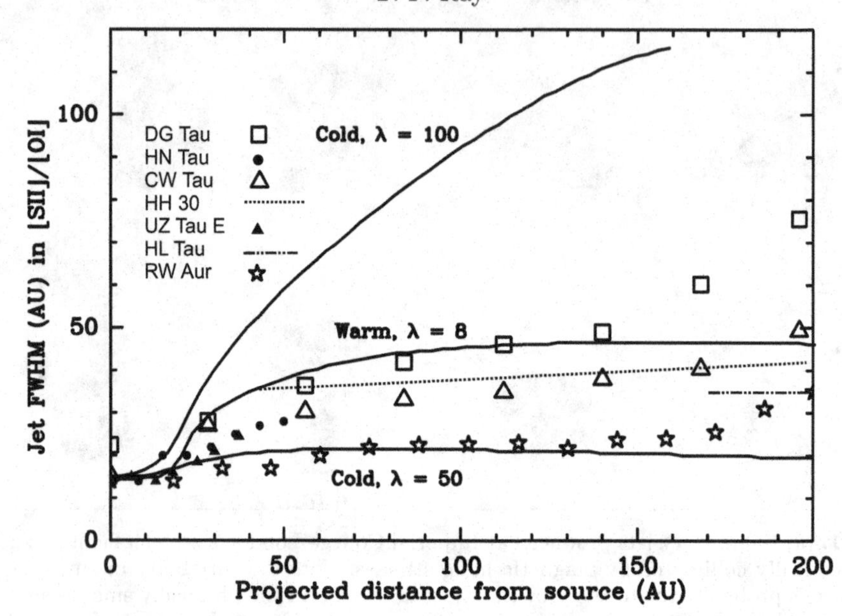

Figure 2. Jet width (FWHM) as a function of distance from the source using [SII] and [OI] images. Solid lines are predicted variations based on cold, low efficiency (high λ) models and warm, high efficiency (low λ) models. Efficiency is measured as the ratio of the mass outflow to the mass inflow (accretion) rates. Although cold solutions can be found to fit the data, the resultant poloidal velocity is much higher than observed favouring warm, moderately efficient, solutions. From Dougados *et al.* (2004) with additional data from Hartigan *et al.* (2004).

3. The challenge of observing the jet engine

While there is some uncertainty as to the origins of jets (see for example Ferreira, Dougados & Cabrit 2006 and Shu, this volume) the general consensus is that the jet is initially generated in a region less than 1 AU in diameter and almost certainly arises from the circumstellar disk. Even for the nearest star formation regions, for example Taurus Auriga or Ophiuchus, this scale corresponds to less than 10 milliarcseconds on the sky. Such a small region cannot be observed using conventional ground based instruments although interferometry (see below) is beginning to observe such zones. The situation however is not quite as bad as it initially looks. Almost all models (see, for example, Pudritz *et al.* 2007) predict jets are accelerated and focussed over larger scales, we thus stand a chance of probing them in the region where they are reaching their asymptotic values (i.e., within 1″ for the nearest star formation regions).

Studying the jet line emission close to the source with sufficient angular resolution usually means observing in the optical. Many jet sources however (e.g. HH1/2 VLA-1) are highly embedded and thus we must turn to a number of Classical T Tauri stars (CTTS) with jets if we are to probe their base. Examples include DG Tau (Dougados *et al.* 2000), CW Tau and HN Tau (Hartigan, Edwards & Pearson 2004). Observations of these jets have not only been done from the ground using adaptive optics (AO) but also from space using the novel technique of slit-less spectroscopy (Hartigan *et al.* 2004). Such observations reveal the variation in jet width as a function of distance from the source (see Figure 2). Comparison with MHD models (Dougados *et al.* 2004) show that the jets must be launched warm. Although 'cold' solutions can be made to fit (again see Figure 2) they predict too high a poloidal jet velocity in comparison to what is observed.

There are a number of other questions that observations close to the source are beginning to address. For example do molecules and dust survive the jet launching process?

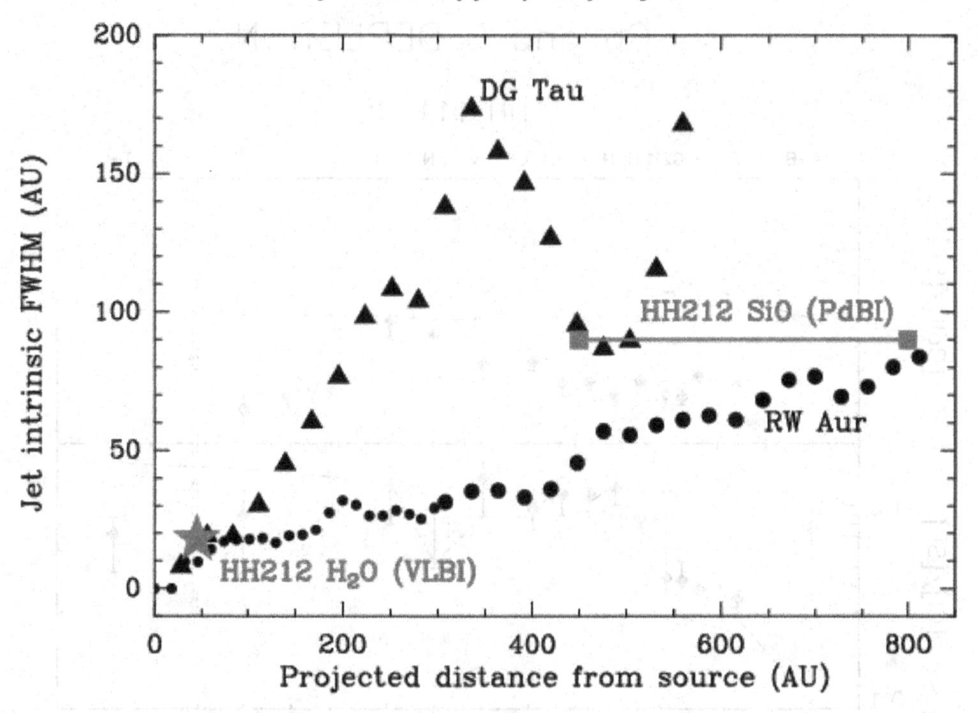

Figure 3. Comparison plot of jet width (FWHM) as a function of distance from the source for DG Tau, RW Aur and HH 212. Whereas the observations for the two classical T Tauri jets were made in optical forbidden lines, those for HH 212 were made in SiO lines using the Plateau de Bure Interferometer. Widths, at a given distance from the source, are seen to be similar in all cases suggesting SiO molecules are launched with the jet rather than entrained. The width of the cone defining the H_2O maser emission of HH 212 is also shown. Note how the masers seem confined to the SiO jet again suggesting they are part of the initial flow. Figure from Cabrit *et al.* (2007).

Clearly molecules are part of the outflow as witnessed by the well-known, often bipolar, lobes seen in low order CO rotational transitions. As stated above however this gas is clearly pushed ambient material and has not travelled in towards the young star and been ejected from the disk. The situation is less clear for example with high velocity CO (which appears much more collimated) seen in higher order transitions, shocked H_2 and SiO emission. Are such molecules entrained in the jet or are they launched with it and therefore constituent part of the initial flow? Recent high spatial resolution SiO observations ($0\farcs34$) by Codella *et al.* (2007) using the Plateau de Bure (PdB) interferometer in its extended configurations certainly suggests that some molecules are launched as part of the jet. HH 212 is a well known outflow with a dramatic large-scale bipolar jet first seen in shocked H_2 emission (Zinnecker, McCaughrean & Rayner 1998). The PdB observations show that the SiO emission is confined to a highly-collimated bipolar jet (with a FWHM width of $0\farcs35$ close to the source) in the same direction as the H_2 jet and that it has very similar kinematics to the latter. Such jet widths are comparable to those seen in outflows from classical T Tauri stars (see Figure 3 and previous Figure). Moreover in the case of HH 212, the H_2O masers also seem confined to a narrow cone (see Figure 3).

Evidence is also emerging that dust may at least partially survive the launching process. Figure 4 is from Podio *et al.* (2006). It shows for example in the HH 111 jet, Ca is depleted close to the source presumably onto dust grains. At larger distances the expected

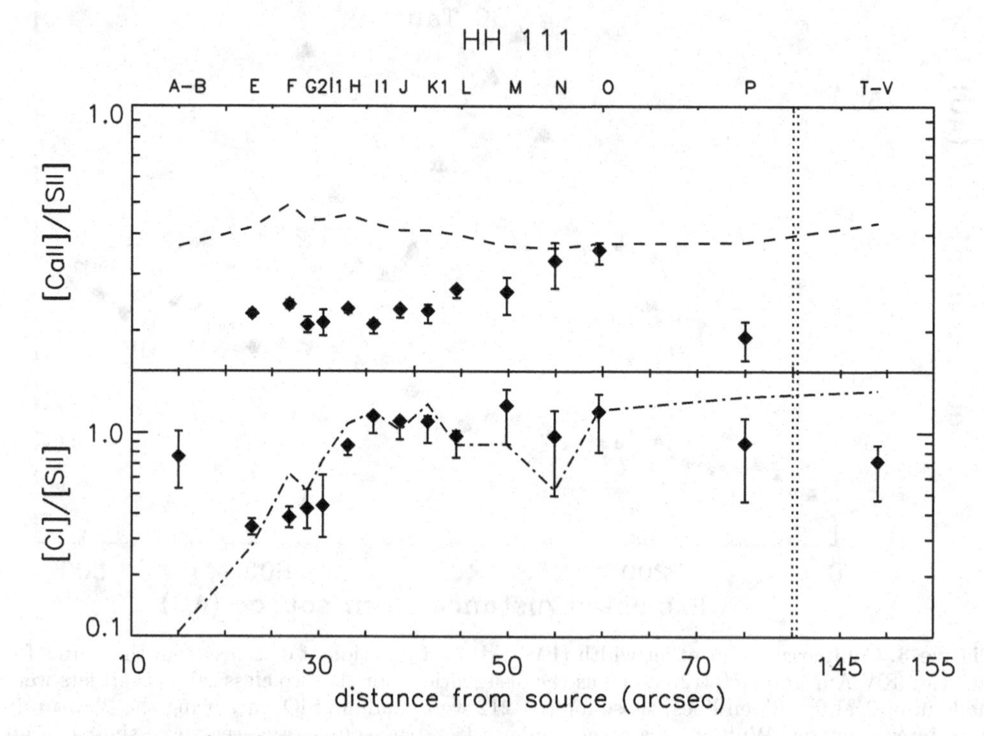

Figure 4. Plots of [CaII]/[SII] and [CI]/[SII] for the HH 111 jet taken from Podio *et al.* (2006). These show that Ca is depleted onto dust grains at least for the initial part of the flow. At larger distances however the dust may be destroyed, presumably as a result of passing through multiple weak shocks in the jet.

abundance of Ca is recovered perhaps because the dust has passed through a number of shocks before being destroyed. The effect is also seen in HH 34 and in Fe as well as Ca.

One of the most interesting findings in recent years from high spatial resolution studies is that YSO jets may be rotating (Bacciotti *et al.* 2002; Coffey *et al.* 2004; Coffey *et al.* 2007; Ray *et al.* 2007). Of course MHD models, which assume jet material is launched centrifugally along magnetic field lines, expect jets to rotate (Pudritz *et al.* 2007). The difficulty in detecting such rotation is that at distances of about 200 AU from the source, corresponding to 1-2″ for the nearest star formation regions, we expect rotational velocities of around 10-20 km s^{-1}. This compares to jet (poloidal) velocities that are typically 10 times larger. Moreover as the jet widens, with increasing distance from the source, rotational velocities are expected to decrease even further in order to conserve angular momentum.

The earliest observations to discover possible rotation close to the source were made using the Space Telescope Imaging Spectrograph (STIS) (Bacciotti *et al.* 2002). A series of overlapping narrow (0.″1) slits were placed parallel to the DG Tau jet. Using data from these observations it was then possible to reconstruct "images" of the jet not only in individual optical lines but also in different velocity channels. Interestingly differences in radial velocity were observed between opposing slit pairs on either side of the outflow axis. These differences were typically found to be 5 to 25 km s^{-1} with errors of around ±5 km s^{-1}, i.e., more or less in line with expected values based on models that launch the jet from extended parts of the disk (around 0.1–1 AU).

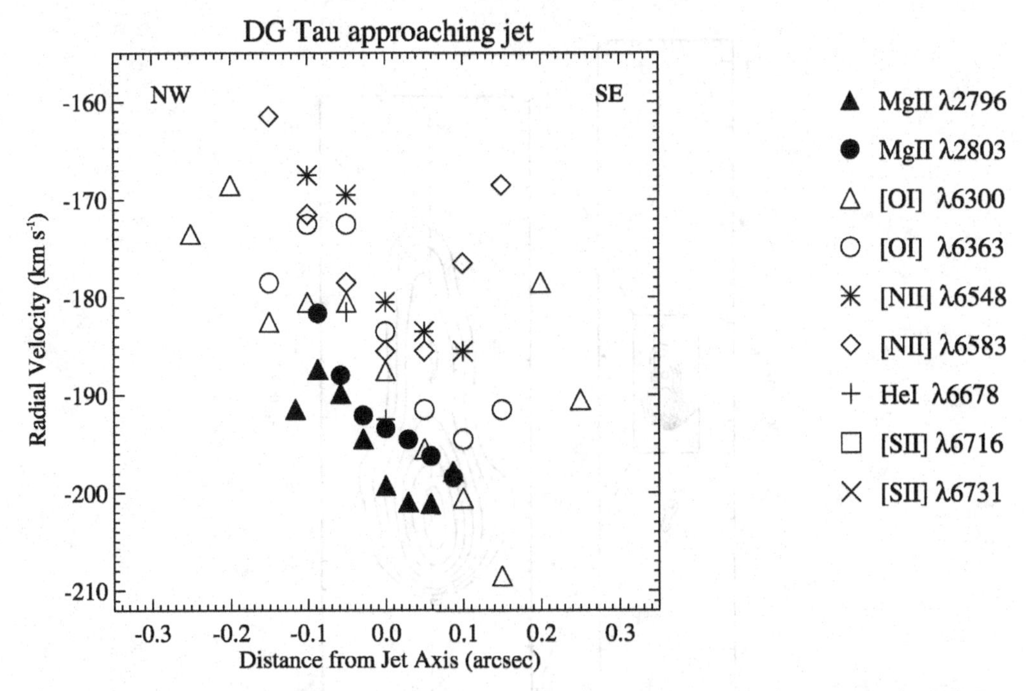

Figure 5. Comparison of observed radial velocities, in optical and UV lines, *across* the DG Tau blueshifted jet as a function of distance from the jet axis. The Mg UV lines are filled in circles and triangles (see key). Note the consistency between the optical and UV data. Plot taken from Coffey *et al.* (2007).

Further confirmation of the observed latitudinal velocity differences came from UV line studies again using STIS (Coffey *et al.* 2007). The magnitude of the differences were seen to be similar to those observed optically (see Figure 5). Velocity differences, again compatible with jet rotation, have also been seen from the ground in the H_2 2.12μm line (Chrysostomou *et al.* 2007, in preparation).

In almost all cases, consistency is seen in the system as a whole, e.g. the disk and jet lobes of HH 212 and DG Tau rotate in the same sense. This is as one might expect if the jet is launched from a rotating disk. Moreover in the case of bipolar jets, e.g. RW Aur, both red and blueshifted jets are seen to rotate in the same sense, i.e., with opposite helicity. Again this is as expected. There appears however to be a difficulty in one case: the disk of the RW Aur jet seems to rotate in the opposite sense to its bipolar jet (Cabrit *et al.* 2006). This system is a hierarchical one and thus it dynamics may be complicated. Nevertheless further study is required to understand this anomaly.

To summarise jets from young stars (both the atomic and molecular components) seem to rotate. Since line diagnostics can also be used to derive mass outflow rates, momentum transfer rates, etc., rotational velocities estimates can be used in conjunction with this data to derive approximate angular momentum transfer rates (e.g. Bacciotti *et al.* 2004). The values derived are consistent with most of the angular momentum, necessary to maintain accretion in the disk, being transported by the jets. Note however that different models have varying levels of jet angular momentum transport: extended disk-wind models have higher specific angular momentum per unit mass than X-wind models. The derived values tend to favour the former over the latter.

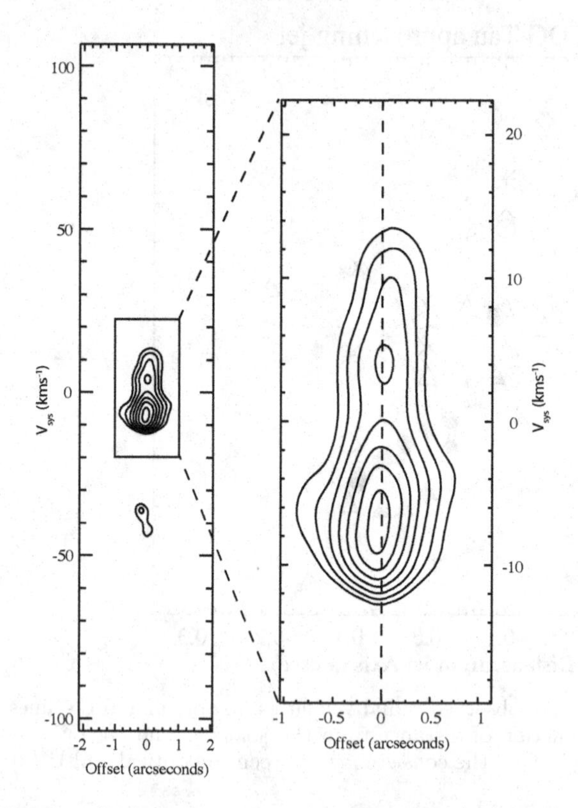

Figure 6. Position Velocity (PV) diagram in the vicinity of the [OI]λ6300 line for the 24 Jupiter mass young brown dwarf 2MASSW J1207334-393254. The data were taken with UVES on the VLT. Velocities are systemic. Both blue and redshifted emission is seen spatially offset either side of the continuum indicating the presence of a bipolar outflow. The offsets are more evident in the spectro-astrometric plot of Figure 7. Plot is from Whelan *et al.* (2007).

4. Spectro-astrometry and interferometry

In recent years the specialised technique of spectro-astrometry has been used to derive data about outflows on scales less than the seeing-disk. The method is explained in detail in Bailey (1998) and Porter, Oudmaijer, & Baines (2004) as a means of separating close binaries, so I will only briefly describe it here.

Although the image of a star is smeared into a disc comparable to the seeing, the accuracy with which we can determine the centroid of the disc, i.e., the position of the star, is a function of the number of collected photons N_p. More precisely

$$\sigma_{Centroid} = \frac{Seeing(mas)}{2.3548\sqrt{N_p}} \tag{4.1}$$

Thus for $N_p = 10^6$, and seeing around $1''$, $\sigma_{Centroid} \approx 1mas$.

Now suppose we have a long-slit spectrum of a YSO which in addition to the continuum from the star, also contains a HH knot close to its source, i.e., well within the seeing disc. Then a centroid fit to the continuum, as a function of wavelength, will deviate at the lines emitted by the HH object. In effect the fit is dragged by the latter. Assuming no *a priori* knowledge of the outflow direction, two preferably orthogonal slit positions are required to specify the direction of the HH emission with respect to its source (note that the HH emission is contained *within* the slit).

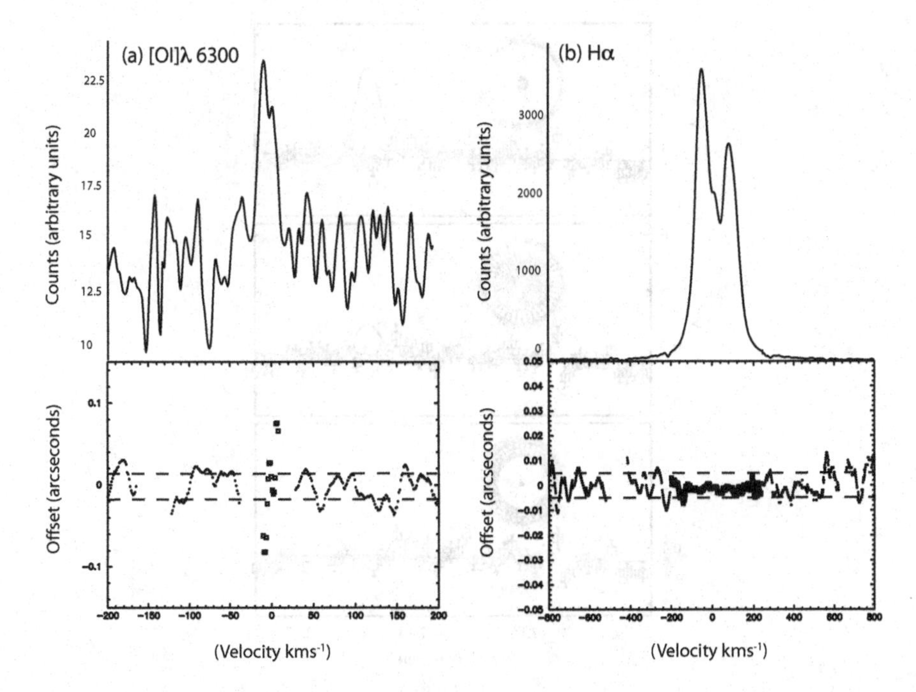

Figure 7. Spectro-astrometric plots for the young brown dwarf 2MASSW J1207334-393254. The [OI]λ6300 line (a) is on the left and Hα (b) on the right. Top panels show line profiles and bottom panels the corresponding offset from the continuum against velocity. Note how the blue and red [OI] wings are offset in the negative and positive direction respectively. In comparison no offset is seen in the Hα line as expected if most of the line originates from accretion.

As the emission we are interested in is within the seeing disc, or more precisely the point spread function (PSF) since the method is obviously applicable to observations from space, the amount of information we can recover is limited. Nevertheless the technique has been used to:

(*a*) Reveal what parts of a line are due to an "extended outflow" as opposed to magnetospheric activity close to the star (Whelan, Ray & Davis 2004),

(*b*) Determine the outflow direction for HH emission that is very close to its source, using either orthogonal slits or an integral field unit,

(*c*) Uncover disk dust holes by being able to distinguish the red and blueshifted jets close to the star in permitted lines but not in forbidden lines (e.g. Takami, Bailey & Chrysostomou 2003). This is because the forbidden emission comes from more extended (distant from the source) regions of the jet where the critical electron density is sufficiently low,

(*d*) Detect outflows from young brown dwarfs (see below).

An example of the power of the technique is the recent detection of outflows from young brown dwarfs (Whelan *et al.* 2005; Whelan *et al.* 2007). Figure 6 shows a long-slit spectrum, in the region of the [OI]λ6300 line, taken with UVES on the VLT of the 24 Jupiter mass young brown dwarf 2MASSW J1207334-393254. The total exposure time was 5 hours. The velocity scale is systemic and both blue and redshifted line emission can be seen. The red and blueshifted emission is spatially offset either side of the continuum position (marked as a dashed line in the figure). This is even more evident in the spectro-astrometric plot shown in Figure 7 which shows the offsets ($\approx 0\farcs1$) of the [OI] line

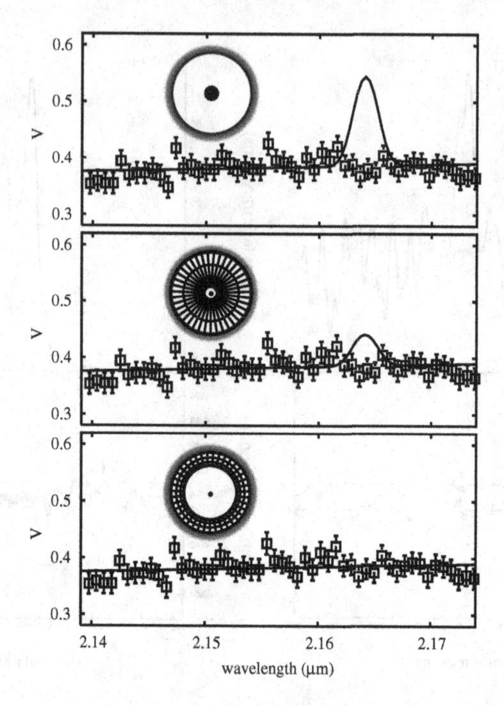

Figure 8. Brγ emission visibility curves from HD 104237 based on data from AMBER/VLTI. Data points are empty squares with error bars. Model fits are solid lines. Three different models are shown in the panels. The upper one is for a magnetospheric accretion model in which the Brγ emission arises from inside the co-rotation radius, i.e., very close to the star. The middle panel assumes the emission comes from the co-rotation radius out to 0.45 AU. Finally the bottom panel, which is the best fit, assumes the emission comes from 0.2 to 0.5 AU. Figure from Tatulli *et al.* (2007).

displaced on either side of the continuum as we pass through its blue and red wings. In contrast to the forbidden emission, most of the Hα line is thought to arise from accretion very close to the BD (Natta *et al.* 2004). This is confirmed in the spectro-astrometric plot (Figure 7) which shows no offset in the Hα line. In fact the scatter is even less than in the continuum itself due to the high S/N ratio of the line.

2MASSW J1207334-393254 appears to be the lowest mass galactic object known with a jet and suggests that even Jupiter mass planets may drive jets when forming. One other remark worth making is that the radial velocity of this outflow appears to be low. This is consistent with the fact that Scholz, Jayawardhana & Brandeker (2005) consider 2MASSW J1207334-393254 to have an edge-on disk and thus the outflow should almost be in the plane of the sky.

A number of optical/near-infrared interferometers are either coming on-stream, such as AMBER/VLTI (Petrov *et al.* 2007), or will be available soon, like the LBT Interferometer (Herbst 2003). Although it is early days, first results are very encouraging and already provide very interesting information. An example is illustrated in Figure 8 which shows the AMBER/VLTI visibility curve against model fits for the origin of the wind from the young Herbig Ae star HD 104237. Full details are given in Tatulli *et al.* (2007) but essentially the interferometer results show the Brγ emission must originate from a zone 0.2 to 0.5 AU from the star, i.e., it is consistent with the wind being launched from an extended region of the accretion disk. Such distances are an order of magnitude greater than expected from the X-wind model (Shu *et al.* 2000) according to which most of the

Brγ emission should arise from close to the co-rotation radius. Finally it should be noted that HD 104237 is a jet source (Grady *et al.* 2004).

Acknowledgements

I wish to acknowledge support through the Marie Curie Research Training Network JETSET (Jet Simulations, Experiments and Theory) under contract MRTN-CT-2004-005592 and from Science Foundation Ireland under contract 04/BRG/P02741. I also would like to thank Immo Appenzeller and Jerome Bouvier for their patience in awaiting this manuscript.

References

Ampleford, D. J., *et al.* 2007, *Astrophys. Space. Sci.*, 307, 51
Arce, H. G., & Sargent, A. I. 2004, *Astrophys. J.*, 612, 342
Bacciotti, F., Ray, T. P., Mundt, R., Eislöffel, J., & Solf, J. 2002, *Astrophys. J.*, 576, 222
Bacciotti, F., & Eislöffel, J. 1999, *Astron. & Astrophys.*, 342, 717
Bacciotti, F., Ray, T. P., Coffey, D., Eislöffel, J., & Woitas, J. 2004, *Astrophys. Space. Sci.*, 292, 651
Bailey, J. 1998, *Mon. Not. Roy. Astron. Soc.*, 301, 161
Bally, J. 2007, *Astrophys. Space. Sci.*, 279
Banerjee, R., Klessen, R. S., & Fendt, C. 2007, *Astrophys. J.*, in press
Böhm, K. H., & Böhm-Vitense, E. 1984, *Astrophys. J.*, 277, 216
Bührke, T., Mundt, R., & Ray, T. P. 1988, *Astron. & Astrophys.*, 200, 99
Cabrit, S., Pety, J., Pesenti, N., & Dougados, C. 2006, *Astron. & Astrophys.*, 452, 897
Cabrit, S., Codella, C., Gueth, F., Nisini, B., Gusdorf, A., Dougados, C., & Bacciotti, F. 2007, *Astron. & Astrophys.*, 468, L29
Ciardi, A., *et al.* 2007, *Astrophys. Space. Sci.*, 307, 17
Codella, C., Cabrit, S., Gueth, F., Cesaroni, R., Bacciotti, F., Lefloch, B., & McCaughrean, M. J. 2007, *Astron. & Astrophys.*, 462, L53
Coffey, D., Bacciotti, F., Woitas, J., Ray, T. P., & Eislöffel, J. 2004, *Astrophys. J.*, 604, 758
Coffey, D., Bacciotti, F., Ray, T. P., Eislöffel, J., & Woitas, J. 2007, *Astrophys. J.*, 663, 350
Cunningham, A. J., Frank, A., Quillen, A. C., & Blackman, E. G. 2006, *Astrophys. J.*, 653, 416
de Colle, F., & Raga, A. C. 2006, *Astron. & Astrophys.*, 449, 1061
Dougados, C., Cabrit, S., Lavalley, C., & Ménard, F. 2000, *Astron. & Astrophys.*, 357, L61
Dougados, C., Cabrit, S., Ferreira, J., Pesenti, N., Garcia, P., & O'Brien, D. 2004, *Astrophys. Space. Sci.*, 292, 643
Downes, T. P., & Cabrit, S. 2003, *Astron. & Astrophys.*, 403, 135
Ferreira, J., Dougados, C., & Cabrit, S. 2006, *Astron. & Astrophys.*, 453, 785
Grady, C. A., *et al.* 2004, *Astrophys. J.*, 608, 809
Haro, G. 1952, *Astrophys. J.*, 115, 572
Hartigan, P., Edwards, S., & Pierson, R. 2004, *Astrophys. J.*, 609, 261
Herbig, G. H. 1950, *Astrophys. J.*, 111, 11
Herbst, T. 2003, *Astrophys. Space. Sci.*, 286, 45
Lebedev, S. V., *et al.* 2004, *Astrophys. J.*, 616, 988
Li, Z.-Y., & Nakamura, F. 2006, *Astrophys. J.*, 640, L187
McGroarty, F., Ray, T. P., & Froebrich, D. 2007, *Astron. & Astrophys.*, 467, 1197
Mundt, R., & Fried, J. W. 1983, *Astrophys. J.*, 274, L83
Natta, A., Testi, L., Muzerolle, J., Randich, S., Comerón, F., & Persi, P. 2004, *Astron. & Astrophys.*, 424, 603
Nisini, B., Bacciotti, F., Giannini, T., Massi, F., Eislöffel, J., Podio, L., & Ray, T. P. 2005, *Astron. & Astrophys.*, 441, 159

Petrov, R. G., *et al.* 2007, *Astron. & Astrophys.*, 464, 1

Podio, L., Bacciotti, F., Nisini, B., Eislöffel, J., Massi, F., Giannini, T., & Ray, T. P. 2006, *Astron. & Astrophys.*, 456, 189

Porter, J. M., Oudmaijer, R. D., & Baines, D. 2004, *Astron. & Astrophys.*, 428, 327

Pudritz, R. E., Ouyed, R., Fendt, C., & Brandenburg, A. 2007, Protostars and Planets V, 277

Raga, A. C., de Colle, F., Kajdič, P., Esquivel, A., & Cantó, J. 2007, *Astron. & Astrophys.*, 465, 879

Ray, T. P. 1987, *Astron. & Astrophys.*, 171, 145

Ray, T. P., Mundt, R., Dyson, J. E., Falle, S. A. E. G., & Raga, A. C. 1996, *Astrophys. J.*, 468, L103

Ray, T., Dougados, C., Bacciotti, F., Eislöffel, J., & Chrysostomou, A. 2007, Protostars and Planets V, 231

Scholz, A., Jayawardhana, R., & Brandeker, A. 2005, *Astrophys. J.*, 629, L41

Shu, F. H., Najita, J. R., Shang, H., & Li, Z.-Y. 2000, Protostars and Planets IV, 789

Snell, R. L., Loren, R. B., & Plambeck, R. L. 1980, *Astrophys. J.*, 239, L17

Stehlé, C., Ciardi, B., Lebedev, S. V., & Lery, T. 2005, SF2A-2005: Semaine de l'Astrophysique Francaise, 355

Sublett, S., Knauer, J. P., Igumenshchev, I. V., Frank, A., & Meyerhofer, D. D. 2007, *Astrophys. Space. Sci.*, 307, 47

Takami, M., Bailey, J., & Chrysostomou, A. 2003, *Astron. & Astrophys.*, 397, 675

Tatulli, E., *et al.* 2007, *Astron. & Astrophys.*, 464, 55

Whelan, E. T., Ray, T. P., & Davis, C. J. 2004, *Astron. & Astrophys.*, 417, 247

Whelan, E. T., Ray, T. P., Bacciotti, F., Natta, A., Testi, L., & Randich, S. 2005, *Nature*, 435, 652

Whelan, E. T., Ray, T. P., Randich, S., Bacciotti, F., Jayawardhana, R., Testi, L., Natta, A., & Mohanty, S. 2007, *Astrophys. J.*, 659, L45

Zinnecker, H., McCaughrean, M. J., & Rayner, J. T. 1998, *Nature*, 394, 862

Star-Disk Interaction in Young Stars
Proceedings IAU Symposium No. 243, 2007
J. Bouvier & I. Appenzeller, eds.

doi:10.1017/S1743921307009556

The role of thermal pressure in jet launching

Noam Soker

Department of Physics, Technion, Haifa 32000, Israel

Abstract. I present and discuss a unified scheme for jet launching that is based on stochastic dissipation of the accretion disk kinetic energy, mainly via shock waves. In this scheme, termed thermally-launched jet model, the kinetic energy of the accreted mass is transferred to internal energy, e.g., heat or magnetic energy. The internal energy accelerates a small fraction of the accreted mass to high speeds and form jets. For example, thermal energy forms a pressure gradient that accelerates the gas. A second acceleration stage is possible wherein the primary outflow stretches magnetic field lines. The field lines then reconnect and accelerate small amount of mass to very high speeds. This double-stage acceleration process might form highly relativistic jets from black holes and neutron stars. The model predicts that detail analysis of accreting brown dwarfs that launch jets will show the mass accretion rate to be $\dot{M}_{\rm BD} \gtrsim 10^{-9} - 10^{-8} M_\odot$ yr^{-1}, which is higher than present claims in the literature.

Keywords. ISM: jets and outflows, stars: formation.

1. Introduction

In many popular models for the formation of astrophysical massive jets (to distinguish from low density hot-plasma jets from radio pulsars) magnetic fields play a dominate role in accelerating the jet's material from the accretion disk. Most models in young stellar objects (YSOs) are based on the operation of large scale magnetic fields driving the flow from the disk. In the "X-wind mechanism" introduced by Shu *et al.* (1988, 1991) the jets are launched from a narrow region in the magnetopause of the stellar field. A different model, although also using open radial magnetic field lines, is based on an outflow from an extended disk region, and does not rely on the stellar magnetic field (Ferreira & Pelletier 1993, 1995; Wardle & Königl 1993; Königl & Pudritz 2000; Shu *et al.* 2000; Ferreira 2002; Krasnopolsky *et al.* 2003; Ferreira & Casse 2004).

Other MHD simulations show that the high post-shock thermal pressure might accelerate gas and form jets and/or winds, e.g., as in the accretion around a black hole (BH) simulations performed by De Villiers *et al.* (2004; also Hawley & Balbus 2002). In simulations of accretion onto a rotating BH De Villiers *et al.* (2005) find that both gas pressure gradients and Lorentz forces in the inner torus play a significant role in launching the jets.

A confining external pressure is required at the edge of accelerated jets (Komissarov *et al.* 2007). Therefore, whether a model is based on large scale magnetic fields or not, a confining external medium is required to form a collimated jet, and there is no advantage to magnetic fields-based models. The magnetic fields can further collimate the internal region, i.e., a self collimation (Komissarov *et al.* 2007).

In Soker & Regev (2003, hereafter SR03) and Soker & Lasota (2004, hereafter SL04) we reexamined the launching of jets from accretion disks by thermal pressure following Torbett (1984; also Torbett & Gilden 1992). This paper describes the basic ingredients of the processes described in those papers, as well as new ideas.

2. Motivation

There are several arguments that point to problems with models for launching jets that are based only on large scale magnetic fields.

(1) Precessing jets. In several YSOs the jets precess on a time scale $\lesssim 100$ years (see e.g. Barsony's poster at this conference). A large scale magnetic field cannot change its symmetry axis on such a short time.

(2) A collimated jets in a planetary nebulae There is a highly collimated clumpy double-jet in the planetary nebulae Hen 2−90 (Sahai & Nyman 2000), very similar in properties to jets from YSOs that form HH objects. The source of the accreted mass in planetary nebulae is thought to be a companion star. In such a system large scale magnetic fields are not expected.

(3) No jets in DQ Her (intermediate polars) systems. Intermediate polars (DQ Her systems) are cataclysmic variables where the magnetic field of the accreting WD is thought to truncate the accretion disk in its inner boundary. This magnetic field geometry is the basis for some jet-launching models in YSOs (e.g., Shu *et al.* 1991). However, no jets are observed in intermediate polars.

(4) Thermal pressure. Thermal pressure seems to be an important ingredient even in MHD models for jet launching (e.g., Ferreira & Casse 2004; Vlahakis *et al.* 2003; Vlahakis & Königl 2003). In particular I note that in the exact solutions for steady relativistic ideal MHD outflows found by Vlahakis & Königl (2003; also Vlahakis *et al.* 2003) the initial acceleration phase is by thermal pressure, and internal (thermal) energy is converted to magnetic energy.

3. Launching jets by thermal pressure gradients

In YSOs the scheme was developed by Torbett (1984; also Torbett & Gilden 1992), and discussed in more detail by SR03. SL04 further discussed it, and include accretion into white dwarfs (WDs) as well. In this model, the accreted disk material is strongly shocked due to large gradients of physical quantities in the boundary layer, and then radiatively cools on a time scale longer than the ejection time from the disk.

The model assumes that hundreds of small blobs are formed in the sheared boundary layer (section 2 of SR03). The blobs occasionally collide with each other, and create shocks which cause the shocked regions to expand in all directions. If the shocked regions continue to expand out into the path of yet more circulating blobs, stronger shocks may be created, as was proposed by Pringle & Savonije (1979) to explain the emission of X-rays out of disk boundary layers in dwarf novae. For the shocked blobs to expand, the radiative cooling time of *individual blobs* must be longer than the adiabatic expansion time of individual blobs. SR03 demand also that the blobs be small, because the dissipation time of disk material to form the strong shocks must be shorter than the jet ejection time (eq. 24 of SR03). SR03 find that the thermal acceleration mechanism works only when the accretion rate in YSO accretion disks is large enough and the α parameter of the disk small enough – otherwise the radiative cooling time is too short and significant ejection does not take place. SR03 term the strong shocks which are formed from the many weakly shocked blobs, 'spatiotemporally localized (but not too small!) accretion shocks', or SPLASHes. Such SPLASHes can be formed by the stochastic behavior of the magnetic fields of the disk itself and of the central object, e.g., as in cases where the inner boundary of the disk is truncated by the stellar magnetic field.

The model then has two conditions. The first condition is that the strongly shocked gas in the boundary layer will cool slowly, such that the thermal pressure will have enough time to accelerate the jet's material. The radiative cooling is via photon diffusion. The constraints translate to a condition on the mass accretion rate to be above a minimum value, depending on the accreting objects, e.g., a WD or a main sequence star. The second condition, which in general is stringent, is that weakly shocked blobs in the boundary layer will expand, and disturb the boundary layer in such a way that a strong shock will develop. This also leads to a minimum value for the mass accretion rate (SR03; SL04 eq. 12)

$$\dot{M} \gtrsim 4.2 \times 10^{-5} \kappa^{-1} \left(\frac{\alpha}{0.1}\right) \left(\frac{R}{R_\odot}\right) M_\odot \text{ yr}^{-1}, \tag{3.1}$$

where R is the disk radius from where the jet is launched, α is the disk's viscosity parameter, and κ is the opacity. Using this criterion SL04 find that the mass accretion rate above which jets could be blown from accretion disks around YSOs, where $R \simeq R_\odot$ and after substituting the opacity, is

$$\dot{M}_b(\text{YSOs}) \gtrsim 7 \times 10^{-7} \left(\frac{R}{R_\odot}\right)^{1.2} \left(\frac{\epsilon}{0.1}\right)^{1.4} \left(\frac{\alpha}{0.1}\right) M_\odot \text{ yr}^{-1}, \tag{3.2}$$

where $\epsilon = H/R$, and H is the vertical disk's scale height. This condition is drawn in the right hand side of Figure 1 for two sets of $(\alpha; \epsilon)$ parameters.

For WDs the post shock temperature is much higher than that YSOs, and the opacity is $\kappa = 0.4$ cm^2 g^{-1}. This gives

$$\dot{M}_b(\text{WD}) \gtrsim 10^{-6} \left(\frac{R}{0.01 R_\odot}\right) \left(\frac{\alpha}{0.1}\right) M_\odot \text{ yr}^{-1}. \tag{3.3}$$

This condition is drawn in the left hand side of Figure 1. SL04 noted that this limit is almost never satisfied in CVs (Warner 1995). However, Retter (2004) suggested that this limit might be met during the transition phase in novae, where a claim for a jet has been made.

4. Jets from brown dwarfs

The thermal launching model presented here can work only when mass accretion rate is high enough. I therefore turn to check the situation with brown dwarfs (BDs), where the claimed low mass accretion rate was presented as an evidence against the thermal-launching model.

Whelan *et al.* (2005, 2007) argued for a BD jet similar to that observed in YSOs. They use a forbidden line which forms at a critical density of 2×10^6 cm^{-3}, or $\rho_c \simeq 3 \times 10^{-18}$ g cm^{-3}, achieved at a distance r_o. Assuming that the half opening angle of each jet is $\alpha \sim 10°$, and the jets speed is v_j, the outflow rate of the two jets combined is

$$2\dot{M}_j = 4\pi \rho_c r_o^2 (1 - \cos\alpha) v_j \simeq 8 \times 10^{-10} \left(\frac{r_o}{10 \text{ AU}}\right)^2 \left(\frac{\alpha}{10°}\right)^2 \left(\frac{v_j}{40 \text{ km s}^{-1}}\right) M_\odot \text{ yr}^{-1}. \tag{4.1}$$

I substitute numbers as given by Whelan *et al.* (2007, Table 1) for two systems. For 2MASS1207-3932 I take for the observed values $v_j = 8$ km s^{-1} and $r_o = 4$ AU, while for ρ–Oph 102 I take $v_j = 40$ km s^{-1} and $r_o = 10$ AU. The unknown inclination implies that both the distance and velocity are larger than the observed values, and the mass outflow rate should be multiply by a number > 2.6; I multiply it by 2.6.

For 2MASS1207-3932 I find $\dot{M}(2MASS1207 - 3932) \simeq 6.6 \times 10^{-11} M_\odot$ yr^{-1}. From Figure 2 of Whelan *et al.* (2007) it seems that the blue-shifted outflow has a large opening angle (or large covering factor), and I expect that for this case $\alpha > 10°$. Over all, the

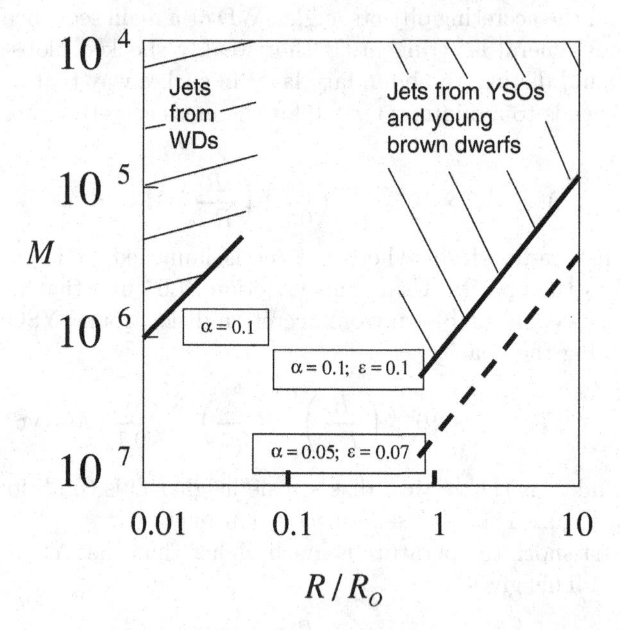

Figure 1. The condition for the radiative cooling time to be longer than mass ejection time, in the accretion rate (in $M_{\odot}$ yr^{-1}) versus jet launching radius (in $R_{\odot}$) plane. Due to several uncertainties, the allowed mass accretions rate can be an order of magnitude lower, but the general constraints are as shown. The radiative cooling time of the postshock gas is dictated by the photon diffusion time (SR03).

mass loss rate is $\dot{M}(2MASS1207-3932) \gtrsim 10^{-10} M_{\odot}$ yr^{-1}. This is an order of magnitude larger than the accretion rate given in Table 1 of Whelan *et al.* (2007). For ρ–Oph 102 I find $\dot{M}(\rho - \text{Oph } 102) \simeq 2 \times 10^{-9} M_{\odot}$ yr^{-1}. This mass outflow rate is 60% higher than the mass accretion rate given in Table 1 of Whelan *et al.* (2007). Jets with opening angle much smaller that 10° are not likely close to their source in YSOs. Considering that mass outflow rates in YSOs are 0.01 – 0.1 times the mass accretion rate, I conclude that there is inconsistency in the data given by Whelan *et al.* (2005, 2007) for outflow from BDs. The resolution to this problem can be one of the following.
(1) The accretion rate is indeed very low ($\sim 10^{-11} M_{\odot}$ yr^{-1}), and the mass outflow rate is much smaller than what I estimated above. In that case, the outflow rate ($\sim 10^{-12} M_{\odot}$ yr^{-1}) can be accounted for by a BD stellar-type wind, with no need for jets.
(2) The mass accretion rate is much higher than that given by Whelan *et al.* (2007) and Mohanty *et al.* (2005), i.e., it is $\dot{M}_{\text{acc}} \simeq 10^{-8} M_{\odot}$ yr^{-1}. In that case jets can be launched according to the model presented here. A higher mass accretion rate is suggested also by the young age of accreting BDs, which is similar to that of YSOs (Mohanty *et al.* 2005). According to Mohanty *et al.* (2005) the number of accreting BDs declines substantially by the age of $\sim 10^7$ yr. For a BD mass of $M_{\text{BD}} > 0.01 M_{\odot}$ the implied average accretion rate should be $\gtrsim 10^{-9} M_{\odot}$ yr^{-1}. A short accretion phase for YSOs is also suggested by the work of Lucas Cieza et al (2007).
(3) There are very large variations on short time scales of the mass accretion rate, as suggested by Scholz & Jayawardhana (2006). Jets are launched then only during the

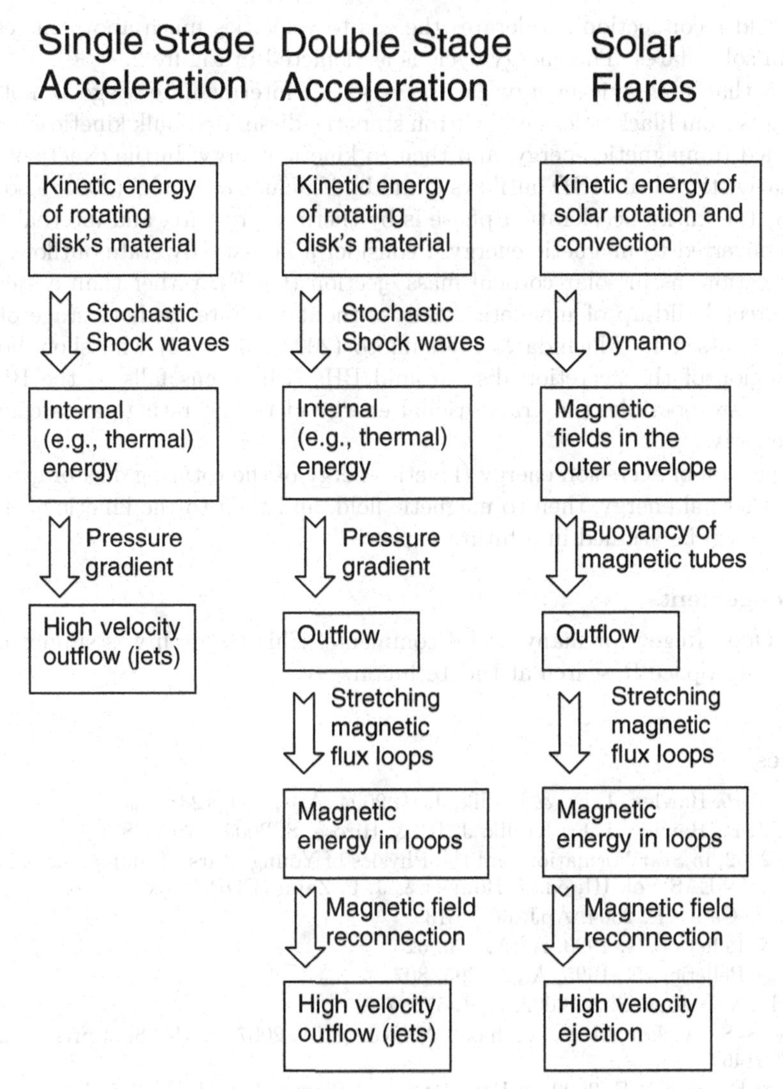

Figure 2. The energy transformations in the single-stage acceleration scenario and in the double-stage acceleration scenario.

very high mass accretion phases. On average, the mass accretion rate can stay low. The outflow is composed of many small clumps, and the *average* mass outflow rate is much smaller than what I calculated above.

I predict that future studies will show that the mass accretion rate of BDs that launch jets (and not stellar-type winds) is $\dot{M}_{\rm acc}(BDs) \gtrsim 10^{-9} - 10^{-8} M_\odot\ {\rm yr}^{-1}$, i.e., larger than current studies show (e.g., Mohanty *et al.* 2005). Such an accretion rate with a low enough value of the disk-viscosity parameter α (see equation 3.2), is compatible with the model presented here.

5. A double-stage acceleration scenario

I propose the possibility that thermal energy is transfered to magnetic energy, via kinetic energy of the outflowing gas, and a second stage of acceleration takes place wherein magnetic field reconnection accelerates the gas to velocities much above the escape velocity, as in solar flares. The energy cycle is as depicted in Figure 2.

The idea that magnetic energy can serve as an intermediate stage is not new. In launching jets from black holes and neutron stars the dissipated bulk kinetic energy might be channelled to magnetic energy, and then to kinetic energy. In the exact solutions for steady, relativistic, ideal MHD outflows found by Vlahakis & Königl (2003; also Vlahakis *et al.* 2003) the initial acceleration phase is by thermal pressure, and internal (thermal) energy is converted to magnetic energy. I consider a non-steady state outflow, based on flare-like ejection, as in solar coronal mass ejection (CME), rather than a steady state outflow. Direct build-up of magnetic fields, without the intermediate stage of thermal energy, is discussed by Machida & Matsumoto (2003, their § 4) who show how in the plunging region of the accretion disk around BHs, where gas falls to the BH and no stable orbits are possible, the gravitational energy of the accreting gas is converted to magnetic energy.

The conversion of accretion energy (kinetic energy of the rotating disk or gravitational energy) to thermal energy, then to magnetic field, and then to the kinetic energy of the ejected wind will be studied in a future paper.

Acknowledgements

I thank Oded Regev for many useful comments. This research was supported by the Asher Fund for Space Research at the Technion.

References

De Villiers, J. P., Hawley, J. F., & Krolik, J. H. 2004, ApJ, 599, 1238
De Villiers, J. P., Hawley, J. F., Krolik, J. H., & Hirose, S. 2005, 620, 878
Ferreira, J. 2002, in Star Formation and the Physics of Young Stars - Summer school on Stellar Physics X - EAS Vol. III, ed. J. Bouvier & J.-P. Zahn (EDP Books)
Ferreira, J., & Casse, F. 2004, ApJ, 601, L139
Ferreira, J. & Pelletier, G. 1993, A&A, 276, 625
Ferreira, J. & Pelletier, G. 1995, A&A, 295, 807
Hawley, J. F., & Balbus, S. A. 2002, ApJ, 573, 738
Komissarov, S. S., Barkov, M. V., Vlahakis, N., Königl, A. 2007. MNRAS, in press (arXiv:astro-ph/0703146)
Königl, A., & Pudritz, R.E. 2000, in Protostars and Planets IV, ed. V. Mannings, A. P. Boss & S. S. Russell, (Univ. of Arizona Press, Tucson), 759
Krasnopolsky, R., Li, Z.-Y., & Blandford, R. D. 2003, ApJ, 595, 631
Lucas Cieza *et al.* 2007, ApJ in press (arXiv:0706.0563
Machida, M., & Matsumoto, R. 2003, ApJ, 585, 429
Mohanty, S., Jayawardhana, R., & Basri, G. 2005, ApJ, 626, 498
Pringle, J. E. & Savonije, G.J. 1979, MNRAS, 187, 777
Retter, A. 2004, ApJ, 615, L125
Sahai, R. & Nyman, L.-A. 2000, *ApJ* 538, L145
Scholz, A., & Jayawardhana, R. 2006, ApJ, 638, 1056
Shu, F. H., Lizano, S., Ruden, S., & Najita, J. 1988, ApJ, 328, L19
Shu, F.H., Najita, J.R., Shang, H. & Li Z.-Y. 2000, in Protostars and Planets IV, ed. V. Mannings, A. P. Boss & S. S. Russell, (Univ. of Arizona Press, Tucson), 789
Shu, F. H., Ruden, S.P., Lada, C.J. & Lizano, S. 1991, ApJ, 370, L31
Soker, N., & Lasota, J.-P. 2004, A&A, 422, 1039 (SL04)
Soker, N., & Regev, O. 2003, A&A, 406, 603 (SR03)
Torbett, M. V. 1984, ApJ, 278, 318
Torbett, M. V., & Gilden, D. L. 1992, A&A, 256, 686 (TG)

Vlahakis, N., & Königl, A. 2003, ApJ, 596, 1104
Vlahakis, N., Fang, P., & Königl, A. 2003, ApJ594, L23
Wardle, M., & Königl, A. 1993, ApJ, 410, 218
Warner, B. 1995, Cambridge Astrophysics Series, Cambridge, New York: Cambridge University Press, 1995,
Whelan, E. T., Ray, T. P., Bacciotti, F., Natta, A., Testi, L., & Randich, S. 2005, Natur, 435, 652
Whelan, E. T., Ray, T. P., Randich, S., Bacciotti, F., Jayawardhana, R, Testi, L., Natta, A., & Mohanty, S. 2007, ApJ, 659, L45

Star-Disk Interaction in Young Stars
Proceedings IAU Symposium No. 243, 2007
J. Bouvier & I. Appenzeller, eds.

doi:10.1017/S1743921307009568

The accretion-ejection connexion in T Tauri stars: jet models vs. observations

S. Cabrit

LERMA, Observatoire de Paris, UPMC, CNRS, 61 Av. de l'Observatoire, 75014 Paris, France
email: Sylvie.Cabrit@obspm.fr

Abstract. Key observational constraints for jet models in T Tauri stars are outlined, including the jet collimation scale, kinematic structure, and ejection/accretion ratio. It is shown that MHD self-collimation is most likely required. The four possible MHD ejection sites (stellar surface, inner disk edge, extended disk region, magnetosphere-disk reconnexion line) are then critically examined against observational constraints, and open issues are discussed.

Keywords. Hydrodynamics, stars: pre–main-sequence, stars: winds, outflows, ISM: jets and outflows.

---

1. Introduction: key properties of spatially resolved T Tauri jets

Modelling of Helium and Hydrogen line profiles in classical T Tauri stars (CTTS) suggests the presence of inner winds from the star or the inner disk edge (see reviews by S. Edwards and by S. Alencar, this volume). The contribution of these winds to the large scale jets observed in forbidden lines and to the angular momentum regulation of CTTS is still unclear, however. In this review, I will approach the problem from the other end and use asymptotic properties beyond 15 AU derived from spatially resolved forbidden line observations (see T. Ray, this volume) to test proposed models for T Tauri jets. These properties include:

• *1. Jet collimation:* the opening angle drops from $20° - 30°$ initially to a few degrees beyond 50 AU of the source, where the apparent jet HWHM radius reaches $\simeq$ 10–20 AU (Hartigan *et al.* 2004; Ray *et al.* 1996; Dougados *et al.* 2000).

• *2. Jet terminal speeds*: the high-velocity component (HVC) typically reaches $\simeq$ 200 to 350 km s^{-1} after deprojection, i.e.,g $1 - 2$ times the stellar keplerian speed, within 15 AU of the source (Bacciotti *et al.* 2000; Woitas *et al.* 2002).

• *3. Transverse velocity decrease:* several jets exhibit an intermediate velocity component (IVC) at 100–10 km s^{-1}, arising from a slower sheath at $\simeq$ 15–30 AU of the jet axis (Lavalley-Fouquet *et al.* 2000; Bacciotti *et al.* 2000; Coffey *et al.* 2004; Coffey *et al.* 2007).

• *4. Jet rotation:* the specific angular momentum at the jet outer edge could be up to $100 - 300$ AU km s^{-1}; it would be lower if other effects than rotation are present (Bacciotti *et al.* 2002; Coffey *et al.* 2004; Cabrit *et al.* 2006; Cerqueira *et al.* 2006).

• *5. Jet ejection/accretion ratio:* current best estimates lead to a *two-sided* value of $\simeq$ 0.1–0.2 (see below).

Recent advances on jet heating and jet mass fluxes are reviewed in § 2, while § 3 presents several arguments in favor of MHD self-collimation of CTTS jets. § 3–6 then confront the possible jet origins (star, inner disk edge, extended disk region, magnetosphere-disk reconnexion site) with the above observational constraints. Concluding remarks are presented in § 8.

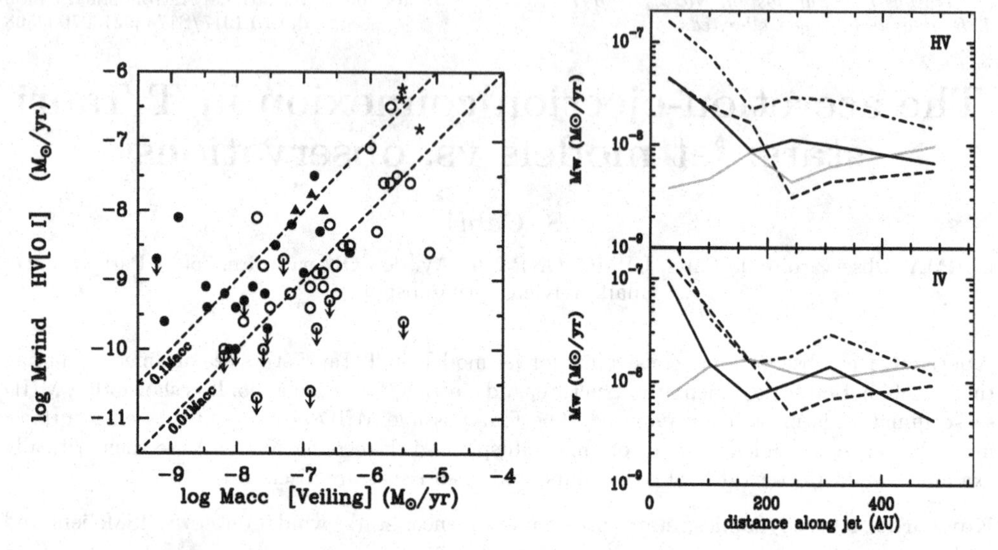

Figure 1. *Left:* Correlation of ejection rate in the blueshifted jet to accretion rate from veiling in CTTS. Open circles are data from HEG95. Filled circles use revised $\dot{M}_{\rm acc}$ from Muzerolle *et al.* 1998. The 3 microjets with updated $\dot{M}_j$ from spectroimaging are denoted as filled triangles, and the 3 Class I jets as asterisks. The mean one-sided $\dot{M}_j/\dot{M}_{\rm acc}$ is 0.05–0.1. From Cabrit *et al.* in prep. *Right:* Mass-loss rates in the HVC and IVC of DG Tau obtained with the 4 methods described in the text (A.1: dash-dot, A.2: solid, B.1: grey, B.2: dashed). From Cabrit (2002).

2. Jet mass-flux and ejection/accretion ratio

The good correlation of [OI] jet brightness with mid-infrared excess from the inner disk and with optical excess from the hot accretion layer reveals that jets are ultimately powered by accretion (Cabrit *et al.* 1990; Hartigan *et al.* 1995, hereafter HEG95). The ejection/accretion ratio is then a key parameter to constrain the jet acceleration mechanism and launch site. HEG95 inferred a mean *one-sided* ratio $\dot{M}_j/\dot{M}_{\rm acc} \simeq 0.01$ (see open circles in left panel of Figure 1), but updated accretion rates using revised bolometric corrections and A_V are on average 10 times smaller (Muzerolle *et al.* 1998). This would suggest a 10 times higher ratio, provided $\dot{M}_j$ does not suffer from a similar bias.

Significant progress on CTTS jet mass-flux estimates have been made recently thanks to sub-arcsecond spectroimaging: as shown in Fig. 2, spatially-resolved line ratios in microjets demonstrate that heating is dominated by shocks beyond 30 AU, and yield estimates of shock parameters and postshock density as a function of distance and velocity. The mass-flux can then be cross-checked using 4 different methods: one may either use the jet mean density and radius (option A), or the [OI] line luminosity (option B); and in each case one may assume either (1) a uniform emissivity within the beam, or (2) a single shock wave (see Cabrit 2002 for a detailed review).

A comparison of the 4 methods in the DG Tau jet is presented in the right panel of Fig. 1. They agree to within a factor 3 beyond 150 AU, but greatly diverge closer in. This could be due to the steeper gradients in physical conditions close to the star, and to the larger uncertainties in A_V and jet radius there. "Asymptotic" mass-loss rate values beyond 150 AU ($t_{dyn} \simeq 3$ yrs) are therefore more reliable.

Comparing with earlier mass-loss rates obtained by HEG95 from integrated [OI] fluxes, the improved asymptotic value is a factor of 10 lower in DG Tau (Lavalley-Fouquet *et al.* 2000), similar in RW Aur (Woitas *et al.* 2002), and a factor 10 higher in RY Tau (Agra-Amboage *et al.*, submitted). Thus, HEG95 mass-loss rates currently do not appear to

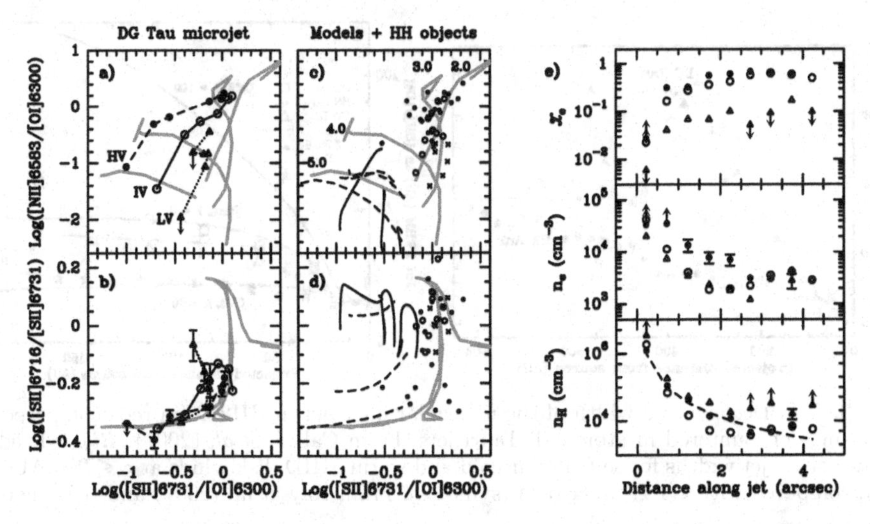

Figure 2. *Left:* Line ratios along the DG Tau microjet in 3 velocity intervals (large connected symbols in a,b) agree much better with predictions for planar shocks (thick green curves), than with ambipolar diffusion heating (thin solid in c,d) or viscous mixing-layers (dashed in c,d). The same is true for HH objects (small symbols in c,d). *Right):* Ionisation fraction x_e, electronic density n_e, and total density $n_H = n_e/x_e$ along the DG Tau jet, inferred from line ratios with the BE99 technique (Bacciotti & Eislöffel 1999). All panels from Lavalley-Fouquet *et al.* (2000). See also Dougados *et al.* (2002) for a similar analysis in the RW Aur microjet.

suffer from a large systematic bias and should remain useful for a statistical analysis. Combining them with updated $\dot{M}_{acc}$ values from Muzerolle *et al.* (1998), and adding the 3 revised $\dot{M}_j$ from spectroimaging, one obtains $\dot{M}_j/\dot{M}_{acc} \simeq 0.05$–$0.1$ (filled symbols in Figure 1). The same is found for 3 resolved Class I jets assuming $L_{bol} = L_{acc}$ (Hartigan *et al.* 1994; Cabrit 2002). Multiplying by 2 to account for the occulted redshifted jet yields $f \equiv (2\dot{M}_j)/\dot{M}_{acc} \simeq 0.1 - 0.2$, and $2L_j \simeq 0.1L_{acc}$ within a factor 2–3.

3. The need for MHD self-collimation of jets in CTTS

External collimation of an isotropic wind is difficult to reconcile with current constraints on the jet collimation scale (Cabrit 2007). The main arguments are summarized below, for a typical mass-loss rate of $\dot{M}_w = 10^{-8}\ M_\odot \mathrm{yr}^{-1}$ and $\dot{M}_{acc} = 10^{-7}\ M_\odot \mathrm{yr}^{-1}$.

3.1. *External hydrodynamical collimation*

An isotropic hydrodynamical wind is refocussed into a polar jet at distance Z_{max} if the ambient pressure P_0 is comparable to the wind ram pressure there: $P_0 \simeq \dot{M}_w V_w/(4\pi Z^2_{max})$ (Barral & Cantó 1981; Cabrit 2007). With V_w = 300km s^{-1}, Z_{max} = 50 AU, and the additional constraint that $n_{coll} \leqslant 4 \times 10^6$ cm^{-3} over this scale ($A_V \leqslant$ 3mag in T Tauri stars), one would need hot material at $T > 6000$ K, only expected in a photoionised flow from the disk surface. Such a thermal flow reaches a speed $v_{evap} \simeq 30$ km s^{-1} (Font *et al.* 2004), ie 1/10th that of the wind. Thus, even including its ram pressure, it would need a mass-flux of $10 \times \dot{M}_w = 10^{-7}\ M_\odot \mathrm{yr}^{-1}$ to refocus it. This is 1000 times more than the typical disk evaporation rate for a CTTS (Font *et al.* 2004) and comparable to the disk accretion rate, which is excluded.

Collimation by external hydrodynamic pressure may thus be safely ruled out. This conclusion is reinforced by the identical width of the molecular microjet of the HH212

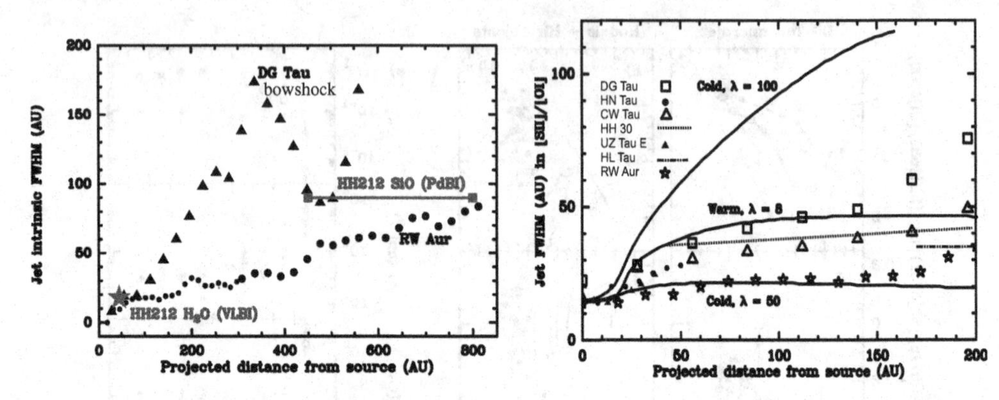

Figure 3. *Left:* Deconvolved width of the molecular microjet in HH212 (in red) compared with the full range encountered in atomic T Tauri jets. From Cabrit *et al.* (2007). *Right:* Predicted beam-convolved jet widths for self-similar cold and warm MHD disk winds at $z \leqslant 200$ AU (solid curves), compared with CTTS microjets (symbols). From Ray *et al.* (2007) and refs. therein

Class 0 source vs. atomic T Tauri jets, demonstrating that a dense infalling Class 0 envelope does not affect jet collimation (Cabrit *et al.* 2007, see left panel of Fig 3).

3.2. *External magnetic collimation*

Replacing P_0 above by $B^2_{\rm coll}/8\pi$ and keeping the same wind parameters one obtains a rough indication of the poloidal field that would focus an isotropic CTTS wind into a jet at z=50 AU: $B_{\rm coll} \simeq 10$ mG. For jet collimation to be effective, the field should be anchored over a disk region of radius $r_D \simeq 100$ AU $\gg r_j$ (Kwan & Tademaru 1988). The corresponding trapped flux would be $(\Phi_B)_{\rm coll} = \pi r_D^2 B_{\rm coll} = 8 \times 10^{28}$G cm^2, i.e., *at least* 2% of the flux present before gravitational collapse $(\Phi_B)_{\rm init} < (\Phi_B)_{\rm crit} = M\sqrt{G}/0.13 = 4 \times 10^{30}(M/1M_\odot)$G cm^2 (Mouschovias & Spitzer 1976). In contrast, 3D numerical MHD simulations of collapse find that only 0.1% of the initial flux remains in the disk, due to ohmic field diffusion (see contribution by Inutsuka, this volume).

Shu *et al.* (2007, and this volume) investigate an alternative scenario where 25% of the critical flux is conserved within 100 AU. External collimation could then occur on observed scales. However, the strong field also causes subkeplerian disk rotation, by 65% in CTTS. The predicted systematic discrepancy of a factor 1/0.4 between theoretical tracks and dynamical masses from disk rotation curves in CO does not appear supported by observations (White *et al.* 1999, Simon *et al.* 2000).

Finally, "active" external magnetic confinement may be provided by a self-collimated outer magnetized disk wind, but the latter would dominate the overall jet mass-flux (Meliani *et al.* 2006). A *purely passive, external* magnetic collimation of T Tauri jets therefore also appears unlikely.

3.3. *MHD self-collimation*

Any MHD wind launched along rotating open field lines tends to undergo self-collimation towards the spin axis (Bogovalov & Tsinganos 1999 and refs. therein). Collimation is achieved by the *toroidal* field component created by the wind inertia and not by the poloidal component. Therefore, it is much more efficient than external magnetic collimation in terms of required flux. For example, in self-similar MHD disk winds, the magnetic flux within the jet launch region $r_{out} \leqslant 1$–10 AU is $\Phi_B \leqslant 10^{26} - 10^{27}$ G cm^2 for $\dot{M}_{\rm acc}$= $10^{-7} M_\odot {\rm yr}^{-1}$ (cf. Eq. (19) in Ferreira *et al.* 2006), ie 2–3 orders of magnitude smaller than for external magnetic collimation, and less than 0.1% of the primordial flux.

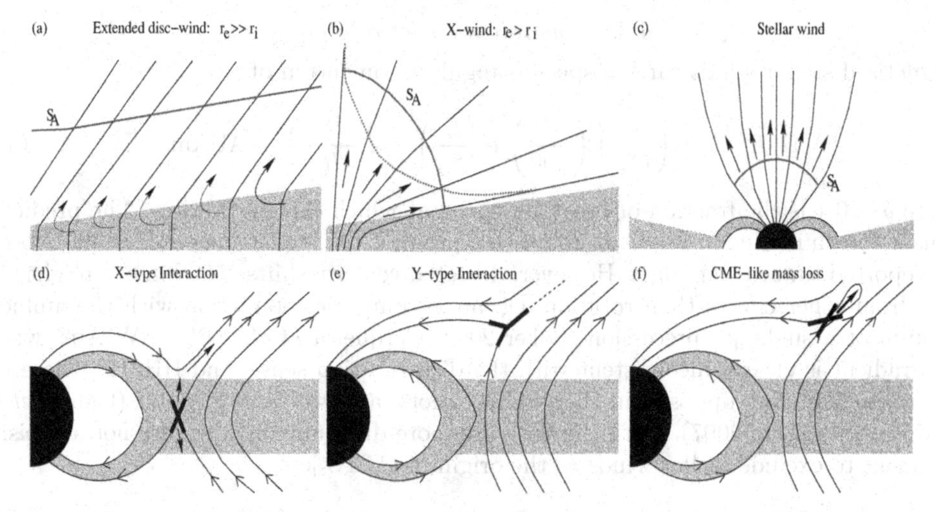

Figure 4. Possible launchsites for MHD winds in young accreting stars. Only those in Panels a) to d) are self-collimated. See Ferreira *et al.* (2006) for details.

Given the large flux loss expected during collapse, it thus seems most probable that CTTS jets trace *self-collimated MHD wind(s)*. Ferreira *et al.* (2006) distinguish four possible launch sites for such winds, illustrated in Fig. 4: (a) extended disk winds, (b) inner disk winds ("X-winds"), (c) stellar winds, (d) reconnexion X-winds ("ReX-winds"). In the following, we confront each of the four options against large-scale jet properties.

4. Pressure-driven MHD stellar winds

Helium line profiles indicate that stellar winds are present in at least 60% of CTTS (see Edwards, this volume; Kwan *et al.* 2007). If they were the main agent regulating angular momentum in CTTS, their mass-loss rate would be $\simeq 0.1$ of $\dot{M}_{\rm acc}$ (Matt & Pudritz 2005). Therefore they are a prime candidate for the origin of T Tauri jets.

4.1. *Jet collimation*

Current analytical and numerical models of MHD stellar winds predict a narrow region of cylindrical flow with radius $\leqslant$ 10–20 Alfvén radii, i.e., $\leqslant 200 R_\star \simeq 2$ AU (eg. Sauty & Tsinganos 1994; Bogovalov & Tsinganos 2001). Possible ways to reproduce the apparent jet radii of 10–20 AU, eg. through density collimation or enhanced mass-flux at intermediate latitudes, remain to be investigated.

4.2. *Jet poloidal speeds along and transverse to the axis*

CTTS stellar winds probed in Helium lines reach a speed similar to that of the large-scale jets (Edwards *et al.* 2003). Since T Tauri stars rotate at only 10% of break-up, centrifugal launching is ineffective and strong pressure gradients are needed to accelerate the gas. Following Ferreira *et al.* (2006), the asymptotic speed may be written $V_j = \sqrt{(\beta - 2) G M_\star / R_\star}$ where $\beta = 2(\Delta H + \mathcal{F})/(G M_\star / R_\star)$ parametrises the amount of energy given to the flow from enthalpy gradients, Alfvén wave pressure, etc.

Observed HVC speeds of 1 – 2 times the stellar keplerian speed thus require $\beta \simeq 3 - 6$. This is only slightly changed by the effect of centrifugal acceleration, as the star rotates slowly (cf. Fig. 7 and Ferreira *et al.* 2006). The strong transverse velocity decrease observed in several jets could be reproduced, eg. if β drops at lower latitudes, or if the last stellar field line is recollimated by a disk field, allowing the development of slower bowshocks/turbulent wakes around the fast jet beam.

4.3. *Constraints on jet rotation*

Magnetised stellar winds carry a specific angular momentum of

$$\lambda_\star \Omega_\star R_\star^2 = 70 \left(\frac{\delta}{0.1}\right) \left(\frac{\lambda_\star}{200}\right) \left(\frac{M_\star}{M_\odot}\right)^{1/2} \left(\frac{R_\star}{3R_\odot}\right)^{1/2} \mathrm{AU\,km\,s^{-1}}, \tag{4.1}$$

where $\delta \simeq 0.1$ is the fraction of break-up speed at which the star rotates. The predicted locus is shown in blue in Fig. 7. It does not reproduce the large values of 200–300 AU km s^{-1} reported towards jet edges. However, detected velocity shifts $\leqslant$ 20 km s^{-1} might also arise from other effects than rotation, eg. an asymmetric interaction with the ambient medium or a slight jet precession (Soker 2005; Cerqueira *et al.* 2006). RW Aur, where the gradient is clearly inconsistent with the disk rotation sense, and HH212, where the transverse shifts are opposite in H_2 and SiO knots, are two cases in point (Cabrit *et al.* 2006; Codella *et al.* 2007). Until the data are more discriminant, they are not a decisive argument to exclude stellar winds as the origin of CTTS jets.

4.4. *Jet ejection to accretion ratio*

The large optical depth and unknown geometry of CTTS stellar winds currently prevent an accurate measure of their contribution to the jet mass-flux. But the following theoretical arguments, from Ferreira et al. (2006), show that an ejection/accretion ratio $\simeq$ 0.2 would be challenging: with $\beta \simeq 3$, the net energy input in the two jets would be:

$$L_\beta = \frac{\beta}{2} \frac{GM_\star}{R_\star} (2\dot{M}_j) = \beta \left(\frac{\dot{M}_j}{\dot{M}_a}\right) L_{\mathrm{acc}} \simeq 30\% L_{\mathrm{acc}}. \tag{4.2}$$

If energy were provided in the form of enthalpy, the true total heating rate *including radiative losses* would then be excessive (cf. Matt, this volume; De Campli 1981). In addition, CTTS stellar winds appear cooler than 20,000 K in their acceleration region (Johns-Krull & Herczeg 2007), also arguing against significant enthalpy gradients.

Non-thermal acceleration by Alfvén wave pressure gradients meets a similar efficiency problem. For $B_\star$ = 150-500 G and a final speed of 300 km s^{-1}, De Campli (1981) found that the required power in *coherent* Alfvén waves is $5 - 10$ times the jet power, i.e., 50%-100% of L_{acc}. This sounds prohibitive, as incoherent Alfvén waves and dissipative waves (acoustic, magnetosonic) will be excited as well. Note that De Campli concluded otherwise because he was comparing the wave power to the total stellar luminosity, not to L_{acc} which is typically much smaller in CTTS.

A more promising pressure-drive for stellar winds is the "magnetic coil" push produced by strongly twisted open field lines in the stellar corona. This effect is observed in numerical simulations (see contributions by Inutsuka and by Romanova, this volume) but may be transient. Unless it proves to be long-lived and efficient enough, pressure-driven stellar winds would provide no more than $\simeq$ 10% of the mass-flux in T Tauri jets.

5. Inner disk winds: The "X-wind" model(s)

In the X-wind scenario, a steady-state "disk-locking" is assumed, where angular momentum accreted through funnel flows is balanced by angular momentum deposited slightly outside corotation by trailing closed stellar field lines. The excess angular momentum deposited in the disk is then assumed to power a centrifugal outflow from a tiny region beyond this point, along field lines that have been disconnected from the star. Using a prescribed mass-loading function, the Alfvén surface and asymptotic collimation

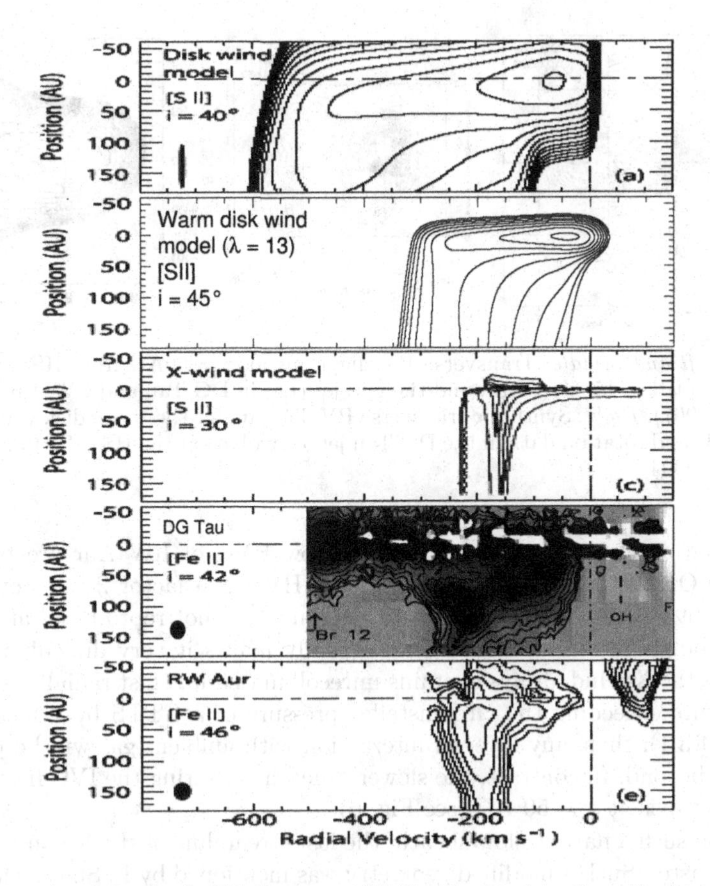

Figure 5. Observed and synthetic PV diagrams of T Tauri microjets along the jet axis (adapted from Pyo *et al.* 2006): the warm disk wind model in (b) is from Cabrit *et al.* (in prep), and the DG Tau PV diagram in (d) is from Pyo *et al.* (2003).

were calculated in the case of no external disk field, making several specific predictions that can be compared with observations (Najita & Shu 1994; Shu *et al.* 1995).

5.1. *Jet collimation*

Despite the presence of unrecollimated radial streamlines at wide angle, the X-wind quickly achieves a cylindrical *density* distribution. The jet beam is then somewhat of an "optical illusion" (Shang *et al.* 1998). Unconvolved synthetic maps yield a power-law transverse intensity distribution with a narrow core of 2 AU. PSF convolution would thus be needed for a definite comparison with observed jet widths.

5.2. *Jet poloidal speeds along and transverse to the axis*

The X-wind predicts a mean magnetic lever arm parameter $\bar{\lambda} = (r_A/r_o)^2 \simeq 3.5$ over most of the flow. Panel (c) of Fig. 5 shows an unconvolved synthetic PV cut along the jet calculated for a corotation radius $R_{\rm cor} = 12R_\odot$ and $V_{K,\rm cor} = 92$ km s^{-1} (Shang *et al.* 1998) corresponding to $M_\star = 0.5M_\odot$ and $P_\star = 6.6$ days, ie. typical of a CTTS like DG Tau. The mean (deprojected) terminal jet speed is then $\simeq 180$ km s^{-1}, with 10% of the mass flux reaching 270 km s^{-1}.

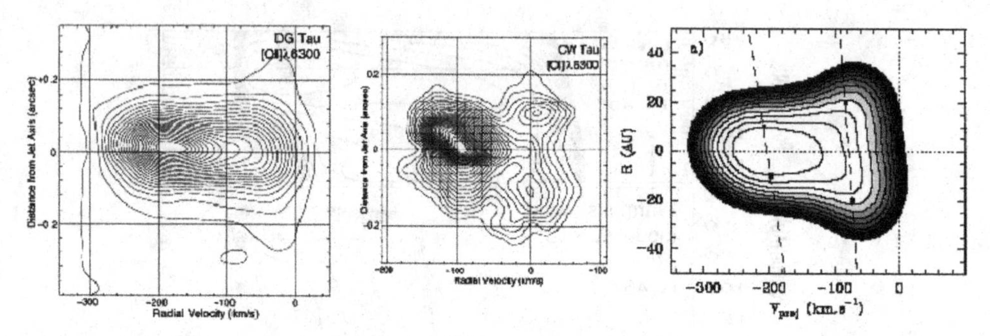

Figure 6. *Left and Middle:* Transverse PV diagrams at z = 50AU from HST illustrating the slower sheath at R = 15–30 AU around the fast jet core in DG Tau and CW Tau (adapted from Coffey *et al.* 2007). *Right:* Synthetic transverse PV diagram for the warm disk wind model fitting both the HVC and rotation data in the DG Tau jet, convolved at the HST/STIS resolution (from Pesenti *et al.* 2004).

The overall acceleration scale and the narrow HVC of RW Aur are both very well reproduced. On the other hand, the predicted HVC is a factor 2 too slow in DG Tau, and the bright IVC at velocities down to -50 km s^{-1} is not reproduced at all.

The wide-angle nature of the X-wind actually makes it very difficult to produce an IVC feature: the X-wind always contains unrecollimated *and* fast radial streamlines, with a ram pressure exceeding the circumstellar pressure of a CTTS by a factor $\simeq$ 1000 at 50 AU (see § 3.1); thus, any layer of interaction with ambient gas will be pushed out to 1000 AU or beyond. In contrast, the slower "sheath" emitting the IVC lies at only 15–30 AU of the jet axis at z = 50 AU (see Fig. 6).

To develop such a narrow slow sheath, the last streamline of the X-wind should recollimate much faster. Such a modified geometry was mentioned by F. Shu at this conference, but the proposed confining disk field predicts subkeplerian disk rotation that does not seem supported by observations (cf. § 3.2). Jets with an IVC thus remain a challenge for X-wind model(s) as currently envisioned.

5.3. *Jet rotation*

The specific angular momentum carried away by the X-wind is

$$\lambda\Omega_\star R_{\rm cor}^2 = \lambda\Omega_{K,\star} R_\star^2 \delta^{-1/3} = 22\text{–}44 \left(\frac{\delta}{0.1}\right)^{-1/3} \left(\frac{M_\star}{M_\odot}\right)^{1/2} \left(\frac{R_\star}{3R_\odot}\right)^{1/2} \text{AU km s}^{-1}. \quad (5.1)$$

The predicted locus in the $rV_\phi - V_p$ plane is indicated in Fig. 7. It falls near stellar winds with $\beta = 2$, $\lambda_\star = 50 - 150$, and is again compatible with current data if considered as upper limits to the true jet rotation.

5.4. *Ejection to accretion ratio*

Current predictions for the X-wind were calculated with a prescribed ejection/accretion ratio $f = 2\dot{M}_j/\dot{M}_{acc} \simeq 0.25$, compatible with the mean value in CTTS jets. The fact that $f\bar{\lambda} \simeq 1$ means that the total angular momentum flux carried away by the X-wind is equal to that extracted from the funnel flow by disk-locking: $\dot{M}_{\rm acc}\Omega_\star R_{\rm cor}^2$ (assuming a truncation radius close to corotation). However, the turbulent viscosity and steady-state disk structure required to transport the extracted angular momentum to the X-wind launch region, and to sustain the high mass-loading, have not yet been calculated.

6. Extended MHD disk winds

Magneto-centrifugal ejection from keplerian accretion disks is a well-understood jet formation process (see Fendt, this volume; Pudritz *et al.* 2007 and refs. therein). Steady, self-similar solutions including full treatment of the mass-loading have been calculated for vertically isothermal disks ("cold" disk winds, Ferreira 1997), and for disks with moderate surface heating ("warm" disk winds, Casse & Ferreira 2000). From these, a complete set of synthetic predictions was produced and tested against observational constraints, with the following results:

6.1. *Jet collimation*

As shown in the right panel of Fig. 3, beam-convolved synthetic maps for self-similar disk winds reproduce very well the observed jet FWHM as a function of distance, for an inner launch radius close to corotation ($r_{in} \simeq$0.07 AU) and a magnetic lever arm parameter $\lambda < 70$ (Cabrit *et al.* 1999; Garcia *et al.* 2001). The outer launch radius has a minor effect, unless ionisation is much higher there than on-axis (Cabrit *et al.* 1999).

6.2. *Jet kinematics*

Disk winds are mainly magneto-centrifugally accelerated, with a negligible effect from enthalpy even in "warm" solutions ($\beta \ll 1$). The asymptotic speed along a streamline with footpoint radius r_o and magnetic lever arm parameter $\lambda \simeq (r_A/r_o)^2$ is then $V_p^\infty(r_o) = \sqrt{GM_\star/r_o}\sqrt{2\lambda - 3}$ (cf. Blandford & Payne 1982). Convolved synthetic PV diagrams along the jet axis are presented in Fig. 5 for an inner launch radius $r_{in} = 0.07$ AU and an outer radius of 1–3 AU. The cold model in Fig. 5a has a large $\lambda \simeq 50$ that is seen to produce excessive jet speeds (cf. Garcia *et al.* 2001; Pyo *et al.* 2006). However, warm disk wind models can reach lower λ and adequate velocities (Casse & Ferreira 2000; Pesenti *et al.* 2004). The warm model with $\lambda = 13$ in Fig. 5b is now in excellent agreement with the HVC in the DG Tau PV diagram. The slower HVC in RW Aur could be reproduced with an even lower $\lambda \simeq 4 - 6$.

The warm model in Fig. 5b is also seen to naturally produce an IVC similar to that of DG Tau, as matter launched from large disk radii of 1– 3 AU achieves lower speeds (in proportion to the kepler speed at the anchoring radius). The transverse PV diagram for the same model is shown in Fig. 6, and fits well the observed transverse velocity decrease in DG Tau (as well as the associated rotation signatures, see Pesenti *et al.* 2004 and below). On the other hand, jets without an IVC, such as in RW Aur, are not as easily explained as they need ad-hoc assumptions (Cabrit *et al.* in prep).

6.3. *Jet rotation*

As illustrated in Fig. 7, the combination of rotation and poloidal speed in an MHD wind from a keplerian disk allows to derive both the launch radius (Anderson *et al.* 2003) and a lower limit λ_ϕ to the wind lever arm parameter λ (Ferreira *et al.* 2006). Current rotation estimates in the IVC of two jets, DG Tau and Th28-Red, would indicate launch radii r_{out} of 1 AU to 3 AU for the slow sheath (cf. Bacciotti *et al.* 2002; Coffey *et al.* 2004; Fendt 2006), and a true magnetic lever arm parameter $\lambda \simeq 13$ (Pesenti *et al.* 2004). This is the same range of λ as inferred independently from HVC maximum speeds (see above). Thus, the HVC and IVC components could be interpreted as inner and outer streamlines of the same extended MHD disk wind, if the rotation interpretation is confirmed.

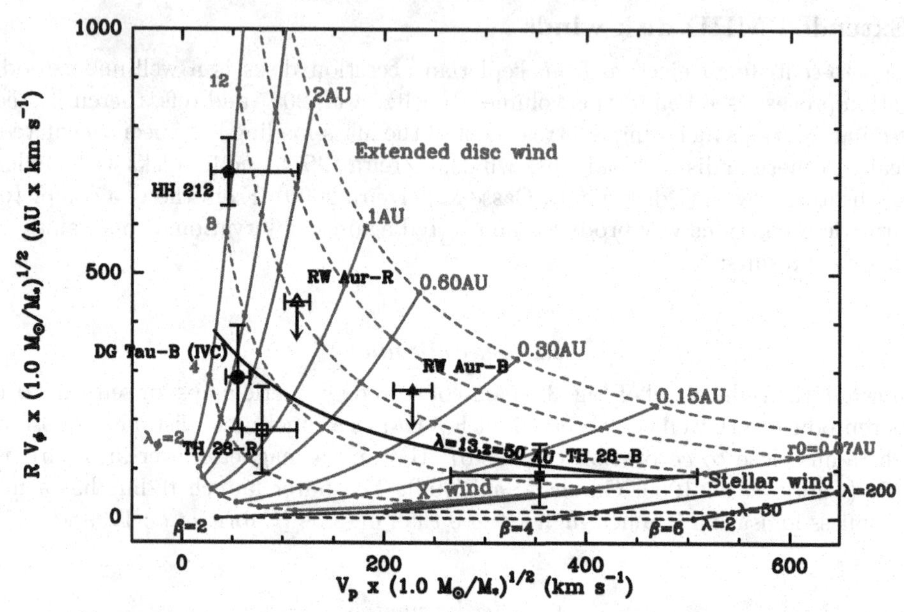

Figure 7. Constraints in the $V_p - rV_\phi$ plane on the launch point r_o and effective magnetic lever arm λ_ϕ of steady extended MHD disk winds. The thick black curve shows a cut at $z = 50$ AU across a warm disk wind solution with $\lambda = 13$. The locus of the X-wind and stellar winds with various pressure parameters β is also indicated. Symbols show current measurements; they are only upper limits for RW Aur and HH212 (see § 4.3). From Ferreira *et al.* (2006).

6.4. *Ejection to accretion ratio*

The mass ejection to accretion ratio in a self-similar extended disk wind is given by

$$2\dot{M}_j/\dot{M}_{acc} \simeq \frac{\ln(r_{out}/r_{in})}{(2\lambda - 2)}. \tag{6.1}$$

With $\lambda \simeq 6 - 13$, and $r_{out}/r_{in} \simeq 10 - 40$ (from IVC rotation data), one could reach a total ejection/accretion ratio $\simeq 0.15 - 0.2$ compatible with the mean observed ratio in CTTS jets (Ferreira *et al.* 2006). Note however that an extended MHD disk wind brakes only the disk, not the star. If the stellar wind is not sufficient for this purpose, excess angular momentum accreted from the inner disk edge would have to be removed by another agent, possibly a reconnexion X-wind (see below).

7. The reconnection X-wind model ("ReX-wind")

When the stellar and disk magnetic moments are parallel — instead of anti-parallel as assumed in the X-wind model — a magnetic X-point forms at the magnetopause. This leads to a fourth type of self-collimated MHD ejection illustrated in panel (d) of Fig. 4, triggered by reconnexion between closed stellar loops and open disk field.

As shown by Ferreira *et al.* (2000; see also Ferreira, this volume), this centrifugal "reconnexion X-wind" (hereafter ReX-wind) flows along newly opened field lines anchored *in the star*, not in the disk. It thus brakes down the star very efficiently, without the need for previous angular momentum extraction through disk-locking as in the X-wind scenario. Assuming a magnetopause close to corotation, the condition to maintain a slow rotation in CTTS despite accretion would again write $f\lambda \simeq 1$. Since matter is lifted up above the magnetic X-point by magnetic pressure, not just by the disk hydrostatic pressure gradient, one might expect a high mass-loading efficiency $f \simeq 0.1 - 0.3$. With $\lambda = 1/f \simeq 3 - 10$, the ReX wind would then reproduce both the HVC mass and speed.

In contrast to the X-wind, the ReX wind would not fan out over a wide-angle but would be automatically confined by the outer poloidal disk field that feeds reconnexion. The observed sheath of intermediate/low velocity material, currently unexplained in X-wind scenarii (see § 5), could naturally develop at this interface and/or through an extended MHD disk wind launched further out.

8. Conclusions

The collimation of jets in CTTS cannot be due to external hydrodynamical pressure and most likely results from MHD self-collimation along rotating open field lines. CTTS jets are thus an important agent of angular momentum removal from the star or disk. A detailed comparison of theoretical model predictions with spatially resolved jets properties reveals open issues with most scenarii for the jet origin:

- Stellar winds are present in CTTS and reach adequate terminal speeds but they do not seem able to provide more than 10% of the jet mass-flux, unless efficient acceleration by a "magnetic coil" is operative. Further research along this line is definitely needed, including insight from both numerical and laboratory experiments (eg. the "magnetic tower" jets studied by Lebedev *et al.* 2005).

- The X-wind is successful in producing a narrow HVC. However, its wide-angle nature prevents the formation of a sheath of lower velocity gas at 15–30 AU of the jet axis. A much tighter recollimation of outer streamlines would be required, with a disk field compatible with the observed keplerian disk rotation.

- Conversely, extended "warm" MHD disk winds agree very well with observational constraints in jets with a low-velocity sheath (Ferreira *et al.* 2006), but they may have difficulties reproducing the properties of jets with an HVC only (Cabrit *et al.* in prep.).

- Reconnexion X-winds are very promising as they seem potentially able at the same time to brake down the star, produce an HVC with sufficient mass and speed, and develop a lower velocity sheath around the jet beam. Modelling of the wind dynamics and geometry would be essential for a closer comparison with observations.

Acknowledgements

Useful discussions with my collaborators, C. Dougados and J. Ferreira, and with S. Edwards, S.-I. Inutsuka, S. Matt, M. Romanova, and F. Shu are gratefully ackowledged.

References

Anderson, J.M., Li, Z.-Y., Krasnopolsky, R., & Blandford, R. 2003, ApJ 590, L107

Bacciotti, F., & Eislöffel, J. 1999, A&A 342, 717 (BE99)

Bacciotti, F., Mundt, R., Ray, T.P., Eislöffel, J., Solf, J., & Camenzind, M. 2000, ApJ 537, L49

Bacciotti, F., Ray, T. P., Mundt, R., Eislöffel, J., & Solf, J. 2002, ApJ 576, 222

Barral, J. F., & Canto, J. 1981, RMxAA 5, 101

Blandford, R.D., & Payne D.G. 1982, MNRAS 199, 883

Bogovalov, S., & Tsinganos, K.1999, MNRAS 305, 211

Bogovalov, S., & Tsinganos, K. 2001, MNRAS 325, 249

Cabrit, S. 2002, in: *Star Formation and the Physics of Young Stars*, ed. by J. Bouvier, J.-P. Zahn (EDP Sciences, Les Ulis 2002) pp. 147–182

Cabrit 2007, in *MHD jets and winds from young stars*, eds. J. Ferreira, C. Dougados et E. Whelan, Lecture Notes in Physics (Spinger), in press.

Cabrit, S., Edwards, S., Strom, S.E., & Strom, K.M. 1990, ApJ 354, 687

Cabrit, S., Ferreira, J., & Raga, A.C. 1999, A&A 343, L61
Cabrit, S., Pety, J., Pesenti, N., & Dougados, C. 2006, A&A, 452, 897
Cabrit, S., *et al.* 2007, A&A 468, L29
Casse, F., & Ferreira, J. 2000, A&A 361, 1178
Cerqueira, A. H., Velázquez, P. F., Raga, A. C., Vasconcelos, M. J., & de Colle, F. 2006, A&A 448, 231
Codella, C. *et al.* 2007, A&A 462, L53
Coffey, D., Bacciotti, F., Woitas, J., Ray, T. P., & Eislöffel, J. 2004, ApJ 604, 758
Coffey, D., Bacciotti, F., Ray, T. P., Eislöffel, J., & Woitas, J. 2007, ApJ 663, 350
De Campli, W.M. 1981, ApJ 244, 124
Dougados, C., Cabrit, S., Lavalley-Fouquet, C., & Ménard, F. 2000, A&A 357, L61
Dougados, C., Cabrit, S., Lavalley-Fouquet, C. 2002, RMxAA Conf. Ser. 13, 43
Edwards, S., *et al.* 2003, ApJ 599, L41
Fendt, C. 2006, ApJ 651, 272
Ferreira, J. 1997, A&A 319, 340
Ferreira, J., & Pelletier, G., Appl, S. 2000, MNRAS 312, 387
Ferreira, J., Dougados, C., & Cabrit, S. 2006, A&A 453, 785
Font, A., McCarthy, I.G., Johnstone, D., & Ballantyne, D.R. 2004, ApJ 607, 890
Garcia, P., Cabrit, S., Ferreira, J., & Binette, L. 2001, A&A 377, 609
Hartigan, P., Morse, J., & Raymond,. J. 1994, ApJ 436, 125
Hartigan, P., Edwards, S., & Gandhour, L. 1995, ApJ 452, 736 (HEG95)
Hartigan, P., Edwards, S., Pierson, R. 2004, ApJ 609, 261
Johns-Krull, C. M., & Herczeg, G. J. 2007, ApJ 655, 345
Kwan, J., Tademaru, E. 1988, ApJ 332, L41
Kwan, J., Edwards, S., & Fischer, W. 2007, ApJ 657, 897
Lavalley-Fouquet, C., Cabrit, S., & Dougados, C. 2000, A&A 356, L41
Lebedev, S.V., *et al.* 2005, MNRAS 361, 97
Matt, S., & Pudritz, R. 2005, ApJ 632, L135
Meliani, Z., Casse, F., & Sauty, C. 2006, A&A 460, 1
Mouschovias, T., & Spitzer, L. 1976, ApJ 210, 326
Muzerolle, J., Hartmann, L., & Calvet, N. 1998, AJ 116, 2965
Najita, J.R., & Shu, F.H. 1994, ApJ 429, 808
Pesenti, N., *et al.* 2004, A&A 416, L9
Pudritz, R.E., Ouyed, R., Fendt, C. & Brandenburg, A. 2007. In *Protostars & Planets V*, ed. by B. Reipurth, D. Jewitt, K. Keil, (University of Arizona Press, Tucson), in press
Pyo, T.-S. *et al.* 2003, ApJ 590, 340
Pyo, T.-S. *et al.* 2006, ApJ 649, 836
Ray, T.P., Mundt, R., Dyson, J.E., Falle, S.A.E., Raga, A. 1996, ApJ 468, L103
Ray, T.P. *et al.* 2007, In *Protostars & Planets V*, ed. by B. Reipurth, D. Jewitt, K. Keil, (University of Arizona Press, Tucson), in press.
Sauty, C., & Tsinganos, K. 1994, A&A 287, 893
Shang, H., Shu, F.H., & Glassgold, A.E. 1998, ApJ 493, L91
Shu, F.H., Najita, J., Ostriker, E.C., & Shang, H. 1995, ApJ 455, L155
Shu, F.H., Galli, D., Lizano, S., Glassgold, A.E., & Diamond, P. 2007, submitted
Simon, M., Dutrey, A., & Guilloteau, S. 2000, ApJ 545, 1034
Soker, N. 2005, A&A 435, 125
Stapelfeldt, K. *et al.* 2003, ApJ 589, 410
White, R. J., Ghez, A. M., Reid, I. N., & Schultz, G. 1999, ApJ 520, 811
Woitas, J., Ray, T.P., Bacciotti, F., Davis, C.J., & Eislöffel, J. 2002, ApJ 580, 336

Star-Disk Interaction in Young Stars
Proceedings IAU Symposium No. 243, 2007
J. Bouvier & I. Appenzeller, eds.

doi:10.1017/S174392130700957X

The structure accompanying young star formation

Vasily A. Demichev[1]† and L. I. Matveyenko[1]
[1]Space Research Institute, Moscow, Russia
email: demichev@iki.rssi.ru

Abstract. We studied the structure of the H_2O super maser region in Orion with VLBI angular resolution of 0.1 mas or 0.05 AU. The maser emission (F $\sim$ 8 MJy) was determined by highly organized structure: accretion disk, bipolar outflow, torus and surrounding shell. The accretion disk, divided into protoplanetary rings, is viewed edge-on. The disk rotates as a rigid body with velocities $V \sim \Omega R$ and the rotation period is $\sim$ 170 yrs. The highly collimated bipolar outflow has a size of 9×0.7 AU, a velocity of $\sim$ 10 km/s. In the center a bright compact ($\leqslant$0.05 AU) source ejector is located, surrounded by a torus 0.6 AU in diameter. The outflow has a helix structure, which is determined by precession with a period of T $\sim$ 10 yrs. Comet-like bullets were observed on distances up to 80 AU.

Keywords. Jets, bipolar outflows, accretion disk, maser.

---

1. Introduction

Gravitational instability in gas-dust complexes gives rise to active regions of protostar formation. The formation of protostars is accompanied by strong maser emission in water-vapor lines, its flux density reaches several tens of kJy. In extremely rare cases, intense H_2O maser flares are observed; their nature and triggering mechanism are not yet completely clear (Burke, Johnston & Efanov 1970; Matveyenko, Kogan & Kostenko 1980). In the Orion Nebula, compact maser sources are concentrated in eight zones whose sizes reach $\sim$ 2000 AU (Genzel *et al.* 1978). The velocities of the maser sources are within several tens of km/s. There was two high activity periods in one of the zones in Orion KL – 1979-1987 and 1998-1999. The radio flux densities of H_2O maser flares reached 8 MJy, and the linewidth of the profile was $\sim$ 0.5 km/s. The velocity of outbursts changed only slightly relative to V = 8 km/s. The coordinates of the active region are $RA = 5^h35^m14^s.121$ and $DEC = -05^0 22' 36''.27$ (2000.0). Below, we analyze in detail the structure of the active supermaser emission region over the period under consideration(Matveyenko, Graham & Diamond 1988; Matveyenko, Diamond & Graham 1998).

2. High activity – epoch 1979-1987

The period of high activity 1979-1987 was accompanied by intense H_2O maser outbursts. The flux densities of the maser outbursts reached F = 1 – 8 MJy, and some of them had a duration of a few days; on average, the enhanced activity lasted for several months (Abraham, Vilas Boas & del Ciampo 1981; Matveyenko 1981; Garay, Moran & Hashick 1989). The supermaser had Gaussian profile with a high-velocity or low-velocity tail.

The profile consists of several features, which are indicative of a complex spatial structure. The observed change of profile velocity V $\simeq$ 8 km/s is determined by the relative

† Present address: Moscow, 117997, Profsojuznaja 84/32.

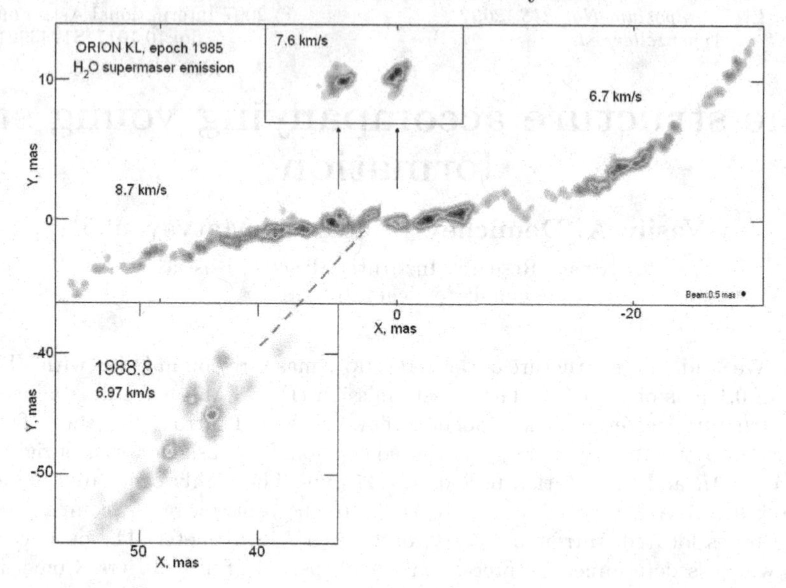

Figure 1. An accretion disk, divided into protoplanetary rings, seen edge-on.

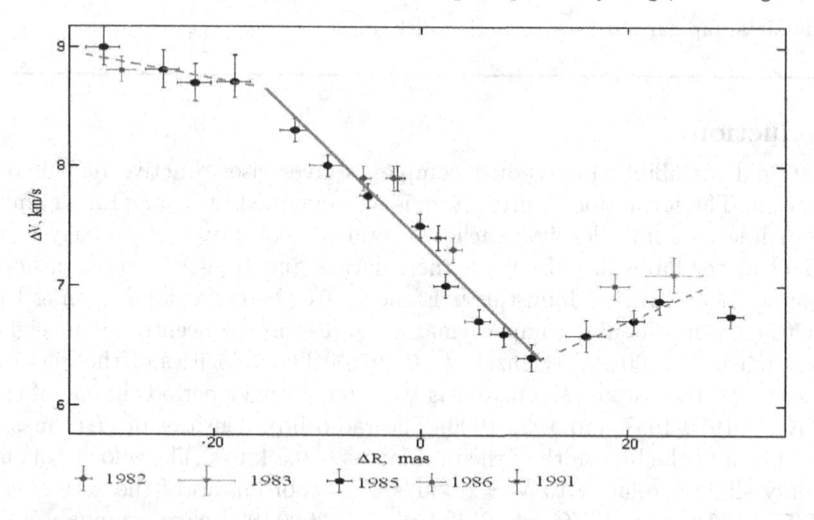

Figure 2. Velocities of the compact components in the disk, $V_{LSR} = 7.6km/s$.

contribution of the features. The linewidth at half maximum (FWHM) of the line profile is $\Delta V \simeq 0.5$ km/s. VLBI measurements of the active region in Orion KL showed that the maser emission in the period under consideration is determined by a highly organized structure, a chain of compact components (Matveyenko 1981). The most complete VLBI measurements were performed in October 1985. A detailed analysis revealed both an extended structure of low brightness, $T_b \sim 10^{11}$K, and a superfine component of high brightness, $T_b \sim 10^{16}$K. The structure consists of a chain of bright compact components distributed along a thin, $\sim$ 0.3 AU, extended S-shaped structure $\simeq$ 27 AU in length (Fig. 1). The velocity of the active region is $V_{LSR} = 7.6$ km/s (Matveyenko, Diamond & Graham 1998). The velocities of the individual components of the structure are also indicated here (Demichev & Matveyenko 2004). The two components located in the central part of the chain have the highest brightness temperatures, reaching $T_b \sim 10^{16}$ K. The radial velocities and brightness temperatures of these components are $V_E =$

7.50 km/s, $T_b \sim 10^{16}$ K and V_W = 7.75 km/s, $T_b \sim 0.6 \times 10^{16}$ K. They contain compact cores ~0.05 AU in size, with brightness temperatures reaching $T_b \sim 4 \times 10^{16}$ K.

The velocity distribution of the components is shown in Fig.2. In the central part of the structure within ≃ 15 AU we have linear velocity change dV/dR = 0.17 km/s/AU. The components correspond to the tangential directions of the concentric rings seen edge-on. Their velocities are equal to the rotation velocity. In this case, the central part of the structure is rotating as a rigid body $V_{rot} = \Omega R$ with period of T ≃ 170 yr.

The brightness of the features in the chain declines with increasing distance from the center to $T_b \sim 10^{12}$ K at its edges. The outer shape of the disk is deformed, probably by interaction with accreting matter. The velocities of the outer parts of the chain deviate from the linear dependence in its central part (Fig. 2). The outer part of the disk is outside the region of rigid body rotation and has a mean rotation velocity of $V_{rot} \simeq 1.1$ km/s, and a radius of R ≃ 14 AU. Its rotation velocity may be determined by Keplerian motion, $V^2R = MG$. In this case, the total mass of the inner part of the structure is $\leqslant 0.02 M_\odot$ (Matveyenko, Demichev & Sivakon 2005). The ambient medium/envelope amplifies maser emission on velocity V ≃ 7.65 km/s in a 0.5 km/s band (Matveyenko, Graham & Diamond 1988).

A fine elongated structure consisting of bullets was detected at a distance of 30 AU from the center of the chain in the direction PA = 135° at the end of the period of high activity 1988-1989.

3. Quescent period – epoch 1995

Observations of the active region in the period of low activity (the epoch 1995.6) showed that the H_2O maser emission at a velocity of 8 km/s does not exceed 1 kJy. VLBI studies with an ultrahigh angular resolution 0.1 mas (0.045 AU) revealed a highly collimated bipolar outflow with a bright compact central source, but the chain of components was absent or its emission was below the detection limit (Fig. 3; Matveyenko, Diamond & Graham 1998). The bipolar outflow 4.5 x 0.5 AU is oriented at PA = -33°, and its brightness temperature is $T_b \sim 10^{12}$ K.

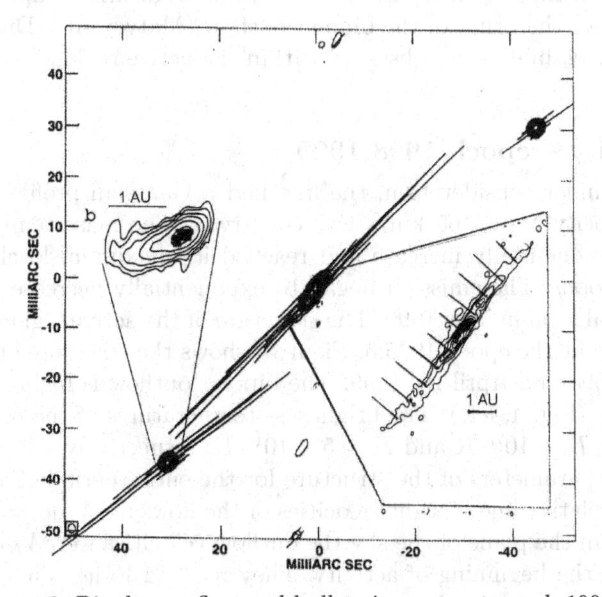

Figure 3. Bipolar outflow and bullets in quescent epoch 1995.6.

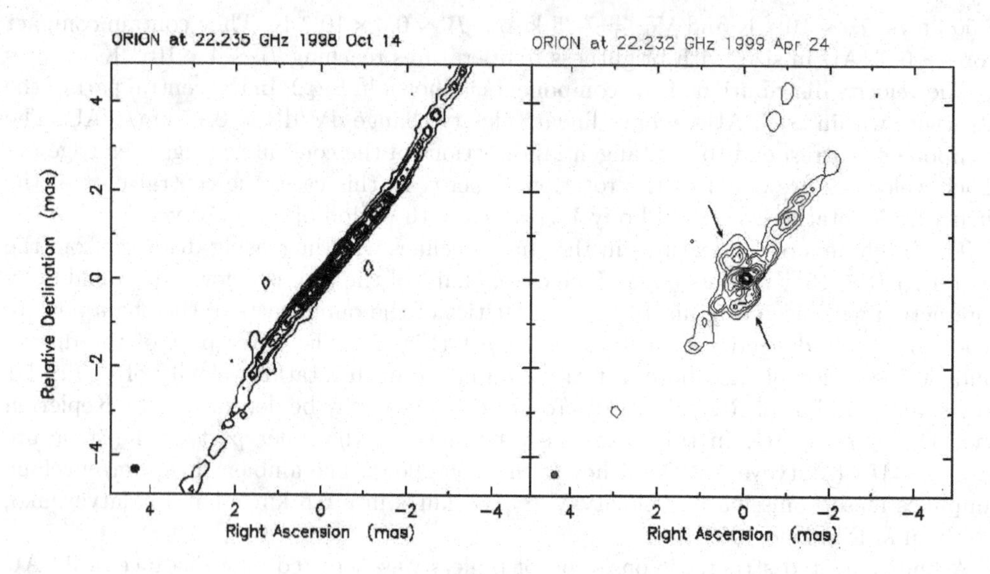

Figure 4. Bipolar outflow on 14 oct 1998 (left) and on 24 apr 1999 (right). The arrows mark a toroidal structure.

The compact bright source in the central part of the bipolar outflow is a nozzle. The bipolar outflow is ejected from the nozzle at PA $= -44°$. The nearest part of the flow has a size not exceeding 0.05×0.15 AU, and its brightness temperature reaches $T_b \sim 10^{13}$ K. The velocity of the injector is V = 7.63 km/s. The maser emission of the structure is linearly polarized, see Fig. 3. The length and orientation of the lines correspond to the degree of polarization and orientation of the polarization plane. Also shown here on an enlarged scale are the bullet and the bipolar outflow with an angular resolution of 0.3 mas. The polarization of the emission from the bipolar outflow near the nozzle is 33 % and increases with distance to 50 %. This may be attributable to partial saturation of the amplification in the region of the nozzle, whose brightness temperature is an order of magnitude higher than that of the bipolar outflow (Matveyenko, Diamond & Graham 1998). As previously, bullets are observed within the active region.

4. High activity – epoch 1998-1999

In the period under consideration, the line had a Gaussian profile with linewidth $\sim$ 0.5 km/s, its velocity V = 7.65 km/s was conserved. The maser emission in February 1998 began to exponentially increase and reached its maximum level of F = 4.3 MJy in August – October. The emission began to exponentially decrease in November and reached its original level in May 1999. The structure of the active region remained almost the same as it was at the epoch 1995.6. Figure 4 shows the structure of the active region on October 14, 1998 and April 24, 1999. The bipolar outflow is highly collimated and is 5 x 0.2 AU in size (Fig. 4, left). The brightness temperatures of the bipolar outflow and the nozzle rose to $T_b \sim 10^{15}$ K and $T_b \sim 5 \times 10^{16}$ K, respectively.

Analysis of the parameters of the structure for the entire period of activity 1998-1999 showed that the relative line of sight velocities of the flows are $V_{NW} = --0.3$ km/s and $V_{SE} = 0.3$ km/s. In the plane of the sky the outflow velocities were $V_{NW} = 10$ km/s and $V_{SE} = 8$ km/s in the beginning of activity. They reached $V_{NW} = 6.8$ km/s and $V_{SE} =$ 5.0 km/s at the peak and then decreased to $V_{NW} = 4.0$ km/s and $V_{SE} = 3.5$ km/s in the period of activity decline. Thus, the high and low velocities precede the high activity and the decline in emission, respectively. The observed correlation of the supermaser emission with the flow velocity suggests collisional pumping, the interaction of the flow with the ambient medium.

The bipolar outflow has a helix structure that is determined by the precession of the rotation axis of the nozzle; the precession period is T $\simeq$ 10 yr, and the precession angle is $\simeq 16°$ (Matveyenko, Zakharin & Diamond 2004; Matveyenko, Demichev & Sivakon 2005).

A toroidal structure was discovered around nozzle at the end of the period of high activity. The torus is 0.6 AU in diameter, and its brightness temperature reaches $T_b \sim 10^{13}$ K. The plane of the torus is oriented perpendicular to the nozzle axis. (Fig.4, right). Probably the torus was present earlier, but its emission was blended with the supermaser. The dynamic range of the measurements was not enough to reveal a structure of relatively low brightness.

5. Bullets

Bullets are observed mostly in periods of low emission, when the disk-jet structure has a low brighness temperature. A first bullet was discovered after the first period of activity in 1988-1989 on a distance of 30 AU from the center of the accretion disk. The bullet had an elongated shape, reaching 6 AU in size. Its brightness temperature was $T_b \sim 10^{13}$ K (Fig. 2, 5) and its line of sight velocity relative to the nozzle was V = −0.46 km/s.

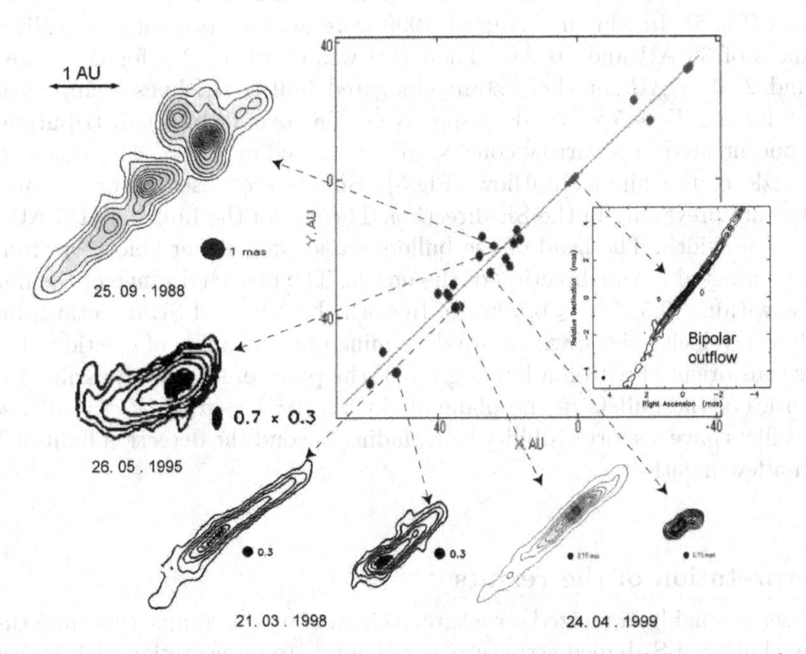

Figure 5. Structure of active region: bullets and outflow at different epoch.

Two bullets were observed during the quescent period of 1995. The brightness temperature of the bullets was $T_b \sim 10^{12}$ K. The southeastern bullet is at a distance of 18.5 AU from the nozzle in the direction PA = 132°. It has a comet-like (headtail) shape, and its line of sight velocity is V_{SE} = 0.32 km/s. However, the head is behind its tail. The northwestern bullet is at a distance of 32.5 AU from the nozzle in the direction PA

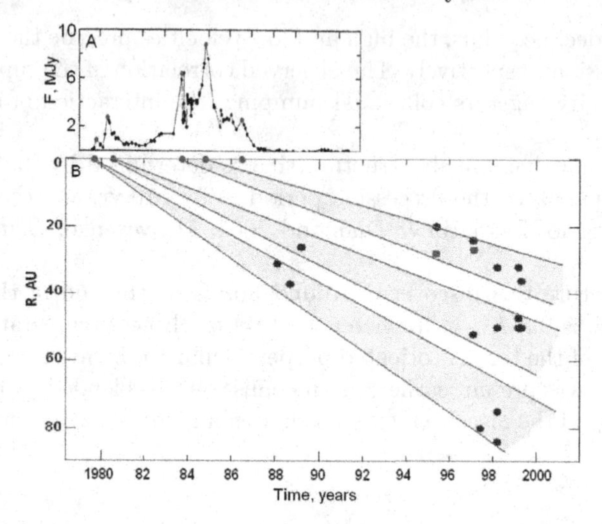

Figure 6. Maser flux distribution during the first active period (Abraham, Vilas Boas & del Ciampo 1981) (up) and bullets movement (down).

= 54°. Its velocity is V_{NW} = 0.18 km/s. The polarization of the emission from the southeastern bullet reaches 45% and 13 % from the northwestern bullet.

During the second active period 1998-1999 bullets were observed many times. In March 1998 two bullets were observed on distances of 47 AU and 68 AU, and heads were in front of the tails (Fig.5). In March – August 1999 were another two bullets in SE direction, on distances of 35 AU and 50 AU. Their size was 0.5×0.15 AU for the close compact bullet, and 2×0.15 AU for the distant elongated bullet. Brithness temperatures were $T_b \sim 10^{13}$ K, and $T_b \sim 5 \times 10^{13}$ K respectively. The overall bullet distribution in 1988-1999 is concentrated in a narrow cone ~ 10° orientated at PA = -44°, that corresponds to the nozzle of the bipolar outflow (Fig.5). Bullets are observed on distances up to 80 AU and are prevailing in the SE direction. The size of the bullets is 1-5 AU in length and 0.2 AU in width. The head of the bullets are located either before or after the tail. Bullets are oriented in the direction of the nozzle. The observed range of the line of sight velocity is within $-0.5 \leqslant V \leqslant 0.5$ km/s. In both the NW and SE directions bullets can have different velocity signs, which are determined by the angle of ejection. The cone of bullets is thus oriented within a few degrees of the plane of sky. It is possible to estimate the velocities of the bullets in the plane of sky $V_{sky} = V/sin(\phi/2)$, $\phi = 10°$, so $V_{sky} \geqslant$ 6 km/s. Bullets have a short visibility time, fading beyond the detection limit of $T_b \sim 10^{11}$ K within a few months.

6. Interpretation of the results

The observed highly organized structure, a chain of bright compact features distributed along the elongated S-shaped structure, corresponds to an accretion disk separated into rings seen edge-on. The disk is 27 AU in diameter and 0.3 AU in thickness. The observed velocity–distance dependence of the components (the rings) corresponds to a rigid-body rotation, $V_{rot} = \Omega R$. The angular velocity Ω corresponds to a rotation period of the disk T = 170 yrs. The maser radiation is concentrated in the azimuthal plane of the rings; its directivity reaches 10^{-3}. The decrease of the rotation velocity towards the center of disk is determined by the transfer of its kinetic energy to the bipolar flow. The energy transfer

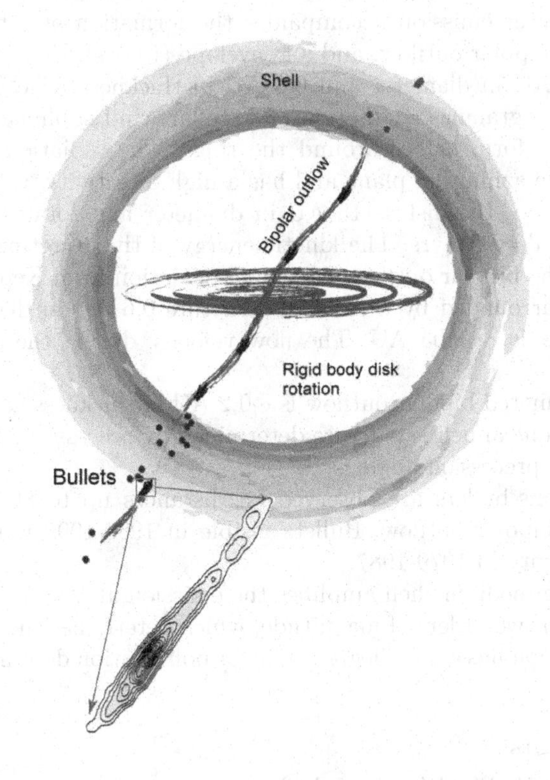

Figure 7. Model of H_2O supermaser in Orion.

causes the decrease of the Keplerian velocity $V^2R = MG$ to the observed rigid-body limit.

The diameter of the outflow on the exit of the nozzle does not exceed 0.05 AU. The outflow is higly collimated, spreading much farther than its visible size up to 80 AU (Fig. 7). Dense fragments of the outflow are visible as bullets. The precession of the nozzle axis produces the conical helix shape of the outflow. The period of precession is T=10 years.

Outflow, bullets and the surronding medium contain ice granules, which become sublimated and produce H_2O molecules. The velocity of compact knots in the outlflow is $\sim$ $4 - 10$ km/s. The interaction of H_2O molecules with the surrounding medium produces collisional pumping of the maser.

The analysis of bullets movement show that they were ejected during the first period of activity in 1979-1987 (Fig. 6). The ejection is accompanied by powerful supermaser outbursts.

The emission of the structure is amplified in the surrounding medium – shell at V = 7.65 km/s in a 0.5 km/s band by more than two orders of magnitude. This increases the brightness temperatures to the supermaser level of 10^{17} K.

7. Conclusion

Our studies of the superfine structure of one of the active star formation regions in Orion KL in H_2O maser emission have shown the following:

Intense H_2O maser emission accompanies the formation of a thin accretion disk, a highly collimated bipolar outflow, and an envelope (Fig. 7).

The disk is 27 AU in diameter and 0.3 AU in thickness. The disk is separated into rings containing ice granules. Radiation and stellar wind sublimate and blow away the water molecules to form haloes around the rings. The radiation of the maser ring is concentrated in the azimuthal plane and has a high directivity, $\leqslant 10^{-3}$.

The central part of the disk is 15 AU in diameter and rotates as a rigid body; the rotation period is T = 170 yrs. The kinetic energy of the accreting matter and the disk is transferred to the bipolar outflow, causing a deviation from Keplerian motion.

The nozzle is surrounded by a toroidal structure 0.6 AU in diameter. The diameter of the ejected flow is $\sim$ 0.05 AU. The flow velocity during the active period reaches 10 km/s.

The highly collimated bipolar outflow is $\sim$0.2 AU in thickness and $\sim$5 AU in apparent length and has a conical helix structure determined by precession. The precession period is $\sim$10 yr, and the precession angle is 16°.

Compact ejections-bullets are observed on distances up to 80 AU. They are dense fragments of the bipolar outflow. Bullets visible in 1988-1999 were ejected during the first period of activity in 1979-1987.

The surrounding medium-shell amplifies the emission at V = 7.65 km/s in a 0.5 km/s band by more than two orders of magnitude, which determines the supermaser emission.

The maser emission has a high degree of linear polarization determined by the pumping directivity.

Acknowledgements

One of authors (V. Demichev) would like to acknowledge RFBR grant 07-02-08192 and IAU for support to attend the conference.

References

Z. Abraham, J.W.S. Vilas Boas, & L.F. del Ciampo 1986, *A&A* 167, 311

B. F. Burke , K. J. Johnston, V. A. Efanov , B. G. Clark , L. R. Kogan, V. I. Kostenko ,K. Y. Lo , L. I. Matveenko , I. G. Moiseev, J. M. Moran, S. H. Knowles, D. C. Papa, G. D. Papadopoulos, A. E. E. Rogers, & P. R. Schwartz 1972, *AZh* 49, 465

V. A. Demichev & L. I. Matveyenko 2004, *Astron. Rep.* 48, 979

G. Garay, J. M. Moran, & A. D. Hashick *et al.* 1989, *ApJ* 338, 244

R. Genzel, D. Downs, J. M. Moran *et al.* 1978, *A&A* 66, 13

L. I. Matveyenko, L. R. Kogan, & V. I. Kostenko 1980, *Astron. Lett.* 6, 279

L. I. Matveyenko 1981, *Astron. Lett.* 7, 54

L. I. Matveyenko, D. A. Graham, & P. J. Diamond 1988, *Astron. Lett.* 14, 468

L. I. Matveyenko, P. J. Diamond & D. A. Graham 1998, *Astron. Lett.* 24, 623

L. I. Matveyenko, K.M. Zakharin, P. J. Diamond & D. A. Graham 2004 *Astron. Lett.* 30, 100

L. I. Matveyenko, V. A. Demichev, S.S. Sivakon, Ph. D. Diamond, & D. A. Graham 2005 *Astron. Lett.* 31, 913

Star-Disk Interaction in Young Stars
Proceedings IAU Symposium No. 243, 2007
J. Bouvier & I. Appenzeller, eds.

doi:10.1017/S1743921307009581

Last gasp of V1647 Ori: a brief post-outburst warm, molecular wind

Sean D. Brittain[1], Theodore Simon[2], Terrence W. Rettig[3], Erika L. Gibb[4], Dinshaw Balsara[3], David Tilley[3], and Kenneth H. Hinkle[5]

[1]Department of Physics and Astronomy, Clemson University, Clemson, SC 29634-0978
email: sbritt@clemson.edu

[2]Institute for Astronomy, University of Hawaii, 2680 Woodlawn Dr, Honolulu, HI 96822

[3]Center for Astrophysics, University of Notre Dame, Notre Dame, IN 46556

[4]University of Missouri at St. Louis, 8001 Natural Bridge Road, St. Louis, MO, 63121

[5]National Optical Astronomy Observatory P.O. Box 26732, Tucson, AZ 85726-6732

Abstract. Followup infrared spectroscopy is reported for V1647 Ori, a young star whose recent eruption illuminated McNeil's Nebula. Lines of H I, H_2, and CO are compared to previous observations. We find that the accretion rate fell two orders of magnitude and the CO bandheads disappeared at the end of the outburst. We also report a striking metamorphosis of the fundamental CO spectrum from centrally peaked profiles to emission lines with superimposed blue-shifted absorption lines and back again one year later. This remarkable change in spectral appearance indicates that the system did not return to equilibrium immediately following the outburst. In this paper we propose a mechanism to explain the emergence of a transient post-outburst outflow.

Keywords. Accretion, accretion disks, stars: V1647 Ori, formation, mass loss, circumstellar matter.

---

1. Introduction

The EXor V1647 Ori recently underwent an outburst that lasted two years (Kóspál *et al.* 2005; Ojha *et al.* 2006; Acosta-Pulido *et al.* 2007). During this event, the source brightened by a factor of 50 in X-rays (Kastner *et al.* 2006), by a factor of 40 in the red (Briceño *et al.* 2004), by a factor of 15 in the near-infrared (Reipurth & Aspin 2004), and by roughly a factor of 15 at wavelengths from 3.6 μm to 70 μm (Muzerolle *et al.* 2005). The rise time for the outburst was roughly 80 days (see Acosta-Pulido *et al.* 2007 for a compilation of *I*-band photometry of this source). This was followed by a plateau phase lasting ~700 days during which time the source faded by one magnitude in the NIR. Beginning in November 2005, the star faded to its preoutburst brightness in ~70 days.

From the brightening of the source, Muzerolle *et al.* (2005) concluded that the bolometric luminosity increased by a factor of 15 (see also Andrews *et al.* 2004) and the stellar accretion rate increased from $\sim 10^{-7} M_\odot$ yr^{-1} to $\sim 10^{-5} M_\odot$ yr^{-1}. Similarly, Gibb *et al.* (2006) inferred a stellar accretion rate of $\sim 5\times 10^{-6} M_\odot$ yr^{-1} from the luminosity of the Brγ emission one year later. This value is somewhat larger than the typical accretion rate of a young low mass star ($10^{-8} - 10^{-7} M_\odot$ yr^{-1}; Bouvier *et al.* 2007), yet lower than is expected for a star of the FUor type ($\sim 10^{-4} M_\odot$ yr^{-1}; Hartman & Kenyon 1996).

During the onset of the outburst of V1647 Ori, observations of atomic lines with P Cygni profiles indicated the presence of a wind (Briceño *et al.* 2004; Reipurth & Aspin

2004; Vacca *et al.* 2004; Walter *et al.* 2004; Ojha *et al.* 2006). The hot ($T \sim 10,000$ K) and fast (-400 km s^{-1}) wind produced during the early phase of this event was escaping at a rate of $\dot{M}_{\rm wind}$=4$\times$10^{-8} M$_\odot$ yr^{-1} (Vacca *et al.* 2004). This mass loss rate is much lower than that of the typical FUor or EXor (Hartmann & Kenyon 1996) and is comparable to that of a classical T Tauri star (cTTS; Hartigan *et al.* 1995). Curiously, the ratio of the mass loss rate to the accretion rate is two orders of magnitude smaller than the typical ratio for cTTSs (Cabrit *et al.* this volume). The absorption component of the P Cygni profile for several lines disappeared within a few months following the peak of the outburst in early 2004 (Gibb *et al.* 2006). The Hα line, however, retained a P Cygni profile throughout the outburst, indicating that a weaker wind continued (Ojha *et al.* 2006; Fedele *et al.* 2007). There were two different absorption components seen in Hα: a highly variable one at a velocity of -400 km s^{-1} and a steady one at -150 km s^{-1}. There was no indication of a lower velocity wind during the outburst.

In contrast to the hydrogen and helium lines, the fundamental NIR ro-vibrational lines of CO observed on 2004 February 27 were broad and centrally peaked (Rettig *et al.* 2005). The temperature of the gas was 2500 K and thus consistent with the CO bandhead emission at 2.3 μm. The width of the emission lines was shown to result from the Keplerian orbital motion of the gas within the inner disk surrounding the central star, similar to the broad emission line profiles that are observed around cTTSs and Herbig Ae/Be stars (HAeBes; Najita *et al.* 2003; Blake & Boogert 2004; Rettig *et al.* 2006; Brittain *et al.* 2007a). Observations obtained on 2004 July 30 showed that the CO lines remained broad but the temperature of the gas fell to 1700 K (Gibb *et al.* 2006).

In this contribution we report followup observations of the NIR spectrum of V1647 Ori, which were acquired immediately following the end of the outburst and one year later. The fading of the source indicates that the accretion rate had dropped and thus we expected the CO emission lines to fade as well (Najita *et al.* 2003). We present a moderate resolution spectrum spanning 1.1–4.1 μm and high resolution spectra centered at 5 μm. We find that the accretion rate dropped two orders of magnitude and the CO emission lines cooled. We also find, however, that the fundamental ro-vibrational CO lines underwent a striking metamorphosis, changing from centrally peaked lines to lines with blue-shifted absorption features and back again. Such a stunning transformation has not been observed from other young stars with similar mass-loss rates. We conclude this article by proposing a mechanism to explain this unique phenomenon.

2. Observations & data reduction

V1647 Ori was observed with the SpeX instrument at the Infrared Telescope Facility on Mauna Kea, Hawaii, in 2006 January. The observations and data reductions were carried out on our behalf by Dr. Wm. Vacca, to whom we are extremely grateful. SpeX is a moderate-resolution ($\lambda/\delta\lambda \sim 1200 - 2500$), near-infrared (1–5 μm), cross-dispersed spectrometer (Rayner *et al.* 2003). The data were acquired using standard observing procedures (see, e.g., Gibb *et al.* 2006) and were reduced using the SpeXtool package (Cushing *et al.* 2004). The telluric correction and flux calibration were performed using standard techniques, as described by Vacca *et al.* (2003).

High resolution spectra of V1647 Ori were also acquired with NIRSPEC at the W. M. Keck Observatory on Mauna Kea in 2006 February and with the Phoenix instrument at the Gemini South telescope in 2006 December and 2007 February. The data analysis and line equivalent width in these spectra are presented in Brittain *et al.* (2007b).

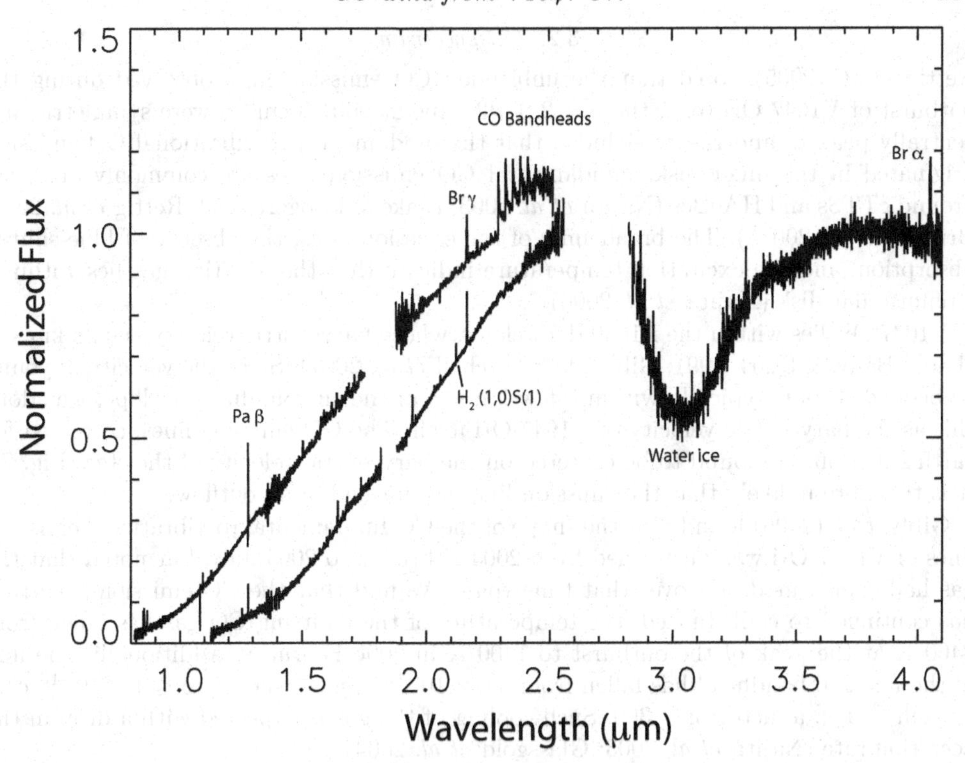

Figure 1. Comparison of the NIR spectrum of V1647 Ori during outburst and in quiescence. The spectra have been normalized to their their *L*-band fluxes. The spectrum that is brighter from 1–2.4 μm was observed in 2004 February, and the fainter spectrum was observed in 2006 January. The continuum faded ~3 magnitudes during this time.

3. Results

3.1. *Atomic emission lines*

Examination of the moderate-resolution spectra taken during the outburst (2004 February) and following outburst (2006 January) reveals several changes (see Fig. 1). For example, Mg I, Fe I, and the CO bandheads disappeared, while H_2 and other unidentified emission lines emerged. Various features also changed shape. At the onset of the outburst, the Paschen series and two He I lines exhibited P Cygni line profiles (Vacca *et al.* 2004; Fig. 1), as did Hα (Briceño *et al.* 2004; Reipurth & Aspin 2004; Walter *et al.* 2004; Ojha *et al.* 2006). By 2004 November these lines had become centrally peaked (Gibb *et al.* 2006) and remain so now. A subsequent spectrum of V1647 Ori taken in 2006 November shows that the emission lines did not undergo significant changes post-outburst.

In contrast to the Balmer and Paschen lines, the Brackett lines revealed no evidence for a blue-shifted absorption component. Brγ has remained centrally peaked throughout the eruption, although its flux has dropped from $\sim10^{-13}$ ergs s^{-1} cm^{-2} during outburst (2004 February; Gibb *et al.* 2006) to 1×10^{-14} ergs s^{-1} cm^{-2} post-outburst (2006 January). If one assumes that the Brγ line forms in the accretion flow and is minimally affected by a stellar wind, its luminosity can be used to infer the stellar accretion rate (Muzerolle *et al.* 1998). Based on the strength of the Brγ line, Gibb *et al.* (2006) determined an accretion rate of $\sim5\times10^{-6}\,M_\odot$ yr^{-1}. Adopting the same stellar parameters as Gibb *et al.* (2006), we find a rate of $\sim10^{-7}\,M_\odot$ yr^{-1} for 2006 January. Thus, in two years the accretion rate has fallen by two orders of magnitude from its peak value of $\sim10^{-5}\,M_\odot$ yr^{-1}.

3.2. *CO spectrum*

Rettig *et al.* (2005) noted that the unblended CO emission lines observed during the outburst of V1647 Ori (e.g., the $v = 2$–1 P27 line at 2001.8 cm^{-1}) were symmetric and centrally peaked, and they concluded that the fundamental ro-vibrational CO emission originated in the inner disk. Fundamental CO emission lines are commonly observed around cTTSs and HAeBes (Najita *et al.* 2003; Blake & Boogert 2004; Rettig *et al.* 2006; Brittain *et al.* 2007a). The broadening of the emission lines, the absence of blue-shifted absorption, and the excitation temperature indicate that the emitting gas lies within a circumstellar disk (Najita *et al.* 2000).

V1647 Ori lies within the L1630 dark cloud, whose barycentric velocity is +26 km s^{-1} (Lada, Bally, & Stark 1991; Gibb 1999; Mitchell *et al.* 2001). Since the velocity of young embedded stars is typically within a few km s^{-1} of the surrounding envelope, we adopt this as the barycentric velocity of V1647 Ori itself. The CO emission lines, corrected for earth's motion, are found to be centered on the barycentric velocity of the star (Fig. 2). It is therefore unlikely that the emission lines are formed in an outflow.

Gibb *et al.* (2006) found that the shape of the CO fundamental ro-vibrational emission lines of V1647 Ori was unchanged from 2004 February to 2004 July, but noted that the gas had cooled modestly over that time span. We find that the CO emission spectrum has continued to cool. Indeed, the temperature of the emitting CO gas has fallen from 2400 K at the peak of the outburst to 1400 K in 2006 February. Additionally, the flux of the $v = 2$–0 bandhead has fallen from 3.6×10^{-13} ergs s^{-1} cm^{-2} to $\leqslant 1 \times 10^{-15}$ ergs s^{-1} cm^{-2}, i.e., a factor of $\gtrsim 300$. Such cooling of the gas is expected with a drop in the accretion rate (Najita *et al.* 2003; Glassgold *et al.* 2004).

The most striking change from the earlier observations is the emergence of a blue-shifted absorption feature in the 2006 February data (Fig. 2). The centroid of the absorption is shifted by -30 km s^{-1} relative to the velocity of the star. Its full width at half-maximum (FWHM) is 20 km s^{-1}. If this absorption is produced by gas that is in Keplerian motion around a $0.5M_{\odot}$ star, and if it is observed against an extended continuum, then the gas must be located at a disk radius of $R = 1.1 \sin^2 i$ (AU). Interestingly, the observed blue-shift corresponds to the projected escape velocity from a $0.5M_{\odot}$ star at a distance of $1.0 \sin^2 i$. Thus the spectrum is consistent with gas being lifted off the surface of the disk at $\sim 1.0 \sin^2 i$ AU from the star. A follow-up observation of the P30 line was acquired one year later with the Phoenix spectrograph on Gemini South. As is apparent in Fig. 2, the emission line is again clearly discernible, but there is no sign of the absorption component.

4. Discussion

Outflows from cTTSs and HAeBes are common, but fundamental ro-vibrational CO lines with blue-shifted absorption components are observed around none of the >300 such sources observed to date (Najita *et al.* 2000, 2003; Blake & Boogert 2004; Rettig *et al.* 2006; Brittain *et al.* 2007a; J. Brown private communication). This fact suggests that the appearance and subsequent disappearance of the warm, absorbing CO in the case of V1647 Ori is not due to the precession of a highly collimated outflow. Indeed the absence of blue-shifted CO absorption in the outflows commonly observed from other young stars indicates that their outflows do not contain warm ($\sim 10^3$ K) CO molecules.

Although the early outburst spectra of V1647 Ori showed no evidence of a 30 km s^{-1} outflow, one might hypothesize that the absorbing CO molecules formed in a decelerating wind that cooled as it expanded. As noted earlier, during the outburst, there were two

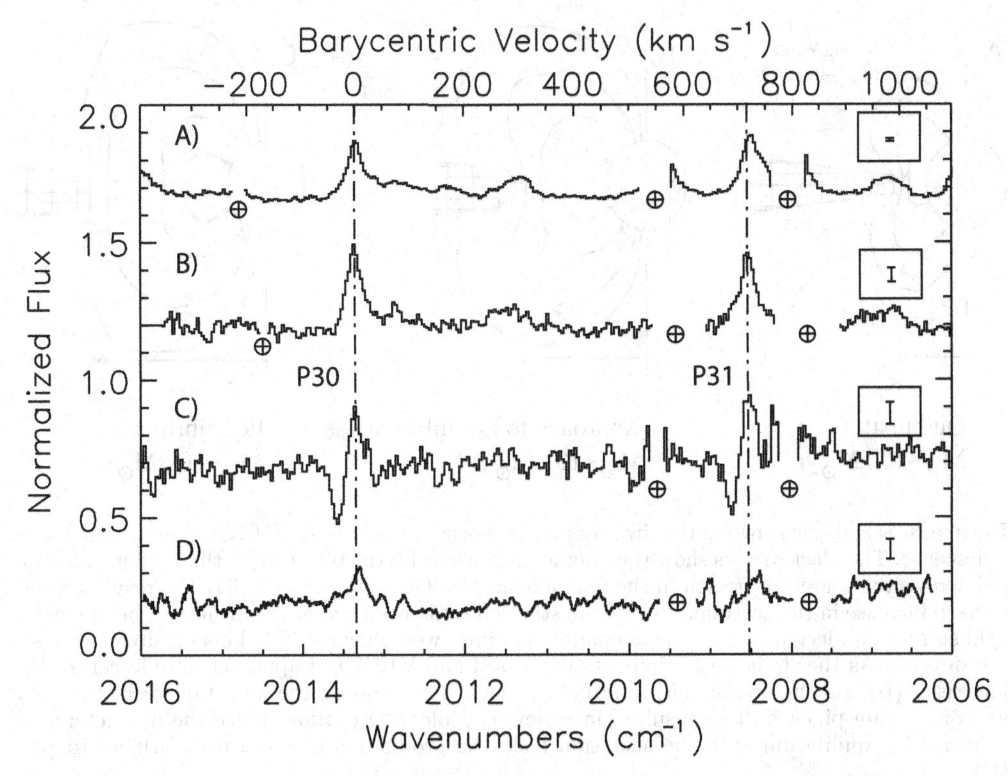

Figure 2. Detailed comparison of the high-J CO lines. The v = 1–0 P30 and P31 lines are plotted over four epochs spanning three years. An error bar has been plotted on the right side of each spectrum representing the standard deviation of the continuum. The spectra have been normalized, offset vertically, and shifted to the rest frame of the molecular cloud in which V1647 Ori is embedded. The spectrum acquired 2004 February is plotted in panel A (from Rettig *et al.* 2005). Six months into the outburst (2004 July) the CO emission lines remain broad and centrally peaked (Panel B; from Gibb *et al.* 2006). However, in 2006 February, after the star had returned to it pre-outburst magnitude, the CO lines have blue-shifted absorption features superimposed on them (Panel C). One year later (2006 December–2007 February) the CO lines return to their original profile (Panel D).

blue-shifted outflow components observed in Hα: a variable high-velocity component (~400 km s^{-1}) and a steady component of moderate velocity (~150 km s^{-1}; Fedele *et al.* 2007). If the lower-velocity gas decelerated at a constant rate to 30 km s^{-1} by 2004 July, then the wind would have expanded to 10AU. However, our observations on that date reveal no evidence of blue-shifted CO absorption (Fig. 2). By 2006 February, when the absorption first appeared, the wind would have expanded to nearly 40 AU. We think it is unlikely that the observed spectrum could be formed 40 AU from the star, and thus we conclude that the CO outflow observed in 2006 February must have emerged post-outburst.

We use a standard magnetospheric accretion model (Bouvier *et al.* 2007 and references therein) to account for the post-outburst outflow. Such models were first developed in the context of accretion onto neutron stars (Ghosh & Lamb 1979). Johns-Krull & Valenti (2000) have found that late-type pre-main-sequence stars typically have kG magnetic fields, which makes it plausible to suppose that the truncation radius of the disk is established by disk-magnetospheric interaction rather than by a disk-star boundary layer. In the standard magnetospheric accretion model, the truncation radius of the disk is

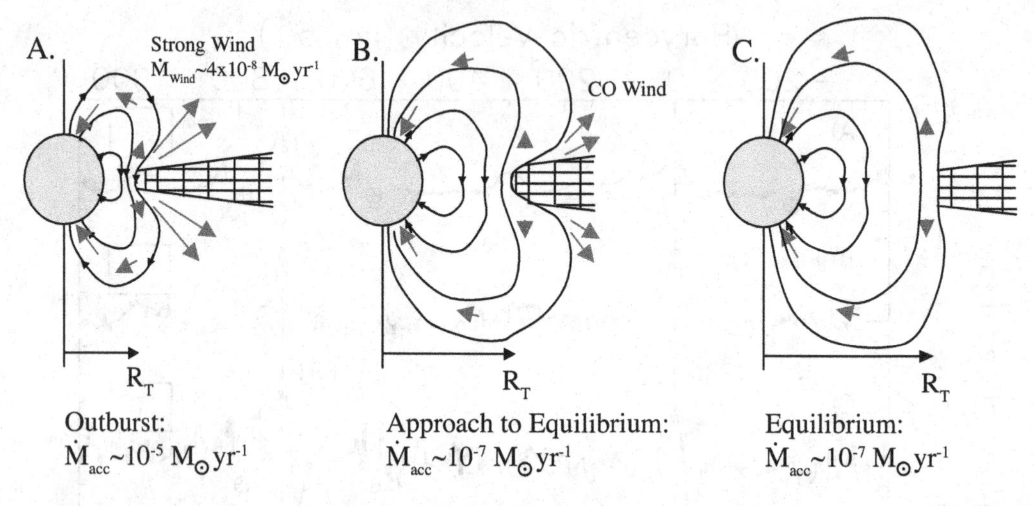

Figure 3. Schematic showing the disk-magnetospheric interaction in V1647 Ori in outburst and quiescence. The black arrows show the magnetospheric field, the blue arrows the outflow, and the red arrows the accretion stream in the magnetosphere. The outburst of V1647 Ori resulted from a sharp increase in the accretion rate of the star. The enhanced accretion pushed in the magnetic field so that R_T decreased and the magnetic field lines were pinched (A). This resulted in a fast, hot outflow. As the stellar accretion rate decreased and V1647 Ori approached quiescence, R_T increased (B). As the stellar magnetic field pushed back against the accretion disk, the field was once again pinched. This resulted in a slower, cooler CO outflow. Once the magnetic field returned to equilibrium with the accretion disk, the production of the warm outflow stopped (C).

related to the accretion rate by the following expression:

$$R_T = 7.1 B_{kG}^{4/7} \dot{M}_{-8}^{-2/7} M_{0.5}^{-1/7} R_2^{5/7}, \tag{4.1}$$

where R_T is the truncation radius in units of a stellar radius, B_{kG} is the magnetic field strength in units of 1 kG, $\dot{M}_{-8}$ is the stellar accretion rate in units of $10^{-8} M_\odot$ yr^{-1}, $M_{0.5}$ is the stellar mass in units of $0.5 M_\odot$, and R_2 is the stellar radius in units of $2R_\odot$ (Bouvier *et al.* 2007). This expression predicts that when the accretion rate decreases by two orders of magnitude, as in the case of V1647 Ori, the truncation radius of the disk should move outward by about a factor of three. Adopting the same parameters used to calculate the accretion rate, $R_{\rm star} = 3R_\odot$ and $M_{\rm star} = 0.5 M_\odot$, and $B_{\rm star} = 1$ kG, the truncation radius moved from 0.02 AU to 0.07 AU. If the movement of the truncation radius was contemporaneous with the optically observed fading of the source, then the stellar magnetic field pushed the inner edge of the disk back at a rate of $\sim$1 km s^{-1}.

We use the following scenario to provide a qualitative explanation of the development of the CO outflow. When the accretion rate increased, the truncation radius moved inward, eventually reaching dynamical equilibrium with the magnetosphere (Fig. 3a). If the pinching of the magnetic field caused it to form an angle of $\leqslant 60°$ with the disk, then the magnetic field could drive an outflow (Blandford & Payne 1982). Evidence from simulations (Balsara 2004; Matsumoto *et al.* 1996) suggests that such disk-magnetosphere systems tend to form outflows when the system is undergoing the greatest amount of dynamical rearrangement. Consistent with simulations and disk wind theory (e.g. Ferreira *et al.* 2006), much of this warm outflow should emanate from the innermost radii of the disk. The accretion rate of V1647 Ori decreased by two orders of magnitude at the end of the outburst, and the truncation radius moved outward in response. During this readjustment, the disk and magnetosphere were not in dynamical equilibrium. As the magnetic field pushed the truncation radius of the disk back, the field once again entered a pinched configuration that resulted in a slower outflow at a larger radius (Fig. 3b). When the disk and the magnetosphere settled into dynamical equilibrium consistent with the reduced accretion rate, the angle between the magnetic field and the disk likely increased and the cooler CO outflow diminished (Fig. 3c). Thus the blue-shifted absorption is not observed in the post-outburst data (Fig. 2). This state is consistent with other cTTSs systems.

5. Conclusions

While accreting disks around low-mass stars are expected to interact with stellar magnetospheres and generate outflows (Hartmann 1998), the transformation of centrally peaked ro-vibrational CO emission lines to CO emission lines with blue-shifted absorption is unique. We suggest that the mechanism responsible for producing this outflow is distinct from the one that drives the outflow of the typical cTTSs. The rapid and dramatic change in the accretion rate that characterizes the EXor phenomenon provides an important opportunity to study the interplay between stellar accretion, the inner disk and outflows. This insight is key to reaching a satisfactory theoretical understanding of these events.

Acknowledgements

The data presented herein were obtained in part at the W.M. Keck Observatory. The Observatory was made possible by the generous financial support of the W.M. Keck Foundation. Also based in part on observations obtained at the Gemini Observatory. The Phoenix infrared spectrograph was developed and is operated by the National Optical Astronomy Observatory. The Phoenix spectra were obtained as part of program GS-2006A-DD-1 and GS-2006B-DD-1.

References

Acosta-Pulido, J. A., *et al.* 2007, *AJ*, 133, 2020
Andrews, S. M., Rothberg, B., & Simon, T. 2004, *Ap. Lett.*, 610, L45
Balsara, D. S. 2004, *ApJS*, 151, 149
Blake, G. A., & Boogert, A. C. A. 2004, *Ap. Lett.*, 606, L73
Blandford, R. D., & Payne, D. G. 1982, *MNRAS*, 199, 883
Bouvier, J., Alencar, S. H. P., Harries, T. J., Johns-Krull, C. M., & Romanova, M. M. 2007, Protostars and Planets V, 479
Briceño, C., *et al.* 2004, *Ap. Lett.*, 606, L123
Brittain, S. Simon, T., Najita, J. R., & Rettig, T .W. 2007a, *ApJ*, 659, 685
Brittain, S. *et al.* 2007b, *Ap. Lett.*, submitted
Cushing, M. C., Vacca, W. D., & Rayner, J. T. 2004, *PASP*, 116, 362
Fedele. D. van den Ancker, M. E., Petr-Gotzens, M. G. & Rafanelli, P. 2007, *A&A*, in press
Ferreira, J., Dougados, C., & Cabrit, S. 2006, *A&A*, 453, 785
Gibb, A. G. 1999, *MNRAS*, 304, 1
Gibb, E. L., Rettig, T. W., Brittain, S. D., Wasikowski, D., Simon, T., Vacca, W. D., Cushing, M. C., & Kulesa, C. 2006, *ApJ*, 641, 383
Glassgold, A. E., Najita, J., & Igea, J. 2004, *ApJ*, 615, 972
Ghosh, P., & Lamb, F. K. 1979, *ApJ*, 234, 296
Hartigan, P., Edwards, S., & Ghandour, L. 1995, *ApJ*, 452, 736
Hartmann, L. 1998, Accretion processes in star formation, Cambridge University Press, 1998
Hartmann, L., & Kenyon, S. J. 1996, *ARAA*, 34, 207

Johns-Krull, C. M., & Valenti, J. A. 2000, ASP Conf. Ser. 198: Stellar Clusters and Associations: Convection, Rotation, and Dynamos, 198, 371
Kastner, J. H., *et al.* 2006, *Ap. Lett.*, 648, L43
Kóspál, A., Abraham, P. A.-P. J., Csizmadia, S., Eredics, M., Kun, M., & Racz, M. 2005, Informational Bulletin on Variable Stars, 5661, 1
Lada, E. A., Bally, J., & Stark, A. A. 1991, *ApJ*, 368, 432
Matsumoto, R., Uchida, Y., Hirose, S., Shibata, K., Hayashi, M. R., Ferrari, A., Bodo, G., & Norman, C. 1996, *ApJ*, 461, 115
Mitchell, G. F., Johnstone, D., Moriarty-Schieven, G., Fich, M., & Tothill, N. F. H. 2001, *ApJ*, 556, 215
Muzerolle, J., Megeath, S. T., Flaherty, K. M., Gordon, K. D., Rieke, G. H., Young, E. T., & Lada, C. J. 2005, *Ap. Lett.*, 620, L107
Najita, J. R., Edwards, S., Basri, G., & Carr, J. 2000, Protostars and Planets IV, 457
Najita, J., Carr, J. S., & Mathieu, R. D. 2003, *ApJ*, 589, 931
Ojha, D. K., *et al.* 2006, *MNRAS*, 368, 825
Rayner, J. T., Toomey, D. W., Onaka, P. M., Denault, A. J., Stahlberger, W. E., Vacca, W. D., Cushing, M. C., & Wang, S. 2003, *PASP*, 115, 362
Reipurth, B., & Aspin, C. 2004, *Ap. Lett.*, 606, L119
Rettig, T., Brittain, S., Simon, T., Gibb, E., Balsara, D. S., Tilley, D. A., & Kulesa, C. 2006, *ApJ*, 646, 342
Rettig, T. W., Brittain, S. D., Gibb, E. L., Simon, T., & Kulesa, C. 2005, *ApJ*, 626, 245
Vacca, W. D., Cushing, M. C., & Simon, T. 2004, *Ap. Lett.*, 609, L29
Walter, F. M., Stringfellow, G. S., Sherry, W. H., & Field-Pollatou, A. 2004, *AJ*, 128, 1872

Star-Disk Interaction in Young Stars
Proceedings IAU Symposium No. 243, 2007
J. Bouvier & I. Appenzeller, eds.

doi:10.1017/S1743921307009593

The rotational evolution of young low mass stars

Jérôme Bouvier

Laboratoire d'Astrophysique, Observatoire de Grenoble, Université Joseph Fourier,
B.P.53, 38041 Grenoble, Cedex 9, France
email: jbouvier@obs.ujf-grenoble.fr

Abstract. Star-disk interaction is thought to drive the angular momentum evolution of young stars. In this review, I present the latest results obtained on the rotational properties of low mass and very low mass pre-main sequence stars. I discuss the evidence for extremely efficient angular momentum removal over the first few Myr of pre-main sequence evolution and describe recent results that support an accretion-driven braking mechanism. Angular momentum evolution models are presented and their implication for accretion disk lifetimes discussed.

Keywords. Stars: pre–main-sequence, stars: rotation, accretion, accretion disks.

1. Introduction

Star-disk interaction in young systems involves mass and angular momentum transfer. Mass transfer manifests itself by accretion onto the star and wind/jet outflows while angular momentum transfer is thought to drive the rotational evolution of young stars and is also seen in jet rotation (cf. T. Ray's review in this volume). The purpose of this review is to address the following issues : i) what do observations tell us about the rotational evolution of young stars, ii) what is the impact of star-disk interaction on angular momentum evolution during the pre-main sequence, and iii) how successful are models in accounting for the observed rotational evolution of young stars?

Section 2 describes the rotational properties of low mass (0.2-1.0 $M_{\odot}$) pre-main sequence stars. Recent results include the determination of rotation periods for hundreds of stars in young clusters spanning an age range from $\leqslant$1 Myr to the zero-age main sequence and beyond. For the first time, these results provide a relatively well sampled age sequence, from which the rotational evolution of young stars can be read.

Since the early 90's, it is commonly thought that the star-disk interaction is somehow reponsible for the low rotation rate of young stars. Section 3 discusses the issue of accretion-related angular momentum loss, a process which should manifest itself by accreting stars being, on average, slower rotators than non accreting ones. Conflicting evidence for such a connection has been reported and is revisited here in the light of new results.

Section 4 provides a brief status of current angular momentum evolution models. Most models rely on the same simplified assumptions and still await to be upgraded to an actual physical description of the angular momentum loss processes acting during the pre-main sequence. The implications of these models for the lifetime of accretion disks are discussed.

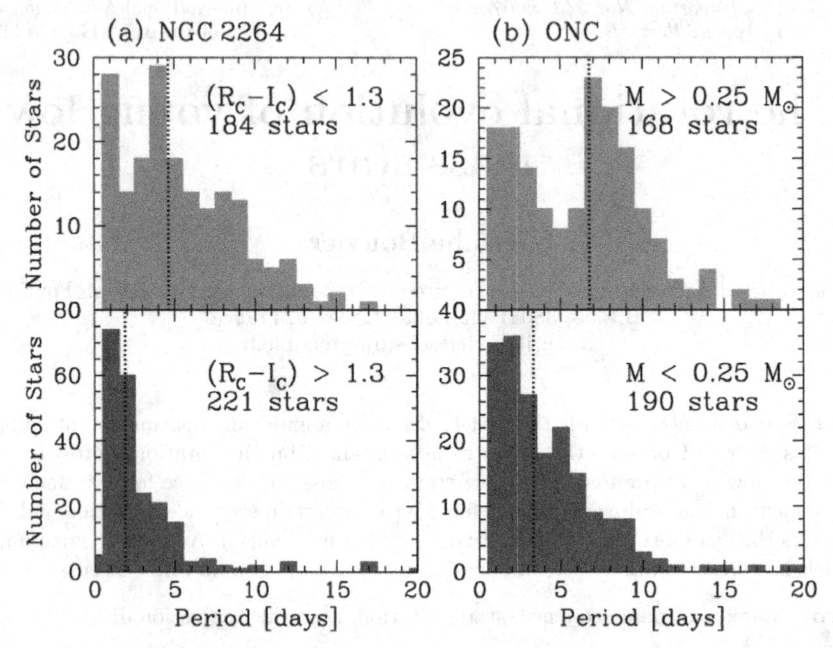

Figure 1. The rotational period distributions of low mass (*top*) and very low mass (*bottom*) stars in the ~2-4 Myr NGC 2264 cluster (*left*) and in the ~1 Myr Orion Nebulae Cluster (*right*). From Lamm *et al.* (2005).

2. The rotational periods of young low mass stars

In the last 10 years, rotational periods have been measured for hundreds of low mass pre-main sequence stars in various star forming regions and young open clusters. Rotational periods are derived by detecting a periodic component in the star's photometric variability, which results from the modulation of the star's brightness by surface spots. While this observational approach is time consuming, requiring the photometric monitoring of hundreds of stars over a timescale of several weeks, it is also far superior to spectroscopic $v \sin i$ measurements as the photometrically-derived rotational period is not affected by inclination effects. Moreover, the star's angular velocity, a key physical parameter of star-disk interaction models, relates directly to the rotational period ($\Omega = 2\pi/P$), without having to resort to the poorly known stellar radius. These properties have been prime motivations for lauching large scale photometric monitoring campaigns targetting star forming regions with the hope to derive the rotational period distribution of statistically significant samples of pre-main sequence stars at various stages of their evolution.

The first large scale photometric monitoring campaign by Choi & Herbst (1996) revealed the peculiar *bimodal* distribution of periods for young stars in the Orion Nebulae Cluster (ONC). The distribution of periods at an age of about 1 Myr exhibits two peaks, one located at a period of about 8 days, the other at a period of about 2-3 days, with approximately twice as many slow rotators as fast ones. While this result was disputed a few years later (Stassun *et al.* 1999), the bimodal distribution was eventually confirmed, with peaks near 2 and 8 days, for ONC stars more massive than 0.25 $M_\odot$ (Herbst *et al.* 2002). Interestingly, the same study also showed that the bimodality does not extend to lower mass stars whose period distribution exhibits a single peak at fast rotation with a mean period of about 2 days. This was the first hint at possibly different rotational properties between low mass (0.3-1.0 $M_\odot$) and very low mass ($\leqslant$0.3 $M_\odot$) young stars.

One of the first comparative studies between 2 clusters of different ages came from the measurement of hundreds of rotational periods for low mass stars in NGC 2264 (Lamm *et al.* 2005). With an estimated age of 2-4 Myr, NGC 2264 is slightly older than ONC, thus allowing a first attempt to trace the evolution of rotation at the start of pre-main sequence evolution. The distribution of periods for ONC and NGC 2264 are shown in Figure 1 for low mass and very low mass stars, with a dividing line around 0.25 $M_\odot$. In each mass bin, the rotational distributions for the 2 clusters have a similar shape, bimodal for low mass stars and single-peaked for very low mass stars. However, on a timescale of a few Myr from ONC to NGC 2264, the peaks of the distributions have shifted towards shorter periods, indicative of stellar spin up by a factor of about 1.5 to 2. Within the uncertainties on the age of the 2 clusters and assuming that the initial period distributions were similar in both clusters, this amount of spin up is consistent with stellar contraction and angular momentum conservation (Lamm *et al.* 2005). While most stars appear to spin up between ONC and NGC 2264, a tail of slow rotators remains, which suggests efficient braking for a fraction of low mass stars during the first 2 Myr of their evolution.

A major leap in the study of angular momentum evolution of young stars recently came from the derivation of hundreds of rotational periods for low mass and very low mass stars in various clusters spanning an age range from 5 to 200 Myr. The Monitor project (Aigrain *et al.* 2007; Irwin *et al.* 2007a) aims at detecting eclipsing binaries and planetary transits in young low mass stars. As a by-product of intense photometric monitoring campaigns, precise rotational periods are derived for the young populations of these clusters (Irwin *et al.* 2006, 2007b). The period distributions derived for stars in the mass range 0.1-1.0 $M_\odot$ are shown in Figure 2 over the age range 1-200 Myr. The shape of the period distribution evolves drastically over the first 40 Myr. Starting from ONC at 1 Myr, the very low mass stars (M$\leqslant$0.25 $M_\odot$) appear to continuously spin up, from a median initial period of 2-3 days at 1 Myr to a median period of 0.5-0.7 days at 40 Myr and even shorter at 150 Myr. The rapid convergence of very low mass stars towards fast rotation, with no slow rotators left, is quite remarkable indeed.

As a group, higher mass stars (M$\geqslant$0.5 $M_\odot$) also tend to spin up during PMS evolution, though not as much as very low mass stars, with a median initial period of about 6 days at 1 Myr shortening to about 3 days at 40 Myr. However, the initial bimodal distribution of periods seen for stars in this mass range remains visible over the whole PMS evolution. A fraction of initially slowly rotating stars, with periods of 8-15 days at 1 Myr, retain similar periods up to 5 Myr and experience only mild spin up later on, with periods in the range 5-10 days at 40 Myr. Meanwhile, the initially fast rotators, with periods of 1-3 days at 1 Myr, spin up continuously with the shortest periods decreasing from 0.9d at 1 Myr, to 0.5d at 5 Myr and 0.2d at 40 Myr.

Recent results thus converge in indicating a somewhat different rotational evolution between low mass and very low mass stars during the pre-main sequence. Most very low mass stars (M$\leqslant$0.25 $M_\odot$) start their evolution as fast rotators and spin up continuously as they descend their Hayashi tracks and approach the ZAMS. In contrast, a significant fraction of low mass stars (0.3-1.0 $M_\odot$) experience milder spin up during their pre-main sequence evolution. Indeed, some appear to evolve at constant angular velocity for the first 5 Myr at least (see also Rebull *et al.* 2004). Clearly, a very efficient brake must be at work to extract angular momentum in these young stars in order to prevent them from spinning up in spite of stellar contraction.

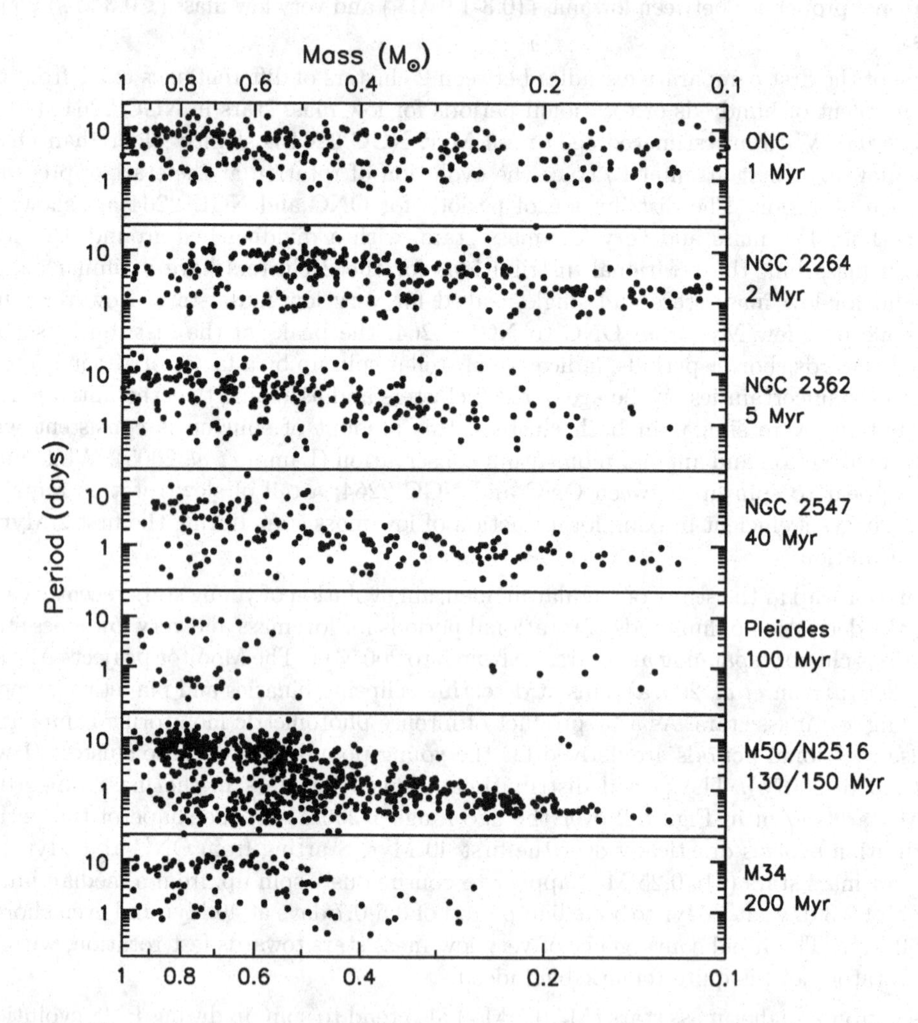

Figure 2. The rotational period distribution of low mass and very low mass stars in various clusters in the age range from 1 to 200 Myr. Note the evolution of the shape of the period distribution as a function of time, especially at very low masses. Adapted from Irwin *et al.* (2007b) and references therein.

3. Is PMS braking related to the accretion process?

On the main sequence, solar-type stars are braked at a low pace by their magnetized winds. Even though magnified versions of solar-type winds probably exist in the magnetically active low mass PMS stars, the associated braking timescale is much longer than the Kelvin-Helmotz contraction timescale (Bouvier *et al.* 1997). As a result, these winds cannot prevent the star from spinning up as it contracts towards the ZAMS (see Matt, this volume).

As an alternative, following models developped for X-ray binaries, Königl (1991) suggested that the magnetic star-disk interaction could regulate the angular momentum of the star. This idea has since been developped in a variety of MHD models where angular momentum is extracted from the star by the magnetic field and carried away by the disk or by an accretion-driven wind (see the contributions by Shu, Fendt, Romanova, and

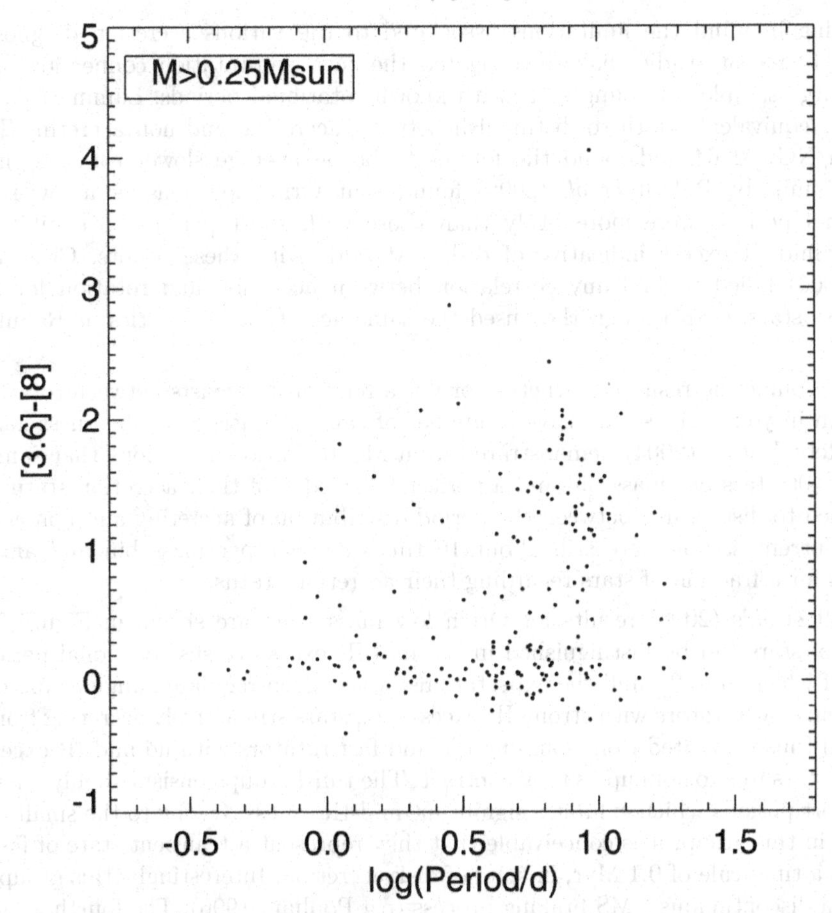

Figure 3. Spitzer [3.6]-[8.0] color excess as a function of rotational period for Orion low mass stars (M⩾0.25 $M_\odot$). Stars with disks have [3.6]-[8.0]⩾1. From Rebull *et al.* (2006).

Ferreira in this volume). In these models, the star is thus braked as long as it accretes from its disk.

Irrespective of the specific underlying physical model (X-wind, disk locking, accretion-driven disk winds or stellar winds), any accretion-related angular momentum loss process should reveal itself as accreting stars being, on average, slower rotators than non accreting ones. Such a relationship between rotation and accretion has therefore been actively searched for. A first hint that such a correlation may exist was reported for a limited sample of T Tauri stars in Taurus with known rotational period and IR excess (Edwards *et al.*, 1993). A larger scale study of ONC stars however revealed no such correlation (Stassun *et al.* 1999).

One difficulty in searching for rotation-accretion connection, besides statistical robustness, is to identify an unambiguous diagnostics of accretion onto the star. The often used near-IR and mid-IR excesses are somewhat ambiguous in this respect as they can arise from passive, i.e., non accreting, disks. The UV continuum excess is a more direct measurement of accretion onto the star but has been measured for too few young stars to be useful. H_α line emission is a proxy of accretion onto the stellar surface, as long as it can be shown that the line flux or width exceeds the chromospheric emission component. In some regions with strong nebular H_α emission, like ONC, the interpretation of the H_α line is however not straighforward.

Keeping in mind the limitations associated to the various accretion diagnostics, a number of recent studies have investigated the accretion-rotation connection for relatively large samples of young stars with known rotational periods. Lamm *et al.* (2005) used H_α equivalent width to distinguish between accreting and non accreting T Tauri stars in NGC 2264 and found the former to be on average slower rotators than the latter. Similarly, Rebull *et al.* (2006) found that Orion low mass stars with longer rotational periods were more likely than those with short periods to exhibit a continuum mid-IR excess indicative of disks. At odds with these results, Cieza & Baliber (2006) failed to find any correlation between accretion and rotation for IC 348 low mass stars, even though they used the same accretion diagnostics as Rebull *et al.* (2006).

These conflicting results on whether or not a relationship exists between rotation and accretion in young stars may have a number of causes. Statistical robustness is still an issue. Rebull *et al.* (2004) demonstrated from Monte Carlo simulations that samples of at least 400 stars per mass bin and a perfect knowledge of their accretion status would be needed to distinguish between the period distribution of accreting and non accreting stars. Current samples are still about 10 times smaller per mass bin and ambiguity remains for a fraction of stars regarding their accretion status.

Rebull *et al.*'s (2006) results for Orion low mass stars are shown in Figure 3. Four groups of stars can be distinguished in the mid-IR excess versus rotational period diagram, of which only 2 fulfill the expectations of accretion-regulated angular momentum evolution: slow rotators with strong IR excess, i.e., stars still actively accreting from their disk and thus prevented from spinning up, and fast rotators with no mid-IR excess, i.e., diskless stars free to spin up as they contract. The third group consists of only a few stars with short periods which exhibit a significant mid-IR excess. Owing to the small number of stars in this group, it is conceivable that they represent a transient state of fast rotation, on a timescale of 0.1 Myr, in spite of disk accretion. Interestingly, this group might hint at a discontinuous PMS braking process (e.g Popham 1996). The fourth group consists of a significant fraction of slow rotators with no mid-IR excess. Thus, among stars without IR excess, about half have periods longer than 5 days.

This latter group is the most puzzling as these stars lack evidence for a disk and yet rotate slowly. They are often interpreted as having dissipated their disk recently, and thus have not had time yet to spin up. However, similar groups of slowly rotating and apparently non accreting stars are observed in other star forming regions, over the age range from $\leqslant$1 Myr (e.g. ONC) to $\sim$5 Myr (e.g. Taurus). The spin up rate scales as R^2_{star} during the first few Myr, as long as the star remains fully convective, and increases later on as the radiative core develops. Thus, assuming an average initial period of 8 days at 1 Myr, a star would spin up to periods shorter than 5 days in about 1 Myr. One thus has to assume that, in each of the observed regions, a significant fraction of stars ($\sim$30%) were released from their disk less than a Myr ago over the age range 1-5 Myr.

Overall, signatures of the accretion-rotation connection, as expected from accretion-regulated angular momentum evolution, have been recently reported. However, some intriguing results remain. In particular, a significant subgroup of apparently non accreting stars have long periods, which does not fit the accretion-regulated angular momentum scenario. Clearly, further characterization of these slow rotators is needed to assess their actual accretion status. Also, additional Monte Carlo simulations would help to clarify the interpretation of the accretion-rotation diagrams over PMS evolution timescales.

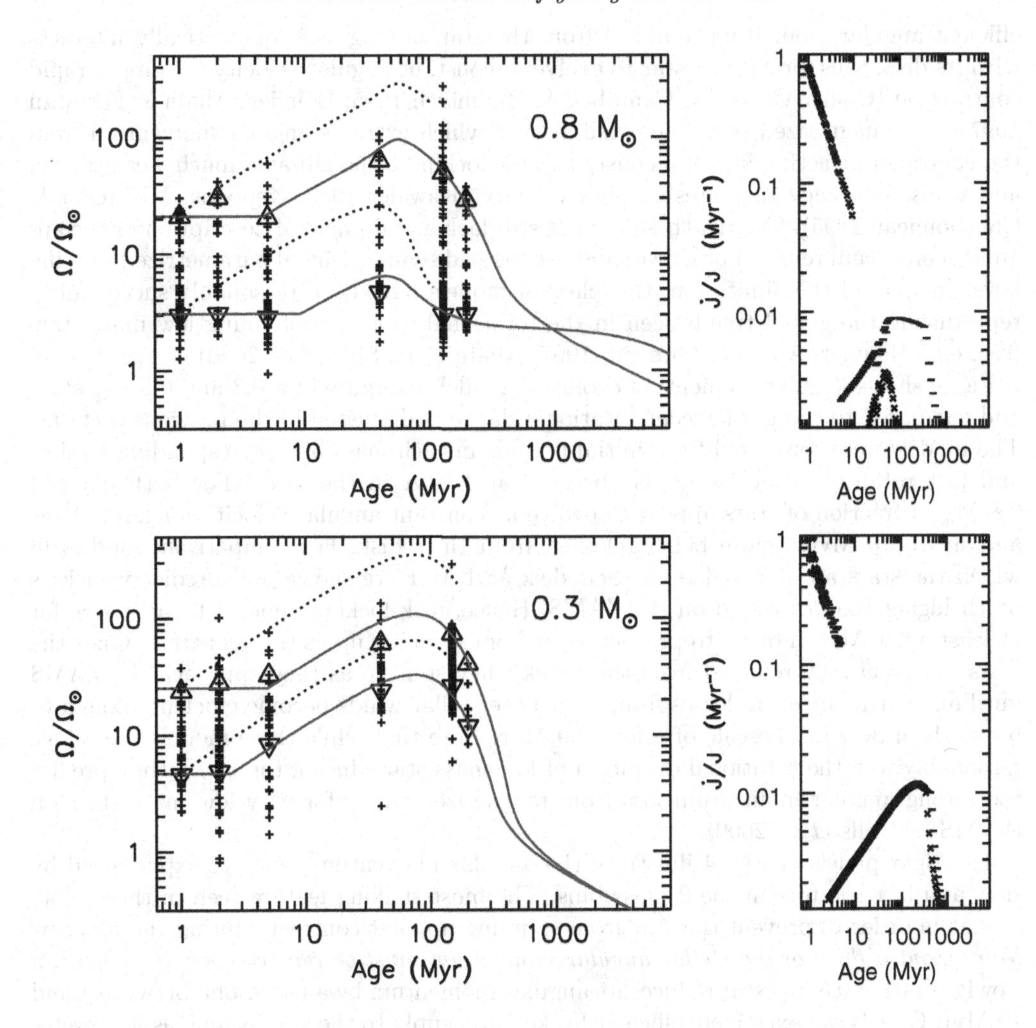

Figure 4. Angular momentum evolution models for 0.8 $M_\odot$ and 0.3 $M_\odot$ stars. *Left :* The distribution of rotational periods for stars in the mass range 0.7-0.9 $M_\odot$ (*top*) and 0.20-0.35 $M_\odot$ (*bottom*) are shown by crosses. Large triangles indicate the 25th and 90th percentiles of the distribution in each cluster. Models for slow and fast rotators are shown by solid lines. The initial rotational periods at 1 Myr are 1.2 and 8.3 days for 0.8 $M_\odot$ stars (*top*) and 1.0 and 5.4 days for 0.3 $M_\odot$ stars (*bottom*). Disk locking timescales range from 2.5 to 10 Myr. Once the star is eventually released from its disk, solar-type winds are the only source of angular momentum loss. Dotted lines show similar models over the first 200 Myr for stars released from their disk at 1 Myr. See Bouvier, Forestini & Allain (1997) and Allain (1998) for the details of the models. *Right :* Angular momentum loss rate (per Myr) for the models shown in the left panels. In order to evolve at constant angular velocity during the first 5-10 Myr in spite of contraction, the star must lose, on average, about a third of its angular momentum per Myr. The angular momentum loss rate from "disk locking" is over a hundred times more efficient than solar-type winds to brake the star during the early PMS.

4. Models of angular momentum evolution and disk lifetimes

Models of angular momentum evolution during the pre-main sequence (PMS) and on the zero-age main sequence (ZAMS) have been developped in the mid- and late-90's and have not much progressed since then. These models rely on 2 processes to remove angular momentum from the star : i) the so-called "disk-locking", which assumes extremely efficient angular momentum removal from the star as long as it magnetically interacts with its disk, thus forcing the star to evolve at constant angular velocity in spite of rapid contraction (Collier Cameron, Campbell & Quaintrell 1995; Bouvier, Allain & Forestini 1997), ii) a magnetized solar-type stellar wind which extracts angular momentum from the central star, acting simultaneously as disk locking although at a much weaker rate, and whose efficiency saturates at high velocity (Kawaler 1988; Keppens, McGregor & Charbonneau 1995). Clearly, these models still lack a real physical description of the angular momentum removal processes and use instead semi-empirical parametrized braking laws. In spite of this limitation, this class of models have been reasonably successful in reproducing the global trends seen in the rotational evolution of young low mass stars (see, e.g., Bouvier, Allain & Forestini 1997; Allain 1998; Sills *et al.* 2000).

Fig. 4 shows angular momentum evolution models computed for 0.3 and 0.8 $M_{\odot}$ stars and compared to the most recent rotational datasets discussed in the previous sections. The models were computed for 2 initial periods in each mass bin, corresponding to slow and fast rotators, respectively, as observed at 1 Myr in the ONC. For both 0.3 and 0.8 $M_{\odot}$, a fraction of stars appear to evolve at constant angular velocity for a few Myr, and up to 10 Myr, before being released from their disk. For comparison, models in which the stars are released from their disk at 1 Myr are shown and predict velocities much higher than observed on the ZAMS. Hence, disk locking appears to be active for at least a few Myr in order to prevent stars from spinning up as they contract. Once the stars are eventually released from their disk, they spin up as they approach the ZAMS until angular momentum losses from solar-type stellar winds become efficient enough to brake them over a timescale of a few 100 Myr. Note that while these models reproduce reasonably well the rotational evolution of low mass stars during the PMS, they predict too strong angular momentum loss from magnetized winds for very low mass stars on the MS (cf. Sills *et al.* 2000).

The right panels of Fig. 4 illustrate the angular momentum loss rate experienced by slow and fast rotators in the 2 mass bins. The most striking feature seen in these plots is that in order to prevent the star from spinning up as it contracts during the first few Myr, *about a third of the stellar angular momentum must be removed per Myr.* Thus, a slowly rotating 0.8 $M_{\odot}$ star reduces its angular momentum by a factor of 5 between 1 and 10 Myr. Clearly, an extremely efficient brake must apply to the star as long as it accretes from its disk, far more efficient indeed than angular momentum losses due to solar-type winds. MHD star-disk interaction models have to face this challenge : not only the star-disk interaction does not spin the star up, but it actively brakes it. In other words, it is not enough to balance positive accretion torques by negative magnetic torques to reach a zero net flux of angular momentum onto the star. Instead, it is mandatory that the star-disk interaction process actually removes angular momentum from the star at a high rate, i.e. that the net torque on the star be strongly negative, if the star is to evolve at constant angular velocity in spite of contraction.

Finally, the comparison of models with observations suggests that the rotational velocity of low mass PMS stars is regulated over a timescale of a few Myr, typically from 2 to 10 Myr. Since the strong brake is thought to result from active star-disk interaction, this requires that disk accretion lasts over at least this timescale. Current estimates of disk lifetimes have been obtained from a variety of diagnostics, including near- and mid-IR excess, and H_{α} emission line width. While the former probe both active and passive disks, the latter provides more direct evidence for accretion onto the star. The disk fraction derived from near-IR excess amounts to about 40-60% at 1 Myr, with no disk remaining past 5 Myr (Hillenbrand 2005). When diagnosed from the more sensitive mid-IR excess and/or H_{α} width measurements, longer disk survival times are obtained, with about 40-60% of stars still surrounded by circumstellar disks at 2 Myr, a fraction which decreases to about 10-25% at 10 Myr (e.g. Damjanov *et al.* 2007; Lyo & Lawson 2005; Jayawardhana *et al.* 2006). Interestingly, a recent study of the 8 Myr η Cha cluster suggests a mean disk lifetime of 9 Myr for single low mass stars while binary systems seem to dissipate their disks on a timescale of 5 Myr (Bouwman *et al.* 2007). No primordial disks appear to survive past 30 Myr (Gorlova *et al.* 2007).

5. Conclusions

New datasets now provide hundreds of rotational periods measured for low mass and very low mass stars in young clusters over the age range from 1 Myr to 0.2 Gyr. The distribution of rotational periods shows a clear evolution over the pre-main sequence contraction timescale. While most stars tend to spin up as they descend their Hayashi tracks, a fraction retain constant angular velocity for a few million years. This provides clear evidence for a strong brake acting on the stars on a timescale of 2-10 Myr. The new data also confirm that very low mass stars ($M \leqslant 0.3\ M_\odot$) tend to suffer lower angular momentum losses than low mass ones (0.3-1.0 $M_\odot$). The early evolution at nearly constant angular velocity implies angular momentum loss rates much larger that those achievable by solar-type magnetized stellar winds. New empirical evidence has been recently reported which confirms that the magnetic star-disk interaction is indeed responsible for the braking of low mass stars at the start of their PMS evolution. These new results challenge the ability of current MHD star-disk interaction models to extract as much angular momentum from the young star as needed to prevent it from spinning up in spite of accretion and contraction. On the observational side, a critical time step still to be sampled in order to better constrain accretion disk lifetimes from rotational evolution is the 5-40 Myr range as most disks appear to dissipate on these timescales.

Acknowledgements

I would like to thank Jonathan Irwin for providing some of the Monitor data prior to publication.

References

Aigrain, S., Hodgkin, S., Irwin, J., Hebb, L., Irwin, M., Favata, F., Moraux, E., & Pont, F. 2007, *MNRAS*, 375, 29

Allain, S. 1998, *A&A*, 333, 629

Bouvier, J., Forestini, M., & Allain, S. 1997, *A&A*, 326, 1023

Bouwman, J., Lawson, W. A., Dominik, C., Feigelson, E. D., Henning, T., Tielens, A. G. G. M., & Waters, L. B. F. M. 2006, *ApJ (Letters)*, 653, L57

Choi, P. I., & Herbst, W. 1996, *AJ*, 111, 283

Cieza, L., & Baliber, N. 2006, *ApJ*, 649, 862

Collier Cameron, A., Campbell, C. G., & Quaintrell, H. 1995, *A&A*, 298, 133

Damjanov, I., Jayawardhana, R., Scholz, A., Ahmic, M., Nguyen, D. C., Brandeker, A., & van Kerkwijk, M. H. 2007, ArXiv e-prints, 708, arXiv:0708.0266

Edwards, S., *et al.* 1993, *AJ*, 106, 372

Gorlova, N., Balog, Z., Rieke, G. H., Muzerolle, J., Su, K. Y. L., Ivanov, V. D., & Young, E. T. 2007, ArXiv e-prints, 707, arXiv:0707.2827

Herbst, W., Bailer-Jones, C. A. L., Mundt, R., Meisenheimer, K., & Wackermann, R. 2002, *A&A*, 396, 513

Hillenbrand, L. A. 2005, ArXiv Astrophysics e-prints, arXiv:astro-ph/0511083

Irwin, J., Irwin, M., Aigrain, S., Hodgkin, S., Hebb, L., & Moraux, E. 2007a, *MNRAS*, 375, 1449

Irwin, J., Aigrain, S., Hodgkin, S., Irwin, M., Bouvier, J., Clarke, C., Hebb, L., & Moraux, E. 2006, *MNRAS*, 370, 954
Irwin, J., Hodgkin, S., Aigrain, S., Hebb, L., Bouvier, J., Clarke, C., Moraux, E., & Bramich, D. M. 2007b, *MNRAS*, 377, 741
Jayawardhana, R., Coffey, J., Scholz, A., Brandeker, A., & van Kerkwijk, M. H. 2006, *ApJ*, 648, 1206
Kawaler, S. D. 1988, *ApJ*, 333, 236
Keppens, R., MacGregor, K. B., & Charbonneau, P. 1995, *A&A*, 294, 469
Königl, A. 1991, *ApJ (Letters)*, 370, L39
Lamm, M. H., Mundt, R., Bailer-Jones, C. A. L., & Herbst, W. 2005, *A&A*, 430, 1005
Popham, R. 1996, *ApJ*, 467, 749
Rebull, L. M., Wolff, S. C., & Strom, S. E. 2004, *AJ*, 127, 1029
Rebull, L. M., Stauffer, J. R., Megeath, S. T., Hora, J. L., & Hartmann, L. 2006, *ApJ*, 646, 297
Sills, A., Pinsonneault, M. H., & Terndrup, D. M. 2000, *ApJ*, 534, 335
Stassun, K. G., Mathieu, R. D., Mazeh, T., & Vrba, F. J. 1999, *AJ*, 117, 2941

Star-Disk Interaction in Young Stars
Proceedings IAU Symposium No. IAU 243, 2007
J. Bouvier & I. Appenzeller, eds.

doi:10.1017/S174392130700960X

The rotation of very low-mass stars and brown dwarfs

Jochen Eislöffel[1] and Alexander Scholz[2]
[1]Thüringer Landessternwarte, Sternwarte 5, D-07778 Tautenburg, Germany
[2]SUPA, School of Physics & Astronomy, University of St. Andrews, North Haugh, St. Andrews, Fife KY16 9SS, United Kingdom

Abstract. The evolution of angular momentum is a key to our understanding of star formation and stellar evolution. The rotational evolution of solar-mass stars is mostly controlled by magnetic interaction with the circumstellar disc and angular momentum loss through stellar winds. Major differences in the internal structure of very low-mass stars and brown dwarfs – they are believed to be fully convective throughout their lives, and thus should not operate a solar-type dynamo – may lead to major differences in the rotation and activity of these objects. Here, we report on observational studies to understand the rotational evolution of the very low-mass stars and brown dwarfs.

Keywords. Stars: activity, stars: evolution, stars: formation, stars: low-mass, brown dwarfs, stars: magnetic fields, stars: rotation.

---

1. Introduction

The study of stellar rotation during the pre-main sequence (PMS) and the main sequence (MS) phase has provided us with many new insights into their formation and evolution (Bodenheimer 1995; Bouvier *et al.* 1997; Stassun & Terndrup 2003; Herbst *et al.* 2007). The so-called angular momentum problem of star formation asks how the specific angular momentum (angular momentum/mass) of dense molecular cloud cores, from which low-mass stars form, gets reduced by 5 – 6 orders of magnitude compared to what is left in solar-type stars on the zero-age main sequence (ZAMS). An additional 1 – 2 orders of magnitude is lost from the ZAMS to the age of the Sun (~ 5 Gyr).

At the early formation stages, magnetic torques between the collapsing cloud core and the surrounding interstellar medium, the deposition of a large amount of angular momentum into the orbital motion of a circumstellar disc, a planetary system, and/or a binary star system play important roles. It is known that solar-mass stars already in their T Tauri phase rotate slowly, although they are still accreting matter from their disc. Magnetic coupling between the star and its circumstellar disc, and the consequent removal of angular momentum in a highly collimated bipolar jet are considered to be the agent for this rotational braking (e.g., Camenzind 1990; Königl 1991; Shu *et al.* 1994; Matt & Pudritz 2005). After the cessation of accretion and the following dispersal of the disc this braking mechanism obviously comes to an end, and the rotation is observed to accelerate as the PMS stars contract towards the ZAMS. On the main sequence then, the rotation rates of solar-mass stars decrease again, because now angular momentum loss through stellar winds takes over as the dominant process.

The rotation of stars can either be measured spectroscopically from the line-broadening of photospheric spectral lines, or it can be derived from periodic variability in the light curves of stars seen in photometric time series observations. While the spectroscopic method suffers from projection effects – the inclination angle of the rotation axis with respect to our line of sight is unknown – the photometric method allows us to determine the rotation period with high precision and independent of inclination angle.

Whereas a large amount of rotation periods are now available in the literature for low-mass stars in clusters younger than about 3 Myr (ONC: Herbst *et al.* 2001; Herbst *et al.* 2002; NGC2264: Lamm 2003; Lamm *et al.* 2004), and most recently in the much older M34: Irwin *et al.* (2006) and NGC2516: Irwin *et al.* (2007), not much is known about the early evolution of very low-mass (VLM) stars and brown dwarfs up to the age of a few Gyr, when they are found as field object in the solar neighbourhood (Bailer-Jones & Mundt 2001; Clarke *et al.* 2002). This has lead us to initiate a programme to study the rotation periods of the VLM stars and brown dwarfs, and to compare them to solar-mass stars as well as to evolutionary models. For our monitoring programme we decided to follow the photometric time series approach to obtain precise rotation periods.

In this text, we first present our results on rotation rates and variability of the sources in Section 2. The observed rotation rates are then compared to various models of rotational evolution in Section 3. In Section 4 we discuss some first attempts to characterise the spots on VLM objects, followed by some comments about accretion and time variability in these objects in Section 5. Finally, Section 6 contains our conclusions.

2. Rotation and variability of VLM objects

In the course of our ongoing monitoring programme, we have so far created a database of rotation periods for 23 VLM objects in the cluster around sigma Ori (Scholz & Eislöffel 2004a), for 30 in the field around epsilon Ori (Scholz & Eislöffel 2005), which are belonging to the Ori OBIb association, and for 9 objects in the Pleiades open cluster (Scholz & Eislöffel 2004b). At ages of about 3, 5, and 125 Myr these three groups of VLM objects form an age sequence that allows us some insights into a relevant part of their PMS evolution.

In general, the observed periodic variability in the light curves of our VLM targets is attributed to surface features, which are asymmetrically distributed on the surface and are co-rotating with the objects. Such surface features may arise from dust condensations in the form of "clouds", or from magnetic activity in the form of cool "spots". Because of their youth all our objects have surface temperatures $T_{eff} > 2700$ K (Baraffe *et al.* 1998), which corresponds to spectral types earlier than M8. These temperatures are higher than the dust condensation limits, so that we are most likely observing cool, magnetically induced spots.

It is interesting to compare the mass dependence of the rotation periods for the VLM and solar-mass objects. For the Pleiades, we find that their period distributions are different. While periods of up to ten days are known for the solar-mass objects, Fig. 1 shows that among the VLM objects we are lacking members with rotation periods of more than about two days. Although our photometric time series extends over a time span of 18 days, slow rotators might have been missed among the VLM objects, if their spot patterns evolved on a much shorter time scale, or if they did not show any significant spots.

These possibilities can be checked by converting the spectroscopically derived lower limits for rotational velocities from Terndrup *et al.* (1999) and references therein into upper limits for the rotation periods of the VLM objects using the radii from the models by Chabrier & Baraffe (1997). Such spectroscopically derived rotational velocities should not be affected by the evolution of spot patterns on the objects. The derived upper period limits are shown in Fig. 1 as a solid line. With a single exception, all our data points fall below this line. Hence, they are in good agreement with the spectroscopic

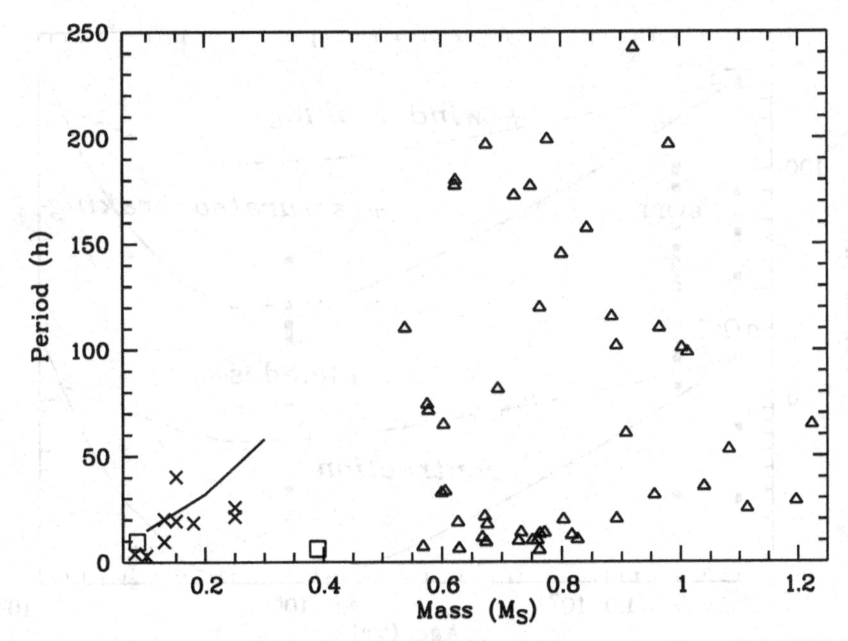

Figure 1. Rotation periods versus mass in the Pleiades. Our rotation periods for VLM objects are shown as crosses. Triangles mark the periods for solar-mass stars from the Open Cluster Database. The two squares show periods from Terndrup *et al.* (1999) The solid line marks the upper limit to the observed $v \sin i$ values of Terndrup *et al.* (2000).

rotation velocities. Both complementary data sets indicate the absence of slow rotators among the VLM objects. Looking at them in detail, they even show a trend towards faster rotation even within the VLM regime from higher to lower masses. Such a trend is also seen in our data of the epsilon Ori cluster, and in the Orion Nebula Cluster data by Herbst *et al.* (2001).

3. Rotational evolution of VLM objects

We now want to combine the periods for all three clusters that we observed, namely sigma Ori (Scholz & Eislöffel 2004a), epsilon Ori (Scholz & Eislöffel 2005), and the Pleiades (Scholz & Eislöffel 2004b), to try to reproduce their period distributions with simple models. These models should include the essential physics of star formation and evolution that we described in Section 1. A practical way of doing this is to project the period distribution for sigma Ori forward in time and then compare the model predictions with our observations for epsilon Ori and the Pleiades.

It is obvious, that the hydrostatic contraction of the newly formed VLM objects has to be taken into account in a first step. Changes in their internal structure may be negligible for these fully convective objects (Sills *et al.* 2000). Then, their rotation periods should evolve from the initial rotation period at the age of sigma Ori strictly following the evolution of the radii (which were taken from the models by Chabrier & Baraffe 1997). Therefore, the rotation accelerates, and should only level out for ages older than the Pleiades, when the objects have settled (dotted lines in Fig. 2). This model, however, is obviously in conflict with the observed Pleiades rotation periods. Half of the sigma Ori objects would get accelerated to rotation periods below the fastest ones observed in the Pleiades of about 3 h. Furthermore, even the slowest rotators in sigma Ori would get

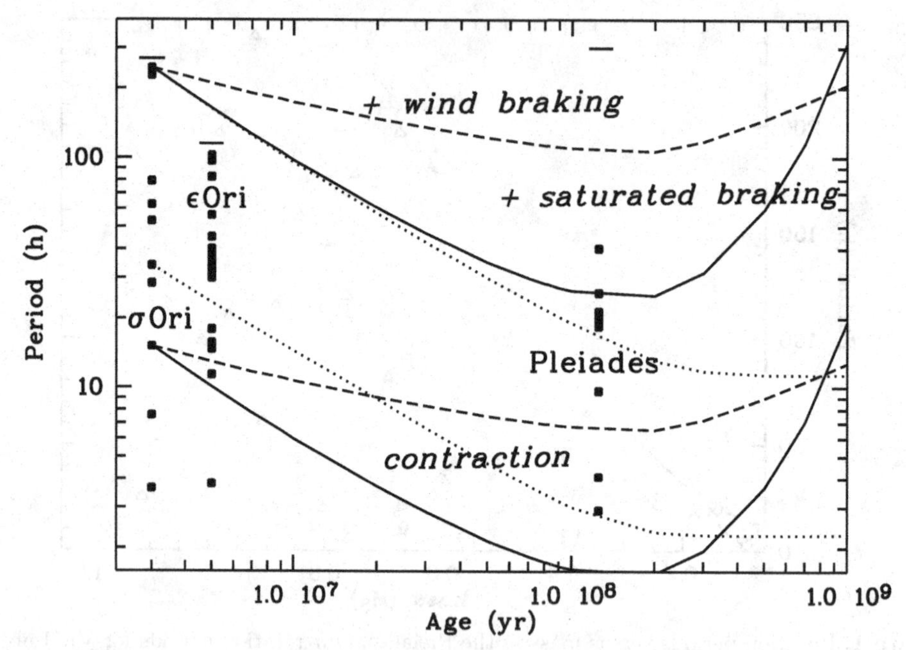

Figure 2. Rotational evolution of VLM objects. The dotted lines show the evolution of the rotation periods for a couple of objects for a model of hydrostatic contraction only. A model with additional Skumanich type braking through stellar winds is shown as dashed lines, and models that use a saturated wind braking instead are shown as solid lines.

spun up to velocities much faster than the slower rotators in the Pleiades, which would remain unexplained then. This teaches us that significant rotational braking must occur until the objects reach the age of the Pleiades, because it is clear that the sigma Ori VLM objects will undergo a significant contraction process.

In a second model we now add a Skumanich type braking through stellar winds, as it is seen in solar-type MS stars (Skumanich 1972). This wind braking increases the rotation periods $\sim t^{1/2}$, as shown by the dashed lines in Fig. 2. However, following this model some of the sigma Ori slow rotators would get braked far to strongly. For a Skumanich type wind braking they would become clearly slower rotators than are observed in the Pleiades (see also above). A possible solution to this problem is to assume that even the slowest sigma Ori rotators seem to rotate so fast, that they are beyond the saturation limit of stellar winds (Chaboyer *et al.* 1995; Terndrup *et al.* 2000; Barnes 2003). In this saturated regime, angular momentum loss is assumed to depend only linearly on angular rotational velocity, and therefore rotation periods increase exponentially with time. The solid lines in Fig. 2 show a model of contraction and saturated wind braking. The period evolution of this model is the most consistent with our data.

It is interesting to explore if disc-locking – as an angular momentum regulator active only at very young ages – may also play a role for the evolution of rotation periods. The sigma Ori cluster would be young enough for this process to play a role, and indeed we found evidence that some of our objects in this cluster may possess an accretion disc. Therefore, we investigate a scenario in which disc-locking is active for an age up to 5 Myr, typical for the occurrence of accretion discs in solar-mass stars. During this time rotation periods would remain constant. This disc-locking scenario we combined with the saturated wind braking, with an adapted spin-down time scale. In Fig. 3 dashed lines

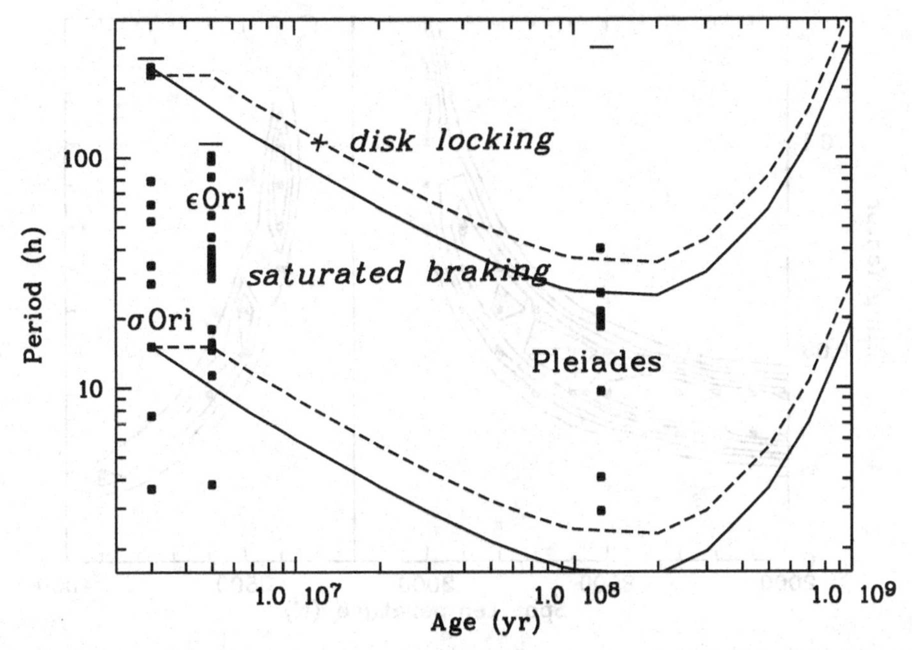

Figure 3. Rotational evolution of VLM objects. The evolution of the rotation periods for a couple of objects for a model with hydrostatic contraction and saturated wind braking are shown as solid lines, as in Fig. 2, while a model with added disc-locking up to an age of 5 Myr is shown as dashed lines.

are shown for two objects following this model, together with the pure saturated wind braking model discussed above (the solid lines, as in Fig. 2). The period evolution for both models is nearly indistinguishable. It turns out that from our currently available rotation periods for these three clusters, strong constraints for or against disc-locking on VLM objects cannot yet be placed.

4. The spots of VLM objects

Not much is known about the properties of the spots on VLM stars and brown dwarfs. First clues on the physical properties of the spots may be obtained from multi-filter time series observations. In principle, they allow us to measure the photometric amplitudes in the light curves at various wavelengths, and from this information to derive the (asymmetric part of the) spot filling factor and the temperature difference between the spots and the average atmosphere – although this method is not capable of delivering unique solutions.

Therefore, in parallel to a photometric time series campaign of the Pleiades in the I-band, we measured in the J- and H-band simultaneously on a second telescope (Scholz *et al.* 2005). Only one VLM star (BPL 129) showed a period in all three wavelength bands at a signal-to-noise high enough so that we could derive spot properties. For several other Pleiades VLM stars and brown dwarfs only limits could be placed.

The best agreement between the observations and a one-spot model is reached for a cool spot with a temperature 18 to 31% below the average photospheric temperature and a filling factor of 4 – 5% (see Fig. 4). These results indicate that spots on VLM stars may have a similar temperature contrast between spot and average atmosphere, but a rather low spot filling factor compared to solar-mass stars. This might be a consequence of a

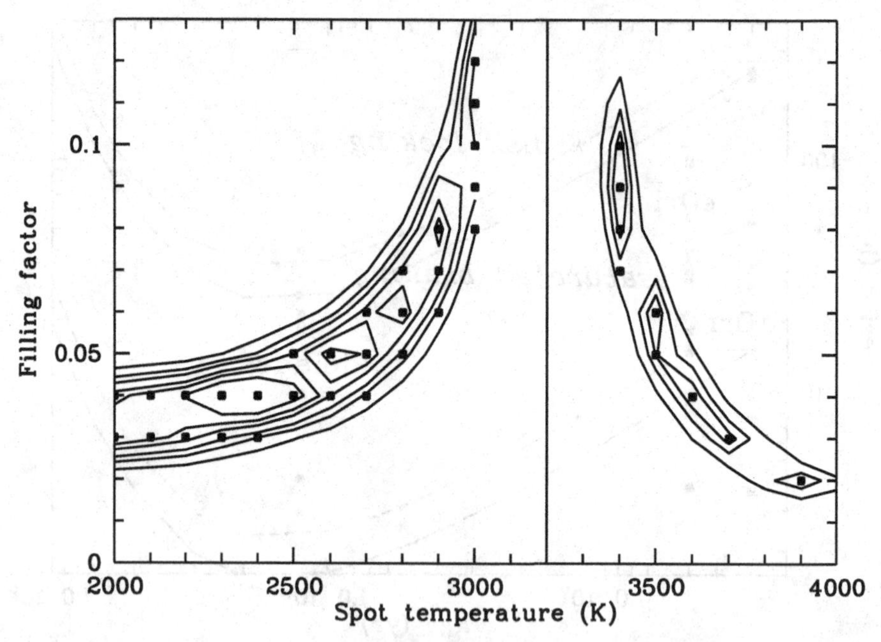

Figure 4. Contour plot for the χ^2 values from the comparison between observed and simulated spot amplitudes in the Pleiades VLM star BPL 129. Contour lines start at χ^2 =3.0 and are plotted for χ^2 =3.0, 2.0, 1.5, 1.0, 0.75, 0.5, indicating increasing quality of the fit. Filled squares show all combinations of spot temperature and filling factor which would produce amplitudes within the error bars of our observations. The vertical solid line indicates the photospheric temperature of BPL 129. Note that the hot spot solutions on the right side never reach to the 4th contour line, and χ^2 is larger than 0.9 everywhere. They are thus significantly worse than the cool spot solutions on the left side. (One data point with χ^2 = 0.92 at T_S = 3300 K and f = 19% is not shown in the figure.)

change in the dynamo from a solar-type shell dynamo to a small-scale turbulent dynamo in these fully convective stars.

5. Accretion and time variability

In the course of our data analysis we noted that a few of the VLM objects in the two Orion regions show large amplitudes of up to 0.6 mag (see Fig. 5). These variations are, however, much of an irregular character. Because of the large amplitudes, it is most likely that they result from hot spots originating from accretion of circumstellar disc matter onto the object surface (see also Fernández & Eiroa 1996). Emission lines in Hα, the far-red Calcium triplet, and – in some cases – even forbidden emission lines of [OI]$\lambda\lambda$6300,6363 and [SII]]$\lambda\lambda$6716,6731 are seen in optical spectra that we obtained of some of these objects in sigma Ori. These spectra thus show signatures typical of accretion, much like those of classical T Tauri stars. In addition, in a near-infrared colour-colour-diagram the high-amplitude variables mostly lie in the reddening path or even redward of it, thus showing near-infrared excess emission that is usually taken as evidence for a circumstellar disc. With their photometric variability, spectral accretion signatures, and indications for near-infrared excess emission from discs, they appear to be the low-mass and substellar counterparts to solar-mass T Tauri stars.

It is interesting that the high-amplitude T Tauri analogs on average are slower rotators than their low-amplitude non-accreting siblings. A similar tendency that brown dwarfs

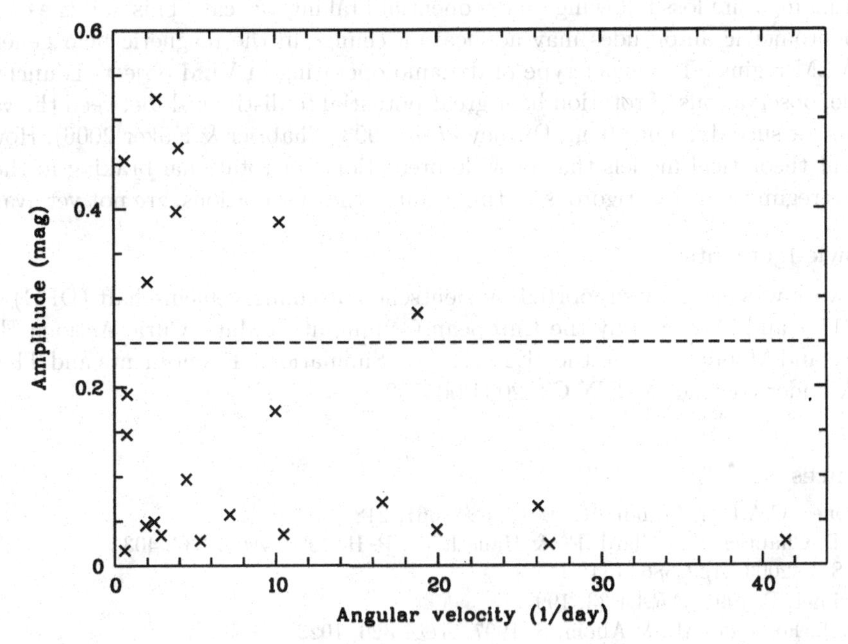

Figure 5. Angular velocity versus amplitude of VLM stars and brown dwarfs in the sigma Ori cluster. The dashed line delineates the separation between low-amplitude and high-amplitude objects. The high-amplitude objects are mostly active accretors, and on average rotate slower that the non-accreting low-amplitude objects.

with discs seem to rotate more slowly is also seen in spectroscopic measurements of $v \sin i$ by Mohanty *et al.* (2005). It thus seems that even in the substellar regime a connection between accretion and rotation exist, possibly implying rotational braking due to the interaction between the object and the disc (Scholz & Eislöffel 2004a).

6. Conclusions

We report results from our ongoing photometric monitoring of VLM objects in the clusters around sigma Ori, epsilon Ori, and the Pleiades, and our first attempts to model their rotational evolution.

It is very likely that the observed periodic variability of many VLM objects originates from magnetically induced cool spots on the surface of the objects. In particular in the Pleiades, photometric amplitudes in VLM objects indicate either less asymmetric spot distribution, smaller relative spotted area, or lower contrast between spots and average photosphere than in solar-mass stars. VLM objects show shorter rotation periods with decreasing mass. This effect is observed already at the youngest ages, and therefore should have its origin in the earliest phases of their evolution.

Combining the rotation periods for all our objects, we find that their evolution does not follow hydrostatic contraction alone. Some kind of braking mechanism, e.g. wind braking similar to the one observed in solar-mass stars, is required as well. Such a wind braking is intimately connected to stellar activity and magnetic dynamo action (Schatzman 1962). Nonetheless, since all the investigated VLM objects are expected to be fully convective, they should not be able to sustain a solar-type large-scale dynamo, which is at the heart of the Skumanich type angular momentum loss of solar-mass stars. In fact, our modelling shows that such a Skumanich type wind braking cannot explain our data, while saturated angular momentum loss following an exponential braking law can. This and the observed small photometric amplitudes may advocate a change in the magnetic field generation in the VLM regime. The exact type of dynamo operating in VLM objects is unclear. In principle, observations of rotation bear great potential to distinguish between the various scenarios for such dynamos (e.g., Durney *et al.* 1993, Chabrier & Küker 2006). However, consistent theoretical models that provide predictions for rotational braking in the very low-mass regime and thus rigorous testing against the observations, are not yet available.

Acknowledgements

This work was partially supported by Deutsche Forschungsgemeinschaft (DFG) grants Ei 409/11-1 and 11-2, and by the European Community's Marie Curie Actions–Human Resource and Mobility within the JETSET (Jet Simulations, Experiments and Theories) network under contract MRTN-CT-2004 005592.

References

Bailer-Jones, C.A.L. & Mundt, R. 2001, *A&A* 367, 218
Baraffe, I., Chabrier, G., Allard, F., & Hauschildt, P. H. 1998, *A&A* 337, 403
Barnes, S.A. 2003, *ApJ* 586, 464
Bodenheimer, P. 1995, *ARAA* 33, 199
Bouvier, J., Forestini, M., & Allain, S. 1997, *A&A* 326, 1023
Camenzind, M. 1990, *RvMA* 3, 234
Chaboyer, B., Demarque, P., & Pinsonneault, M.H. 1995, *ApJ* 441,876
Chabrier, G. & Baraffe, I. 1997, *A&A* 327, 1039
Chabrier, G. & Küker, M. 2006, *A&A* 446, 1027
Clarke, F.J., Tinney, C.G., & Covey, K.R. 2002, *MNRAS* 332, 361
Durney, B.R., DeYoung, D.S., & Roxburgh, I.W. 1993, *Solar Physics* 145, 207
Fernández, M. & Eiroa, C. 1996, *A&A* 310, 143
Herbst, W., Bailer-Jones, C.A.L., & Mundt, R. 2001, *ApJ* 554, 197
Herbst, W., Bailer-Jones, C.A.L., Mundt, R., Meisenheimer, K., & Wackermann, R. 2002, *A&A* 396, 513
Herbst, W., Eislöffel, J., Mundt, R., & Scholz, A. 2007, *Protostars and Planets V*, B. Reipurth, D. Jewitt, and K. Keil (eds.), University of Arizona Press, Tucson, p.297
Irwin, J., Aigrain, S., Hodgkin, S., Irwin, M., Bouvier, J., Clarke, C., Hebb, L., & Moraux, E. 2006, *MNRAS* 370, 954
Irwin, J., Hodgkin, S., Aigrain, S., Hebb, L., Bouvier, J., Clarke, C., Moraux, E., & Bramich, D.M. 2007, *MNRAS* 377, 741
Königl, A. 1991, *ApJ* 370, 39
Lamm, M.H. 2003, Ph.D. thesis, University of Heidelberg
Lamm, M.H., Bailer-Jones, C.A.L., Mundt, R., Herbst, W., & Scholz, A. 2004, *A&A* 417, 557
Matt, S. & Pudritz, R.E. 2005, *ApJ* 632, L135
Mohanty, S., Jayawardhana, R., & Basri, G. 2005, *MmSAI* 76, 303
Schatzman, E. 1962, *Ann. d'Astrophys.* 25, 18
Scholz, A. & Eislöffel, J. 2004a, *A&A* 419, 249
Scholz, A. & Eislöffel, J. 2004b, *A&A* 421, 259
Scholz, A. & Eislöffel, J. 2005, *A&A* 429, 1007
Scholz, A., Eislöffel, J., & Froebrich, D. 2005, *A&A* 438, 675
Shu, F., Najita, J., Ostriker, E., Wilkin, F., Ruden, S., & Lizano, S. 1994, *ApJ* 429, 781
Sills, A., Pinsonneault, M.H., & Terndrup, D.M. 2000, *ApJ* 534, 335
Skumanich, A. 1972, *ApJ* 171, 565
Stassun, K.G. & Terndrup, D. 2003, *PASP* 115, 505
Terndrup D.M., Krishnamurthi A., Pinsonneault M.H., & Stauffer J.R. 1999, *AJ* 118, 1814
Terndrup, D.M., Stauffer, J.R., Pinsonneault, M.H. *et al.* 2000, *AJ* 119, 1303

Star-Disk Interaction in Young Stars
Proceedings IAU Symposium No. 243, 2007
J. Bouvier & I. Appenzeller, eds.

doi:10.1017/S1743921307009611

Magnetization, accretion, and outflows in young stellar objects

Frank H. Shu[1], **Daniele Galli**[2], **Susana Lizano**[3], **Mike J. Cai**[4]

[1]Department of Physics, University of California, San Diego, CA 92093
[2]INAF-Osservatorio Astrofisico di Arcetri, Largo E. Fermi 5, Firenze I-50125, Italy
[3]CRyA, Universidad Nacional Autónoma de México, Apdo. Postal 72-3, 58089 Morelia, Mexico
[4]Academia Sinica, Institute of Astronomy and Astrophysics, Taiwan

Abstract. We review the theory of the formation and gravitational collapse of magnetized molecular cloud cores, leading to the birth of T Tauri stars surrounded by quasi-Keplerian disks whose accretion is driven by the magnetorotational instability (MRI). Some loss of magnetic flux during the collapse results typically in a dimensionless mass-to-flux ratio for the star plus disk of $\lambda_0 \approx 4$. Most of the mass ends up in the star, while almost all of the flux and the angular momentum ends up in the disk; therefore, a known mass for the central star implies a computable flux in the surrounding disk. A self-contained theory of the MRI that drives the viscous/resistive spreading in such circumstances then yields the disk radius needed to contain the flux trapped in the disk as a function of the age t. This theory yields analytic predictions of the distributions with distance ϖ from the central star of the surface density $\Sigma(\varpi)$, the vertical magnetic field $B_z(\varpi)$, and the (sub-Keplerian) angular rotation rate $\Omega(\varpi)$. We discuss the implications of this picture for disk-winds, X-winds, and funnel flows, and we summarize the global situation by giving the energy and angular-momentum budget for the overall problem.

Keywords. Accretion disks; MHD; stars: formation, magnetic fields; ISM: jets and outflows.

1. Introduction

Contemporary star formation occurs in dense cores found in giant molecular clouds (Shu *et al.* 1987; Evans 1999). Two viable but divergent opinions hold concerning the mechanism that produces such cores. The first relies on the leakage of magnetic support by the action of ambipolar diffusion (e.g., Mestel & Spitzer 1956; Nakano 1979; Lizano & Shu 1989; Basu & Mouschovias 1994). The second invokes transient compression by converging turbulent flows (e.g., Elmegreen 1993; Padoan 1995; Klessen, Heitsch, & MacLow 2001; Vázquez-Semadeni 2005). Recent observations show that early-stage, loosely aggregated, cores in the Pipe Nebula (Lada et al. 2007) and late-stage, tightly aggregated, cores in the Rho Oph region (André *et al.* 2007) are both internally quiet and have little core-to-core relative motion. These facts demonstrate that hypersonic turbulence of the variety embraced by the early enthusiasts plays little role in the evolution of dense cores into stars and planetary systems. Turbulence may yet enter in initiating core formation through its decay or in triggering the formation of the most massive stars under very crowded conditions, but these possibilities are not of interest in the present review.

Many studies, conducted under a variety of assumptions and boundary conditions, show that the end product of the leakage of magnetic fields from a slowly condensing, lightly ionized, dense pocket of gas and dust by ambipolar diffusion is a gravomagneto catastrophe, whereby the central regions of the condensing core formally tries to reach infinite density in finite time. A useful convention is to define $t = 0$ as the moment of catastrophe, with $t < 0$ being the condensation phase driving starless cores toward gravomagneto catastrophe, and $t > 0$ being the subsequent, inside-out, dynamical collapse to form a star plus disk. In a spherical coordinate system (r, θ, φ) with origin at the core center, a nonlinear attractor involving magnetized isothermal support against self-gravity apparently exists that produces solutions in which the density ρ and magnetic flux function Φ at the critical instant $t = 0$ approach the form of a singular isothermal toroid (Li & Shu 1996):

$$\rho \to \frac{a^2}{2\pi G r^2} R(\theta), \qquad \Phi(r, \theta) \to \frac{4\pi a^2 r}{G^{1/2}} \phi(\theta), \tag{1.1}$$

where a is the isothermal sound speed and $R(\theta)$ and $\phi(\theta)$ are well-defined functions of the polar angle θ once the dimensionless mass-to-flux ratio λ along flux tubes achieves a critical constant value λ_0 for the inner regions of the core. The likely value of λ_0 varies from 2 to 1 for regions of average to high ionization-fraction (perhaps characteristic of regions of low- and high-mass star formation; see Crutcher & Troland 2006).

The subsequent evolution for $t > 0$ yields a mass infall-rate onto a growing central protostar plus disk that has the form,

$$\dot{M} = \frac{m_0 (\Theta^{1/2} a)^3}{(1 - \lambda_0^{-2}) G}, \tag{1.2}$$

where m_0 is a dimensionless coefficient of order unity and $\Theta \geqslant 1$ is a factor accounting for the enhancement of the effective isothermal sound speed a squared associated with the effects of magnetic pressure and possibly turbulence. The natural development of head-start velocities (see Lee *et al.* 2001; Harvey *et al.* 2002) and over-densities (compared to the static singular isothermal sphere; see Harvey *et al.* 2001) from the previous epoch of ambipolar diffusion yield typical values of m_0 that are a factor of 2 or 3 larger than the classical value $m_0 = 0.975$ (Shu 1977, Adams & Shu 2007). For conditions measured in dense molecular cloud cores, $\dot{M}$ might vary from a few times 10^{-6} $M_\odot$ yr^{-1} (low-mass star formation) to $\sim 10^{-4}$-10^{-3} $M_\odot$ yr^{-1} (high-mass star formation).

2. Catastrophic magnetic braking if field freezing applies

Because the dynamical collapse occurs quickly compared with ambipolar diffusion, field freezing holds as a rough approximation for the epoch $t > 0$. Numerical calculations of the collapse phase using the ZEUS-2D code (Stone & Norman 1992) reveals difficulties, however, some that were anticipated, and others, not (Allen, Li & Shu 2003). An anticipated difficulty is that if gravitational collapse occurs to a virtual point because of the absence of rotation, then the central star would end with a mass-to-flux ratio given approximately by $2\pi G^{1/2} M_*/\Phi_* \sim \lambda_0 \sim 2$. For a star with mass $M_* \sim 1 M_\odot$ and radius $R_* \sim 3 R_\odot$, this would imply a surface field $B_* \sim \Phi_*/\pi R_*^2 \sim 10^7$ G, about four orders of magnitude larger than measured fields on the surfaces of T Tauri stars. Instead of being dragged in from the interstellar medium, the observed kG fields in these stars probably result from dynamo action.

The above difficulty was already known to Mestel & Spitzer (1956) and prompted them to propose that ambipolar diffusion would prevent the discrepancy: the interstellar flux would be left behind in the interstellar medium during the collapse to form a star. The modern situation reviewed above shows that λ_0 at the end of the ambipolar diffusion epoch typically amounts to only 2, and not 10^4 times larger.

From observations we now know that stars of all masses form via the intermediary of an accretion disk (for recent evidence concerning high-mass stars, see Rodríguez, Zapata & Ho 2007). Accretion disks have cross-sectional areas that are much larger than stars, so they can easily contain the same flux without having absurdly high levels of magnetic field. However, an unexpected surprise in the simulations of Allen, Li & Shu (2003) was that accretion disks do *not* form if λ_0 is anything like 2 and field freezing were to apply. Galli *et al.* (2006) give an analytical demonstration that the magnetic braking associated with the collapse to a central point, which produces a magnetic configuration that is a so-called "split monopole," transfers angular momentum away from the central regions so efficiently by the radiation of torsional Alfvén waves that Keplerian disks cannot form. High-resolution numerical simulations conducted by Fromang *et al.* (2006) and Price & Bate (2007) have confirmed these results.

Fortunately, a little bit of flux slippage, perhaps due to electrical resistivity rather than ambipolar diffusion, can go a long way toward promoting disk formation (Shu *et al.* 2006). From a combination of theory and observations (see Girart *et al.* 2006), Shu *et al.* (2007) argue that an additional loss of flux by a factor of 2 on the way down to forming a star plus disk is plausible, yielding the estimate $\lambda_0 \approx 4$ for the value appropriate for such configurations. Using different starting configurations, Hennebelle & Fromang (2007) find that disk formation is possible even with field freezing if $\lambda_0 = 20$, but not if $\lambda_0 = 5$. Below, we adopt a value $\lambda_0 = 4$ to compute the flux trapped in the circumstellar disks of YSOs, but an uncertainty of a factor of a few exists in this estimate.

3. Mean-field MHD of accretion disks

A net magnetization of the circumstellar disks surrounding YSOs makes the MRI (Hawley & Balbus 1991; Balbus & Hawley 1998) a natural candidate for the mechanism of inward transport of matter and outward transport of angular momentum. Such a process is needed to explain why objects like the mature solar system have almost all the mass in the central stars and almost all the angular momentum in the companions (planets) that supposedly formed out of their nebular disks. Unfortunately, no previous MRI simulation is both global and has nonzero net flux threading the disk, perhaps explaining why most MRI simulations give too low an effective viscosity compared to the requirements of real systems (King, Pringle & Livio 2007; Fromang & Papaloizou 2007).

3.1. *Turbulent viscosity*

In the absence of relevant numerical simulations, we use mixing length theory to estimate the transport coefficients associated with the turbulence induced by the MRI. Consider the schematic diagram in Figure 1, adapted from Shu *et al.* (2007). We imagine that accretion in the cylindrically radial direction ϖ stretches the field lines that would otherwise stick vertically through the disk so that the poloidal $\mathbf{B}$ has not only a vertical component B_z, but also a radial component that reverses sign from $B_\varpi^- < 0$ at the lower surface to $B_\varpi^+ > 0$ at the upper surface. We then suppose that turbulent fluctuations occur that occasionally create a loop that pinches and disconnects from its parent field line (see steps 2 and 3 of top panel). Because a detached loop is easy to shear, differential rotation would stretch the radial fluctuation δB_ϖ into a toroidal field δB_φ, where the two fields are related (including the proper sign) by the order of magnitude estimate,

$$\delta u \delta B_\varphi \sim \delta B_\varpi \varpi \frac{\partial \Omega}{\partial \varpi} \delta \varpi. \tag{3.1}$$

For the fluctuations most effective in transport, the velocity perturbation δu is related to the mixing length $\delta \varpi$ by the natural inverse correlation time, Ω,

$$\delta u \sim \Omega \delta \varpi. \tag{3.2}$$

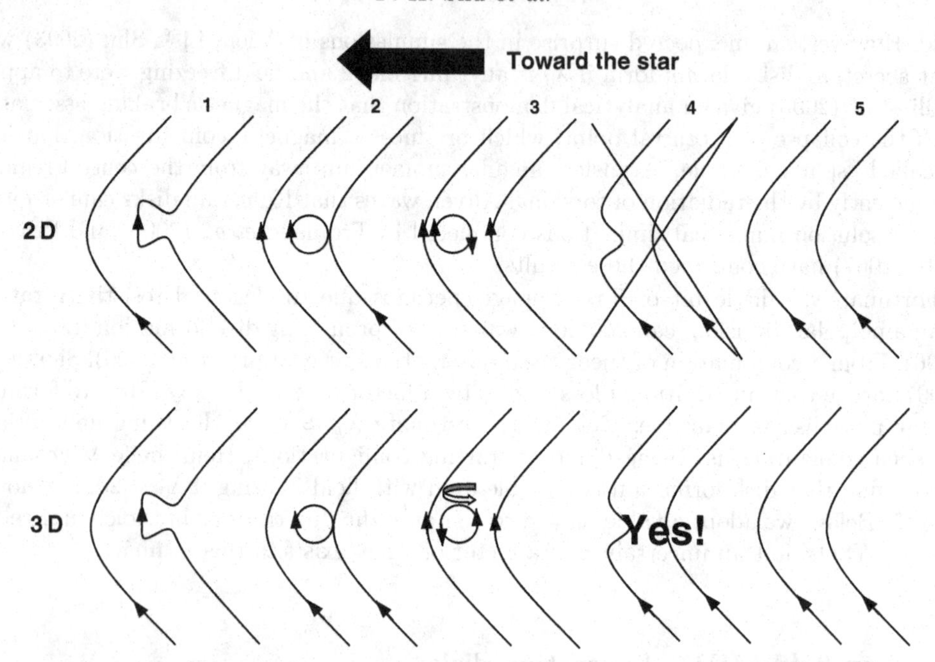

Figure 1. Schematic diagram of scenarios by which field loops are created and destroyed by magnetohydrodynamic turbulence when the mean field is strong: (top) in 2-D by stretch, pinch, disconnect and (bottom) in 3-D by stretch, pinch, disconnect and twist, reconnect, relax. The depiction is the meridional plane (ϖ, z), except for the twist associated with differential rotation indicated by the block arrow, which occurs out of the plane of the page. The twist in the bottom diagram gets the fields oriented in opposite directions at the target reconnection point, which results in the "yes" sign to proceed to steps 4 and 5. (adapted from Shu *et al.* 2007)

Solving for δB_φ, we get

$$\delta B_\varphi \sim \frac{\varpi}{\Omega}\frac{\partial \Omega}{\partial \varpi}\delta B_\varpi. \tag{3.3}$$

Using B_ϖ^+ as the natural scale for δB_ϖ, we estimate the associated Maxwell stress as

$$\frac{\delta B_\varpi \delta B_\varphi}{4\pi} \sim \frac{(B_\varpi^+)^2}{4\pi}\frac{\varpi}{\Omega}\frac{\partial \Omega}{\partial \varpi}. \tag{3.4}$$

The above expression should be compared with the tangential momentum-transport term usually modeled as a viscous stress:

$$\rho\nu\varpi\frac{\partial \Omega}{\partial \varpi} \sim \frac{\Sigma}{2z_0}\nu\varpi\frac{\partial \Omega}{\partial \varpi}, \tag{3.5}$$

where ν is the coefficient of (turbulent) kinematic viscosity, Σ is the surface density of the disk (gas plus dust), and z_0 is its effective half-height. Comparing the two expressions, we may identify

$$\nu = F\left[\frac{(B_\varpi^+)^2 z_0}{2\pi\Sigma\Omega}\right], \tag{3.6}$$

where F is a dimensionless "form factor" of order unity (or somewhat smaller) that corrects for the uncertainties in the twiddles and an effective integration over the thickness of the disk.

In a steady accretion disk where F is a spatial constant, the radial component of the field at the upper surface of the disk is related to the vertical component B_z in the mid-plane by the equation (Shu *et al.* 2007):

$$B_{\varpi}^{+} = I_{\ell} B_z, \tag{3.7}$$

where I_ℓ is an integral that has a well-determined, order-unity, value once the disk flare is specified (i.e., the power-law dependence of z_0 on ϖ, which is related to the index ℓ). The substitution of equation (3.7) into equation (3.6) yields the form:

$$\nu = D \frac{B_z^2 z_0}{2\pi \Sigma \Omega}, \tag{3.8}$$

where $D = I_\ell^2 F$ is an adjustable dimensionless parameter of order unity if the MRI is fully developed. Reassuringly, equation (3.8) is the viscosity expression of Shakura & Sunyaev (1973) if we substitute magnetic pressure in the mid-plane for gas pressure.

3.2. *Turbulent resistivity*

In order for accretion to occur, not only must outward angular-momentum transport occur diffusively at an enhanced rate relative to friction at a molecular level, but so must inward transport of matter occur by crossing stationary mean-field lines faster than the resistive diffusion associated with knocking ions and electrons off field lines by atomic and molecular collisions. Inward matter transport does not happen if loop dynamics involve only the steps indicated in the top series of picturess in Figure 1. The reason follows.

Consider the matter trapped inside the magnetic loop as the latter tries to move radially inward from the position where it disconnected from the upstream mean field line. If only the 2D degrees of freedom are considered, this loop cannot attach onto the downstream mean field line because its field orientation is parallel to the target reconnection point. Magnetic reconnection will not occur under such circumstances. The field loop in the sequence of the top panel is thus doomed to bounce (random-walk $\equiv$ diffuse) between the mean field lines sandwiching it, until it accidentally, sooner or later, reattaches to its original parent (upstream) field line. Although the stretching of the field line in the azimuthal direction can diffusively transport some angular momentum by the afore-computed Maxwell stresses, no transport of matter across field lines will occur above and beyond what collisions at a microscopic level allow.

Consider the situation, however, if an extra 3D step of twisting the field loop occurs out of the (meridional) plane of the page, as indicated by the block arrow in panel 3 of the bottom sequence of pictures in Figure 1. Such a vortical twist reverses the inner and outer field directions of the original loop, allowing the loop to attach onto the downstream mean field line by magnetic reconnection. In the process, the enclosed matter will have moved from the upstream field line to the downstream field line at a rate that potentially much exceeds molecular diffusion. We may estimate the associated turbulent diffusion coefficient (turbulent resistivity) by the following product of terms:

$$\eta = F \left(\frac{B_{\varpi}^{+} B_z z_0}{2\pi \Sigma \Omega} \right) \left(-\frac{z_0}{\Omega} \frac{\partial \Omega}{\partial \varpi} \right). \tag{3.9}$$

The term in the first parenthesis is the same as its bracketed counterpart in equation (3.6) except that one factor of B_{ϖ}^{+} has been replaced by B_z. Viscous transport depends only on there being a $(B_{\varpi}^{+})^2$, whereas resistive transport via the formation of loops requires the poloidal field to have mean curvature $B_{\varpi}^{+} B_z$. As discussed previously, however, the creation of field loops by itself is not enough to give resistive diffusion. The created field loops must also be able to reattach to downstream field lines, and this requires an added step. For loops of typical vertical size z_0, the fraction of all formed loops that will have the right orientation for the downstream reconnection can be estimated to be the ratio of z_0 to the radial length associated with the angular shear $-\Omega(\partial\Omega/\partial\varpi)^{-1}$. This ratio is the term inside the second parenthesis. Finally, a dimensionless coefficient of order unity exists in front, which must be taken to equal the same F as appears in equation (3.6) if steady disk accretion is to be possible (Shu *et al.* 2007).

When equation (3.7) holds and when the rotation law of the disk is quasi-Keplerian, $-(\varpi/\Omega)(\partial\Omega/\partial\varpi) = 3/2$, the turbulent resistivity is smaller than the turbulent viscosity,

$$\eta = \frac{3z_0}{2I_\ell \varpi}\nu. \tag{3.10}$$

by a factor $\sim$ the aspect ratio $z_0/\varpi \ll 1$ for a thin disk. First found by Lubow, Papaloizou & Pringle (1994) via a set of arguments motivated by the desire to drive disk winds, this result is surprising if we use a naive picture of turbulent mixing, but it is a natural outcome when viewed via the model of "loop dynamics" depicted in Figure 1.

4. Steady-state solution

When substituted into the governing equations for a steady-state, magnetized, disk with accretion rate $\dot{M}_\mathrm{d}$, the viscosity and resistivity expressions yield the following solutions for the rotation rate Ω, surface density Σ, and vertical component of the magnetic field B_z as functions of radial position ϖ in a disk surrounding a star of mass M_*:

$$\Omega(\varpi) = f\left(\frac{GM_*}{\varpi}\right)^{1/2}, \tag{4.1}$$

$$\Sigma(\varpi) = \frac{f}{1-f^2}\left(\frac{I_\ell}{3\pi DA}\right)\frac{\dot{M}_\mathrm{d}}{(GM_*\varpi)^{1/2}}, \tag{4.2}$$

$$B_z(\varpi) = \left(\frac{2f}{3DA}\right)^{1/2}\left(\frac{GM_*\dot{M}_\mathrm{d}^2}{\varpi^5}\right)^{1/4}, \tag{4.3}$$

where $f \leqslant 1$ is a constant accounting for the partial support against stellar gravity by the magnetic tension and A is the disk aspect ratio, $A \equiv z_0/\varpi$. The above results are derived in an approximation that ignores the self-gravity of the disk relative to the gravity of the star, and that assumes the effects of gas pressure and magnetic pressure are negligible compared with magnetic tension force in modifying the centrifugal balance of the disk.

The disk aspect ratio is computed from the equation for vertical hydrostatic equilibrium, which yields a relationship between A and the square of the thermal speed $a_\mathrm{m}^2 \equiv kT_\mathrm{m}/m$ in the mid-plane of the disk of the form (see Appendix C of Shu *et al.* 2007):

$$a_\mathrm{m}^2 = \frac{1}{2}\left[I_\ell(1-f^2)A + A^2\right]\frac{GM_*}{\varpi}. \tag{4.4}$$

In principle, we should compute the mid-plane temperature T_m for a thin disk from the condition that radiative conduction and thermal convection (if any) carry the energy of viscous and resistive dissipation to the disk's upper and lower surfaces where this heat is radiated away per unit area at some effective black-body rate σT_e^4. In practice, for quick modeling purposes, we shall assume $A(\varpi)$ to be given typically and empirically as a power-law, which we will take here to be (see, e.g., D'Alessio *et al.* 1999):

$$A(\varpi) = 0.1\left(\frac{\varpi}{100\ \mathrm{AU}}\right)^{1/4}. \tag{4.5}$$

With 1/4 as the power-law exponent, Shu *et al.* (2007) compute the self-consistent value for I_ℓ to be 1.742. They also argue that the plausible theoretical domain for this exponent ranges from 1/8 to 1/2, implying that the power-law for Σ varies from $\Sigma \propto \varpi^{-5/8}$ to $\Sigma \propto \varpi^{-1}$, in agreement with the empirical findings of Andrews & Williams (2007).

In addition, we suppose that we can ascribe a fiducial age $t_{\rm age}$ to a disk based on the time it would take to completely drain the disk mass $M_{\rm d}$ at the accretion rate $\dot{M}_{\rm d}$:

$$\int_0^{R_\Phi} \Sigma(\varpi)\, 2\pi\varpi d\varpi \equiv M_{\rm d}(R_\Phi) = \dot{M}_{\rm d} t_{\rm age}. \tag{4.6}$$

In the above we have taken the outer radius of the disk to equal the value R_Φ needed to contain the flux $\Phi_{\rm d}$,

$$\int_0^{R_\Phi} B_z\, 2\pi\varpi d\varpi = \Phi_{\rm d} = 2\pi G^{1/2} M_*/\lambda_0, \tag{4.7}$$

where λ_0 is the dimensionless mass-to-flux ratio brought into the system by the gravitational collapse from a molecular cloud core. When the indicated integrals are performed with the integrands given by the steady-state solutions for $\Sigma(\varpi)$ and $B_z(\varpi)$, we are implicitly assuming a definite value for the total angular momentum of the system,

$$\mathcal{J}_{\rm d} = \int_0^{R_\Phi} \varpi^2 \Omega \Sigma\, 2\pi\varpi d\varpi. \tag{4.8}$$

This value will generally not equal the actual amount. At late times, the actual system will not have enough angular momentum to maintain the system in steady state all the way out to the flux radius R_Φ, and the disk will experience time-dependent spreading.

Other problems may also arise, such as dead zones because of insufficient ionization in the central layers (Gammie 1996). Disregarding such difficulties, in steady state, we can write the requirement of centrifugal balance against the joint attraction of the stellar gravity and the magnetic repulsion (due to tension) acting inside the disk as

$$\varpi\Omega^2 = \frac{GM_*}{\varpi^2} - \frac{B_\varpi^+ B_z}{2\pi\Sigma}. \tag{4.9}$$

Using equations (3.7), (4.1), (4.2), (4.3), (4.6), and (4.7), we now transform the above requirement to the following formula for the departure from Keplerian rotation:

$$1 - f^2 = \left(\frac{0.5444}{\lambda_0^2}\right) \frac{M_*}{M_{\rm d}(R_\Phi)}, \tag{4.10}$$

where we have specialized the numerical coefficient to the value applicable to the flaring law (4.5). Because the star's mass M_* increases as the disk accretion proceeds, whereas the disk's mass $M_{\rm d}(R_\Phi)$ decreases, equation (4.10) states that the departure from Keplerian rotation, $1 - f^2$, increases with time. With a fixed total mass and flux for the system, the disk becomes increasingly magnetized with time, leading to a growing role for magnetic repulsion relative to gravitational attraction of the disk.

4.1. *Four astronomical models*

The four columns following the object in Table 1 gives the input parameters when we assume that $\lambda_0 = 4$ in the models for a T Tauri star, a low-mass protostar, an FU Orionis star, and a high-mass protostar. Notice that the input values are ordinary except for the choice $D = 10^{-2.5}$ for the T Tauri model, which is to be contrasted with the more natural selection $D = 1$ for the other three models. The small value for D is made so that the T Tauri disk would not have spread to ridiculously large radii R_Φ, and it can be partially justified physically on the grounds that T Tauri stars may have dead zones.

Table 1. Parameters of four model systems

Object	M_*	$\dot{M}_d$	t_{age}	D	f	R_Φ	$M_d(R_\Phi)$	$\mathcal{J}_d(R_\Phi)$
	$(M_\odot)$	$(M_\odot/\text{yr})$	(yr)			(AU)	$(M_\odot)$	($M_\odot$ AU km/s)
T Tauri star	0.5	1×10^{-8}	3×10^{6}	$10^{-2.5}$	0.658	298	0.0300	5.12
Low-mass Protostar	0.5	2×10^{-6}	1×10^{5}	1	0.957	318	0.200	51.4
FU Ori	0.5	2×10^{-4}	1×10^{2}	1	0.386	16.5	0.0200	0.473
High-mass Protostar	25	1×10^{-4}	1×10^{5}	1	0.957	1,520	10.0	39,700

The last four columns in Table 1 yield the output parameters of the analytic theory. A detailed discussion of these results can be found in Shu *et al.* (2007). For here, we merely note that the predicted sub-Keplerianity $f = 0.658$ of T Tauri disks is surprising and extreme. If we were to make a less extreme choice for λ_0, say, $\lambda_0 = 8$, the quantity $1 - f^2$ in equation (4.10) would decrease by a factor of 4, and f would equal (a perhaps more acceptable) 0.926, with magnetic fields a factor of 2 lower than discussed below.

4.2. *Expected magnetic fields*

Figure 2 shows the magnetic field distribution from equation (4.3) when we adopt the parameter choices of Table 1. The solid line applies to the $M_* = 0.5M_\odot$ models and the dashed line to the $M_* = 25M_\odot$ model, with f chosen to be 0.386 for the former (appropriate to FU Ori) and 0.957 for the latter (appropriate to the high-mass protostar). The disk accretion rate $\dot{M}_d$ and stellar mass M_* have been scaled according to the formula that labels the vertical axis. The data points come from Donati *et al.*(2005) for FU Ori, Winnberg *et al.*(1981) for V1057 Cyg, Hutawarakorn & Cohen (1999) for G35.2-0.74N, Hutawarakorn *et al.*(2002) for W75N, and Edris *et al.*(2005) for IRAS 20126.

In general, the theoretical expectations are in line with the empirical measurements. The agreement is impressive when we realize that the magnetic fields measured from spectropolarimetry and the Zeeman effect in maser emission lines span almost six orders of magnitude in field strength and over four magnitudes in spatial scale. From Figure 2, however, we see that a vast observational desert exists in the measurements made at the extremes of this range. In this desert, there exists empirically only the meteoritic measurement of roughly 1 G from chondritic materials that come to us presumably from about 3 AU in the asteroid belt (Levy & Sonett 1978). Fortunately or fortuitously, this measurement also lies on the theoretical line (see the discussion in Shu *et al.* 2007). Much work awaits observers in the ALMA era to fill in the gaps of Figure 2.

5. Implications for disk winds

The expectation that strong magnetization makes the inner parts of accretion disks sub-Keplerian in their rotation rates has important consequences for the feasibility of disk winds. It is well-known that one criterion for a cold, magnetocentrifugally driven, disk wind to arise is that the **B** field should emerge from the disk surface at an outward angle of more than 30° with respect to the vertical direction (Blandford & Payne 1982). Translated to our language, this requires $B_\varpi^+/B_z > 1/\sqrt{3}$, which is amply satisfied by equation (3.7) where $B_\varpi^+/B_z = I_\ell = 1.742$.

Magnetocentrifugally driven disk winds need to satisfy a second criterion, namely, that the acoustic speed at the surface a_s must make up for any departure from Keplerian

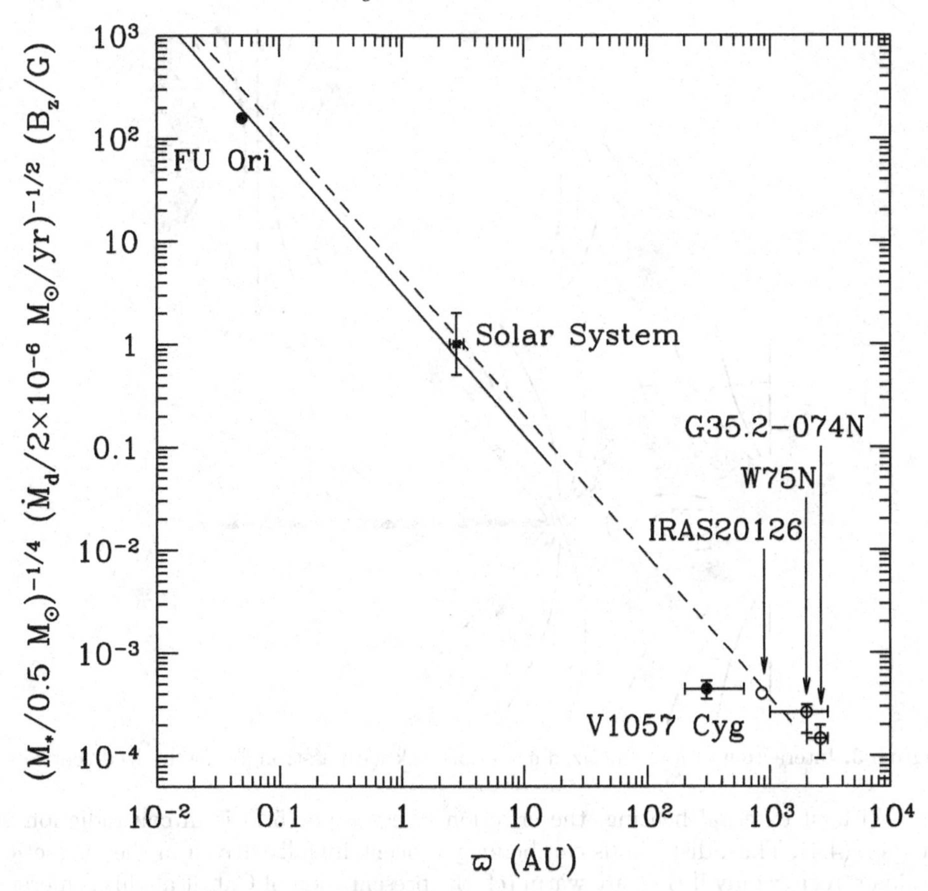

Figure 2. Scaled magnetic field B_z plotted against radial distance ϖ in the accretion disk from the central star.

rotation at the launch point ϖ, a condition that Shu *et al.* (2007) express as

$$a_s^2 > \frac{1}{4}(1 - f^2)\frac{GM_*}{\varpi}. \tag{5.1}$$

If the departure from Keplerian rotation is zero, $1 - f^2 = 0$, then even cold winds with $a_s = 0$ can be magnetocentrifugally launched from the disk surface provided that the first criterion $B_\varpi^+/B_z > 1/\sqrt{3}$ is satisfied. However, if $1 - f^2 > 0$, then thermal help is needed to overcome the gravitational potential GM_*/ϖ. The necessary coefficient is estimated as 1/4, because Parker's (1963) theory for the unmagnetized, non-rotating, solar wind produces one factor of 1/2, and another 1/2 can be justified on the grounds that rotation at Kepler speeds brings the gas within a factor of 1/2 of having enough energy to escape.

Comparison with equation (4.4) now reveals the potential problem. The first term in that equation is smaller than the term on the right-hand side of equation (5.1) by the factor $2I_\ell A$, which is a small number in the inner regions of the disk where A has a value of only a few percent. The first term in equation (4.4) dominates in the case of strongly magnetized disks where the vertical thickness of the disk is compressed more by the gradient of the magnetic pressure associated with B_ϖ^2 than it is by the vertical component of the stellar gravity (effect of the second term). But the sound speed in the surface layers cannot exceed the sound speed in the mid-plane, unless the disk surface is heated externally, for example, by high-energy photons (X-rays, UV) coming from the central

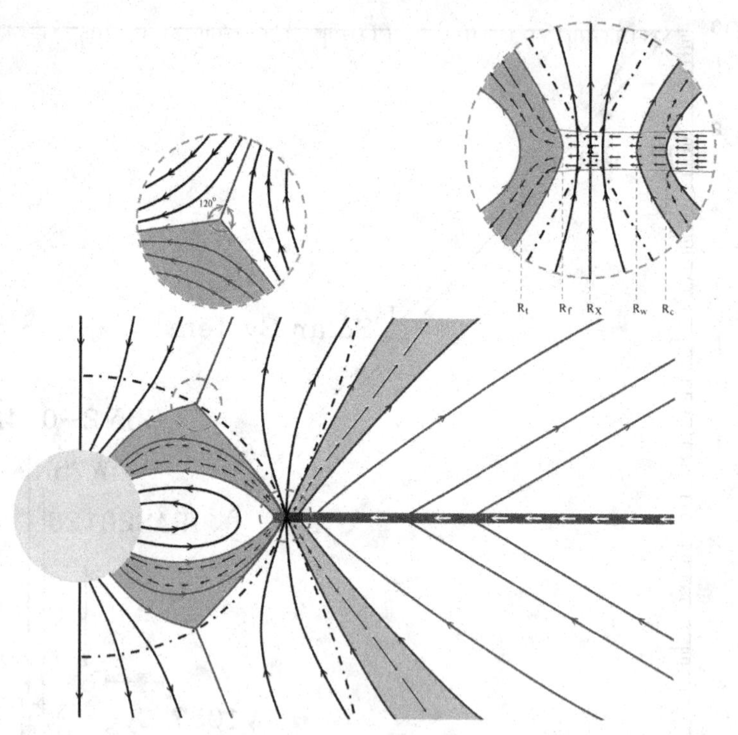

Figure 3. Interaction of a magnetized accretion disk with a strongly magnetized central star.

star. Without external heating, the criterion of equation (5.1) is in contradiction with equation (4.4). Thus, disk winds can be magnetocentrifugally driven in the magnetically dominant regime only if they are warm (cf. the presentation of Cabrit at this conference). In this case the rate-limiting factor in determining the wind mass-loss rate is the stellar photons reaching the disk surface, a condition that would make disk winds closely related to the process of photo-evaporation (e.g., Font *et al.* 2004). Conversely, if the second term in equation (4.4) dominates, then the magnetization of the disk has to be weak to make $1 - f^2 < 2A^2$, severely hampering the ability of the disk to drive powerful, fast, cold jets.

6. Implications for X-winds

Fast stellar jets almost certainly require an X-wind (Shu *et al.* 1994), and only X-winds can make sense of the peculiar interpretation given to the He I observations by Suzan Edwards in her presentation at this conference. The observational results reviewed by Chris Johns-Krull at this conference similarly have no plausible explanation other than X-wind theory. However, why does not sub-Keplerian rotation of the disk have the equally ominous implication for X-winds that it does for disk winds?

6.1. *Influence of disk fields*

Equation (4.9) combined with Figure 3 supplies the answer to the last question. Equation (4.9) states that sub-Keplerian rotation is expected only as long as the product $B_\varpi^+ B_z$ is *positive*, i.e., only when the disk field makes an outward bend. But the sense of the bend must reverse as we come to the inner edge of the disk and the accretion flow attaches onto the stellar magnetosphere in a funnel flow onto the central star (Fig. 3). The reversal in sign of $B_\varpi^+ B_z$ as this happens makes the disk rotation rate transition from sub-Keplerian values (in the disk proper), to Keplerian values (in the X-region), to super-Keplerian values (in the forbidden region interior to the inner disk edge). Before the latter happens, the disk (dark shade) is truncated and the matter flows onto the star via a magnetospheric funnel (Shu *et al.* 1994; see also the simulations of Romanova at this conference). The reversed pressure gradients at a sharp inner edge help with making the transition from sub-Keplerian to super-Keplerian values.

To visualize what is going on in Figure 3, imagine the inward press of the disk field to be resisted by the outward press of the squeezed magnetosphere (assumed to have a dipole moment in the picture that is anti-aligned with the disk field; for a more general multipole treatment, see Mohanty & Shu 2007). The squeezing pops open some of the previously closed stellar fields, and these open fields become available to form the regions that are dead or alive to the magnetocentrifugal flow (field angles at the surface of the disk less than or greater than 30° from the vertical). The inward press on the stellar magnetosphere also increases the disk rotation rate near the inner edge of the disk. Because of the added presence of the disk fields (with comparable strengths as the compressed stellar field in the X-region), X-winds will be better focused than the case with no disk field. Moreover, the fraction f_{w} of the disk accretion rate $\dot{M}_{\mathrm{d}}$ in Figure 3 that ends up in outflow (lightly shaded X-wind) rather than inflow (lightly shaded funnel-flow) will be less than 1/3 because of the pushing up of the trapped stellar flux toward a more vertical direction. The lighter loading of field lines will make the terminal speeds of X–winds faster than computed under the assumption that the disk is itself unmagnetized.

6.2. *X-winds in action*

The basic equations for steady, axisymmetric, X-winds were written down by Shu *et al.* (1988) – all that has changed in the interim are the boundary conditions that are attached to the two governing partial-differential equations. For a cold flow, the Grad-Shafranov equation (Grad & Rubin 1958; Shafranov 1966) for the stream function $\psi(\varpi, z)$ reads:

$$\nabla \cdot (\mathcal{A}\nabla\psi) + \frac{1}{\mathcal{A}}\left(\frac{J}{\varpi^2} - 1\right)\frac{J'}{\varpi^2} + \frac{2\beta\beta' V_{\mathrm{eff}}}{(\beta^2 - \varpi^2\mathcal{A})^2} = 0, \tag{6.1}$$

where $V_{\mathrm{eff}} = -(\varpi^2 + z^2)^{-1/2} - \varpi^2/2 + 3/2$ is the dimensionless effective potential in the corotating frame. Bernoulli's equation for the Alfvén discriminant $\mathcal{A} \equiv (\beta^2\rho - 1)/\varpi^2\rho$ is

$$|\nabla\psi|^2 + \frac{1}{\mathcal{A}^2}\left(\frac{J}{\varpi^2} - 1\right)^2 + \frac{2\varpi^2 V_{\mathrm{eff}}}{(\beta^2 - \varpi^2\mathcal{A})^2} = 0. \tag{6.2}$$

In the above $J(\psi)$ and $\beta(\psi)$ are functions of ψ associated, respectively, with the conservation of angular momentum carried in matter and field and with the freezing of gas loaded onto the field at the base of the flow from the surface of the disk.

The difficulty with the above formulation is that equation (6.2) substituted into equation (6.1) yields a partial-differential of mixed type, changing from elliptic to hyperbolic as the flow velocity crosses certain signal speeds (Heinemann & Olbert 1978). In the cold limit, the relevant signal speed is the fast MHD speed, with the Alfvén surface where $\mathcal{A} = 0$ being only an apparent singularity where $J(\psi) = \varpi^2$. These equations have been attacked in various guises by Najita & Shu (1994, X-wind in the sub-Alfvénic region), Ostriker & Shu (1995, funnel flow), Shu *et al.* (1995, X-wind far asymptotics), and Shang *et al.* (1998, full X-wind solution by interpolation of near and far asymptotics). This piecemeal treatment has been criticized by other workers (cf. Ferreira *et al.* 2006), but it has produced global solutions of adequate accuracy for most astronomical applications.

Nevertheless, a more systematic approach is desirable. Such an approach has been found and implemented by Cai *et al.* (2007). It begins by defining an action integral:

$$S \equiv \int\int \left[\frac{\mathcal{A}}{2}|\nabla\psi|^2 - \frac{1}{2\mathcal{A}}\left(\frac{J}{\varpi^2} - 1\right)^2 + \frac{V_{\text{eff}}}{\beta^2 - \varpi^2 \mathcal{A}} \right] 2\pi\varpi d\varpi\, dz. \tag{6.3}$$

The action S gives equation (6.1) when extremized with respect to variations of ψ, while it gives equation (6.2) when extremized with respect to variations of $\mathcal{A}$. Instead of attacking the two relations as partial differential equations, we may then reduce the problem to looking for trial functions in a variational approach. Details are given in Cai *et al.* (2007). For here, we merely note that the solutions so obtained are in good agreement with the earlier ones obtained by less sophisticated means.

7. Budget for energy and angular momentum

7.1. *Photon Luminosity*

To summarize the outlook of this paper, let us examine the global budget for energy and angular momentum as the gas spirals from large distances in the disk to land in hot-spots at the end of the funnel flow (see also Shu 1995). The hot-spot luminosity is given dimensionally by

$$L_{\text{hot}} = (1 - f_{\text{w}})\frac{GM_*\dot{M}_{\text{d}}}{R_*}\left(1 + \frac{R_*^3 \sin^2\theta_{\text{h}}}{2R_X^3} - \frac{3R_*}{2R_X}\right), \tag{7.1}$$

where θ_{h} is the mean co-latitude of the hot-spot. The term inside the parenthesis is R_*/R_X times the dimensionless specific kinetic energy of gas falling freely from rest at the X-point in the corotating frame,

$$\frac{1}{2}u^2 = \frac{1}{r} + \frac{1}{2}\varpi^2 - \frac{3}{2} \equiv -V_{\text{eff}}, \tag{7.2}$$

evaluated at the position of the hot-spot on the surface of the star, $r = R_*/R_X$ and $\varpi = r\sin\theta_{\text{h}}$. To get the system photon-luminosity, L_{sys}, the amount from the hot spot is to be added to the amount L_* liberated by the star and the disk-accretion luminosity:

$$L_{\text{d}} = \left(\mathcal{T} + \frac{1}{2}\right)\frac{GM_*\dot{M}_{\text{d}}}{R_X}, \tag{7.3}$$

where $\mathcal{T}$ is the dimensionless viscous torque just outside R_X. Thus,

$$L_{\text{sys}} = L_* + L_{\text{hot}} + L_{\text{d}}. \tag{7.4}$$

7.2. *Flow and Field Luminosity*

In a cold gas, even in the presence of a magnetic field, the dimensionless specific energy of the gas in the corotating frame, $H \equiv |\mathbf{u}|^2/2 + V_{\text{eff}} = 0$ is a constant of motion. In an inertial frame, H is the Jacobi constant, $H = E^g - J^g$, where E^g and J^g are the specific energy and specific angular momentum of the gas measured in an inertial frame, with $E^g = 1 = J^g$ at the X-point where $V_{\text{eff}} = 0$ by definition. Thus, the time rate of change of the dimensionless specific energy E^g of a parcel of gas as we follow its motion is given by the time rate of change J^g along a streamline. There is also a Poynting flux per unit mass flux given by the analogous quantity J^B. Thus, in a convention where we define the gravitational potential energy in an inertial frame to be zero at infinity, the mechanical luminosity flowing out of the system in the form of an X-wind is given by

$$L_{\text{w}} = f_{\text{w}} \left(\bar{J}_{\text{w}} - \frac{3}{2} \right) \frac{GM_* \dot{M}_{\text{d}}}{R_X}, \tag{7.5}$$

where $\bar{J}_{\text{w}}$ is the dimensionless specific angular-momentum carried on average in wind gas and field. The constant streamline-averaged quantity $\bar{J}_{\text{w}} = \bar{J}^g_{\text{w}} + \bar{J}^B_{\text{w}}$ is the sum of two variables where the gaseous contribution $\bar{J}^g_{\text{w}}$ starts at 1 at R_X and becomes $\bar{J}_{\text{w}} > 3/2$ at infinity, whereas the field contribution $\bar{J}^B_{\text{w}}$ starts at $\bar{J}_{\text{w}} - 1$ at R_X and goes to zero at infinity. Similarly, the mechanical luminosity contained in the funnel flow equals

$$L_{\text{fun}} = (1 - f_{\text{w}}) \left(\bar{J}_* - \frac{3}{2} \right) \frac{GM_* \dot{M}_{\text{d}}}{R_X}. \tag{7.6}$$

The constant streamline-averaged quantity $\bar{J}_* = \bar{J}^g_* + \bar{J}^B_*$ is the sum of two variables where the gaseous contribution $\bar{J}^g_*$ starts at 1 at R_X and becomes $\bar{J}_*$ at the hot-spot, whereas the field contribution $\bar{J}^B_*$ starts at $\bar{J}_* - 1$ at R_X and goes to zero at the hot spot (if the hot-spot field is very strong).

If we add the two mechanical luminosities to the power dissipated in the disk to accrete the gas from infinity to R_X, we get

$$L_{\text{w}} + L_{\text{fun}} + L_{\text{d}} = \frac{GM_* \dot{M}_{\text{d}}}{R_X} \left[f_{\text{w}} \bar{J}_{\text{w}} + (1 - f_{\text{w}}) \bar{J}_* - 1 + \mathcal{T} \right], \tag{7.7}$$

which must equal zero since the gas and field in the star formation process starts with zero specific energy at infinity and energy is conserved. This requires the expression in the square bracket to vanish, which is equivalent to the conservation of angular momentum because what leaves the X-region in the X-wind, $f_{\text{w}} \bar{J}_{\text{w}}$, and in the funnel flow, $(1 - f_{\text{w}}) \bar{J}_*$, must come from what was originally there, 1, minus any angular momentum per unit mass removed by the viscous torque, $\mathcal{T}$. The conservation of energy is equivalent to the conservation of angular momentum because, by the argument involving the Jacobi integral, the only way to transfer energy (in the inertial frame) is to exert torque. In any case, the conservation of angular momentum/energy in the X-region requires

$$f_{\text{w}} = \frac{1 - \mathcal{T} - \bar{J}_*}{\bar{J}_{\text{w}} - \bar{J}_*}. \tag{7.8}$$

The fact that the sum of the three luminosities,

$$L_{\text{fun}} + L_{\text{w}} + L_{\text{d}} = 0, \tag{7.9}$$

vanishes means that L_{fun} must be negative if L_{w} and L_{d} are positive. Mechanical power flows *out* of the star because it gives up its right to rotate faster and faster if it becomes locked to the rotation rate of the inner edge of the disk. Although this happens formally for arbitrary $\mathcal{T}$ and $\bar{J}_*$ as long as $\bar{J}_* < 3/2$ and $\mathcal{T} < 1 - \bar{J}_*$; in fact we expect $\bar{J}_*$ and $\mathcal{T}$ to be small compared to unity if the square of the stellar radius is small in comparison to R_X^2 and if the MRI torque just outside R_X cannot compete with wind or funnel flows that remove or add angular momentum at dynamical rates.

The physical significance of L_{fun} being negative cannot be overstated. Angular momentum is transferred by the funnel flow to the inner edge of the disk, except for a small part that is needed to keep the star locked to the disk. Otherwise, the star would begin to rotate faster and faster, becoming highly flattened as a result. Disk truncation occurs not by ram pressure effects (Ghosh & Lamb 1978), but by considerations of angular-momentum transport. Funnel flows are the reason why stars are spheres and not disks.

This fact is the ultimate tribute to the proposal made by Claude Bertout and Gibor Basri many years ago that magnetospheric infall underlies the accretion luminosity of T Tauri stars, not boundary-layer emission.

References

Adams, F. C., & Shu, F. H. 2007, ApJ, submitted
Allen, A, Li, Z.-Y., & Shu, F. H. 2003, ApJ, 599, 363
André, P., Belloche, A., Motte, F., & Peretto, N. 2007, A&A, in press
Andrews, S. M., & Williams, J. P. 2007, ApJ, 659, 705
Balbus, S. A., & Hawley, J. F. 1998, Rev. Mod. Phys., 70, 1
Basu, S., & Mouschovias, T. Ch., 1994, ApJ, 432, 720
Blandford, R. D., & Payne, D. G. 1982, MNRAS, 199, 883
Cai, M. J., Shang, H., Lin, H.-H., & Shu, F. H. 2007, ApJ, submitted
Crutcher, R. M., & Troland, T. H. 2006, in Triggered Star Formation in a Turbulent ISM, IAU Symp. No. 237, 25
D'Alessio, P., Calvet, N., Hartmann, L., Lizano, S., & Canto, J. 1999, 527, 893
Donati, J. F., Paletou, F., Bouvier, J., & Ferreira, J. 2005, Nature, 438, 466
Edris, K. A., Fuller, G. A., Cohen, R. J., & Etoka, S. 2005, A&A, 434, 213
Elmegreen, B. G. 1993, ApJL, 419, L29
Evans, N. J. 1999, ARAA, 37, 311
Ferreira, J., Dougados, C., & Cabrit, S. 2006, A&A, 453, 785
Font, A. S., McCarthy, I. G., Johnstone, D., & Ballantyne, D. R. 2004, ApJ, 607, 890
Fromang, S., Hennebelle, P., & Teyssier, R. 2006, A&A, 457, 371
Fromang, S., & Papaloizou, J. 2007, Astro-ph 0705.3621
Galli, D., Lizano, S., Shu, F. H., & Allen, A. 2006, ApJ, 647, 374
Gammie, C. F. 1996, ApJ, 457, 355
Ghosh, P., & Lamb, F. K. 1978, ApJ, 223, L83
Girart, J. M., Rao, R., & Marrone, D. 2006, Science, 313, 812
Grad, H., & Rubin, H. 1958, in Proc. Conf. Internat. Atomic Energy Agency 31, (Geneva: Internat. Atomic Energy Agency)
Harvey, D.W.A., Wilner, D.J., Lada, C.J., Myers, P.C., Alves, J.F., & Chen, H. 2001, ApJ, 563, 903
Harvey, D.W.A., Wilner, D. J., DiFrancesco, J., Lee, C. W., Myers, P. C., & Williams, J. P. 2002, AJ, 123, 3325
Hawley, J. F., & Balbus, S. A. 1991, ApJ, 381, 496
Heinemann, M., & Olbert, S. 1978, JGR, 83, 2457
Hennebelle, P., & Fromang, S. 2007, A&A, submitted
Hutawarakorn, B., & Cohen, R. J. 1999, MNRAS, 303, 845
Hutawarakorn, B., Cohen, R. J., & Brebner, G. C. 2002, MNRAS, 330, 349
King, A. R., Pringle, J. E., & Livio, M. 2007, MNRAS, 376, 1740
Klessen, R. S., Heitsch, F., & MacLow, M. 2001, ApJ, 535, 887
Lada, C. J., Muensch, A. A., Rathborne, J., Alves, J. F., & Lombardi, M. 2007, ApJ, submitted
Lee, C. W., Myers, P. C., & Tafalla, M. 2001, ApJS, 136, 703
Levy, E. H., & Sonett, C. P. 1978, in Protostars & Planets, ed. T. Gehrels (Tucson: University of Arizona Press), p. 516
Li, Z.-Y., & Shu, F. H. 1996, ApJ, 472, 211
Lizano, S., & Shu, F. H. 1989, ApJ, 342, 834
Lubow, S. H., Papaloizou, J. C. B., & Pringle, J. E. 1994, MNRAS, 267, 235
Mestel, L., & Spitzer, L.1956, MNRAS, 116, 503
Mohanty, S., & Shu, F. H. 2007, ApJ, in preparation
Najita, J. R., & Shu, F. H. 1994, ApJ, 429, 808
Nakano, T. 1979, PASJ, 31, 697
Ostriker, E. C., & Shu, F. H. 1995, ApJ, 447, 813

Padoan, P. 1995, MNRAS, 277, 377
Parker, E. N. 1963, Interplanetary Dynamical Processes (New York: Interscience Pub)
Price, D. J., & Bate, M.R. 2007, MNRAS, 337, 77
Rodríguez, L. F., Zapata, L. A., & Ho, P. T. P. 2007, ApJL, L143
Shafranov, V. D. 1966, Rev. Plasma Phys, 2, 103
Shakura, N. I., & Sunyaev, R. A. 1973, A&A, 24, 337
Shu, F. H. 1977, ApJ, 214, 488
Shang, H., Shu, F. H., & Glassgold, A. E. 1998, ApJ, 439, 91
Shu, F. H. 1995, Rev Mex AA, 1, 375
Shu, F. H., Adams, F. C., & Lizano, S. 1987, ARAA, 25, 23
Shu, F. H., Galli, D., Lizano, S., & Cai, M. 2006, ApJ, 647, 382
Shu, F. H., Gallli, D., Lizano, S., Glassgold, A. E., & Diamond, P. 2007, ApJ, 665, 535
Shu, F. H., Lizano, S., Ruden, S. P., & Najita, J. 1988, ApJ, 328, L19
Shu, F., Najita, J., Ostriker, E., Wilkin, F., Ruden, S., & Lizano, S. 1994, ApJ, 429, 781
Shu, F. H., Najita, J., Ostriker, E. C., & Shang, H. 1995 ApJ, 455, L155
Stone, J. M., & Norman, M. L. 1992, ApJS, 80, 791
Vázquez-Semadeni, E. 2005, in The Initial Mass Function 50 Years Later, eds. E. Corbelli, F. Palla, & H. Zinnecker (Dordrecht: Springer), p. 371
Winnberg, A., Graham, D., Walmsley, C. M., & Booth, R. S. 1981, A&A, 93, 79

Star-Disk Interaction in Young Stars
Proceedings IAU Symposium No. 243, 2007
J. Bouvier & I. Appenzeller, eds.

doi:10.1017/S1743921307009623

MHD simulations of star-disk magnetospheres and the formation of outflows and jets

Christian Fendt

Max Planck Institute for Astronomy, Königstuhl 17, 69117 Heidelberg, Germany
email: fendt@mpia.de

Abstract. In this review the recent development concerning the large-scale evolution of stellar magnetospheres in interaction with the accretion disk is discussed. I put emphasis on the generation of outflows and jets from the disk and/or the star. In fact, tremendous progress has occurred over the last decade in the numerical simulation of the star-disk interaction. The role of numerical simulations is essential in this area because the processes involved are complex, strongly interrelated, and often highly time-dependent. Recent MHD simulations suggest that outflows launched from a very concentrated region tend to be un-collimated. I present preliminary results of simulations of large-scale star-disk magnetospheres loaded with matter from the stellar, resp. the disk surface demonstrating how a disk jet collimates the wind from the star and also how the stellar wind lowers the collimation degree of the disk outflow.

Keywords. Accretion disks, MHD, methods: numerical, stars: formation, stars: magnetic fields, stars: mass loss, stars: winds, outflows, ISM: jets and outflows.

---

1. Introduction

Highly collimated jets and outflows are one of the most striking signatures of young stars. There is general agreement that these jets are collimated disk/stellar winds, being launched, accelerated, and collimated by magnetic forces (see reviews by Pudritz *et al.* 2007; Shang *et al.* 2007). However, the details of the physical processes involved are not completely understood. In fact, young stellar objects may carry a strong stellar field which is important for the angular momentum exchange between disk and star (Bouvier *et al.* 2007), but will presumable affect the outflow formation as well.

This review discusses some of the essential aspects of outflow formation from magnetized young stars, in particular the application of numerical magnetohydrodynamic (MHD) simulations. I concentrate on the *global*, large-scale picture, i.e. on the question, how outflows are launched in the disk-star environment and how they probably look alike. I will not discuss details of the proposed "disk locking" mechanism in such systems, and not the issues of dipolar accretion or the evolution of the magnetized accretion disk itself. For this, I refer to the contributions by Shu, Romanova, Matt, and Ferreira.

2. Jet formation – the standard model

The principal processes involved in jet formation may be summarized as follows. The underlying hypothesis is that jets can only be formed in a system with a high degree of axi-symmetry (e.g. Fendt & Zinnecker 1998).

- Magnetic field is generated by the star-disk system.
- The star-disk system also drives an electric current.

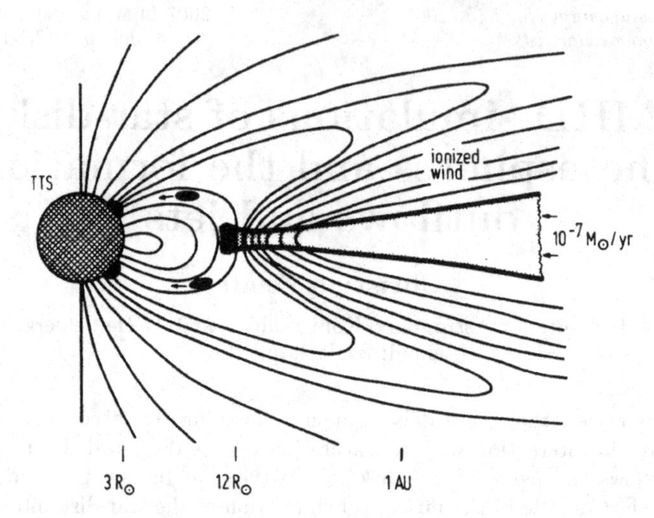

Figure 1. Magnetic star-disk interaction in young star and the generation of protostellar outflows and jets as proposed by Camenzind (1990).

- Accreting matter is launched as a plasma wind (either from the stellar or disk surface), couples to the magnetic field, and is flung out magnetocentrifugally.
- Plasma inertia leads to bending of the poloidal field (i.e. the field along the meridional plane including the jet axis).
- The plasma becomes accelerated magnetically, i.e. by conversion of Poynting flux to kinetic energy.
- Pinching forces of the induced toroidal field component eventually collimate the wind flow, forming a collimated jet structure.
- The plasma velocities subsequently exceed the speed of the magnetosonic waves. The fast magnetosonic regime is causally decoupled from outer boundary conditions.
- Where the outflow meets the interstellar medium (ISM), a bow shock develops, thermalizing the jet energy. Also, the electric current is closed via the bow shock, and the jet net current returns to the source of the current via the ISM.

Historically, we note that the model topology of dipole-plus-disk field were introduced for protostellar jet formation 25 years ago by Uchida & Low (1981) discussing possible magnetic field configurations in such systems. Amazingly, first MHD simulations of such a configuration were performed already in the early 80's by Uchida & Shibata (1984), Uchida & Shibata (1985) following-up these early models. However, probably due to the success of MHD disk-jet models by Blandford & Payne (1982), Pudritz & Norman (1983) and the limitations (spatial and time resolution) of the early numerical simulations, this concept was somewhat repressed until the early 90's when it became evident that young stars do carry a substantial and most probably large-scale magnetic field.

To my knowledge, the first who considered the detailed physical processes involved in disk truncation and channeling the matter along the dipolar field lines in the context of protostars has been Camenzind (1990) followed by Königl (1991), Collier Cameron & Campbell (1993), Hartmann *et al.* (1994) and Shu *et al.* (1994). This sudden boost of conforming papers from competing groups added up to the breakthrough of these ideas to the protostellar jet community. Figure 1 shows the model scenario suggested by Camenzind (1990) proposing the formation of protostellar jets from star-disk magnetospheres and including details of the star-disk interaction as angular momentum exchange ("disk locking"), dipolar accretion, or the turbulent diffusive boundary layer. The inner disk radius was derived as

$$r_{\rm in} = 2.4\, r_\star \left(\frac{\alpha}{2}\right)^{\frac{2}{7}} \left(\frac{B_\star\, r_\star}{10^3{\rm G}\ 3\, r_\odot}\right)^{\frac{4}{7}} \left(\frac{\dot{M}_{\rm acc}}{10^{-7} M_\odot/{\rm yr}}\right)^{-\frac{2}{7}} \left(\frac{h(r)}{r}\right)^{\frac{2}{7}} \left(\frac{r_\star c^2}{10^6 {\rm GM}_\star}\right)^{\frac{1}{7}} \tag{2.1}$$

with the disk viscosity $\alpha < 1$, the disk height $h(r)$, and the stellar radius $r_\star$ mass $M_\star$, and magnetic field strength $B_\star$. Follow-up numerical calculations provided stationary state solutions of the magnetospheric structure and the flow dynamics on a global scale, however, still resolving the central disk-jet-star geometry (see Fendt *et al.* 1995; Fendt & Camenzind 1996). The magnetic field distribution was calculated in a force-free approach and showed a rapid collimation to a cylindrical collimation. The solution could only be obtained for stellar wind type geometry, thus forming a jet from open stellar field lines. Models like this are nowadays discussed as "accretion powered stellar winds" (Matt & Pudritz 2005; Keppens & Goedbloed 1999; see also Matt this proceedings).

I remind of the general difficulty in solving the stationary state MHD equations. The exact solution is determined by matching a smooth transition of the flow at the (singular) critical MHD surfaces which are at an initially unknown location. All stationary state solutions published so far were derived applying essential simplifications, as self-similarity (e.g. Blandford & Payne 1982; Sauty & Tsinganos 1994; Li 1995; Sauty *et al.* 2002; Contopoulos & Lovelace 1994; Ferreira 1997), non-global solutions (e.g. Shu *et al.* 1994; Pelletier & Pudritz 1992), non local force-balance (e.g. Lovelace *et al.* 1991), force-free freeness (see above), or very low rotation (e.g. Sakurai 1995).

Numerical simulations of jet formation became feasible since the mid 90's and did overcome the difficulties of the stationary state approach. Of course, the disadvantage is the lack of spatial resolution and the limitation in some physical parameters like plasma-beta. In general, two approaches were made. One is prescribing the accretion disk properties as a boundary condition for the jet flow (see e.g. Ustyugova *et al.* 1995; Romanova *et al.* 1997; Ouyed & Pudritz 1997; Krasnopolsky *et al.* 1999; Fendt & Elstner 2000; Fendt & Čemeljić 2002; Vitorino *et al.* 2003). The other one is to include the accretion disk structure in the numerical treatment and evolve the accretion-ejection system self-consistently (see e.g. Hayashi *et al.* 1996; Hirose *et al.* 1997; Miller & Stone 1997; Goodson *et al.* 1997; Kudoh *et al.* 1998; Casse & Keppens 2002; von Rekowski *et al.* 2003; Kuwabara *et al.* 2005) for either pure disk systems and/or the dipole-disk interaction.

3. Stellar magnetosphere and large-scale outflow

Here I discuss how the presence of a central stellar magnetic field may affect the overall jet formation process.

Additional magnetic flux. In comparison to the situation of a pure disk magnetic field, the stellar magnetic field adds substantial magnetic flux to the system. For a polar field strength B_0 and a stellar radius $R_\star$, the large-scale stellar dipolar field

$$B_{{\rm p},\star}(r) \simeq 40\,{\rm G} \left(\frac{B_0}{1\,{\rm kG}}\right) \left(\frac{r}{3\,R_\star}\right)^{-3} \tag{3.1}$$

is to be compared to the disk poloidal magnetic field provided either by a dynamo or by advection of ambient interstellar field, both limited by equipartition arguments,

$$B_{\rm p,disk} < B_{\rm eq}(r) = 20\,{\rm G}\,\alpha^{-\frac{1}{2}} \left(\frac{\dot{M}_{\rm a}}{10^{-6}\, M_\odot/{\rm yr}}\right)^{\frac{1}{2}} \left(\frac{M_\star}{M_\odot}\right)^{\frac{1}{4}} \left(\frac{H/r}{0.1}\right)^{-\frac{1}{2}} \left(\frac{r}{10\, R_\odot}\right)^{-\frac{5}{4}} \tag{3.2}$$

The stellar magnetic flux will not remain closed, but will inflate and open up as the poloidal magnetic field is sheared (e.g. Uchida & Shibata 1984; Lovelace *et al.* 1995; Fendt & Elstner 2000; Uzdensky *et al.* 2002; Matt & Pudritz 2005). Some of these field lines may effectively become a disk field, and therefore follow the same processes as for disk winds. The additional Poynting flux that threads the disk may assist the jet launching by MHD forces and may serve as an additional energy reservoir for the jet kinetic energy, thus implying a greater asymptotic jet speed (Michel scaling; Michel 1969; Fendt & Camenzind 1996).

Additional magnetic pressure. However, the stellar field also provides additional central magnetic pressure which may implicate a de-collimation of the overall outflow. The central stellar magnetic field may launch a strong stellar wind which will remove stellar angular momentum. Such an outflow will interact with the surrounding disk wind. If true, observed protostellar jets may consist of two components – the stellar wind and the disk wind, with strength depending on intrinsic (yet unknown) parameters. Note that so far this argument is "ad-hoc" and numerical simulations are needed to figure out the actual dynamical evolution (see § 8). For a further discussion of stellar winds we refer to the contribution by S.Matt.

Angular momentum exchange by the stellar field. In the scenario of magnetic "disk locking", the stellar field which threads the disk will re-arrange the global angular momentum budget. If the star looses angular momentum to the disk (this is not yet decided by the simulations, see below), both disk accretion and outflow formation is affected. The angular momentum flow from the star is transfered by the dipolar field and is deposited close to the inner disk radius (not further out than the last closed field line). Therefore, the matter in this region may be accelerated to slightly super-Keplerian rotation which has two interesting aspects: (i) due to the super-Keplerian speed this disk material could be easily expelled into the corona by magneto-centrifugal launching (Blandford & Payne 1982; Ferreira 1997) and form a disk wind, and (ii) the excess angular momentum will stop accretion unless it is removed by some further (unknown) process. Again, a disk outflow launched from the very inner part of the disk can be an efficient way to do this. This scenario is similar to the X-wind models (Shu *et al.* 1994; Ferreira *et al.* 2000).

The torque on the star by the accretion of disk matter is $\tau_{\rm acc} = \dot{M}_{\rm acc}\,(GM_{\star}r_{\rm in})^{1/2}$ (e.g. Matt & Pudritz 2005; Pudritz *et al.* 2007), with the disk accretion rate $\dot{M}_{\rm acc}$, the stellar mass $M_{\star}$ and the disk inner radius $r_{\rm in}$ inside the co-rotation radius. For "disk locking", the star may be braked-down by the magnetic torque due to stellar field lines connecting the star with the accretion disk outside the co-rotation radius. The differential magnetic torque acting on a disk annulus of dr width is $d\tau_{\rm mag} = r^2 B_{\phi} B_z dr$. However, while B_z may be derived by assuming a central dipolar field, the induction of toroidal magnetic fields (electric currents) is model dependent. This is why numerical simulations of the dipole-disk interaction that evaluate simultaneously the poloidal and toroidal field components are essential. For further discussion I refer to the contributions by Romanova and Matt.

Non-axisymmetric effects from a tipped magnetic dipole. A central dipolar field inclined to the rotation axis of star and disk may strongly disturb the axisymmetry of the system. In extreme cases this may hinder jet formation at all, while weaker non-axisymmetric perturbation may lead to warping of the inner disk, and thus a precession of the outflow launched from this area. A rotating inclined dipole also implies a time-variation of the magnetic field which may lead to a time-variation in the mass flow rates for both the accretion disk and the outflow.

Indeed, photometric and spectroscopic variability studies of AA Tau give evidence for time-dependent magnetospheric accretion on time scales of the order a month. Monte-Carlo models of scattered light by O'Sullivan *et al.* (2005) were able to reproduce the observed photo-polarimetric variability which may arise by the warping of the disk being induced by a tipped magnetic dipole of 5 kG strength.

Investigations of the warping process by Pfeiffer & Lai (2004) using numerical simulations show that the warp could evolve into a steady state precessing rigidly. Disks can be warped by the magnetic torque that arises from the a slight misalignment between the disk and star's rotation axis (Lai 1999). This disk warping mechanism may also operate in the absence of a stellar magnetosphere as purely induced by the interaction between a large-scale magnetic field and the disk electric current and, thus, may lead to the precession of magnetic jets/outflows (Lai 2003).

Three-dimensional radiative transfer models of the magnetospheric emission line profile by Symington *et al.* (2005) based on the warped disk density and velocity distribution obtained by numerical MHD simulations give gross agreement with observations with a variability somewhat larger than observed (see also Harries, this proceedings).

4. Reminder on magnetohydrodynamics

The magnetohydrodynamic concept considers an ionized neutral fluid with averaged particle quantities as fluid quantities (e.g. mass density, current density). For example, the MHD Lorentz force $\vec{F_L} \sim \vec{j} \times \vec{B}$ is defined by the electric current density $\vec{j}$. In MHD, electric fields are negligible small in the rest frame of the fluid. For infinite conductivity matter is frozen into the field (or vice versa), this is the limit of ideal MHD. Resistive MHD allows for a slight slip of matter across the field.

If we consider axisymmetric flows the field components can be decomposed into a poloidal component and a toroidal component, e.g. $\vec{B} = \vec{B_p} \times \vec{B_\phi}$ The helical magnetic field lines follow (and define) magnetic flux surfaces Ψ - axisymmetric surfaces of constant magnetic flux,

$$\Psi(r,z) \equiv \frac{1}{2\pi}\int \vec{B}_{\rm p} \cdot d\vec{A}, \tag{4.1}$$

where $d\vec{A}$ is the area element of a circular area perpendicular to the symmetry axis. With that, the Lorentz force could be de-composed into components parallel and perpendicular to the flux surfaces, $\vec{F_L} \equiv \vec{F}_{L,||} + \vec{F}_{L,\perp}$ with

$$\vec{F}_{L,||} \equiv \vec{j}_\perp \times \vec{B}_\phi \quad \text{and} \quad \vec{F}_{L,\perp} \equiv \vec{j}_{||} \times \vec{B}. \tag{4.2}$$

This implies that a certain configuration of electric current and magnetic field distribution may accelerate the matter along the field (parallel force component) and collimate the flow across the poloidal field (perpendicular force component). Of course, also the opposite might be true under a different field configuration – de-collimation or deceleration. Note that for both magnetic acceleration and collimation the presence of a *toroidal* field component is essential.

Another view of how the Lorentz force do act in jets is to rewrite the force in terms of magnetic pressure and magnetic tension applying Ampère's law, $\vec{F_L} \sim (\nabla \times \vec{B}) \times \vec{B}$, and the well known vector identities,

$$\vec{F_L} = \nabla\left(\frac{|\vec{B}|^2}{8\pi}\right) + \frac{1}{4\pi}\left(\vec{B}\cdot\nabla\right)\vec{B}. \tag{4.3}$$

The first term on the right hand side is the gradient of the "magnetic pressure", the second one represents the magnetic tension due to the field curvature. Both, magnetic pressure force and tension force may contribute to acceleration and collimation. Magnetic pressure acceleration can be the main driving force for dipolar magnetospheres if the differential rotation between the foot-points in the star and the disk will wind up the poloidal field lines, thus introducing a strong vertical toroidal pressure gradient (Uchida & Shibata 1984; Lovelace *et al.* 1995). Note that a pure magnetic dipole is force-free as magnetic tension and pressure forces cancel.

What is denoted by "magneto-centrifugal acceleration" (Blandford & Payne 1982) is in that sense not an MHD effect. What happens is that the magnetic field co-rotating with the star or the disk is so strong that it dominates the matter inertia. The magnetic field lines can be considered as "wires" with "beads" on them. For a sufficiently low inclination of the field lines, the effective potential along the field becomes instable, and any perturbation will centrifugally expel the matter along the field lines. Outflow collimation by self-generated MHD forces happens when the matter inertia becomes so strong that the magnetic field cannot dominate the matter flow any longer. Matter continues to flow under conservation of angular momentum, now dragging the field with it and bending the field lines, thus inducing a toroidal magnetic field component.

5. Self-collimated MHD jets

Essentially, the numerical proof of MHD jet self-collimation came by two pioneering papers investigating the time evolution of a prescribed wind launched from the accretion disk surface (Ustyugova *et al.* 1995; Ouyed & Pudritz 1997). These models assume a fixed-in-time equatorial boundary condition defined by a disk in Keplerian rotation and a prescribed mass flow rate from that "disk surface" into the computational domain (the "outflow"). This allows for long-term simulations and to find potential stationary state solutions. In fact, starting from the initial condition of a magnetohydro-*static* equilibrium, a collimated outflow evolves, proving the stationary state models discussed before.

Further application of this approach has been done e.g. for a change of the simulation box geometry (Ustyugova *et al.* 1999), feedback from the outflow to the boundary condition (Krasnopolsky *et al.* 1999), dipolar magnetospheres (Fendt & Elstner 2000), resistive MHD jets (Fendt & Čemeljić 2002), time-dependent perturbations (Vitorino *et al.* 2003), 3D non-axisymmetric perturbations (Ouyed *et al.* 2003), or different disk magnetic field profiles (Pudritz *et al.* 2006; Fendt 2006).

However, future simulations will clearly treat the large-scale time evolution of star-jet magnetospheres including the disk evolution self-consistently. In fact, future has already begun. Simulations of jet formation including the disk evolution have been presented by Casse & Keppens (2002), Casse & Keppens (2004), von Rekowski & Brandenburg (2004), Meliani *et al.* (2007), Zanni *et al.* (2007) while disk-star interaction simulations were performed by Romanova *et al.* (2002), Romanova *et al.* (2004), Küker *et al.* (2003). For details on the latter topic I refer to the contributions by Romanova and Ferreira.

6. MHD simulations: stellar magnetosphere – disk interaction

Numerical simulation of the magnetospheric star-disk interaction are technically most demanding. The nature of the object requires to treat a complex model geometry in combination with strong gradients in magnetic field strength, density and resistivity, implying a large variation in physical time scales for the three components of disk, jet, and magnetosphere, which all have to be resolved numerically.

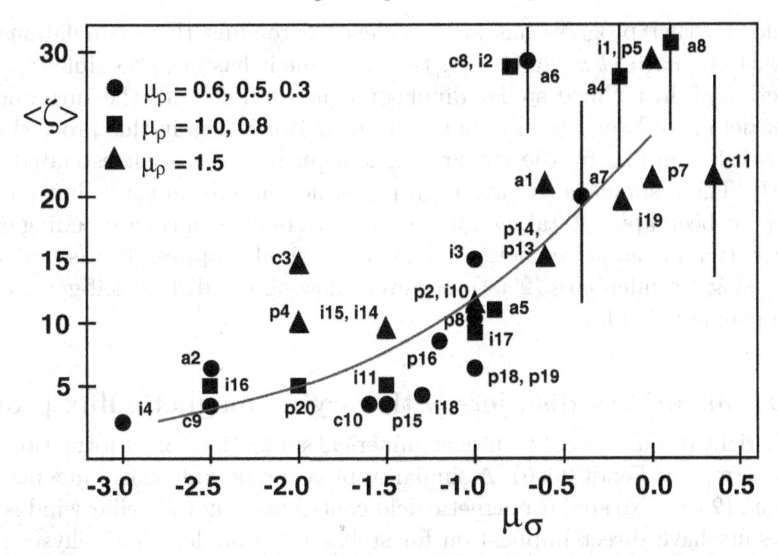

Figure 2. Collimation degree $< \zeta >$ (averaged over differently sized regions in the computational box), plotted against the exponent of the disk wind magnetization power law profile (see Fendt 2006). The symbols trace different disk wind density profiles μ_ρ. The different parameter runs are denoted by "a1", ..., "p20". The "error bars" for the very flat profiles indicate that here the simulation did not reach an overall quasi-steady state and, thus, the mass fluxes slightly vary in time. The line indicates the trend observed in our simulations.

Today huge progress has been made with several groups (and also codes) competing in the field. Early simulations were able to follow the evolution only for a few rotations of the inner disk (note that 100 rotations at the co-rotation radius correspond to 0.3 Keplerian rotations at 10 co-rotation radii) as e.g. Hayashi *et al.* (1996), Miller & Stone (1997), Goodson *et al.* (1997). The next step was to increase the grid resolution and treat several hundreds of (inner) disk rotations on a global grid of 50 AU extension (Goodson *et al.* 1999). The main result of these simulations is a two component flow consisting of a fast and narrow axial jet and a slow disk wind, both launched in the inner part of the disk. Close to the inner disk radius repetitive reconnection processes are seen on time scales of a couple of rotation periods. The dipolar field inflates and an expanding current sheet builds up. After field reconnection along the current sheet, the process starts all over again.

Another approach was taken by Fendt & Elstner (2000). In order to perform long-term simulations, the evolution of the disk structure was neglected and the disk instead taken as a fixed boundary condition for the outflow. After 2000 rotations a quasi-steady state was obtained with a two-component outflow from disk and star. The outflow expands without signature of collimation on the spatial scale investigated (20×20 inner disk radii). One result of this very long simulation is that the axial narrow jet observed in other simulations is shown to be an intermittent feature launched in the early phase of the simulation.

In a series of ideal MHD simulations Romanova and collaborators succeeded in working out a detailed and sufficiently stable numerical model of magnetospheric disk interaction. They were the first to simulate the axisymmetric funnel flow from the disk inner radius onto the stellar surface (Romanova *et al.* 2002) on a global scale ($R_{\rm max} = 50 R_{\rm in}$) and for a sufficiently long period of time in order to reach a steady state in the accretion funnel. Strong outflows have not been observed for the parameter space investigated, probably due to the matter dominated corona which does not allow for opening-up the dipolar field. Further progress has been achieved extending these simulations to three dimensions (Romanova *et al.* 2004). For the first time it has been possible to investigate the interaction of an inclined stellar dipolar magnetosphere with the surrounding disk. For further details of I refer to the contribution by Romanova in this proceedings.

The star-disk coupling by the stellar magnetosphere was also investigated by Küker *et al.* (2003). These simulations have been performed in axisymmetry, but an advanced disk model has been applied, taking into account α-viscosity, a corresponding eddy magnetic diffusivity and radiative energy transport. A similar approach was undertaken by von Rekowski & Brandenburg (2004), however allowing the disk to self-generate its own large-scale magnetic field.

7. MHD simulations: disk jets with varying magnetic flux profile

Here we briefly discuss recent results of numerical simulations of jet formation from disk winds (for details see Fendt 2006). A similar approach was undertaken independently by Pudritz *et al.* (2006). No stellar magnetic field contribution nor a stellar wind is involved, still the results have direct implication for stellar wind models. The physical grid size corresponds to $(r \times z) = (150 \times 300)\, r_{\rm in} = 7 \times 14$AU, which allows to resolve e.g. the velocity structure in micro jets (DG Tau).

Again we start from a force-free initial field distribution in hydrostatic equilibrium and let it evolve in time under the condition of a mass inflow from the accretion disk boundary condition. However, we extended this approach over a huge parameter run, covering a wide range of disk magnetic field profiles and disk wind mass flux profiles, both parameterized by a power law, $B_{\rm p,wind}(r) \sim r^{-\mu}, \rho_{\rm wind}(r) \sim r^{-\mu_\rho}$. Both quantities could be combined in the disk wind magnetization parameter (Michel 1969), $\sigma_{\rm wind} \sim B_{\rm p}^2 r^4 \Omega_F^2 / \dot{M}_{wind} \sim r^{\mu_\sigma}$. We quantify the collimation degree by comparing the mass flux in axial and lateral direction (see Fendt & Čemeljić 2002; Fendt 2006). Figure 2 shows the degree of collimation measured by the parameter ζ plotted against the power law exponent of the disk wind magnetization μ_σ. The main result is that steep magnetization profiles (resp. disk magnetic field strength profiles) are unlikely to generate highly collimated outflows.

This important conclusion holds in particular for outflows launched as stellar winds, or X-winds, where the magnetic flux of the outflow originates in a very small region. In turn, one may say that the existence of collimated jets would require a certain disk magnetic field profile and, thus, may put some constraints on the origin of the disk, i.e. a disk dynamo-generated field versus a field advected from the interstellar medium.

8. MHD simulations: outflows from disk-star magnetospheres

Here I present preliminary results of MHD simulations considering a stellar magnetosphere surrounded by a disk magnetic field where both field components are fed by a mass flux from the underlying boundary condition – representing the stellar surface and the accretion disk.

The essential point is that both cases are treated, a stellar dipole aligned and anti-aligned with the ambient disk magnetic field. Such field geometries were considered already by Uchida & Low (1981) and have recently been reconsidered in the form of reconnection X-winds (see contributions by S. Cabrit and J. Ferreira).

Our model setup is the following. Applying cylindrical coordinates (r, ϕ, z), the equatorial plane is divided in three components - the stellar surface with $r < r_\star = 0.5 r_{\rm in}$, the accretion disk at radii $r > r_{\rm in} = 1.0$ and the gap between star and disk. A mass

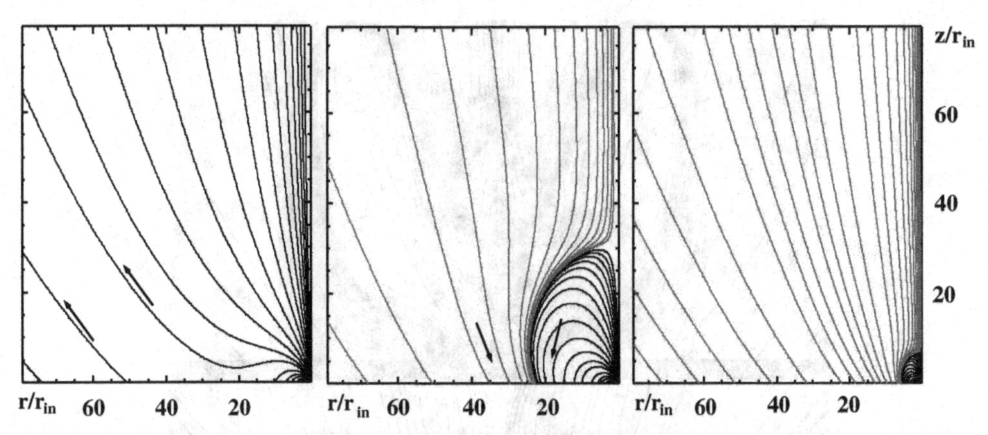

Figure 3. Initial magnetic field distribution for star-disk jet formation simulations, shown are poloidal magnetic field lines. Arrows indicate the magnetic field direction. Different strength and orientation of the superposed stellar and disk magnetic field component, $\Psi_{0,\mathrm{disk}} = 0.01, -0.01, -0.1$, resp. $\Psi_{0,\mathrm{star}} = 5.0, 5.0, 3.0$ (from left to right).

flux is prescribed for both wind components - star and disk. The initial magnetic field distribution is taken as a superposition of the stellar (dipolar) field and the disk field (force-free potential field),

$$\Psi_{\mathrm{total}} = \Psi_{0,\mathrm{disk}} f_{\mathrm{disk}}(r,z) + \Psi_{0,\mathrm{star}} f_{\mathrm{star}}(r,z) \tag{8.1}$$

where $\Psi_{0,\mathrm{disk}}$ and $\Psi_{0,\mathrm{star}}$ measure the strength of both components and the functions $f(r,z)$ describe the initial (force-free) magnetic field distribution of both components (see Fendt 2007, to be submitted). Figure 3 shows the different initial field configurations. The magnetic flux as prescribed by the initial condition remains frozen into to the disk-star boundary. The coronal magnetic field evolves in time. The central star is rotating with a magnetospheric co-rotation radius equal to the disk inner radius. The grid size is $(r \times z) = (80 \times 80)$ inner disk radii.

Figure 4 shows how the coronal field structure evolves in time for the example simulation with $\Psi_{0,\mathrm{disk}} = -0.1$ and $\Psi_{0,\mathrm{star}} = 3.0$. Note that in this case the disk magnetic field and the equatorial dipolar field are aligned. This field structure has been discussed already by Uchida & Low (1981) and is discussed nowadays in the framework of "reconnection X-winds" (see contribution by Ferreira, this proceedings). We evolve the simulations for 2800 rotations at the inner disk radius corresponding to 4 rotations at the outer disk radius. At intermediate time scales (about 700 inner disk rotations) a quasi-stationary state emerges. One clearly sees the de-collimating effect of the central stellar wind component. Note, however, that at this time the outer disk has rotated only about 0.15 times and the coronal structure above the outer disk will further evolve in time and disturb the quasi-steady state. Over the long run such quasi-stationary states may be approached again, what we observed is a cyclic behavior of the opening angle with a periodicity of about 500 (inner disk) rotation periods. We also observe that reconnection processes close to the remaining inner dipole leads to sudden flares (see also Goodson *et al.* 1999) which seem to trigger the large-scale cyclic behavior. The propagation of these flares is very fast, reconnection islands propagate across the jet magnetosphere within a few rotation time steps.

Figure 4. Time evolution of a star-disk magnetosphere from initial state of Fig. 3, right. Time step is 50, 400, 2606, 2700 rotations of the inner disk (from top to bottom). Colors show logarithmic density contours, black lines are poloidal field lines (magnetic flux contours).

9. Conclusions

The stellar magnetic field has an important impact on the jet formation process. It provides an additional magnetic flux component which allows for higher outflows velocities as more magnetic energy is available to be converted into kinetic energy, and an additional central (magnetic) pressure component, which may de-collimate the outflow. It may also provide excess angular momentum in the jet launching region by disk locking. It may disturb the outflow axisymmetry, and/or trigger a time-variation in outflow rate if the central stellar magnetosphere is inclined to the rotation axis of the star-disk system. The traditional difficulties of stationary state MHD solutions have now been overcome by means of numerical simulations. Having a large number of numerical MHD codes available today, it is essential to prove the model suggestions from the past by performing numerical simulations.

Preliminary results of MHD simulations of a superposed stellar and disk magnetosphere demonstrate the de-collimation of the disk jet, respectively the collimation of the stellar wind by the surrounding disk jet. Disk jet simulations with different magnetic field and mass flux profiles provide a unique relation between disk wind magnetization and degree of collimation. Better collimation is achieved for flat magnetic field profiles. Thus, stellar winds and X-winds are unlikely to launch highly collimated outflows.

I like to remind the reader that for a good number of codes applied in this field (as e.g. ZEUS, VAC, FLASH, or PLUTO), the numerical schemes have been published including standard tests and reference simulations. There are, however, a number of interesting publications with important and indeed convincing results, but the underlying code has never made public. Thus, it might not always be clear what the code is actually doing and what the limits of the numerical scheme are. I like both to encourage the authors of these papers to publish their codes and to advice the reader to always check the publications for their numerical background.

I conclude this review noting that tremendous progress has occurred in the numerical simulation of star-disk interaction and jet formation during the last decade. The role of numerical simulations is essential in this field because the physical processes involved are complex, strongly interrelated, and often highly time-dependent. However, it is fair to state that numerical simulations of the star-disk interaction have not yet shown the launching of a collimated jet flow comparable to the observations.

References

Blandford, R.D. & Payne, D.G. 1982, *MNRAS* 199, 883

Bouvier, J., Alencar, S.H.P., Harries, T.J., Johns-Krull, C.M., & Romanova, M.M. 2007, in: B. Reipurth, D. Jewitt, & K. Keil (eds.), *Protostars & Planets V*, University of Arizona Press, Tucson, 2007, p.479

Camenzind, M. 1990, *Reviews in Modern Astronomy* 3, 234

Casse, F., & Keppens, R. 2002, *ApJ* 581, 988

Casse, F., & Keppens, R. 2004, *ApJ* 601, 90

Collier Cameron, A., & Campbell, C.G. 1993, *A&A* 274, 309

Contopoulos, J., & Lovelace, R.. 1994, *ApJ* 429, 139

Fendt, Ch., Camenzind, M., & Appl, S. 1995, *A&A* 300, 791

Fendt, Ch., & Camenzind, M. 1996, *A&A* 313, 591

Fendt, Ch., & Zinnecker, H. 1998, *A&A* 334, 750

Fendt, Ch., & Elstner, D. 2000, *A&A* 363, 208

Fendt, Ch., & Čemeljić, M. 2002, *A&A* 395, 1045

Fendt, Ch. 2006, *ApJ* 651, 272

Ferreira, J. 1997, *A&A* 319, 340

Ferreira, J., Pelletier, G., & Appl, S. 2000, *MNRAS* 312, 387
Goodson, A.P., Winglee, R.M., & Böhm, K-.H. 1997, *ApJ* 489, 199
Goodson, A.P., Bohm, K.-H., & Winglee, R.M. 1999, *ApJ* 524, 142
Hartmann, L., Hewett, R., & Calvet, N. 1994, *ApJ* 426, 669
Hayashi, M.R., Shibata, K., & Matsumoto, R. 1996, *ApJ* 468, L37
Hirose, S., Uchida, Y., Shibata, K., & Matsumoto, R. 1997, *PASJ* 49,193
Keppens, R., & Goedbloed, J. P. 1999, *A&A* 343, 251
Königl, A. 1991, *ApJ* 370, L39
Krasnopolsky, R., Li, Z.-Y., & Blandford, R.D. 1999, *ApJ* 526,631
Kudoh, T., Matsumoto, R., & Shibata, K. 1998, *ApJ* 508, 186
Küker, M., Henning, T., & Rüdiger, G. 2003, *ApJ* 589, 397; erratum: *ApJ* 614, 526
Kuwabara, T., Shibata, K., Kudoh, T., & Matsumoto, R. 2005, *ApJ* 621, 921
Lai, D. 1999, *ApJ* 524, 1030
Lai, D. 2003, *ApJ* 591, L119
Li, Z.-Y. 1995, *ApJ* 444, 848
Lovelace, R., Berk, H., & Contopoulos, J. 1991, *ApJ* 379, 696
Lovelace, R., Romanova, M., & Bisnovatyi-Kogan, G. 1995, *MNRAS* 275, 244
Matt, S., & Pudritz, R.E. 2005, *ApJ* 632, L135
Meliani, Z., Casse, F., & Sauty, C. 2007, *A&A* 460, 1
Michel, F.C. 1969, *ApJ* 158, 727
Miller, K.A., & Stone, J.M. 1997, *ApJ* 489, 890
Ouyed, R., & Pudritz, R.E. 1997, *ApJ* 482, 712
Ouyed, R., Clarke, D.A., & Pudritz, R.E. 2003, *ApJ* 582, 292
O'Sullivan, M., Truss, M., Walker, C., Wood, K., Matthews, O., Whitney, B., & Bjorkman, J.E. 2005, *MNRAS* 358, 632
Pelletier, G., & Pudritz, R.E. 1992, *ApJ* 394, 117
Pfeiffer, H.P., & Lai, D. 2004, *ApJ* 604, 766
Pudritz, R.E., & Norman, C.A. 1983, *ApJ* 274, 677
Pudritz, R.E., Rogers, C., & Ouyed, R. 2006, *MNRAS* 365, 1131
Pudritz, R.E., Ouyed, R., Fendt, Ch., Brandenburg, A. 2007, in: B. Reipurth, D. Jewitt, & K. Keil (eds.), *Protostars & Planets V*, University of Arizona Press, Tucson, 2007, p.277
von Rekowski, B., Brandenburg, A., Dobler, W., & Shukurov, A. 2003, *A&A* 398, 825
von Rekowski, B., & Brandenburg, A. 2004, *A&A* 420, 17
Romanova, M., Ustyugova, G., Koldoba, A., Chechetkin, V., & Lovelace, R. 1997, *ApJ* 482, 708
Romanova, M., Ustyugova, G., Koldoba, A., & Lovelace, R. 2002, *ApJ* 578, 420
Romanova, M., Ustyugova, G., Koldoba, A., & Lovelace, R. 2004, *ApJ* 610, 920
Sakurai, T. 1985, *A&A* 152, 121
Sauty, C., & Tsinganos, K. 1994, *A&A* 287, 893
Sauty, C., Trussoni, E.; & Tsinganos, K. 2002, *A&A* 389, 1068
Shang, H., Li, Z.-Y., Hirano, N. 2007, in: B. Reipurth, D. Jewitt, & K. Keil (eds.), *Protostars & Planets V*, University of Arizona Press, Tucson, 2007, p.261
Shu, F., Najita, J., Ostriker, E., Wilkin, F,. Ruden, S., & Lizano, S. 1994, *ApJ* 429, 781
Symington, N.H., Harries, T.J., & Kurosawa, R. 2005, *MNRAS* 356, 1489
Uchida, Y., & Low, B.C. 1981, *Journal of Astroph. and Astron.* 2, 405
Uchida, Y., & Shibata, K. 1984, *PASJ* 36, 105
Uchida, Y., & Shibata, K. 1985, *PASJ* 37, 515
Ustyugova, G., Koldoba, A., Romanova, M., Chechetkin, V., & Lovelace, R. 1995, *ApJ* 439, 39
Ustyugova, G., Koldoba, A., Romanova, M., Chechetkin, V., & Lovelace, R. 1999, *ApJ* 516, 221
Uzdensky, D.A., Königl, A., & Litwin, C. 2002, *ApJ* 565, 1191
Vitorino, B.F., Jatenco-Pereira, V., & Opher, R. 2003, *ApJ* 592, 332
Zanni, C., Ferrari, A., Rosner, R., Bodo, G., & Massaglia, S. 2007, *A&A* 469, 811

Star-disk Interaction in Young Stars
Proceedings IAU Symposium No. 243, 2007
J. Bouvier & I. Appenzeller, eds.

doi:10.1017/S1743921307009635

MHD simulations of disk-star interaction

Marina M. Romanova[1], M. Long[1], A. K. Kulkarni[1], R. Kurosawa[2]
G. V. Ustyugova[3], A. K. Koldoba[3] and R. V. E. Lovelace[1]

[1]Astronomy Department, Cornell University, Ithaca, NY 14853, USA;
email: romanova@astro.cornell.edu

[2]Department of Physics and Astronomy, University of Nevada Las Vegas, Box 454002,
4505 Maryland Pkwy, Las Vegas, NV 89154-4002, USA; email: rk@physics.unlv.edu

[3]Keldysh Institute of Applied Mathematics, Miusskaya sq. 4, Moscow, 125047, Russia,
email: ustyugg@rambler.ru

Abstract. We discuss a number of topics relevant to disk-magnetosphere interaction and how numerical simulations illuminate them. The topics include: (1) disk-magnetosphere interaction and the problem of disk-locking; (2) the wind problem; (3) structure of the magnetospheric flow, hot spots at the star's surface, and the inner disk region; (4) modeling of spectra from 3D funnel streams; (5) accretion to a star with a complex magnetic field; (6) accretion through 3D instabilities; (7) magnetospheric gap and survival of protoplanets. Results of both 2D and 3D simulations are discussed.

Keywords. Stars: magnetic fields, stars: early-type, accretion, accretion disks.

---

1. Introduction

Disk accretion to a rotating star with a dipole magnetic field has been investigated theoretically by a number of authors (e.g., Ghosh & Lamb 1978; Camenzind 1990; Königl 1991; Ostriker & Shu 1995; Lovelace, Romanova & Bisnovatyi-Kogan 1995). Recently a number of numerical simulations have been performed assuming axisymmetry (e.g., Hayashi, Shibata & Matsumoto 1996; Miller & Stone 1997; Goodson *et al.* 1997; Fendt & Elstner 2000; Matt *et al.* 2002; von Rekowski & Brandenburg 2004; Yelenina, Ustyugova & Koldoba 2006). In some of them, accretion through funnel streams has been clearly observed and investigated (e.g., Romanova *et al.* 2002; Long, Romanova & Lovelace 2005; Bessolaz *et al.* 2007; Zanni, Bessolaz & Ferreira 2007). Full three-dimensional (3D) MHD simulations have been done by Romanova *et al.* (2003, 2004), for which a special Godunov-type MHD code based on the "cubed sphere" grid has been developed (Koldoba *et al.* 2002). Longer 3D simulations have been done by Kulkarni & Romanova (2005) and for accretion to a star with a non-dipole magnetic field by Long, Romanova & Lovelace (2007a,b). Accretion through 3D instabilities has been recently observed in 3D simulations (Romanova & Lovelace 2006; Kulkarni & Romanova 2007). Spectral lines from 3D funnel streams were calculated by Kurosawa, Romanova & Harries (2007) where a 3D radiative transfer grid has been projected onto the 3D MHD grid. A magnetospheric gap may stop the inward migration of protoplanets at the CTTSs stage (Lin, Bodenheimer & Richardson 1996), unless different processes increase the density inside the gap (Romanova & Lovelace 2006). We discuss these topics in greater detail in the following sections.

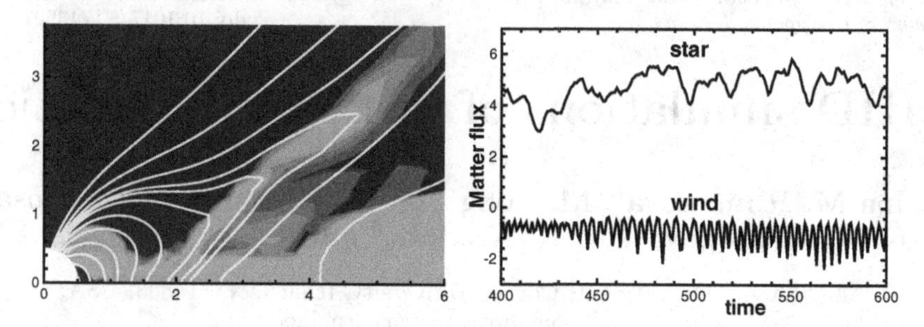

Figure 1. Axisymmetric simulations of winds in the case where the field lines are gathered in an X-type configuration. Left panel shows the distribution of matter flux ρv and sample field lines. Right panel shows temporal evolution of the accretion rate to the star and to the wind, integrated over the $r = 6$ surface. Time is measured in rotational periods at $r = 1$. From Romanova *et al.* 2007.

2. Disk-magnetosphere interaction and disk-locking

2.1. *Disk-magnetosphere interaction through the closed field lines and through the funnel stream*

The accretion disk is disrupted by the stellar magnetosphere and is stopped at the distance R_m from the star where the magnetic pressure is approximately equal to the total matter pressure: $B^2/8\pi = p + \rho v^2$ (e.g., Ghosh & Lamb 1978; see also results of 2D and 3D simulations, Romanova *et al.* 2002, 2003; Long, Romanova & Lovelace 2005). The matter of the disrupted disk accumulates near R_m and is lifted out of the equatorial plane into a "funnel stream" by the pressure gradient force. The matter then follows the star's dipole field lines and is accelerated by gravity until it hits the star's surface (Romanova *et al.* 2002; 2003; Zanni *et al.* 2007; Bessolaz *et al.* 2007). Initially, the angular momentum is carried by the funnel stream matter, but later it is converted to angular momentum carried by the field, so that when funnel matter arrives to the surface of the star, almost all angular momentum is carried by magnetic field lines and only few percent by matter (Ghosh & Lamb 1978, 1979; see simulations by Romanova *et al.* 2002; Long *et al.* 2005). Whether a star spins-up or spins-down is determined by the ratio between the magnetospheric radius R_m and the co-rotation radius $R_{cr} = (GM/\Omega_*^2)^{1/3}$. When the star rotates slowly, and $R_{cr} > R_m$, then the disk matter spins it up (see the main case in Romanova *et al.* 2002; and Zanni *et al.* 2007; Bessolaz *et al.* 2007). In this case the inner disk rotates faster than the star, and closed magnetic field lines connecting the star and the disk form a leading spiral which helps to spin up the star. All of the aforementioned numerical simulations show a direct correlation between the accretion rate $\dot{M}$ (at the surface of the star) and the spin-up rate: the higher the accretion rate, the stronger the spin-up. This correlation may result from the increased twist of closed magnetic field lines connecting the inner disk and the star. If the star rotates fast, so that $R_{cr} < R_m$, then matter is still pulled to the funnel flow by gravity, but the star *spins down* (see e.g., Fig. 6 of Long *et al.* 2005 for large $\boldsymbol{\mu}$) because closed field lines form a trailing spiral, so that the star loses angular momentum. We have evidence from simulations that a significant part of the disk-magnetosphere interaction occurs through closed field lines connecting the star and the disk. Opening of external field lines *does not* lead to disconnection of the star from the disk.

2.2. *Where does the angular momentum go?*

The matter of the disk carries significant specific angular momentum, $j_m = \Omega(R_m)R_m^2$, which is much larger than that of the star, $j_* = \Omega(R_*)R_*^2$, if the angular velocities are

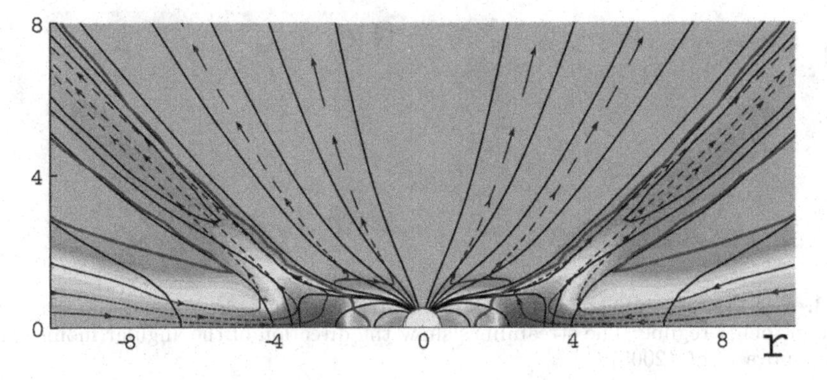

Figure 2. Axisymmetric simulations of a star in the propeller regime. The background shows the density, the thin solid lines are the magnetic field lines, the thick solid lines show the $\beta = 1$ line (magnetic energy-density dominates in the corona and in the wind region above the disk). The dashed lines show the streamlines of matter flow. From Romanova *et al.* (2005).

comparable, $\Omega(R_m) \sim \Omega(R_*)$. The question is whether all the angular momentum of the disk, which is carried initially by matter, goes to the star, or whether a significant part of it returns to the disk via the magnetic field, as suggested by Ostriker & Shu (1995). Numerical simulations have not given a final answer to this question yet, but the last possibility appears probable. In this case angular momentum should be constantly removed from the disk by viscosity, some type of waves propagating through the disk, or by winds. X-type winds may be an efficient mechanism of angular momentum removal from inner regions of the disk (Shu *et al.* 1994) or, winds may flow from the disk at any distance (Pudritz & Norman 1986; Lovelace, Romanova & Bisnovatyi-Kogan 1995; Lamzin *et al.* 2004; Ferreira, Dougados & Cabrit 2006). If angular momentum does not return back to the disk, then the star will spin up and other mechanisms are required to spin it down. This question needs further investigation.

2.3. *Inflation of field lines and spinning-down*

Differential rotation between the star and the disk leads to the inflation of field lines (Aly 1980; Shu *et al.* 1994; Lovelace, *et al.* 1995). Some field lines are strongly inflated and do not transport information between the star and the disk. Other field lines are only partially open and may transport angular momentum between the star and the disk (e.g., the "dead zone" of Ostriker and Shu 1995). Field lines may be only partially open because, for example, the magnetic pressure force leading to opening is balanced by the magnetic tension force opposing the opening. In addition, inflated field lines may reconnect and then inflate again, so that the magnetosphere may not be stationary, but may experience quasi-periodic or episodic restructuring (e.g. Aly & Kuijpers 1990; Uzdensky 2002). Strongly inflated field lines connect the star with the slowly rotating corona, so that angular momentum always flows out from the star to the corona. The corona has a low density and is probably magnetically dominated, so that angular momentum may flow from the star to the corona due to the magnetic stress, $\dot{L}_f = \int dS \cdot rB_\phi B_p/4\pi$. A star may also lose angular momentum due to major matter outflow from its surface or due to re-direction into the outflow of some of the matter coming to the star through the funnel flow (Matt & Pudritz 2005) where a Weber & Davis (1967) type of angular momentum loss is investigated. In these models it is important to understand the force which lifts the matter (up to 10% of the accretion rate) from the vicinity of the star. Compared to the "material" wind of Matt & Pudritz (2005), the angular momentum

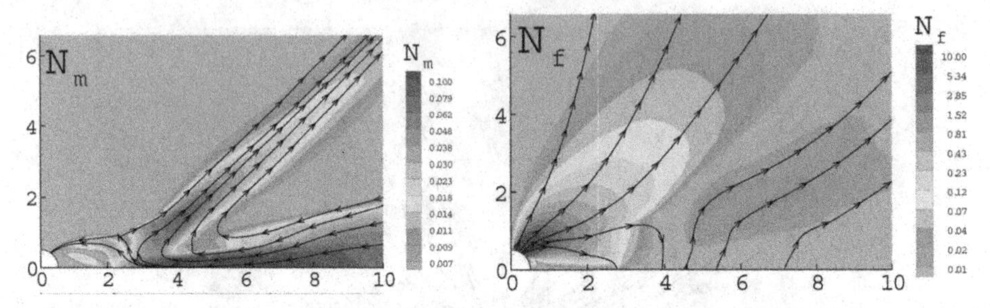

Figure 3. The background shows torques associated with matter N_m and the magnetic field N_f in the propeller regime. The streamlines show the direction of the angular momentum flow. From Romanova *et al.* (2005).

carried by the *magnetic stress* does not require much matter flowing from the star and does not require dense matter in the corona. Estimates show that only $n = 10^5$ cm^{-3} is required to support a current associated with the twist of magnetic field lines. Thus, angular momentum always flows to the "magnetic tower" (see also Lynden-Bell 2003) and the torque depends on the magnetic field strength and angular velocity of the star. In Long et al. (2005), matter dominates in most of the corona, and spin-down through open field lines does depend on the coronal density. However in the propeller regime (Romanova *et al.* 2005; Ustyugova *et al.* 2006), the corona is magnetically-dominated and a significant torque is associated with the magnetic tower.

2.4. *Rotational equilibrium state and disk-locking*

If a magnetized star accretes for a sufficiently long time, it is expected to be in the *rotational equilibrium state*, that is when the total positive spin acting on the star equals the total negative spin. Many CTTSs are expected to be in the rotational equilibrium state (Bouvier 2007). If $R_m = R_{cr}$, then both inflated and partially-open field lines (which thread the disk at $r > R_{cr}$) will lead to spin-down of the star. So in the rotational equilibrium state, $R_{cr} = kR_m$, where $k > 1$. Numerical simulations have shown that $k = 1.2 - 1.5$ (Romanova *et al.* 2002; Long *et al.* 2005). However in both cases matter dominates in the corona and this leads to quite large angular momentum outflows associated with inflated and partially open field lines. In the best cases of Long *et al.* (2005) the coronal density is 3×10^{-4} of that in the disk. If the corona above the disk has this density (say, due to winds from the disk or the star) then the estimations done in this paper are correct.

3. Modeling of jets and winds from CTTSs

Jets and winds are observed from a number of CTTSs (e.g Cabrit et al. 2007). In some cases only slow jets are observed (Brittain *et al.* 2007). Clear outflow signatures are also seen as absorption features in the blue wings of some lines, such as the He I line which shows outflow of the order of up to 10% of the disk mass (Edwards *et al.* 2003; Kwan *et al.* 2007; see also Hartmann 1998).

3.1. *Winds from slowly rotating stars*

In early simulations of slowly rotating stars we did not observe any significant outflows from the disk or from the disk-magnetosphere boundary (Romanova *et al.* 2002, Long *et al.* 2005). It is clear that an X-type configuration is favorable for launching outflows from the disk-magnetosphere boundary (Shu *et al.* 1994). To obtain this configuration,

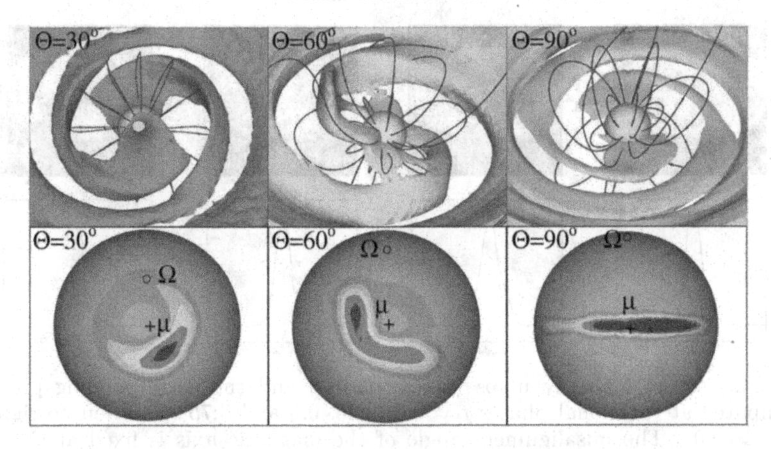

Figure 4. 3D simulations of accretion to a star with a misaligned dipole magnetic field at misalignment angles $\Theta = 30°$, $\Theta = 60°$, $\Theta = 90°$. Top panels show matter flow in the inner part of the simulation region and magnetic field lines. Bottom panels show corresponding hot spots on the surface of the star, where the color background shows kinetic energy density. From Romanova *et al.* (2004).

we suggested that the accretion rate is initially low but later increases, and matter comes to the simulation region bunching field lines into an X-type configuration. In addition we suggested that the $\alpha-$parameters regulating viscosity, α_v, and diffusivity, α_d, are not very small, $\alpha_d = 0.1$ (compared to earlier research, where we typically chose $\alpha \approx 0.02$). In addition, we chose the viscosity to be several times larger than the diffusivity, so that the bunching of field lines would be supported by the flow. An increase of α_d from 0.02 to 0.1 helped obtain a reasonable rate of penetration of incoming matter through field lines. In this case we obtained a quasi-periodic, X-type wind (see an example of simulations in Fig. 1, where $\alpha_d = 0.1$ and $\alpha_v = 0.3$). We observed multiple outbursts of matter into the wind, driven by a combination of magnetic and centrifugal forces (Romanova *et al.* 2007, in preparation). Typical velocities in the outflow are $v = 30 - 60$ km s^{-1}, for typical parameters of CTTS and up to 20% of the disk matter flows to the wind. The simulations run for about 800 rotations at $r = 1$. Similar conical X-type outflows have recently been observed with more general initial conditions. It looks like some outflows are also observed by Bessolaz *et al.* (2007) and Zanni *et al.* (2007). We should note that in their runs diffusivity is even higher. It seems that a relatively high diffusivity is a necessary condition for getting outflows from the disk-magnetosphere boundary.

3.2. *Winds from rapidly rotating stars*

If $R_m > R_{cr}$ then the star is in the "propeller" regime (e.g., Illarionov & Sunyaev 1975; Lovelace, Romanova & Bisnovatyi-Kogan 1999). This regime may occur if the accretion rate decreases, causing R_m to increase. Or, it may be a typical stage during the early years of the CTTSs evolution. Axisymmetric simulations of the propeller stage have shown that once again, at low diffusivity $\alpha_d = 0.02$, there are no outflows, while at larger diffusivity, $\alpha_d > 0.1$, conical quasi-periodic or episodic outflows form and carry away a significant part of the disk matter as wind (see Fig. 2, also Romanova *et al.* 2005; Ustyugova *et al.* 2006). Fig 3 (left panel) shows that a significant part of the angular momentum carried by matter is redirected into conical outflows. Fig 3 (right panel) shows that a large part of the angular momentum flows into the corona along the open magnetic field lines (through the magnetic tower) and some flows from the disk to the corona. In the propeller regime we were able to obtain a magnetically-dominated corona, and Fig. 3 shows an example

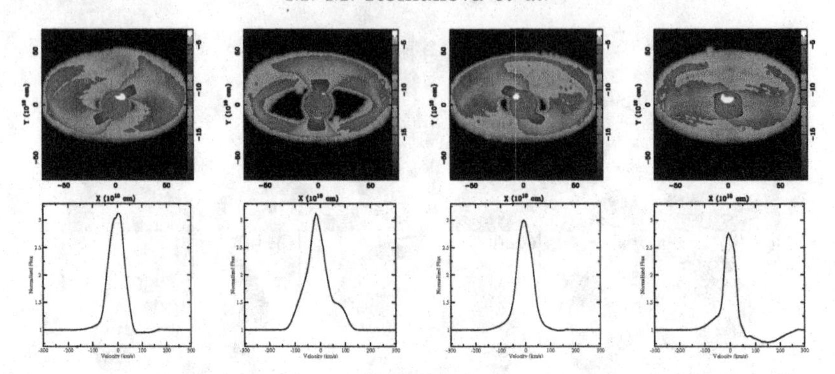

Figure 5. Paβ model intensity maps (upper panels) and the corresponding profiles (lower panels) computed at rotational phases $t = 0.0$, 0.25, 0.5 and 0.75 (from left to right) and for inclination $i = 60°$. The misalignment angle of the magnetic axis is fixed at $\Theta = 15°$. The intensity is shown on a logarithmic scale with an arbitrary units. From Kurosawa *et al.* (2007).

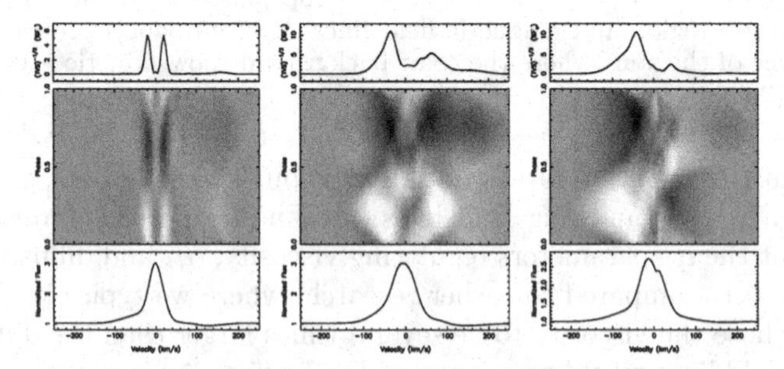

Figure 6. The summary of the Paβ spectra computed for the MHD model with $\Theta = 15°$ and $i = 10°, 60°$ and $80°$ respectively from left to right. The spectra were computed at 50 different rotational phases. In the bottom panels, the mean spectra of all rotational phases are shown. In the middle panels, the quotient spectra (each spectrum divided by the mean spectrum) are shown as grayscale images with increasing rotational phases in the upward vertical direction. The temporal variance spectra $(\mathrm{TVS})^{-1/2}$ are shown in the top panels. From Kurosawa *et al.* (2007).

in which the star loses about half of its angular momentum through the magnetic wind. The propeller stage may be important for fast spin-down of young CTTSs.

4. Properties of the funnel streams, hot spots and the inner disk

A wide variety of different features were found in our full 3D MHD simulations of disk accretion to a rotating magnetized star with a *misaligned* dipole magnetic field (Romanova *et al.* 2003, 2004). Simulations were done for a variety of misalignment angles from $\Theta = 0°$ to $\Theta = 90°$, where Θ is the angle between the magnetic axis of the dipole μ and the angular velocity of the star Ω (which was aligned with the angular velocity of the disk). Below we summarize some interesting features observed in 3D simulations: **(1)** 3D simulations have shown that the system becomes noticeably non-axisymmetric even at very small misalignment angles $\Theta \sim 2° - 5°$. Thus, most magnetized stars are expected to be non-axisymmetric and will form funnel flows and non-axisymmetric hot spots. **(2)** Matter accretes to the star through funnel streams which are not homogeneous. The density is largest in the interior regions of the stream, and decreases outwards, so that the appearance of the streams depends on the density: at the largest densities they look like thin streams. At lower densities the stream is wider. The matter covers the

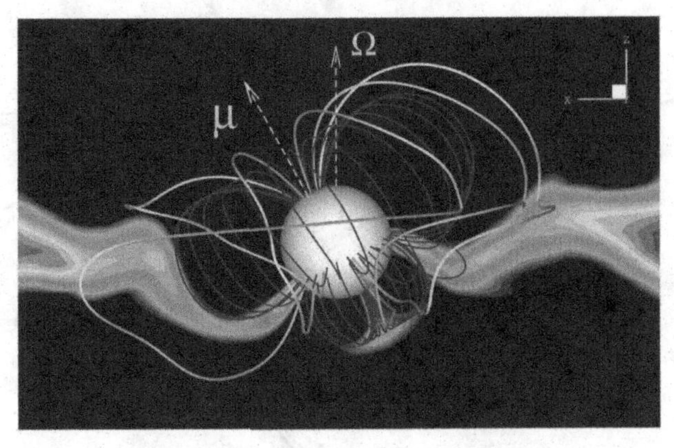

Figure 7. 3D simulations of accretion to a star with aligned dipole plus quadrupole fields. Both dipole and quadrupole magnetic moments are misaligned relative to the rotation axis at $\Theta = 30^\circ$. Here $B_{dipole} \approx 0.45 B_{quad}$. From Long *et al.* (2007a).

whole magnetosphere at the lowest densities. **(3)** There are usually two streams which flow to the nearest poles. Some matter flows to the further pole which often leads to a second, weaker set of streams. In the case of high Θ, one wide stream often splits into two streams. **(4)** The structure of the hot spots reflects the structure of the funnel streams: the density and specific kinetic energy are larger in the central regions of spots. Thus, hot spots are "hotter" in the center, and cooler outside. They have different shapes, ranging from bow-shaped for small misalignment angles to bar-shaped for large misalignment angles (see Fig. 4). **(5)** Variability curves from the hot spots obtained from numerical simulations for different misalignment angles Θ and different inclination angles i of the disk show a variety of shapes, which may be used to constrain Θ and i. **(6)** The density distribution in the disk around the magnetized star is inhomogeneous. It has always a shape of spirals (see Fig. 4). **(7)** The inner regions of the disk are slightly warped which reflects the tendency of matter to flow in the magnetic equatorial plane. This warped disk may obscure the light from the star, and this may lead to quasi-periodic variations of light (Bouvier *et al.* 2003).

5. Radiative transport and calculation of spectral lines

We calculated the line profiles from 3D funnel streams using the radiative transfer code TORUS (Harries 2000; Kurosawa *et al.* 2004, 2005, 2006; Symington *et al.* 2005) which was modified to incorporate the density, velocity and gas temperature structures from the 3D MHD simulations (Romanova *et al.* 2003, 2004). The radiative transfer code uses a three-dimensional (3D) adaptive mesh refinement (AMR) grid. We take 3D MHD simulations which come to quasi-stationary state, take parameters of the flow for some moment of time and calculate spectrum from the funnel streams using TORUS code.

We study the dependence of the observed line variability on two main parameters: (1) the inclination angle (i) and (2) the misalignment angle (Θ). Here, we show sample results for $\Theta = 15^\circ$. As Θ is rather small, the accretion occurs in two arms, creating two hot spots on the surface of the star (Romanova *et al.* 2004; see also Fig. 4). As one can see from Fig. 5 ($i = 60^\circ$) the largest amount of red wing absorption occurs at the rotational phase at which a hot spot is facing the observer and when the spot-funnel-observer alignment favorable.

The line variability behavior is summarized in Fig. 6 which shows the phase averaged spectra, the quotient spectra as a function of rotational phase (in the gray scale images),

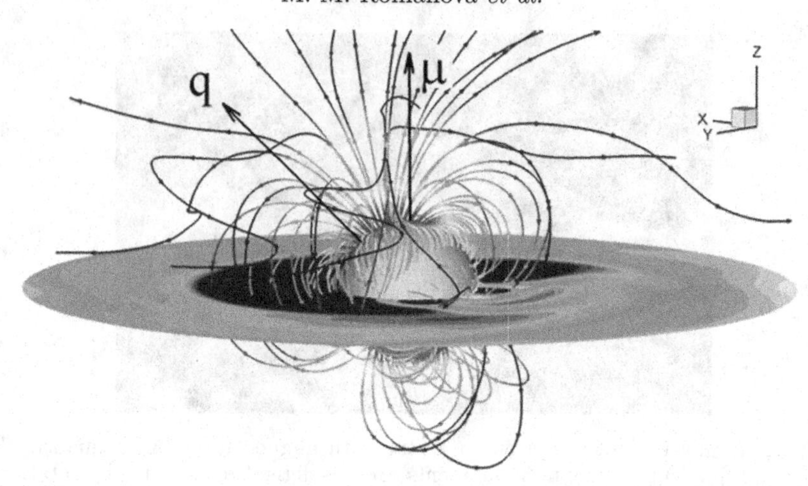

Figure 8. 3D simulations of accretion to a star with a complex magnetic field: a combination of a dipole moment along the z-axis and a quadrupole moment misaligned at $\chi = 45°$. The equatorial slice shows the density distribution. From Long *et al.* (2007b).

and the temporal variance spectrum (TVS), which is similar to the root-mean-square spectra (cf. Fullerton, Gies, & Bolton 1996; Kurosawa *et al.* 2005). The mean spectra of three models are fairly symmetric about the line center; however, a very weak but noticeable amount of absorption in the red wings can be seen in the spectra at all i. For $i = 10°$ and $60°$ cases, the flux levels in their red wing become slightly below the continuum level, but the level remains above the continuum for $i = 80°$ case. Although the line equivalent width of the mean spectra for $i = 10°$ is slightly smaller than that of the $i = 60°$ and $80°$ cases, no major difference is seen between the three models.

In general, the line profile shapes and strength predicted by the radiative transfer model based on the MHD simulations are similar to those seen in the atlas of the observed Paβ and Brγ given by Folha & Emerson (2001). The level of the line variability seen in the time-series spectroscopic observation (Paβ) of SU Aur (Kurosawa *et al.* 2005, see their Fig. 1) is comparable to our models (Fig. 6). The double-peaked variability pattern (TVS) seen in the low inclination angle model (i.e. $i = 10°$ model in Fig. 6) is very similar to that of Hα and Hβ from the CTTS TW Hydra observed by Alencar & Batalha (2002). The system has a low inclination angle ($i = 18° \pm 10°$, Alencar & Batalha 2002) which is consistent with our model ($i = 10°$). Although not shown here, the line variability predicted from a model with $i = 75°$ and $\Theta = 60°$ resembles that of Hβ from the CTTS AA Tau observed by Bouvier et al. (2007b). More careful analysis and tailored model fits of observations are required for deriving fundamental physical parameters of a particular system.

6. Accretion to a star with a complex magnetic field

The magnetic field of a young T Tauri star may have a complex structure (see e.g. Johns-Krull *et al.* 1999; Smirnov *et al.* 2003; Gregory *et al.* 2006; Jardine *et al.* 2006). Observational properties of magnetospheric accretion lead to the conclusion that the magnetic field of many CTTSs may consist of a combination of dipole and multipole fields (e.g., Bouvier *et al.* 2007a). If a star has several magnetic poles, then matter may flow to the star in multiple streams, choosing the shortest path to the nearby pole. Some magnetic poles are expected to be closer to the equator of the star compared with the pure dipole case. The observational properties of hot spots will also be different. A picture of

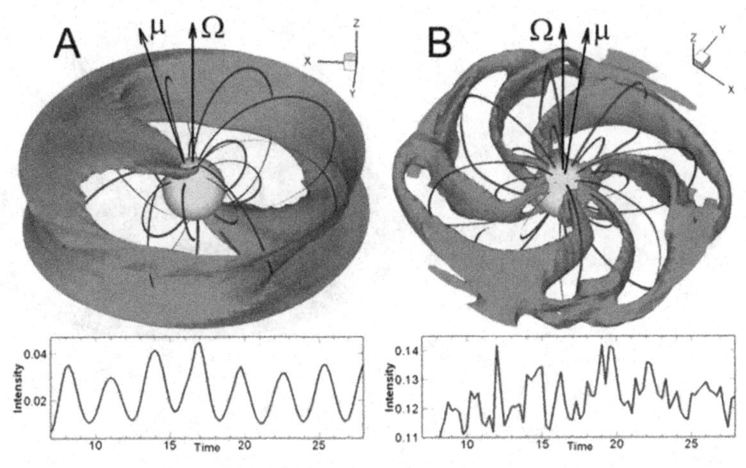

Figure 9. Left panel: in stable accretion matter accretes in two ordered funnel streams and the light-curve has an ordered, almost sinusoidal shape. Right panel: in the unstable regime matter accretes through multiple tongues and the light-curve is irregular. From Kulkarni & Romanova (2007).

matter flow to a star with a complex magnetic field (deduced from Doppler tomography, e.g. Jardine *et al.* 2006) has been calculated by Gregory *et al.* (2006). It was shown that matter accretes through several funnel streams choosing the nearby magnetic poles to accrete.

Recently we were able to perform the first 3D MHD simulations of accretion to a star with a combination of the dipole and quadrupole magnetic fields (Long, Romanova & Lovelace 2007a,b). Simulations of accretion to a star with a *pure quadrupole* field have shown that most of the matter flows to the star through the quadrupole "belt", forming a ring-shaped hot spot at the magnetic equator. In the case of a *dipole plus quadrupole* field, the magnetic flux in the northern hemisphere is larger than that in the southern hemisphere and the quadrupole belt and the ring are displaced to the south. Fig. 7 shows the case when the magnetic moment of the dipole $\boldsymbol{\mu}$ and that of the quadrupole $\mathbf{D}$ are aligned and both are misaligned relative the the rotational axis $\boldsymbol{\Omega}$ at an angle Θ. One can see that the disk is disrupted by the dipole component of the field but the quadrupole component is strong enough to guide matter streams to the poles. At different Θ the light curves have a variety of different features, but many of these features are observed in the pure dipole cases at somewhat different Θ and inclination angle of the system i. So, it would be challenging to deduce the magnetic field configuration from the light curves (Long *et al.* 2007a).

As the next step we investigated accretion to a star with a dipole plus quadrupole field, when the magnetic axis of the quadrupole is misaligned relative to magnetic axis of the dipole at an angle χ (Long *et al.* 2007b). Fig. 8 shows an example of such a accretion for $\chi = 45°$.

7. Accretion through instabilities

Recent 3D simulations have shown that in many cases, disk matter accretes to the star through a 3D Rayleigh-Taylor instability (Romanova & Lovelace 2006; Kulkarni & Romanova 2007). Penetration of matter through 3D instabilities has been predicted theoretically (e.g., Arons & Lea 1976) and has been observed in 2D simulations by Wang & Robertson (1985). In the unstable regime, matter accretes through the magnetosphere

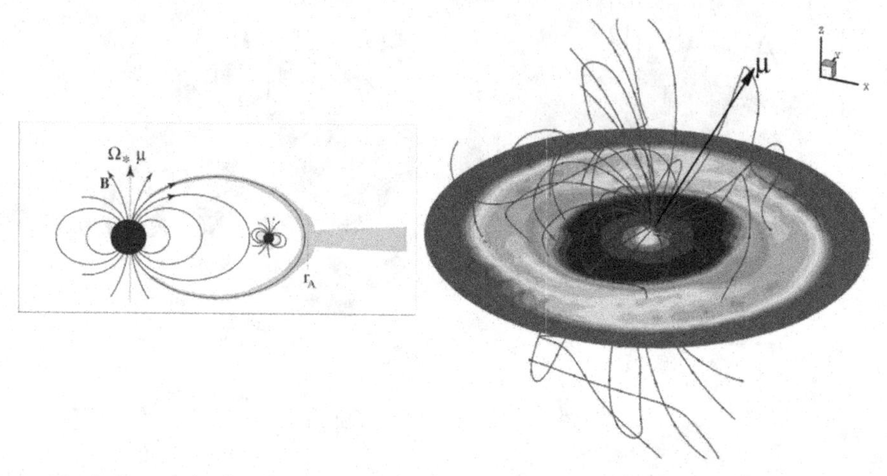

Figure 10. Left panel: sketch of the magnetospheric gap. Right panel: result of 3D MHD simulations, where dark blue shows low density, and red - high density with a contrast of 300. Red lines are sample magnetic field lines. From Romanova & Lovelace (2006).

forming transient but frequently appearing equatorial tongues (see Fig. 9, right panel). The tongues penetrate deep into the magnetosphere and deposit material onto the star much further away from the magnetic poles than funnel streams do. The structure of the tongues is opposite to that of the magnetospheric funnel streams: they are narrow in the longitudinal direction, but wide in latitude (see Fig. 9). The number and location of the tongues varies with time. The number varies between $m = 2$ and $m = 7$ and depends on parameters of the model. For example, $m = 2$ dominates if the misalignment angle of the dipole is not very small, $\Theta \sim 15° - 30°$. The tongues hit the surface of the star and form randomly distributed hot spots, so that the light curve looks irregular. In some cases quasi-periodic oscillations are observed when a definite number of tongues dominates. Simulations have shown that this type of accretion occurs in cases when the accretion rate is relatively high (Kulkarni & Romanova 2007).

8. Magnetospheric gaps and survival of protoplanets

A young star with a strong dipole magnetic field is surrounded by a low-density *magnetospheric gap* where planets may survive longer than in the disk, due to the low density of matter inside the gap (if their orbit is inside the 2:1 resonance with the inner radius of the disk) (Lin *et al.* 1996; Romanova & Lovelace 2006; see Fig. 10). For typical CTTS the inner edge of the truncated disk approximately coincides with the peak in distribution of close-in planets ($P \approx 3$ days). We investigated in 3D simulations the emptiness of magnetospheric gaps in different situations.

Magnetospheric gaps around stars with a dipole magnetic field. We observed from 3D simulations that in case of the dipole field and relatively small misalignment angles, $\Theta < 30°$, the magnetospheric gap has quite a low density (see Fig. 10, right panel). However, for sufficiently large Θ, $\Theta > 45°$, part of the funnel stream crosses the equatorial plane, so that magnetospheric gap is not empty (Romanova & Lovelace 2006).

Magnetospheric gaps around stars with complex magnetic fields. The magnetic field of a young T Tauri star may have a non-dipole, complex structure (see e.g. Gregory, *et al.* 2006). If the dipole component dominates at large distances, and other components of the field are weaker multipoles, then a magnetospheric gap may form, as in the case of the dipole field (e.g., Bouvier *et al.* 2007a). However, if multipoles are sufficiently strong compared with the dipole field, then matter may accrete through several streams and will chose a path to the nearby poles which are closest to the equatorial plane. In this situation some of funnels may cross the equatorial plane and increase the density in the magnetospheric gap (see example in Fig. 8).

Accretion through 3D instabilities and magnetospheric gaps. At sufficiently high accretion rates in the disk matter accretes to the star through 3D instabilities (Romanova & Lovelace 2006; Kulkarni & Romanova 2007, see Fig. 9, right panel) and the magnetospheric gap is not empty. Such accretion is expected at the early stages of evolution of the CTTSs, when the accretion rate is expected to be higher. Probably, at this early stage many protoplanets form, migrate inward, and are absorbed by the star.

Acknowledgements

We would like to thank organizers for excellent meeting. We also acknowledge the useful discussion of the disk-locking problem with Frank Shu, Sean Matt, Jerome Bouvier, and other participants. This research was partially supported by the NSF grants AST-0507760 and AST-0607135, and NASA grants NNG05GL49G and NAG5-13060. NASA provided access to high performance computing facilities.

References

Alencar, S. H. P. & Batalha, C. 2002, *ApJ*, 571, 378

Aly, J.J. 1980, *A&A*, 86, 192

Aly, J.J., & Kuijpers, J. 1990, *A&A*, 227, 473

Arons, J. & Lea, S. M. 1976, *ApJ*, 207, 914

Bessolaz, N., Zanni, C., Ferreira, J., Keppens, R., Bouvier, J., & Dougados, C. 2007, *These Proceedings*

Bouvier, J., Grankin, K. N., Alencar, S. H. P., Dougados, C., Ferndez, M., Basri, G., *et al.* 2003, *A&A*, 409, 169

Bouvier, J., Alencar, S.H.P., Harries, T.J., Johns-Krull, C.M., Romanova, M.M. 2007a, in: B. Reipurth, D. Jewitt, & K. Keil (eds.), *Protostars and Planets V*, University of Arizona Press, Tucson, vol. 951, p. 479

Bouvier, J., Alencar, S. H. P., Boutelier, T., Dougados, C., Balog, Z., Grankin, K., Hodgkin, S. T., Ibrahimov, M. A., Kun, M., Magakian, T. Y., & Pinte, C. 2007b, *A&A*, 463, 1017

Bouvier, J. 2007, *These Proceedings*

Brittain, S., Rettig, T., Balsara, D., Tilley, D., Simon, T., Gibb, E., & Hinke, K. 2007, *These Proceedings*

Cabrit, S., Dougados, C., Ferreira, J., Garcia , P., Raga, A., Agra-Amboade, V., Lavalley, C., & Pesenti, M. 2007, *These Proceedings*

Camenzind, M. 1990, *Rev. Mod. Astron.*, 3, 234

Edwards, S., Fischer, W., Kwan, J., Hillenbrand, L., & Dupree, A. K. 2003, *ApJ Lett.*, 599, L41

Fendt, C., & Elstner, D. 2000, *A&A* 363, 208

Ferreira, J., Dougados, C., & Cabrit, S. 2006, *A&A*, 453, 785

Folha, D. F. M. & Emerson, J. P. 2001, *A&A*, 365, 90

Fullerton, A. W., Gies, D. R., & Bolton, C. T. 1996, *ApJS*, 103, 475

Ghosh, P. & Lamb, F. K. 1978, *ApJ Letters* 223, L83

Ghosh, P. & Lamb, F. K. 1979, *ApJ* 234, 296

Goodson, A.P., Winglee, R., & Böhm, K.H. 1997, *ApJ* 489, 199

Gregory, S. G., Jardine, M., Simpson, I., & Donati, J.-F. 2006, *MNRAS*, 371, 999

Harries, T. J. 2000, *MNRAS*, 315, 722

Hayashi, M.R., Shibata, K., & Matsumoto, R. 1996, *ApJ Letters* 468, L37

Hartmann, L. 1998, *Cambridge University Press*

Illarionov, A.F., & Sunyaev, R.A. 1975, *A&A* 39, 185

Jardine, M., Cameron, A.C., Donati, J.-F., Gregory, S.G., & Wood, K., 2006, *MNRAS*, 367, 917

Johns-Krull, C.M., Valenti, J.A. & Koresko, C. 1999, *ApJ* 516, 900
Koldoba, A.V., Romanova, M.M., Ustyugova, G.V., & Lovelace, R.V.E. 2002, *ApJ Letters* 576, L53
Konigl, A. 1991, *ApJ* 370, L39
Kulkarni, A.K. & Romanova, M.M. 2005, *ApJ* 633, 349
Kulkarni, A.K. & Romanova, M.M. 2007, *in preparation*
Kurosawa, R., Harries, T. J., Bate, M. R., & Symington, N. H. 2004, *MNRAS*, 351, 1134
Kurosawa, R., Harries, T. J., & Symington, N. H. 2005, *MNRAS*, 358, 671
Kurosawa, R., Harries, T. J., & Symington, N. H. 2006, *MNRAS*, 370, 580
Kurosawa, R., Romanova, M.M. & Harries, T. J. 2007, *in preparation*
Kwan, J., Edwards, S. & Fischer, W. 2007, *ApJ* 657, 897
Lamzin, S. A., Kravtsova, A. S., Romanova, M. M. & Batalha, C. 2004, *Astron. Lett.* 30, 413
Lin, D.N.C., Bodenheimer, P., & Richardson, D.C. 1996, *Nature* 380, 606
Long, M., Romanova, M.M. & Lovelace, R.V.E. 2005, *ApJ* 634, 1214
Long, M., Romanova, M.M. & Lovelace, R.V.E. 2007a, *MNRAS* 374, 436
Long, M., Romanova, M.M. & Lovelace, R.V.E. 2007b, *in preparation*
Lovelace, R.V.E., Romanova, M.M., & Bisnovatyi-Kogan, G.S. 1995, *MNRAS* 374, 436
Lovelace, R.V.E., Romanova, M.M., & Bisnovatyi-Kogan, G.S. 1999, *MNRAS* 514, 368
Lynden-Bell, D. 2003, *MNRAS* 341, 1360
Matt, S., Goodson, A.P., Winglee, R.M., & Böhm, K.-H. 2002, *ApJ* 574, 232
Matt, S.& Pudritz, R.E. 2005, *ApJ Letters* 632, L135
Miller, K.A. & Stone, J.M. 1997, *ApJ* 489, 890
Ostriker, E.C. & Shu, F.H. 1995, *ApJ* 447, 813
Pudritz, R.E. & Norman, C.A. 1986, *ApJ* 301, 571
Romanova, M.M. & Lovelace, R.V.E. 2006, *ApJ Lett.* 645, L73
Romanova, M.M., Ustyugova, G.V., Koldoba, A.V., Wick, J.V. & Lovelace, R.V.E. 2003, *ApJ* 595, 1009
Romanova, M.M., Ustyugova, G.V., Koldoba, A.V. & Lovelace, R.V.E. 2002, *ApJ* 578, 420
Romanova, M.M., Ustyugova, G.V., Koldoba, A.V. & Lovelace, R.V.E. 2004, *ApJ* 610, 920
Romanova, M.M., Ustyugova, G.V., Koldoba, A.V. & Lovelace, R.V.E. 2005, *ApJ Lett.* 635, L165
Romanova, M.M., Ustyugova, G.V., Koldoba, A.V. & Lovelace, R.V.E. 2007, *in preparation*
Shu, F., Najita, J., Ostriker, E., Wilkin, F., Ruden, S. & Lizano, S 1994, *ApJ* 429, 781
Smirnov, D.A., Lamzin, S.A., Fabrika, S.N. & Valyavin, G.G. 2003, *A&A*, 401, 1057
Symington, N. H., Harries, T. J., & Kurosawa, R. 2005, *MNRAS*, 356, 1489
Ustyugova, G.V., Koldoba, A.V., Romanova, M.M., & Lovelace, R.V.E. 2006, *ApJ* 646, 304
Uzdensky D.A. 2002, *ApJ*, 572, 432
von Rekowski, B., & Brandenburg, A., *A&A* 2004, 420, 17
Wang, Y.-M. & Robertson, J. A. 1985, *ApJ*, 299, 85
Weber, E.J. & Davis, L.J. 1967, *ApJ*, 148, 217
Yelenina, T. G., Ustyugova, G. V., & Koldoba, A. V., *A&A* 2006, 458, 679
Zanni, C., Bessolaz, N. & Ferreira, J. 2007, *These Proceedings*

Discussion

SHU: I think that your coronal density is too high and this determines angular momentum outflow to corona. What is the density in the corona ?

ROMANOVA: In typical simulation run the density in the corona is about 10^{-3} of the density in the disk. In experimental runs we got density 3×10^{-4}. The question is: what is the real density in the corona in the vicinity of the disk-magnetosphere boundary, where both, stellar and disk winds may contribute to coronal density.

SHU: But angular momentum which flows to the magnetic tower strongly depends on the coronal density which is high in your simulations.

ROMANOVA: The magnetic tower carries angular momentum due to the twisting of the magnetic field. In the propeller case magnetic winds carry away half or more of the angular momentum of the star. In the propeller case we were able to obtain magnetically-dominated corona. In the case of slowly rotating stars magnetic wind carries less angular momentum, and also we are not in a coronal regime yet to give the final answer about this number.

MATT: How much angular momentum flows to radially stretched field lines ?

ROMANOVA: Our simulations show that about 1/3 of the angular momentum flows along these field lines.

Star-Disk Interaction in Young Stars
Proceedings IAU Symposium No. 243, 2007
J. Bouvier & I. Appenzeller, eds.

doi:10.1017/S1743921307009647

MHD instabilities at the disk-magnetosphere boundary: 3D simulations

Akshay K. Kulkarni and Marina M. Romanova
Department of Astronomy, Cornell University, Ithaca, NY 14853, USA
email:{akshay,romanova}@astro.cornell.edu

Abstract. We present results of 3D simulations of MHD instabilities at the accretion disk-magnetosphere boundary. The instability is Rayleigh-Taylor, and develops for a large range of parameter values. It manifests itself in the form of tall, thin tongues of plasma that reach the star by penetrating through the magnetosphere in the equatorial plane. The tongues rotate around the star in the equatorial plane, and their shape and number changes with time on inner-disk dynamical timescales. In contrast with funnel flows, which deposit matter mainly in the polar region, the tongues deposit matter much closer to the stellar equator.

Keywords. Accretion, accretion disks, instabilities, (magnetohydrodynamics:) MHD, stars: circumstellar matter, stars: magnetic fields, stars: oscillations, stars: pre-main-sequence, stars: rotation, stars: spots.

1. Introduction

The geometry of the accretion flow around magnetized stars is a problem of long-standing interest. It is an important factor in determining the observed spectral and variability properties of the accreting system. An accretion disk around a magnetized central object is truncated approximately at the distance from the central star where the magnetic energy density becomes comparable to the matter energy density. Beyond that point, there are two ways in which the gas can accrete to the star: (1) through funnels, or magnetospheric accretion (Ghosh & Lamb 1978, 1979); (2) through plasma instabilities at the disk-magnetosphere interface (Arons & Lea 1976; Elser & Lamb 1977; Spruit & Taam 1990; Rastätter & Schindler 1999; Li & Narayan 2004). Two- and three-dimensional simulations have shown accretion through funnel streams (Romanova *et al.* 2002, 2003, 2004; Kulkarni & Romanova 2005). Plasma instabilities at the disk-magnetosphere interface, when present, will completely alter the geometry of the accretion flow. In general, the Rayleigh-Taylor instability is expected to develop at the disk-magnetosphere interface because of the high-density disk matter being supported by the low-density magnetospheric plasma. The Kelvin-Helmholtz instability is also expected to develop because of the discontinuity in the angular velocity of the matter at the boundary. The inner disk matter is expected to rotate at the local Keplerian velocity, while the magnetospheric plasma corotates with the star. There have been numerous studies of the development of such instabilities.

We report on the discovery of instabilities at the disk-magnetosphere interface in full three-dimensional simulations of disk accretion to a magnetized star. The instabilities occur for a large range of parameter values. We discuss our numerical model in § 2, and present our results in § 3.

2. The numerical model

The model we use is the same as in our earlier 3D MHD simulations (Koldoba *et al.* 2002; Romanova *et al.* 2003, 2004). The star has a dipole magnetic field, the axis of which makes an angle Θ with the star's rotation axis. The rotation axes of the star and the accretion disk are aligned. There is a low-density corona around the star and the disk which also rotates about the same axis. To model stationary accretion, the disk is chosen to initially be in a quasi-equilibrium state, where the gravitational, centrifugal and pressure gradient forces are in balance (Romanova *et al.* 2002). Viscosity is modelled using the α-model (Shakura & Sunyaev 1973; Novikov & Thorne 1973). To model accretion, the ideal MHD equations are solved numerically in three dimensions, using a Godunov-type numerical code, written in a "cubed-sphere" coordinate system rotating with the star (Koldoba *et al.* 2002; Romanova *et al.* 2003). The boundary conditions at the star's surface amount to assuming that the infalling matter passes through the surface of the star. So the dynamics of the matter after it falls on the star is ignored. The inward motion of the accretion disk is found to be stopped by the star's magnetosphere at the Alfvén radius, where the magnetic and matter energy densities become equal. The subsequent evolution depends on the parameters of the model.

The simulations are done using the following dimensionless variables: the radial coordinate $r' = r/R_0$, the fluid velocity $\mathbf{v}' = \mathbf{v}/v_0$, the density $\rho' = \rho/\rho_0$, the magnetic field $\mathbf{B}' = \mathbf{B}/B_0$, the pressure $p' = p/p_0$, the temperature $T' = T/T_0$, and the time $t' = t/t_0$. The variables with subscript 0 are dimensional reference values and the unprimed variables are the dimensional variables. Because of the use of dimensionless variables, the results are applicable to a wide range of objects and physical conditions, each with its own set of reference values. To apply our simulation results to a particular situation, we have the freedom to choose three parameters, and all the reference values are calculated from those. We choose the mass, radius and surface magnetic field of the star as the three independent parameters.

For protostars, we take the mass of the star to be $M = 0.8M_\odot = 1.6 \times 10^{33}$ g and its radius $R = 2R_\odot = 1.4 \times 10^6$ km $= 1.4 \times 10^{11}$ cm. The reference length scale is $R_0 = R/0.35 = 4 \times 10^{11}$ cm ≈ 0.03 AU. The reference velocity is $v_0 = 1.6 \times 10^7$ cm s^{-1}. The reference angular velocity is $\omega_0 = 4 \times 10^{-5}$ s^{-1}. The reference time is $t_0 = 2.5 \times 10^4$ s $= 0.3$ days. The reference rotation period is $P_0 = 1.5 \times 10^5$ s $= 1.8$ days. The reference surface magnetic field is $B_{\star_0} = 10^3$ G, which is a typical value for protostars. Then the reference magnetic field is $B_0 = B_{\star_0}(R/R_0)^3 = 43$ G. The reference density is $\rho_0 = 7 \times 10^{-12}$ g cm^{-3}. The reference pressure is $p_0 = 1.8 \times 10^3$ dynes cm^{-2}. The reference temperature is $T_0 = 1.6 \times 10^6$ K. The reference accretion rate is $\dot{M}_0 = 1.8 \times 10^{19}$ g s^{-1} $= 2.8 \times 10^{-7} M_\odot$ yr^{-1}.

Subsequently, we drop the primes on the dimensionless variables and show dimensionless values in the figures. The dimensionless time in the remainder of this paper is in units of the Keplerian orbital period P_0 at $r' = 1$.

3. Simulation results

We chose the following parameters for our main case: misalignment angle $\Theta = 5°$, stellar surface magnetic field $B_\star = 2$, viscosity parameter $\alpha = 0.1$, stellar angular velocity $\Omega_\star = 0.35$, initial disk radius $= 2.1$.

Fig. 1 shows two views of the accretion flow around the star through instabilities (top row) and two views of a magnetospheric (funnel) flow (bottom row). The growth of unstable perturbations at the disk-magnetosphere boundary results in penetration of the

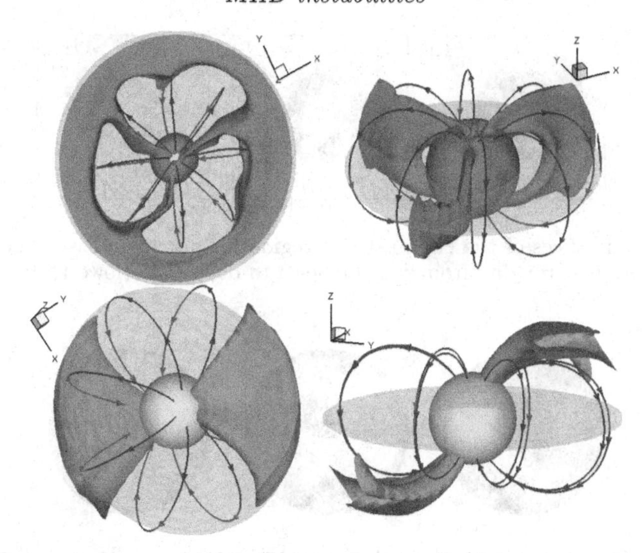

Figure 1. Geometry of the accretion flow around a star through instabilities (top row), contrasted with a traditional funnel flow (bottom row). A constant density surface is shown. The lines are magnetospheric magnetic field lines. The translucent disc denotes the equatorial plane. The star's rotational axis is in the z-direction.

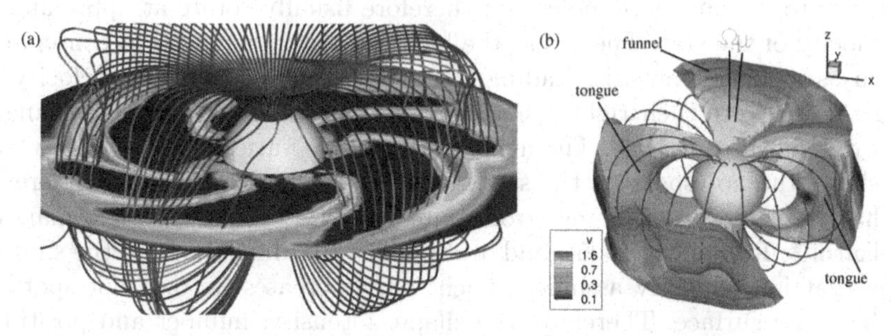

Figure 2. (a) A tongue of gas, shown by density contours in the equatorial plane, pushing aside magnetic field lines on its way to the star. Note that the "hole" in the magnetosphere is not artificially depicted – the field lines start out uniformly spaced on the star's surface, and twist aside around the tongue. (b) Cutaway view of the region around the star showing the matter velocity profile in the funnels and tongues, in a reference frame rotating with the star. A constant density surface is shown, overlaid with velocity contours.

magnetosphere by the disk matter, in the form of tongues of gas travelling through the equatorial plane. The tongues are threaded by the magnetic field lines closer to the star and turn into miniature funnel-like flows which deposit matter much closer to the star's equator than true funnel flows do.

The tongues are tall and thin, as opposed to the funnels which are flat and wide. This is because when the tongues penetrate the magnetosphere, they pry the field lines aside (Fig. 2a), since this is energetically more favourable than bending the field lines inward.

The matter velocity profile in the tongues is very similar to that in the funnels, as Fig. 2b shows. It is also seen that the funnels and tongues are not mutually exclusive. We discuss this last point in more detail in § 3.5.

Fig. 3 shows some disk-plane slices of the circumstellar region at various times, in which the tongues are visible. The density enhancements which result in the formation of the tongues can be seen at the base of the tongues. The tongues rotate around the star with the angular velocity of the gas at the inner edge of the disk. This is in contrast

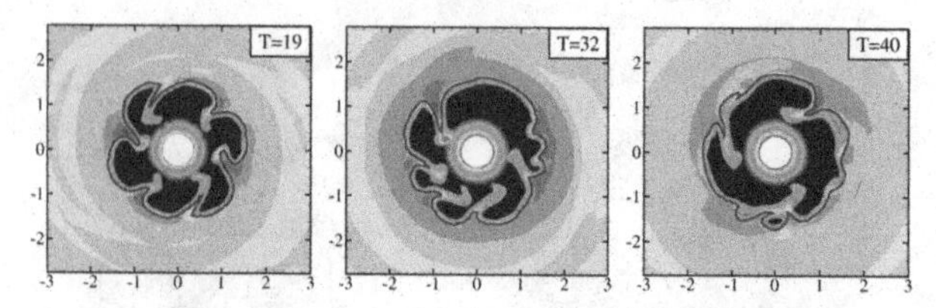

Figure 3. Disk-plane slices of the circumstellar region for our main case. The colors represent plasma density contours, ranging from red (highest) to deep blue (lowest). The solid line is the $\beta = 1$ line.

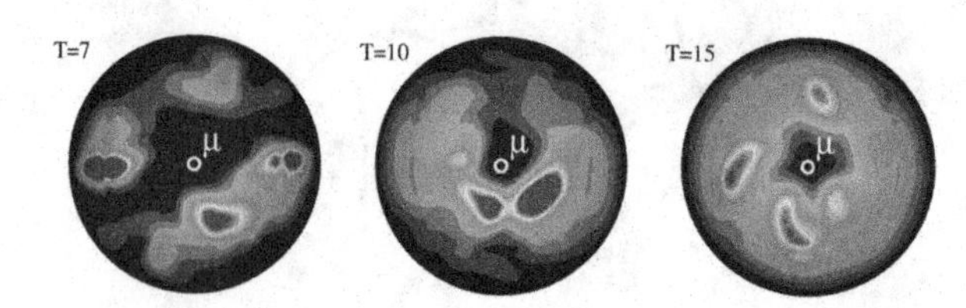

Figure 4. Hot spots on the star's surface at various times for our main case. The colors represent contours of the matter flux onto the star's surface, ranging from 0.2 (deep blue) to 3 (red).

with funnel streams, which flow towards the star's polar region from the part of the disk that is closest to the magnetic pole, and therefore usually rotate at approximately the angular velocity of the star. The gas at the base of the tongues tries to conserve angular momentum as it moves inwards, leading to an increase in its angular velocity, causing the tongues to curve to their right. The shape and number of the tongues change on the inner-disk dynamical timescale. The number of tongues varies between about 2 and 7.

Fig. 4 shows the hot spots on the star's surface for our main case at different times. We see that the spots are different from pure funnel-flow hot spots (Romanova *et al.* 2004, Kulkarni & Romanova 2005), and are significantly different from the simple polar-cap shape that is frequently assumed. Each tongue creates its own hot spots when it reaches the star's surface. Therefore, the shape, intensity, number and position of the spots change on the order of the inner-disk dynamical timescale.

In the following subsections, we investigate the dependence of the instability on different parameters, starting from the main case and varying one parameter at a time.

3.1. *Dependence on the accretion rate*

The accretion rate in our code is regulated by the viscosity coefficient ν which is proportional to the α-parameter (Shakura & Sunyaev 1973). The radial velocity of inward flow in the disk at a distance r from the star is $v_r \approx \nu/r \approx \alpha c_s h/r$, where h is the thickness of the disk and c_s is the sound speed. Thus the accretion rate through the disk is approximately proportional to α: $\dot{M} \approx 4\pi r h \rho v_r \sim \alpha$.

We performed simulations for $\alpha = 0.02, 0.03, 0.04, 0.06, 0.08, 0.1, 0.2, 0.3$. At very small α ($\leqslant 0.03$), the instability does not appear. At larger α, the instability appears, and when α is increased, the instability starts earlier and more matter accretes through it. Fig. 5 shows equatorial slices of the plasma density distribution at different α at $t = 25$. One can see that there are no tongues at $\alpha = 0.02$. The tongues are quite weak at $\alpha = 0.04$, but much stronger at larger α, when more matter comes to the inner region of the disk, and the plasma density in the inner region of the disk is higher than in the low-α cases. This shows that increased accumulation of mass at the inner edge of the disk leads to enhancement of the instability, producing tongues with higher velocities that propagate deeper into the magnetosphere of the star. We should note that in spite

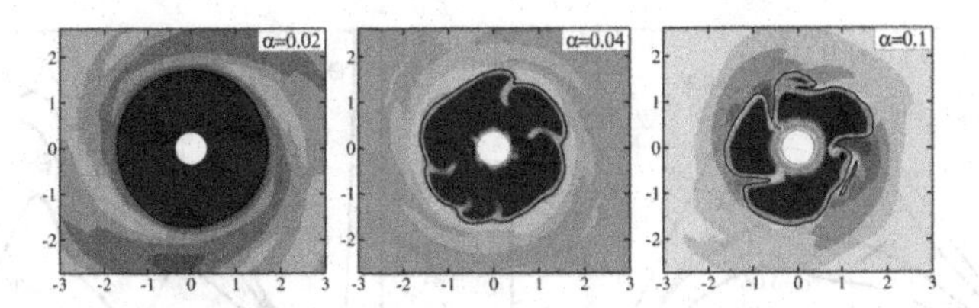

Figure 5. Plasma density distribution in the equatorial plane for different α values at $t = 25$. The colors represent plasma density contours, ranging from red (highest) to deep blue (lowest). The solid line is the $\beta = 1$ line.

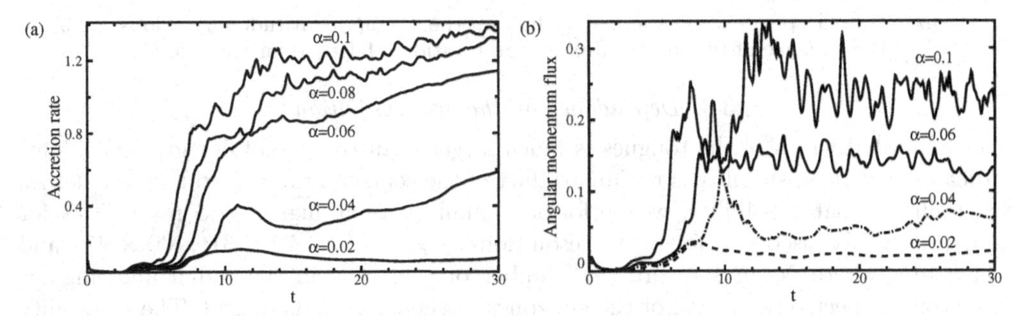

Figure 6. (a) Variation of the accretion rate onto the star's surface with time, for different values of the α-viscosity. Notice the sudden increase in the accretion rate for $\alpha \geqslant 0.04$ between $t = 5$ and $t = 10$, which occurs at the onset of the instability. (b) Variation with time of the angular momentum flux onto the star about the star's rotational axis, for different values of the α-viscosity.

of different conditions at the inner region of the disk (much higher density at larger α), the number and behaviour of the tongues is approximately the same in all cases.

The accretion rate onto the star's surface is higher during accretion through instability, as Fig. 6a shows. The increase the accretion rate with α is not merely due to increase in the amount of matter transported inwards by the accretion disk. The sudden increase in accretion rate between $t = 5$ and $t = 10$ for the unstable cases ($\alpha \geqslant 0.04$) occurs at the onset of the instability, which suggests that the instability, when present, is very efficient at transporting matter onto the star's surface. The higher accretion rate is accompanied by a higher angular momentum flux to the star (Fig. 6b).

Fig. 7a shows the dependence on the magnetic latitude θ of the matter flux onto the star's surface, for different α. For $\alpha = 0.02$ and $\alpha = 0.03$ the matter flux peaks at $\theta \approx 73°$ with a half-width of $\approx 75° - 65° = 10°$. For $\alpha = 0.06 - 0.1$, the peak is at much lower latitudes, $\theta \approx 53°$, with half-width $\approx 70° - 25° = 45°$. It is surprising to see that at the largest viscosity coefficients, $\alpha = 0.2$ and 0.3, the hot spots do not move closer to equator, but have a maximum at $\theta \approx 50°$, like for $\alpha = 0.1$.

3.2. *Dependence on the misalignment angle*

We find that the instability shuts off for $\Theta \gtrsim 30°$. The reason for this is that for large misalignment angles ($\Theta \gtrsim 30°$), the magnetic poles are closer to the disk plane. Therefore, the gravitational energy barrier that the gas in the inner disk region has to overcome in order to form funnel flows is reduced, making funnel flows energetically more favourable.

Fig. 7b shows the accretion rate onto the star's surface as a function of rotational latitude. When the accretion is through instability ($\Theta \leqslant 30°$), most of the matter accretes onto the mid-latitude ($\theta \sim 50°$) region of the star, *independent of the misalignment angle.*

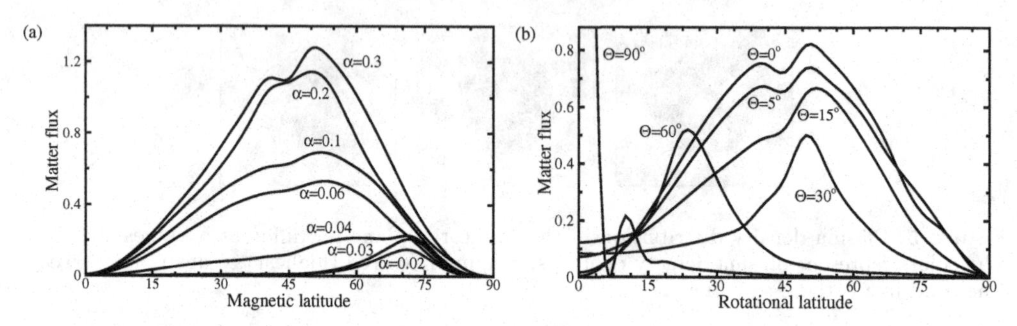

Figure 7. (a) Dependence of the matter flux on the magnetic latitude for various α. (b) Dependence of the matter flux on the rotational latitude for various Θ.

3.3. *Dependence on the grid resolution*

The azimuthal width of the tongues is much larger than the size of our grid cells. This indicates that the instability is not an artefact of the coarseness of the grid. Nevertheless, to eliminate that possibility, we performed simulations at higher grid resolutions for comparison. We used the following resolutions: $N_r \times N^2 = 72 \times 31^2, 100 \times 41^2$ and 144×61^2, where N_r and N are the number of grid cells in the radial and angular directions respectively in each of the six zones of the cubed sphere grid. The instability exists in all these cases, with similar tongue behaviour.

3.4. *Possible perturbation mechanisms*

Different sources of perturbation are expected at the disk-magnetosphere boundary in real accretion disks. There are always natural density and pressure inhomogeneities in the disk, which may act as perturbations. Also, if the magnetic and rotational axes are not aligned, there will always be some density enhancement near the disk foot-points of the funnel streams (Romanova *et al.* 2003, 2004). This would be a constant source of inhomogeneity in the disk. Another source of perturbation is associated with the magnetic field lines which are trapped inside the inner regions of the disk and are azimuthally wrapped by the disk matter. This leads to increase of magnetic energy in some parts of the disk and to partial expulsion of matter from these regions, and thus to inhomogeneous distribution of matter. This mechanism is expected to operate in real astronomical objects as well.

Concerning the role of the grid, it is unlikely, as noted above, that the discrete nature of the grid by itself leads to perturbations. But another perturbing element is the boundaries between the sectors of the cubed sphere grid. Four of these boundaries cross the disk. They produce initial density and pressure perturbations at the 5% level near the disk-magnetosphere boundary, and at even larger levels at larger distances from the star, where the grid is coarser. At later times these perturbations become less important. So at early times in the simulations, this boundary effect is the most important contributor to the perturbations. That is why we often see four tongues initially. However, at later times, we often observe anywhere between 2 and 7 tongues, which shows that there is no direct influence of these boundaries on the perturbations at later times.

3.5. *Empirical conditions for the existence of the instability*

For a given mass, radius, magnetic field strength and misalignment angle Θ of the star, the presence of the instability is correlated with the star's rotation period P and the accretion rate $\dot{M}$ at the star's surface. Fig. 8a shows the regimes of stable and unstable accretion in the $\dot{M} - P$ plane for $\Theta = 5°$, obtained empirically from the simulations.

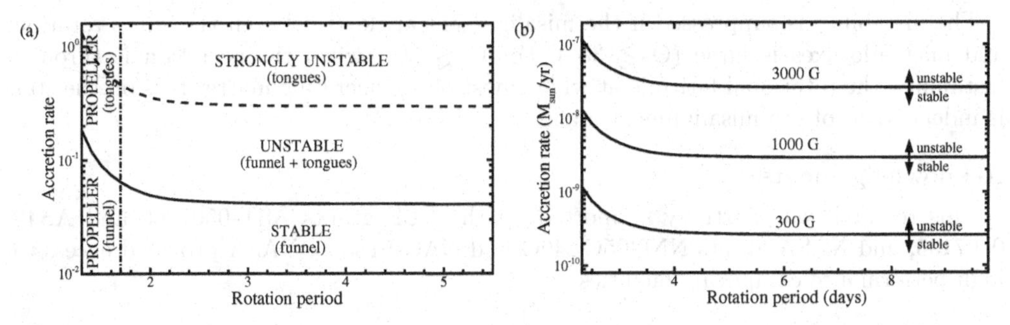

Figure 8. Nature of the accretion flow for different stellar rotation periods and accretion rates, (a) in dimensionless values, and (b) for a protostar with $M = 0.8M_\odot$, $R = 2R_\odot$. The accretion rate is calculated at the star's surface.

As noted in § 3.1, the presence of the instability is accompanied by high accretion rates. Also, unless the star rotates very rapidly (i.e., is close to the propeller stage), the presence of the instability is very weakly correlated with the rotation rate. The transition region between the regimes of pure funnel accretion and accretion solely through instabilities is fairly broad. In this region, simultaneous accretion through funnel streams and instability tongues is observed.

Fig. 8 is in dimensionless units, and therefore, as noted in § 2, can be used for a wide variety of physical situations with appropriately chosen reference values. The following are the reference values for an accretion disk around a protostar of mass M, radius R and surface magnetic field B:

$$P_0 = 1.8 \text{ days} \left(\frac{R}{2R_\odot}\right)^{3/2} \left(\frac{M}{0.8M\odot}\right)^{-1/2} \tag{3.1}$$

$$\dot{M}_0 = 7 \times 10^{-8} M_\odot \text{ yr}^{-1} \left(\frac{B}{10^3 \text{ G}}\right)^2 \left(\frac{R}{2R_\odot}\right)^{5/2} \left(\frac{M}{0.8M\odot}\right)^{-1/2} \tag{3.2}$$

Fig. 8b shows the boundary between the stable and unstable accretion regimes for protostars, for the fiducial values of mass and radius shown in equations 3.1 and 3.2, and different surface magnetic field strengths.

4. Summary

Accretion through instabilities at the disk-magnetosphere interface occurs for a wide range of physical parameters. It results in tall, thin tongues of gas penetrating the magnetosphere and travelling in the equatorial plane. The tongues are very transient, and grow and rotate around the star on inner-disk dynamical timescales. The number of tongues at a given time is of the order of a few. Near the star, the tongues are threaded by the magnetic field lines, and form miniature funnel-like flows, which deposit matter much closer to the star's equator than true funnel flows do. Each tongue produces its own hot spots on the star's surface, and as a result, the hot spots also change on inner-disk dynamical timescales. The resulting light curves often show no clear periodicity. Sometimes, when a certain number of tongues dominate, we see quasi-periodic oscillations in the lightcurves.

The instability is associated with high accretion rates, and coexists with funnel flows for quite a broad range of accretion rates. Protostars with the fiducial mass, radius, surface magnetic fields and accretion rates of $0.8M_\odot$, $2R_\odot$, 10^3 G and $10^{-8} M_\odot$ yr^{-1} respectively are expected to be in this transition region.

The instability is suppressed if the misalignment angle Θ between the star's rotation and magnetic axes is large ($\Theta \gtrsim 30°$). For $\Theta \lesssim 30°$, when the accretion is through instability, the rotational latitude at which most of the accreting matter falls on the star is independent of the misalignment angle.

Acknowledgements

This research was partially supported by the NSF grants AST-0507760 and AST-0607135, and NASA grants NNG05GL49G and NAG-513060. NASA provided access to high performance computing facilities.

References

Arons, J. & Lea, S. M. 1976, *ApJ* 207, 914
Elsner, R. F. & Lamb, F. K. 1977, *ApJ* 215, 897
Ghosh, P. & Lamb, F. K. 1978, *ApJ* 223, L83
Ghosh, P. & Lamb, F. K. 1979, *ApJ* 232, 259
Koldoba, A. V., Romanova, M. M., Ustyugova, G. V., & Lovelace, R. V. E. 2002, *ApJ* 576, L53
Kulkarni, A. K. & Romanova, M. M. 2005, *ApJ* 633, 349
Li, L.-X. & Narayan, R. 2004, *ApJ* 601, 414
Novikov, I., & Thorne, K. S. 1973, in: B. DeWitt and C. DeWitt (eds.), *Black Holes* (New York: Gordon and Breach), p. 409
Rastätter, L. & Schindler, K. 1999, *ApJ* 524, 361
Romanova, M. M., Ustyugova, G. V., Koldoba, A. V., & Lovelace, R.V.E. 2002, *ApJ*, 578, 420
Romanova, M. M., Ustyugova, G. V., Koldoba, A. V., Wick, J. V., & Lovelace, R. V. E. 2003, *ApJ* 595, 1009
Romanova, M. M., Ustyugova, G. V., Koldoba, A. V., & Lovelace, R. V. E. 2004, *ApJ* 610, 920
Shakura, N. I., & Sunyaev, R. A. 1973, *A&A* 24, 337
Spruit, H. C. & Taam, R. E. 1990, *A&A* 229, 475

Discussion

MATT: How does the velocity of matter in the tongues compare with that in the funnels?

KULKARNI: They are of the same order.

JOHNS-KRULL: What is the filling factor of the hot spots during accretion through instability?

KULKARNI: The filling factor depends on the temperature that is taken to define the boundary of the hot spot. It is larger than for funnel flows, though.

HARRIES: Is the magnetic field on the star's surface weaker at the place where the tongues reach the star?

KULKARNI: No, because in the inner magnetosphere, the tongues are threaded by the stronger magnetic field and travel along the field lines.

FENDT: Is it possible that the instability seen in the simulations could be a result of diffusivity?

KULKARNI: That is unlikely, because our model assumes ideal MHD. The numerical diffusivity is relatively small, and decreases when the grid resolution is improved. Also, the instability progresses at much shorter timescales than the diffusive timescale.

Star-Disk Interaction in Young Stars
Proceedings IAU Symposium No. 243, 2007
J. Bouvier & I. Appenzeller, eds.

doi:10.1017/S1743921307009659

The nature of stellar winds in the star-disk interaction

Sean Matt[1] and Ralph E. Pudritz[2]

[1]Dept. of Astronomy, U. of Virginia PO Box 400325, Charlottesville, VA 22904-4325, USA
email: spm5x@virginia.edu

[2]Physics and Astronomy Dept., McMaster University, Hamilton, ON, L8S 4M1, Canada
email: pudritz@physics.mcmaster.ca

Abstract. Stellar winds may be important for angular momentum transport from accreting T Tauri stars, but the nature of these winds is still not well-constrained. We present some simulation results for hypothetical, hot ($\sim 10^6$ K) coronal winds from T Tauri stars, and we calculate the expected emission properties. For the high mass loss rates required to solve the angular momentum problem, we find that the radiative losses will be much greater than can be powered by the accretion process. We place an upper limit to the mass loss rate from accretion-powered coronal winds of $\sim 10^{-11} M_\odot$ yr^{-1}. We conclude that accretion powered stellar winds are still a promising scenario for solving the stellar angular momentum problem, but the winds must be cool (e.g., 10^4 K) and thus are not driven by thermal pressure.

Keywords. MHD, stars: coronae, stars: magnetic fields, stars: pre–main-sequence, stars: rotation, stars: winds, outflows.

---

1. Introduction

Observations (e.g., Herbst *et al.* 2007) reveal that a large fraction of accreting T Tauri stars (CTTSs) spin slowly, that is at $\sim 10\%$ of breakup speed. This is surprising because the accretion of disk material adds angular momentum to the star (e.g., Matt & Pudritz 2007). One promising scenario to explain how the slowly spinning stars rid themselves of this accreted angular momentum, proposed by Hartmann & Stauffer (1989), is that a stellar wind carries it off. For this to work, the mass outflow rate should be approximately proportional to the accretion rate. Depending on the stellar magnetic field strength (among other things), in order to solve the stellar angular momentum problem, the wind outflow rate needs to be of the order of 10% of the accretion rate (Matt & Pudritz 2005).

Since the "typical" mass accretion rate observed in the CTTSs is $\dot{M}_{\rm a} \sim 10^{-8} M_\odot$ yr^{-1} (Johns-Krull & Gafford 2002), this means the stellar wind should have a mass outflow rate of $\dot{M}_{\rm w} \sim 10^{-9} M_\odot$ yr^{-1}. A wind this massive requires a lot of power to accelerate it, and Matt & Pudritz (2005) suggested that a fraction of the potential energy liberated by the accretion process goes into driving the wind. In the case of a coronal wind (i.e., $\sim 10^6$ K, thermally driven), for example, this requires $\sim 10\%$ of the accretion power (Matt & Pudritz 2005). There is some observational evidence for accretion-powered stellar winds in these systems (Edwards *et al.* 2006; Kwan *et al.* 2007).

But what is the nature of T Tauri stellar winds? How massive are they, and what drives them? The mass outflow rates of stellar winds is very poorly constrained observationally (e.g., Dupree *et al.* 2005). This is basically due to the extreme difficulty in disentangling the signatures of a stellar wind from signatures of a wind from the inner edge of a disk and a host of other energetic phenomena exhibited by CTTSs. The wind driving mechanism is also not constrained and is the primary focus of this paper.

2. The T Tauri coronal wind hypothesis

T Tauri stars are magnetically active and possess hot, energetic corona (for a review, see Feigelson & Montmerle 1999). They are 4–5 orders of magnitude more luminous in X-rays than the sun. Thus, it stands to reason that they drive solar-like coronal winds, but more powerful. In this case, the wind is primarily thermal pressure-driven, and the wind temperature needs to be $\sim 10^6$ K for the pressure force to overcome gravity. As a first step, we make the hypothesis here that some of the accretion power is transferred to heat in the stellar corona, and thus drives a coronal wind.

There is only one calculation in the literature (that we are aware of) that constrains the mass outflow rate of coronal winds from CTTSs. Specifically, Bisnovatyi-Kogan & Lamzin (1977) calculated the X-ray emission from coronal winds. From these calculations, Decampli (1981) concluded that, in order for the wind emission to be consistent with the observed X-ray luminosities, the outflow rate of a T Tauri coronal wind must be less than $\sim 10^{-9} M_\odot$ yr^{-1}. As discussed above, a wind this massive may still be important for angular momentum transport, and thus we proceed.

3. Coronal wind simulations

To calculate realistic wind solutions, we carried out 2.5D (axisymmetric) ideal magnetohydrodynamic (MHD) simulations of coronal winds. For simplicity, we did not include the accretion disk. We employ the numerical code and method described by Matt & Balick (2004). This allows us to obtain steady-state wind solutions for a Parker-like coronal wind (Parker 1958), as modified by the presence of stellar rotation and a rotation-axis-aligned dipole magnetic field. We assume a polytropic equation of state ($P \propto \rho^\gamma$), with no radiative cooling. The fiducial parameters are given in Table 1, adopted to represent values for a "typical" CTTS.

Table 1. Fiducial stellar wind parameters

Parameter	Value
M_*	$0.5\ M_\odot$
R_*	$2.0\ R_\odot$
B_* (dipole)	200 G
f	0.1
$\dot{M}_\mathrm{w}$	$1.9 \times 10^{-9} M_\odot$ yr^{-1}
T_c	1.3×10^6 K
γ	1.40

In Table 1, M_* and R_* are the stellar mass and radius; B_* is the magnetic field strength of the dipole magnetic field at the surface and equator of the star; f is the spin rate of the star, expressed as a fraction of the breakup rate; T_c is the temperature at the base of the corona; and γ is the polytropic index.

Figure 1 illustrates the steady-state wind solution for the fiducial case. We find that this wind carries away enough angular momentum to counteract the spin up torque from an accretion rate of $\dot{M}_\mathrm{a} \approx 5 \times 10^{-9} M_\odot$ yr^{-1}. We also carried out a parameter study (Matt & Pudritz 2007, in preparation), which generally validates the idea that a stellar wind can indeed remove the accreted angular momentum in CTTSs.

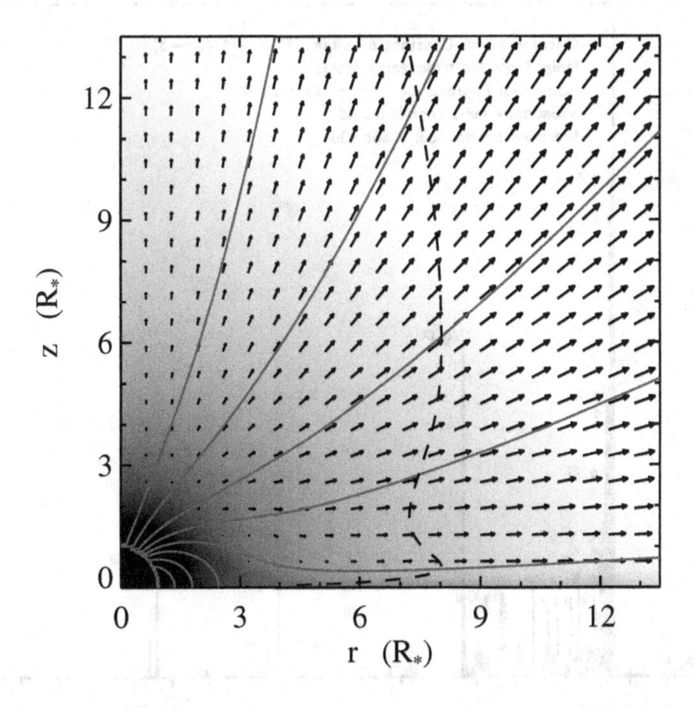

Figure 1. Greyscale of log density, velocity vectors, and magnetic field lines illustrate the structure of the steady-state wind solution for our fiducial case. The dashed line represents the Alfvén surface, where the wind speed equals the local Alfvén speed. The rotation axis is vertical, along the left side of the plot.

4. Emission properties of coronal wind

The simulation results of the previous section provide detailed solutions for the density and temperature in coronal stellar winds. Although the simulations did not include radiative cooling effects, it is instructive to examine the emission properties expected from these winds, ex post facto. For this, we employ the CHIANTI line database and IDL software (Dere *et al.* 1997; Landi *et al.* 2006), which allows us to calculate spectra and total radiative cooling rates in the wind.

The CHIANTI package assumes, among other things, that the ionization and excitation levels in the plasma are in a steady-state; all lines are optically thin; the plasma is in coronal equilibrium, so that the ionization state is in LTE. These assumptions are appropriate for the purposes of this work, and we also adopt cosmic abundances for the gas.

4.1. *Illustrative synthetic spectrum*

For illustrative purposes, Figure 2 shows a spectrum, computed by CHIANTI, of an isothermal plasma with a temperature of 10^6 K. It is clear that the cooling is dominated by line emission. In this case, the three strongest emission lines (of Fe IX 171.1 Å, Fe X 174.5 Å, and Mg IX 368.1 Å) account for approximately 20% of the total luminosity. Furthermore, only about 1% of the total energy is emitted shortward of 30 angstroms (i.e., in X-rays), and the vast majority of the emission is in the extreme UV.

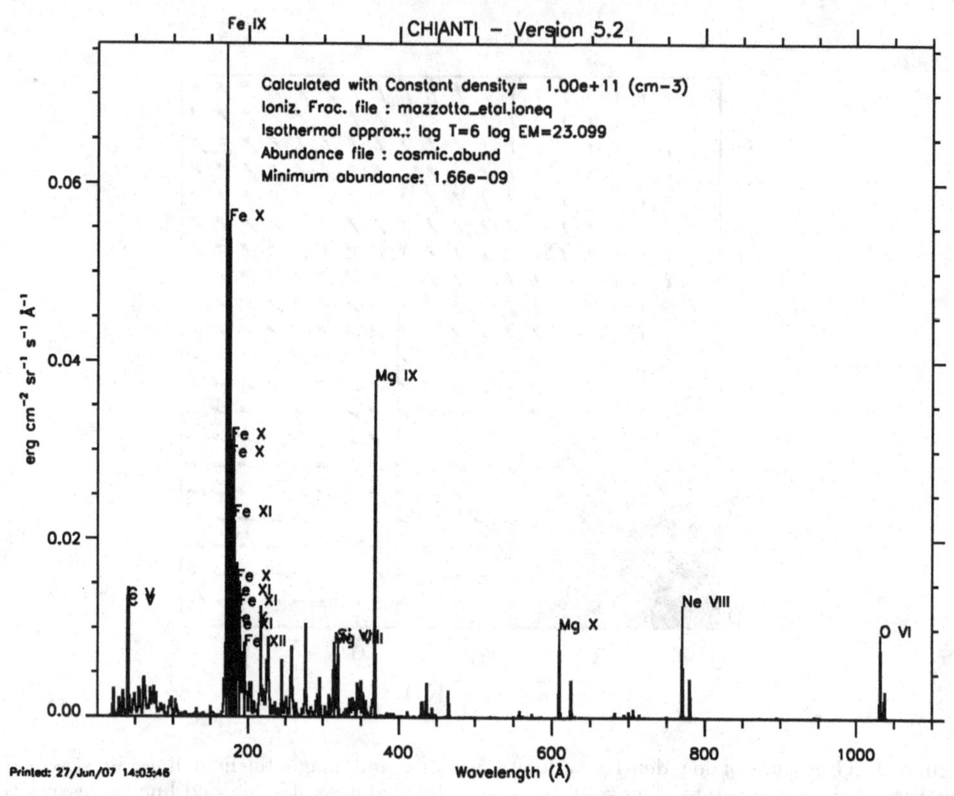

Figure 2. Synthetic spectrum of isothermal, 10^6 K, optically thin coronal plasma. Flux units are arbitrary. The figure is generated by CHIANTI software.

4.2. *Total radiative losses*

CHIANTI also provides a tool to calculate the total cooling rate (i.e., radiated luminosity in erg s^{-1}; which is essentially an integration of the emission spectrum over wavelength) for any given coronal density, temperature, and emitting volume. With this, we calculate the cooling in each computational gridcell of our simulations, and sum over all gridcells, to obtain the total luminosity of the simulated wind solution. For the fiducial case, the total wind luminosity is a few times 10^{34} erg s^{-1}. Since optically thin emission is proportional to density squared, and since the mass outflow rate in the wind is approximately proportional to density, we express the luminosity of the wind as

$$L_{\rm w} \sim 10^{34} \;\; {\rm erg\ s^{-1}} \left(\frac{\dot{M}_{\rm w}}{10^{-9} M_\odot \; {\rm yr}^{-1}} \right)^2 . \tag{4.1}$$

As suggested by the example spectrum (Fig. 2), if $\sim 1\%$ of this emission is emitted in X-rays, the X-ray luminosity of the wind is $\sim 10^{32}$ erg s^{-1}. This is significantly higher than the typically observed X-ray luminosity of CTTSs of $\sim 10^{30}$ erg s^{-1} (Feigelson & Montmerle 1999). Of course, we have calculated the total cooling rate, which is not exactly the observed luminosity. Consider that approximately half of this radiation will be blocked by the star, and there will likely be significant absorption of these soft X-rays. Still, it does not seem avoidable that the predicted X-ray luminosity from the fiducial coronal wind solution is much higher than typically observed.

4.3. *Accretion power*

More importantly, we must consider the energy budget of the wind. The total cooling rate, $L_{\rm w}$, of the fiducial wind is two orders of magnitude larger than the kinetic energy in the wind ($0.5\dot{M}_{\rm w}v_\infty^2$, where v_∞^2 is the wind speed)—this is approximately equivalent to saying that the cooling time is two orders of magnitude shorter than the wind acceleration time. Thus, it takes a lot more energy to keep this plasma hot (while it radiates) than it does to accelerate the matter away from the star.

In the accretion-powered stellar wind scenario, the energy in the wind somehow derives from the gravitational potential energy released by accreting gas ($\sim GM_*\dot{M}_{\rm a}/R_*$). This accretion power, assuming the fiducial stellar mass and radius, can be expressed approximately as

$$L_{\rm a} \sim 10^{32}\ \ {\rm erg\ s^{-1}}\ \left(\frac{\dot{M}_{\rm a}}{10^{-8}M_\odot\ {\rm yr}^{-1}}\right). \tag{4.2}$$

As discussed in § 1, in order for stellar winds to solve the angular momentum problem, torque balance requires $\dot{M}_{\rm w}/\dot{M}_{\rm a} \sim 0.1$. Thus, if we fix this ratio of mass flow rates, it is clear from equations 4.1 and 4.2 that there is not enough accretion energy to keep coronal winds hot, in the fiducial case.

4.4. *An upper limit on T Tauri coronal winds*

If we fix the ratio $\dot{M}_{\rm w}/\dot{M}_{\rm a} \sim 0.1$, it is evident from equations 4.1 and 4.2 that there will be enough accretion power to drive a coronal wind when the wind outflow rate is

$$\dot{M}_{\rm w} \lesssim 10^{-11}\ \ M_\odot\, {\rm yr}^{-1}. \tag{4.3}$$

Thus, in principle, accretion-powered coronal winds can remove the accreted angular momentum for $\dot{M}_{\rm a} \sim 10^{-10}M_\odot\ {\rm yr}^{-1}$. However, for accretion rates this low, the spin up torque from accretion is so small that the time for the star to spin up from this torque is comparable to the entire pre-main-sequence lifetime (e.g., Matt & Pudritz 2007). In other words, for these low accretion rates, there is no angular momentum problem. The logical conclusion is that, in order for accretion-powered stellar winds to solve the angular momentum problem, the winds cannot be as hot as we have considered here.

5. On the validity of our simulated wind solutions

We showed in § 4 that the expected emission properties of our fiducial, coronal winds effectively rules them out. In other words, our assumption in this paper that the wind is driven by thermal-pressure is not realistic. However, it is important to note that the angular momentum carried in the wind does not depend on what drives the wind. Instead, the angular momentum outflow rate depends only on B_*, the rotation rate, $\dot{M}_{\rm w}$, R_*, and the wind velocity. As long as "something" accelerates the wind to speeds similar to what we see in our simulations (of the order of the escape speed), the torque we calculate is approximately correct.

For example, if the wind is cold and driven by Alfvén waves, the driving force can be parameterized as being proportional to $-\nabla\xi$ (where ξ is the wave energy density; Decampli 1981). This has the same functional form as the thermal-pressure force ($-\nabla P$) used in our simulations, so there is some form of ξ that will result in a wind solution with exactly the same density and kinematics as our simulations (but a different temperature).

Thus, while the thermodynamic properties of our simulations have been invalidated, the conclusion that stellar winds are capable of carrying off accreted angular momentum is not affected.

6. Conclusion

Based on the emission properties of $\sim 10^6$ K coronal plasmas, we rule out hot coronal winds as a likely candidate for accretion-powered stellar winds. The coronal wind hypothesis fails. Instead, for mass loss rates comparable to our fiducial value of $10^{-9} M_\odot$ yr^{-1}, the winds must be as cool as $\sim 10^4$ K, where radiative cooling becomes much less efficient than for a coronal plasma. At temperatures this low, the pressure force cannot overcome the gravity of the star, and accretion-powered winds are thus not driven by thermal pressure.

To date, possibly the best observational evidence for accretion-powered stellar winds from CTTSs comes from measurements of blue-shifted absorption features in the He I emission line at 10830 Å (e.g., Edwards *et al.* 2003, 2006; Dupree *et al.* 2005). Furthermore, radiative transfer modeling by Kurosawa *et al.* (2006) suggests that a stellar wind may contribute significantly to the Hα line profile. At densities where collisions between particles are important, both He I and H start to become substantially ionized at a temperature of a few times 10^4 K. If the wind is much hotter than this, it may be difficult to explain the prominence of He I and H I features in observed spectra (see also Johns-Krull & Herczeg 2007). Thus these works also support the conclusion that the winds are much cooler than a coronal plasma.

Accretion-powered stellar winds remain a promising scenario for solving the stellar angular momentum problem. But, the question remains, what is the nature of these winds? What drives them? Possible scenarios include Alfvén wave driving (Decampli 1981), episodic magnetospheric inflation (Goodson *et al.* 1999; Matt *et al.* 2003), and reconnection X-winds (Ferreira *et al.* 2000, 2006).

Acknowledgements

We thank the organizers for a stimulating conference and for the opportunity to present this work. Gibor Basri deserves credit for issuing a friendly challenge to our coronal wind hypothesis, six months prior to this meeting. He wins the challenge, as it turns out, since he correctly surmised there would be too much X-ray emission. Thanks also to Jürgen Schmitt for making us aware of the CHIANTI software and database and for discussion about calculating radiative losses.

CHIANTI is a collaborative project involving the NRL (USA), RAL (UK), and the following Univerisities: College London (UK), Cambridge (UK), George Mason (USA), and Florence (Italy). The research of SM was supported by the University of Virginia through a Levinson/VITA Fellowship partially funded by The Frank Levinson Family Foundation through the Peninsula Community Foundation. REP is supported by a grant from NSERC.

References

Bisnovatyi-Kogan, G. S. & Lamzin, S. A. 1977, Soviet Astronomy, 21, 720

Decampli, W. M. 1981, *ApJ*, 244, 124

Dere, K. P., Landi, E., Mason, H. E., Monsignori Fossi, B. C., & Young, P. R. 1997, *A&AS*, 125, 149

Dupree, A. K., Brickhouse, N. S., Smith, G. H., & Strader, J. 2005, *ApJ*, 625, L131

Edwards, S., Fischer, W., Hillenbrand, L., & Kwan, J. 2006, *ApJ*, 646, 319

Edwards, S., Fischer, W., Kwan, J., Hillenbrand, L., & Dupree, A. K. 2003, *ApJ*, 599, L41

Feigelson, E. D. & Montmerle, T. 1999, *ARA&A*, 37, 363

Ferreira, J., Dougados, C., & Cabrit, S. 2006, *A&A*, 453, 785

Ferreira, J., Pelletier, G., & Appl, S. 2000, *MNRAS*, 312, 387

Goodson, A. P., Böhm, K., & Winglee, R. M. 1999, *ApJ*, 524, 142

Hartmann, L. & Stauffer, J. R. 1989, *AJ*, 97, 873

Herbst, W., Eislöffel, J., Mundt, R., & Scholz, A. 2007, in Protostars and Planets V, ed. B. Reipurth, D. Jewitt, & K. Keil, 297–311

Johns-Krull, C. M. & Gafford, A. D. 2002, *ApJ*, 573, 685

Johns-Krull, C. M. & Herczeg, G. J. 2007, *ApJ*, 655, 345

Kurosawa, R., Harries, T. J., & Symington, N. H. 2006, *MNRAS*, 370, 580

Kwan, J., Edwards, S., & Fischer, W. 2007, *ApJ*, 657, 897

Landi, E., Del Zanna, G., Young, P. R., Dere, K. P., Mason, H. E., & Landini, M. 2006, *ApJS*, 162, 261

Matt, S. & Balick, B. 2004, *ApJ*, 615, 921

Matt, S. & Pudritz, R. E. 2005, *MNRAS*, 356, 167

Matt, S. & Pudritz, R. E. 2007, to appear in proceedings of the 14th Cambridge Workshop on Cool Stars, Stellar Systems, and the Sun, astro-ph/0701648

Matt, S., Winglee, R., & Böhm, K.-H. 2003, *MNRAS*, 345, 660

Parker, E. N. 1958, *ApJ*, 128, 664

Star-disk Interaction in Young Stars
Proceedings IAU Symposium No. 243, 2007
J. Bouvier & I. Appenzeller, eds.

doi:10.1017/S1743921307009660

Large scale magnetic fields in discs: jets and reconnection X-winds

Jonathan Ferreira, Nicolas Bessolaz, Claudio Zanni, and Céline Combet

Laboratoire d'Astrophysique de Grenoble, F-38041 Grenoble, France
email: Jonathan.Ferreira@obs.ujf-grenoble.fr

Abstract. In this contribution we first briefly review our current knowledge on the physics of accretion discs driving self-confined jets. It will be shown that a large scale magnetic field is expected to thread the innermost disc regions, giving rise to a transition from an outer standard accretion disc to an inner jet emitting disc. We then report new progresses on the theory of star-disc interaction, allowing to explain the formation of accretion funnel flows with stellar dipole fields consistent with observational constraints. Such a connection is now not only probed by modern observations but it is also requested for spinning down protostars, which are known to be both actively accreting and contracting. This spin down most probably relies on the angular momentum removal by ejection. Two such scenarios will be addressed here, namely "accretion-powered stellar winds" (Matt & Pudritz 2005) and "Reconnection X-winds" (Ferreira, Pelletier & Appl 2000). The latter can slow down a protostar on time scales shorter or comparable to the embedded phase. It will be shown that these two scenarios are not incompatible and that transitions from one to another may even occur as they mainly depend on the stellar dynamo.

Keywords. Accretion discs, magnetohydrodynamics, stars: pre-main sequence, stars: magnetic fields, stars: rotation, ISM: jets and outflows.

---

1. Introduction

Actively accreting classical T Tauri stars (CTTS) often display supersonic collimated jets on scales of a few 10-100 AU in low excitation optical forbidden lines. Molecular outflows observed in younger Class 0 and I sources may be powered by an inner unobserved optical jet. These jet signatures are correlated with the infrared excess and accretion rate of the circumstellar disc (Cabrit *et al.* 1990; Hartigan *et al.* 1995). It is therefore widely believed that the accretion process is essential to the production of jets. For quite a while, the precise physical connection remained a matter of debate: do the jets emanate from the star, the circumstellar disc or the magnetospheric star-disc interaction? This issue seems now to be almost settled: while all these wind components are probably present, magnetized disc winds would be responsible for most of the mass loss (see Ferreira *et al.* 2006a and S. Cabrit's contribution).

The basic and universal accretion-ejection mechanism would then be the following: an accretion disc around a central object can – under certain conditions and whatever the nature of this object (star or compact object) – drive jets through the action of large scale magnetic fields. These fields would tap the mechanical energy released by mass accretion within the disc and transfer it to an ejected fraction. The smaller the fraction, the larger the final jet velocity. One thing that must be understood is how the presence of these jets modifies the nature of the underlying accretion flow. Many papers in the literature actually assume that the accretion disc resembles a standard accretion disc (hereafter SAD), as first described by Shakura & Sunyaev (1973). Thus, although it was soon recognized that ejection and accretion were tightly related (Blandford & Payne 1982; Konigl 1989), a truly self-consistent model appeared only lately (Ferreira & Pelletier 1995; Ferreira 1997; Casse & Ferreira 2000; Ferreira & Casse 2004). To date, this is the only published magneto-hydrodynamic (MHD) model that describes in a self-consistent way the physics of an accretion disc threaded by a large scale magnetic field and giving rise to self-collimated jets. We term such a disc a Jet Emitting Disc, hereafter a JED. This model is unique in the sense that it provides both the physical conditions within the disc required to steadily launch jets and the distributions of all quantities in space (although the self-similar assumption used introduces some unavoidable biases). Most of its results have been confirmed by numerical MHD simulations using either the VAC code (Casse & Keppens 2002, 2004) or the FLASH code (Zanni *et al.* 2007).

In this contribution, we present some results on how the presence of a large scale vertical magnetic field deeply alters the disc dynamics (Section 2). Section 3 addresses new results on the star-disc interaction obtained with two MHD codes, VAC and PLUTO. It will be argued that such an interaction probably leads to a systematic spin up of the protostar. Section 4 is then devoted to the only model so far that allows a magnetic brake down of a protostar during its embedded phase.

2. Magnetic fields in accretion discs

The necessary condition for launching a self-collimated jet from a Keplerian accretion disc is the presence of a large scale vertical magnetic field close to equipartition (Ferreira & Pelletier 1995), namely

$$B_z \simeq 0.2 \left(\frac{M}{M_\odot}\right)^{1/4} \left(\frac{\dot{M}_a}{10^{-7} M_\odot/yr}\right)^{1/2} \left(\frac{r}{1 \text{ AU}}\right)^{-5/4+\xi/2} \text{G}, \tag{2.1}$$

where ξ is the disc ejection efficiency as measured by a varying disc accretion rate, namely $\dot{M}_a \propto r^\xi$. The value of this magnetic field is far smaller than the one estimated from the interstellar magnetic field assuming either ideal MHD or $B \propto n^{1/2}$ (Heiles *et al.* 1993; Basu & Mouschovias 1994). This implies some decoupling between the infalling/accreting material and the magnetic field in order to get rid off this field. This issue is still under debate. The question is therefore whether accretion discs can build up their own large scale magnetic field (dynamo) or if they can drag in the interstellar magnetic field? Although no large scale fields have been provided by a self-consistent disc dynamo, this scenario cannot be excluded. However the latter scenario (advection) seems a bit more natural (see F. Shu's contribution).

A picture, that can be applied to accretion discs around both young stars and compact objects, is now gradually emerging. A large scale magnetic field is thought to be dragged in by the accretion flow and concentrated in the innermost disc regions. This is consistent with the SAD theory. As a result of the interplay between advection (due to accretion) and turbulent diffusion, the large scale magnetic field scales as $B_z \propto r^{-\mathcal{R}_m}$, where $\mathcal{R}_m$ of order unity is the effective magnetic Reynolds number. As a consequence, one gets a disc magnetization $\mu = B_z^2/\mu_o P$ where P is the gas pressure which naturally increases towards the inner regions (Ferreira *et al.* 2006b). Such a field triggers the magneto-rotational instability (Balbus 2003) in the outer disc regions, producing thereby a standard accretion disc with no ejection (note that a thermally driven or photo-evaporated disc wind is of course clearly possible). When the field reaches equipartition, the accretion flow switches from a SAD to a JED, the latter driving self-confined jets. The physics of this inner disc is thus no longer governed by the radial transport of angular momentum.

This picture puts a strong emphasis on the strength of the disc large scale magnetic field (B_z). Probing inner disc regions where $\mu \sim 1$ (fields given by Eq. 1) may appear too demanding as magnetic fields are very difficult to measure. However, Donati *et al.* (2005) did this *tour de force* using the spectro-polarimeter ESPadOnS and found a ~kG field at 0.05 AU around FU Ori (a field actually larger than equipartition!). Moreover, T Tauri discs may not all reach equipartition fields if these mostly depend on interstellar field advection. This is indeed suggested by the striking results obtained by Ménard & Duchêne (2004). Using a sample of CTTS, these authors found that CTTS are oriented randomly with respect to the local interstellar field. However, sources (discs) with strong outflows are mostly perpendicular to the field (i.e. jets are aligned to it, Strom *et al.* 1986), whereas sources with no jet detection are parallel. That could be a hint that, only in the former case, field dragging leads to the presence of inner JEDs.

An indirect way to probe this region is to look at jet kinematic properties and in particular at the rotation. Current observations clearly favor self-confined jets launched from some radial extension in the disc (say from 0.1 to 0.5-2 AUs, see Ferreira *et al.* 2006a for more details). However, cold models with a low ejection efficiency $\xi \sim 0.01$ hence a magnetic lever arm parameter $\lambda = 1 + 1/2\xi \sim 50$ are excluded. Only "warm" solutions with $\xi \sim 0.1$ ($\lambda \sim 10$) are fully compatible with current observations (mass flux, velocities, collimation). Such models require heat input at the upper disc surface layers in order to allow more mass to be loaded onto the field lines. The origin of this heat deposition remains an open question. It cannot be solely due to illumination by stellar UV and X-ray radiation (Garcia *et al.* to be submitted). Alternatively, the turbulent processes responsible for the required magnetic diffusivity inside the disc might also give rise to a turbulent vertical heat flux, leading to dissipation at the disc surface layers. It is interesting to note that, in current MHD simulations of the magneto-rotational instability, a magnetically active "corona" is quickly established (Stone *et al.* 1996; Miller & Stone 2000). Although no 3D simulation has been run with open magnetic field lines, this result is rather promising. Indeed, it might be an intrinsic property of the MHD turbulence in accretion discs, regardless of the launching of jets (see also arguments developed by Kwan 1997 and Glassgold *et al.* 2004).

From the observational point of view, JED and SAD emission properties are quite different. While in a SAD all the released accretion power is radiated away at the disc surfaces, in a JED this power is feeding the two jets. As a consequence, only a small fraction (of order h/r) of the available power is put into the disc luminosity (Ferreira & Pelletier 1995). This translates right away into a lack of disc emission from the innermost ejecting parts: the spectral energy distribution would thus appear flatter than the usual $-4/3$ scaling. But more interestingly, the disc is much less dense than a corresponding SAD at the same radius (with the same $\dot{M}_a$, Combet & Ferreira, submitted). This is a straightforward consequence of a much larger accretion velocity due to the dominant jet torque. This could lead to optically thin parts in the JED but, in any case, to a sharp decrease of both the disc surface density $\Sigma(r)$ and scale height $h(r)$ at the SAD-JED transition radius r_J. This radius is unknown as it depends on the magnetic flux Φ available in the disc (and may vary from one object to another as discussed above), but may lead to observational investigations.

This last property is very interesting since Masset *et al.* (2006) have shown that a transition of this kind would act as a trap for low mass protoplanetary embryo ($M < 15\ M_{\oplus}$). Indeed, Type I inwards migration is due to the differential Lindblad torque arising from the planetesimal interaction with the viscous disc. But this negative torque is strongly reduced at the transition radius and balanced by the positive corotation torque (which is due to the exchange of angular momentum between the planetesimal and trapped disc material in its vicinity). Thus, these planetesimals would be halted at r_J which may be as large as 1 AU, long before the disc truncation radius due to the star-disc interaction.

3. The star-disc interaction

Once they become visible in the optical, T Tauri stars exhibit rotational periods of the order of 10 days, which is much smaller than expected (Bouvier *et al.* 1997; Rebull *et al.* 2002). This implies a very efficient mechanism of angular momentum removal from the star during its embedded phase. Moreover, a T Tauri star seems to evolve with an almost constant rotational period although it undergoes some contraction and is still actively accreting disc material for roughly a million years. This is a major issue in star formation, unsolved yet, but one solution to this paradox is the star-disc interaction.

3.1. *The Gosh & Lamb picture*

Angular resolution is not yet sufficient to directly image this region (of size 0.1 AU or less: it would require optical interferometry) but there have been mounting spectroscopic and photometric evidences that the disc is truncated by a stellar magnetosphere (assumed to be a dipole) and that accretion proceeds along magnetic funnels or curtains towards the magnetic poles (see Bouvier *et al.* 2007 and references therein). This gave rise to the so-called *disc locking paradigm*, where it is assumed that the stellar angular momentum could be transferred to the disc (Ghosh *et al.* 1977; Collier Cameron & Campbell 1993; Armitage & Clarke 1996). Unfortunately, this idealized picture can probably not be maintained (see a thorough discussion in Matt & Pudritz 2005b).

The simple reason is that accretion onto the star and this "strict" disc locking mechanism are two contradictory requirements. Let Ω_* be the angular velocity of the star. Its magnetosphere will try to make the disc material corotate with the protostar so that the sign of the torque depends directly on the relative angular velocity. Stellar magnetic field lines threading the disc beyond the rotation radius $r_{co} = (GM/\Omega_*^2)^{1/3}$ exert a positive torque, whereas they brake down the disc material below r_{co}. Let us also define the truncation radius r_t below which the stellar magnetic field is strong enough to "truncate" the disc by enforcing the material to flow along the field lines and no longer on the plane of the disc. Now, one can safely realize that accretion onto the star can only proceed if $r_t < r_{co}$. In this situation, the stellar magnetic field can brake down both the disc and the material accreting in the funnel flows. This implies of course a stellar spin up by the disc material located below r_{co}. The disc locking paradigm assumes that stellar field lines remain anchored beyond r_{co}, giving hopefully rise to some angular momentum balance. But within this paradigm, the disc viscosity must be efficient enough so as to radially transport outwards both the disc and stellar angular momentum! This is unrealistic because the stellar angular momentum is far too large. Moreover, all numerical simulations done so far showed a fast opening of the field lines beyond r_{co} (through numerical reconnection), severing the causal link and thereby dramatically reducing this negative torque (Lovelace *et al.* 1995, 1999; Long *et al.* 2005). Although this effect is strongly dependent on the disc magnetic turbulent diffusivity, the main result is to spin up the star whenever $r_t < r_{co}$ (Zanni et al, in prep).

Therefore, an important question is what determines the disc truncation radius r_t? In fact, two constraints must be simultaneously fulfilled for driving steady-state accretion funnel flows (Bessolaz et al, submitted). First, the poloidal stellar magnetic field must be strong enough to halt the accretion motion, namely $B_z^2/\mu_o \sim \rho u_r^2$. Second, the disc thermal pressure must be able to lift material vertically in order to initiate the accretion funnel flow, $B_z^2/\mu_o \sim P$. This last constraint is equivalent to an equipartition field, as first proposed by Pringle & Rees (1972). Putting this two constraints together one derives a ratio of the disc truncation radius to the co-rotation radius

$$\frac{r_t}{r_{co}} \simeq 0.66\ B_*^{4/7} \dot{M}_a^{-2/7} M_*^{-10/21} R_*^{12/7} P_*^{-2/3} \tag{3.1}$$

where the disc accretion rate $\dot{M}_a$ has been normalized to 10^{-8} $M_\odot$ yr^{-1}, stellar dipole field B_* to 150 G, mass to 0.5 $M_\odot$, radius to $3R_\odot$ and period P_* to 8 days. These analytical constraints and estimates have been confirmed using MHD axisymmetric numerical simulations of a star-disc dipole interaction (Bessolaz *et al.* – VAC, Zanni *et al.* – PLUTO, submitted). This is therefore a robust result. It shows that truncating discs can be done with a dipole field of several hundreds of Gauss (not kG !), consistent with both observations of magnetic fields (Donati, private communication) and sizes of inner disc holes (Najita *et al.* 2007). It confirms that for T Tauri parameters $r_t < r_{co}$ and that the "disc locking" picture *a la* Ghosh & Lamb is most probably not working.

3.2. *Stellar spin down via winds*

The obvious way to conciliate this result with the observational constraint that accreting stars are actually being spun down is via *winds that would not exist without the presence of accretion.* This has led Matt & Pudritz (2005a) to propose the name of "accretion powered stellar winds".

Accretion onto the star takes place along closed magnetospheric field lines, shocks the stellar surface and releases there most of its mechanical energy (mostly UV emission). The idea is then that a fraction of this accretion-heated mass diffuses towards the magnetic pole until it reaches open field lines. A warm stellar wind can then be initiated. The problem with T Tauri stars is that they are rotating at about 10% of their break-up speed. This translates into a totally negligible magnetic acceleration (the stellar material is far too deep in the gravitational potential well). One has therefore to rely on pressure-driven winds (see eg. discussion in Ferreira *et al.* 2006a for "enhanced stellar winds"). Now, if that initial pressure is only thermal, then temperatures of several million degrees are required. This raises the critical issue of too strong emission losses due to this inner hot wind (see S. Matt's contribution). The alternative is to rely on a turbulent Alfvén wave pressure and would be less dissipative (Hartmann & MacGregor 1980; DeCampli 1981). Note that the presence of turbulent MHD waves is expected in this context.

It should be noted that current MHD numerical simulations of star-disc interaction (e.g. Long *et al.* 2005; Zanni *et al.* 2007) do show a magnetic braking due to the opened stellar field lines. This has been interpreted as a "magnetic tower" since no real stellar wind was incorporated in the simulations. This is obviously a very promising issue. A thorough investigation should therefore be conducted in order to assess whether or not accretion-powered stellar winds of this kind can indeed (i) be dense enough and with a magnetic lever arm large enough to brake down the protostar, (ii) have radiative losses consistent with observations and (iii) do not pose any energetic problem like e.g. requiring to tap more than 50% of the accretion luminosity.

4. Reconnection X-winds

Accretion-powered stellar winds are somehow designed to explain the "disc locking" paradigm for T Tauri stars, namely to maintain their low rotation rate despite accretion. But how can we explain that T Tauri stars do *already* rotate at 10% of their break-up speed? Numerical simulations of the collapse of rotating magnetized clouds succeed nowadays in explaining the formation of protostellar cores at break-up speeds thanks to magnetic braking (Banerjee & Pudritz 2006; Machida *et al.* 2006). However, it is doubtful that such a braking could provide much lower initial rotation rates. One must then rely on some interaction between the protostar and its disc during the embedded phase (Class 0 and possibly Class I).

To our knowledge, the only model that addresses this issue is the Reconnection X-wind model (Ferreira *et al.* 2000). In this model, it is assumed that the interstellar magnetic field is advected with the infalling material in such a way that a significant magnetic flux Φ is now threading the protostellar core and the inner disc regions (as simulations show). This self-gravitating core will develop a dynamo of some kind but whose outcome is assumed to be the generation of a dipole field with a magnetic moment parallel to the disc magnetic field. This is clearly an assumption as there is no theory of such a constrained dynamo that takes into account both the presence of an initial strong fossil field and the outer disc (see however Moss 2004). The coexistence of this dipolar stellar field with the outer disc field generates an X-type magnetic neutral line, where both fields cancel each other at a radius r_X. Note that such a magnetic configuration has been previously considered by Uchida & Low (1981) and Hirose *et al.* (1997), but without taking into account the stellar rotation.

Let us assume that at $t = 0$ the dipole is emerging from a protostar rotating at break-up speed so that $r_X = R_{*,0}$ with $R_{*,0} = R_*(t=0)$, $M_{*,0} = M_*(t=0)$, $\dot{M}_{a,0} = \dot{M}_a(t=0)$, $\Omega_{*,0} = \sqrt{GM_{*,0}/R^3_{*,0}}$ and $B_{*,0} = B_*(t=0)$. It is further assumed that the field threading the disc is strong enough to drive self-confined disc winds at these early stages. Then Eq. (2.1) applies and provides us the value of the required stellar field. What will be the consequences of this initial state?

From the point of view of the disc, nothing is changed beyond r_X: a disc wind is taking place in the JED and disc material accretes by loosing its angular momentum in the jets. At r_X however, magnetic reconnection converts closed stellar field lines and open disc field lines into open stellar field lines. Accreting material that was already at the disc surface at r_X is now loaded into these newly opened field lines (there is a strong upward Lorentz force above r_X). Since these lines are now rotating at the stellar rotation rate, they exert a strong azimuthal force that drives ejection. This new type of wind has been called "Reconnection X-winds". Although material is ejected along field lines anchored onto the star, this is not a stellar wind since material did not reach the stellar surface and thus did not loose its rotational energy: it is much easier to accelerate matter under these circumstances.

Reconnection X-winds are fed with disc material and powered by the stellar rotational energy. As a consequence, they exert a negative torque on the protostar which leads to a stellar spin down. On the other hand, an increase of the stellar angular velocity Ω_* is expected from both accretion and contraction, with a typical Kelvin-Helmoltz time scale of several 10^5 yrs. Because of the huge stellar inertia, the evolution of Ω_* with time must be followed on these long time scales. One assumption used to compute the angular momentum history of the protostar on those scales is that $r_X \simeq r_{co}$. Such an assumption relies on the possibility for the protostellar magnetosphere to evacuate angular momentum through violent ejection events (Reconnection X-winds) whenever $r_X > r_{co}$, while quasi steady accretion columns form when $r_X < r_{co}$. Consistently with $r_X \simeq r_{co}$, a constant fraction $f = \dot{M}_X/\dot{M}_a$ is assumed on these long time scales, where $\dot{M}_X$ is the ejected mass flux in Reconnection X-winds, as well as a constant magnetic lever arm parameter λ. *These winds are therefore best seen as violent outbursts carrying disc material (blobs?) and stellar angular momentum from the star-disc interaction, channeled* *and confined by the outer disc wind.* Note that a conventional stellar wind would of course take place and fill in the inner field lines with mass, but its effect on the stellar spin evolution has been neglected in this work.

The global picture is then the following. As the protostar is being spun down, the corotation radius r_{co} increases and so must r_X. The stellar dipole field is assumed to follow $B_{star} = B_*(r/R_*)^{-n}$ where the index n describes a deviation from a pure dipole in vacuum. Now, r_X is defined by the cancellation of the stellar and disc field, whose scaling is very different from the former (see Eq. 2.1). The only way to ensure $r_X \simeq r_{co}$ on these long time scales is then to decrease $\dot{M}_a$ in time as well. Note that this is not a surprise as the accretion rate onto the star is controlled by the star-disc interaction. Thus, while computing the stellar spin evolution in time $\Omega_*(t)$, starting from conditions prevailing in Class 0 objects and using f, λ and n as free parameters, one gets also $R_*(t)$, $M_*(t)$ and $\dot{M}_a(t)$. Note that this global process of angular momentum removal is intimately related to the magnetic history of the protostar-disc system. Two additional ingredients are thus necessary: the amount of magnetic flux Φ threading the disc and how the stellar field B_* evolves with time (through dynamo). The calculations reported in Ferreira *et al.* (2000) were performed using simple assumptions about the dynamo and a more realistic modeling is needed. However, the results are already very promising (see Ferreira *et al.* 2000 for more details).

It was found that all low-mass Class 0 objects can indeed be spun down, from the break-up speed to about 10% of it, on a time scale consistent with the duration of the embedded phase for very reasonable values of the free parameters ($n = 3$ or 4, $f\lambda > 0.1$). Stellar period, mass, radius and disc accretion rates were found consistent with values for T Tauri stars with a dipole field smaller than 1 kG. Finally, it is noteworthy that the main difference between "accretion-powered stellar winds" and Reconnection X-winds relies on the stellar magnetic moment. In the former case, it is anti-parallel to the disc field while it is parallel in the latter. If the dynamo action explicitly assumed provides a magnetic field reversal, then recurrent transitions from one wind configuration to another can be expected.

5. Conclusion

The theory of *steady* jet production from Keplerian accretion discs has been completed in the framework of "alpha" discs. The physical conditions required to thermo-magnetically drive jets are constrained and all the relevant physical processes have been included. The role of large scale magnetic fields in discs has gradually emerged and it seems now an unavoidable ingredient of star formation theory as a whole. The progresses in star-disc interaction reported here provide valuable insights but one should remain cautious as stellar magnetic fields are not pure, aligned dipoles.

The amount of magnetic flux Φ in the disc is an unknown parameter but it is reasonable to assume that it scales with the total mass M. If this is verified then two important aspects could be naturally explained:

(1) Reconnection X-winds can brake down a protostar during the embedded phase, explaining that T Tauri stars rotate at about 10% of the break-up speed. Remarkably, the mystery of the low dispersion in angular velocities would be naturally accounted by a low dispersion in the ratio Φ/M (Ferreira *et al.* 2000). These winds are also a very promising means to drive time dependent massive bullets, channeled by the outer steady disc wind.

(2) The transition from disc winds (Class 0, I, II) to stellar winds (Class II, III) would follow the evolution of the disc magnetic flux Φ, with a transition radius between the outer SAD and the inner JED decreasing in time.

References

Armitage, P. J. & Clarke, C. J. 1996, *MNRAS*, 280, 458
Balbus, S. A. 2003, *ARAA*, 41, 555
Banerjee, R. & Pudritz, R. E. 2006, *ApJ*, 641, 949
Basu, S. & Mouschovias, T. C. 1994, *ApJ*, 432, 720
Blandford, R. D. & Payne, D. G. 1982, *MNRAS*, 199, 883
Bouvier, J., Alencar, S. H. P., Harries, T. J., Johns-Krull, C. M., & Romanova, M. M. 2007, in Protostars and Planets V, ed. B. Reipurth, D. Jewitt, & K. Keil, 479–494
Bouvier, J., Wichmann, R., Grankin, K., *et al.* 1997, *A&A*, 318, 495
Cabrit, S., Edwards, S., Strom, S. E., & Strom, K. M. 1990, *ApJ*, 354, 687
Casse, F. & Ferreira, J. 2000, *A&A*, 353, 1115
Casse, F. & Keppens, R. 2002, *ApJ*, 581, 988
Casse, F. & Keppens, R. 2004, *ApJ*, 601, 90
Collier Cameron, A. & Campbell, C. G. 1993, *A&A*, 274, 309
DeCampli, W. M. 1981, *ApJ*, 244, 124
Donati, J.-F., Paletou, F., Bouvier, J., & Ferreira, J. 2005, *Nature*, 438, 466
Ferreira, J. 1997, *A&A*, 319, 340
Ferreira, J. & Casse, F. 2004, *ApJ*l, 601, L139
Ferreira, J., Dougados, C., & Cabrit, S. 2006a, *A&A*, in press
Ferreira, J. & Pelletier, G. 1995, *A&A*, 295, 807
Ferreira, J., Pelletier, G., & Appl, S. 2000, *MNRAS*, 312, 387
Ferreira, J., Petrucci, P.-O., Henri, G., Saugé, L., & Pelletier, G. 2006b, *A&A*, 447, 813
Ghosh, P., Pethick, C. J., & Lamb, F. K. 1977, *ApJ*, 217, 578
Glassgold, A. E., Najita, J., & Igea, J. 2004, *ApJ*, 615, 972
Hartigan, P., Edwards, S., & Ghandour, L. 1995, *ApJ*, 452, 736
Hartmann, L. & MacGregor, K. B. 1980, *ApJ*, 242, 260
Heiles, C., Goodman, A. A., McKee, C. F., & Zweibel, E. G. 1993, in Protostars and Planets III, ed. E. H. Levy & J. I. Lunine, 279–326
Hirose, S., Uchida, Y., Shibata, K., & Matsumoto, R. 1997, *PASJ*, 49, 193
Konigl, A. 1989, *ApJ*, 342, 208
Kwan, J. 1997, *ApJ*, 489, 284
Long, M., Romanova, M. M., & Lovelace, R. V. E. 2005, *ApJ*, 634, 1214
Lovelace, R. V. E., Li, H., Colgate, S. A., & Nelson, A. F. 1999, *ApJ*, 513, 805
Lovelace, R. V. E., Romanova, M. M., & Bisnovatyi-Kogan, G. S. 1995, *MNRAS*, 275, 244
Machida, M. N., Inutsuka, S.-i., & Matsumoto, T. 2006, *ApJ*l, 647, L151
Masset, F. S., Morbidelli, A., Crida, A., & Ferreira, J. 2006, *ApJ*, 642, 478
Matt, S. & Pudritz, R. E. 2005a, *ApJ*l, 632, L135
Matt, S. & Pudritz, R. E. 2005b, *MNRAS*, 356, 167
Ménard, F. & Duchêne, G. 2004, *A&A*, 425, 973
Miller, K. A. & Stone, J. M. 2000, *ApJ*, 534, 398
Moss, D. 2004, *A&A*, 414, 1065
Najita, J. R., Carr, J. S., Glassgold, A. E., & Valenti, J. 2007, ArXiv e-prints, 704
Pringle, J. E. & Rees, M. J. 1972, *A&A*, 21, 1
Rebull, L. M., Wolff, S. C., Strom, S. E., & Makidon, R. B. 2002, *AJ*, 124, 546
Shakura, N. I. & Sunyaev, R. A. 1973, *A&A*, 24, 337
Stone, J. M., Hawley, J. F., Gammie, C. F., & Balbus, S. A. 1996, *ApJ*, 463, 656
Strom, K. M., Strom, S. E., Wolff, S. C., Morgan, J., & Wenz, M. 1986, *ApJ*s, 62, 39
Uchida, Y. & Low, B. C. 1981, Journal of Astrophysics and Astronomy, 2, 405
Zanni, C., Ferrari, A., Rosner, R., Bodo, G., & Massaglia, S. 2007, *A&A*, 469, 811

Star-Disk Interaction in Young Stars
Proceedings IAU Symposium No. 243, 2007
J. Bouvier & I. Appenzeller, eds.

doi:10.1017/S1743921307009672

The implications of close binary stars for star-disk interactions

Robert D. Mathieu
Department of Astronomy, University of Wisconsin – Madison, WI 53706 USA
email: mathieu@astro.wisc.edu

Abstract. The presence of close ($\lesssim$ 0.1 AU) stellar companions must greatly alter the circumstellar environment of classical T Tauri stars, including severe truncation if not elimination of circumstellar disks. It is thus remarkable how little impact the presence of a close companion has on our observable diagnostics for accretion and outflow. Emission line shapes, degrees of continuum veiling, and spectral energy distributions are all indistinguishable between single classical T Tauri stars and classical T Tauri close binaries. Some of the most classical T Tauri stars that laid the foundation for our single-star accretion-disk paradigm have turned out to have close companions. Periodicities in spectral signatures are suggestive of the presence of accretion flows from circumbinary disks to the circumstellar regions; the subsequent flow of material through the circumstellar region to the stellar surface in the presence of a stellar magnetosphere is unstudied. Observations of stellar rotation distributions in close binaries suggest that inner disk regions may act to regulate stellar angular momentum.

Keywords. Binary stars, accretion streams, stellar rotation, UZ Tau E, DQ Tau.

---

1. Introduction

The presence of stellar† companions within protostellar disks raises fascinating dynamical questions regarding disk structures and accretion flows. At the heart of these questions is the expectation that such companions will clear gaps in accretion disks (e.g., Figure 1 of Artymowicz & Lubow 1996), and that the balance of viscous and resonant forces at the inner edge of a circumbinary disk will prevent the flow of circumbinary material to the circumstellar disks and thus to the stellar surfaces.

The accretion disk morphology as a result of this clearing varies qualitatively with the binary semi-major axis. For the widest binaries (a $\gg$ 100 AU) the morphology may be one or two circumstellar disks, likely with properties little different from disks around single stars. For binaries with separations of several tens of AU, the morphology may be one or two truncated circumstellar disks, with radii of 0.2-0.3 times the semi-major axis. For close binaries (a $\lesssim$ 10 AU), the binary may be surrounded by a circumbinary disk and each star may have a truncated circumstellar disk. If the pre-main-sequence binary population is similar to that of the field (e.g., Duquennoy & Mayor 1991), roughly 15% of the population is binaries with semi-major axes of less than 1 AU. In fact, there is evidence that the pre-main-sequence binary frequency may be higher than in the field.

Here I wish to connect these issues to the subject of this symposium. Specifically, the question that I wish to put forward – but alas, not answer – is the following:

† Many of the points of this article also apply to massive planetary companions. At present, the issues are most directly explored in the context of stellar companions.

Given that every indicator of circumstellar accretion and outflow survives in the face of severe (complete?) circumstellar disk truncation within the shortest period classical T Tauri binaries, what does this tell us about star-disk physics?

This question rests on the simple observation that classical T Tauri binary stars with periastron separations of as small as several stellar radii show the same diagnostics for accretion and outflow as do single classical T Tauri stars (or those in wide binaries). Put another way, considering only our observable diagnostics for accretion and outflow, we cannot distinguish those classical T Tauri stars that have very close stellar companions (except perhaps in periodic variation). Given that we expect such companions to greatly modify the properties of circumstellar disks, if not eliminate them completely, this comes as a possibly very significant surprise for our understanding of star-disk interactions.

2. Evidence for accretion in young short-period binaries

One of the most classical of all classical T Tauri stars is a case in point – UZ Tau E. Its properties include: high-amplitude, irregular photometric variability; Hα emission equivalent width in excess of 50 Å; strong ultraviolet excess; and heavily veiled spectra. Indeed, it has been classified as an eruptive variable with behavior akin to FU Ori stars (Herbig 1977). All of these phenomena are taken as evidence for accretion onto the stellar surface; Hartigan, Edwards & Ghandour (1995) determine a large accretion rate of order 10^{-6} $M_{\odot}\,\mathrm{yr}^{-1}$. UZ Tau E also shows signatures of outflows at M $\approx 10^{-8}$ $M_{\odot}\,\mathrm{yr}^{-1}$, and evidence for microjets (Hirth, Mundt & Soif 1997). Finally, UZ Tau E shows large infrared excesses and the canonical Class II spectral energy distribution of a protostellar disk. Millimeter interferometry have shown UZ Tau E to be surrounded by a massive (≈ 0.6 $M_{\odot}$) disk (Jensen, Koerner & Mathieu 1996). In the present paradigm, all of these observational data would be interpreted as – indeed, could be taken as an archetype of – a single star surrounded by an actively accreting disk that is continuous from very near the stellar surface to its outer radius of several hundred AU.

In fact, UZ Tau E is a very close binary with an orbital period of 19.13 days, an eccentricity of 0.33, and a mass ratio of 0.30 (Prato *et al.* 2002; Martín et al. 2005; Jensen *et al.* 2007). Using orbital elements and an inclination estimate from Jensen *et al.* the periastron separation is only 0.10 AU. Using a photometric estimate for the primary stellar radius of 2 $R_{\odot}$, this periastron separation is only of order 10 (primary) stellar radii.

While UZ Tau E is perhaps the most striking case in point, it is not alone. The classical T Tauri stars DQ Tau (periastron separation 0.05 AU; Mathieu *et al.* 1997; Basri, Johns-Krull & Mathieu 1997) and AK Sco (0.08 AU; Alencar *et al.* 2003) also show accretion and outflow diagnostics, and are even closer binary systems. V4046 Sgr might also be included in this sample, albeit likely with a much lower accretion rate (e.g., Stempels & Gahm 2004).

Again, the point of this brief summary is to emphasize that in these young binaries, all of our paradigmatic observables for accretion and outflow appear to be unaffected despite the presence of stellar companions passing within 10 stellar radii or less every 10-20 days. How might, for example, magnetospheric accretion from disk to stellar surface continue unaffected in such an enviroment?

3. Accretion streams

A decade ago Artymowicz & Lubow (1996; AL) found in SPH simulations that for sufficiently warm and viscous circumbinary disks, material can flow across the cleared gaps as streams with mass transfer rates equivalent to those expected from continuous disks around single stars. This idea was further explored by Günther & Kley (2002), who confirmed the presence of such streams using high-resolution numerical simulations (Figure 1).

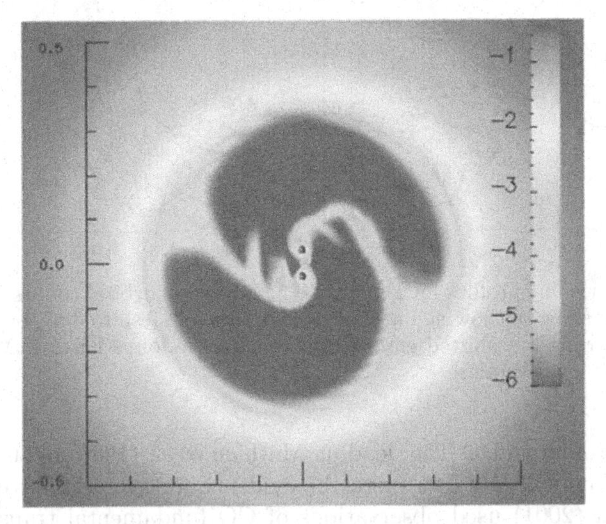

Figure 1. DQ Tau circumbinary disk after 85.5 orbital periods. The binary is shown at periastron, and the size of the stars reflects their actuall stellar radii. The length scales are in AU. (From Günther & Kley 2002, where the figure can be found in color.)

Contemporaneous with AL, my collaborators and I were studying the classical T Tauri binary DQ Tau. This binary has an orbital period of 15.80 days, an eccentricity of 0.56, and a mass ratio of unity (Mathieu *et al.* 1997; see also Mathieu 2001 for a summary discussion). The periastron separation is only 0.05 AU, just a few stellar radii (Figure 2). Nonetheless, the stellar spectra show diagnostics for both accretion and outflow (e.g., Figure 3), albeit at a lower accretion rate (e.g., 5 x 10^{-8} $M_{\odot}\,yr^{-1}$) than for UZ Tau E.

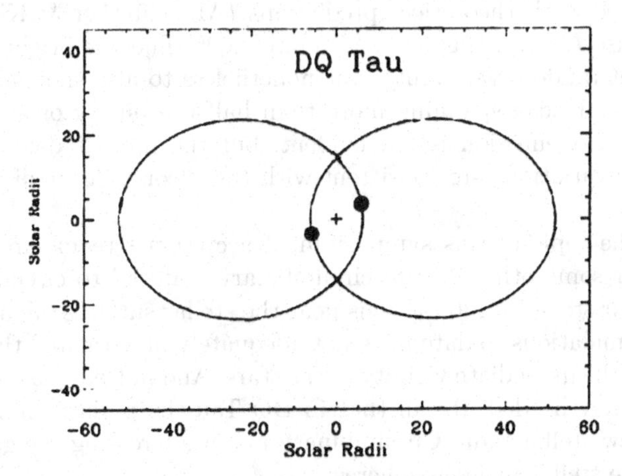

Figure 2. The orbit of the classical T Tauri binary DQ Tau, with the radii of the stars shown to scale. (Adapted from Basri, Johns-Krull & Mathieu 1997.)

Importantly, photometric, veiling, and Hα observations all indicate enhanced accretion rates at periastron passage (Mathieu *et al.* 1997; Basri, Johns-Krull & Mathieu 1997). Such pulsed accretion events were also found by AL in SPH calculations for a binary with

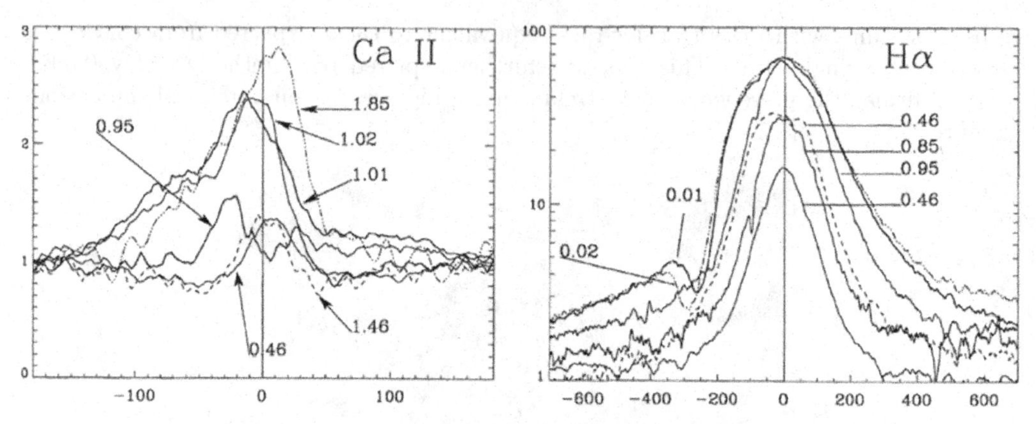

Figure 3. Time-series line profiles for DQ Tau. The labels are orbital counts and phases. During periastron the lines show outflow signatures; Hα develops blueshifted absorption features, and Ca II can display strong blueshifted emission. (From Basri, Johns-Krull & Mathieu 1997. Note that the Hα scale is logarithmic.)

parameters very similar to DQ Tau, leading Mathieu *et al.* (1997) to suggest that the periodic accretion events of DQ Tau might be evidence for accretion streams. Later Carr, Mathieu & Najita (2001) used observations of CO fundamental transitions to demonstrate the presence of 1200 K gaseous material located within the central few tenths of an AU of the binary system. The discovery of residual gas and dust within the regions expected to be cleared by the binary suggests a link between the circumbinary disk and the ongoing accretion at the stellar surfaces. That said, the relation of this material to an accretion stream is not yet established.

More recently, similar evidence for periodic accretion events has been searched for in the short-period classical T Tauri binaries AK Sco and UZ Tau E. Alencar *et al.* (2003) found no evidence of enhanced accretion near the periastron passage in AK Sco. They did find periodicity in the Hα emission with the orbital period, but its phasing is not evidently associated with theoretical predictions (AL, Günther & Kley 2002). Jensen *et al.* (2007; see also Huerta, Hartigan, & White, 2005) find the brightness of UZ Tau E to show significant random variability, but nonetheless to also show an overall periodic pattern with a broad peak spanning more than half the binary orbital period. Similar periodicity in the Hα emission is not evident, but the present data are sparse. They suggest that these variations are consistent with the theoretical predictions of accretion stream models.

Returning to the topic of this symposium, if accretion streams are present in these close binaries – or some other flow of circumbinary material to circumstellar regions – the implications for star-disk interactions near the stellar surfaces are as yet unexplored. The theoretical simulations to date have not adequately investigated the hydrodynamics of the gas flows in the immediate vicinity of the stars. And indeed, magnetohydrodynamic simulations are likely needed. Recall that in DQ Tau the periastron separation is only 0.05 AU, just a few stellar radii. Circumbinary material crossing the gap does so *within* the domains of the stellar magnetospheres.

One clue to how the accreting material organizes itself may be the finding of Basri, Johns-Krull & Mathieu (1997) that while the veiling of DQ Tau does increase at periastron, nonetheless some veiling – and by implication, some accretion – is present throughout the orbital cycle. (Note that this paper also presents an extensive discussion of the possible distribution of line-emitting gas around DQ Tau.)

Another clue may lie in an interesting distinction between the spectral energy distributions (SEDs) of classical T Tauri close binaries and those weak-lined T Tauri stars that still show infrared excesses, as shown in Figure 4. The classical T Tauri close binaries show continuous Class II spectral energy distributions with minimal to no dips in emission at those wavelengths corresponding to the temperature of material missing from cleared regions. The spectral energy distributionss do not preclude such cleared regions (Mathieu *et al.* 1997; Alencar *et al.* 2003), but if gaps are present their signature is subtle. On the other hand, close weak-lined T Tauri binaries show substantial structure in their spectral energy distributions indicative of gaps (e.g., Jensen & Mathieu 1997).

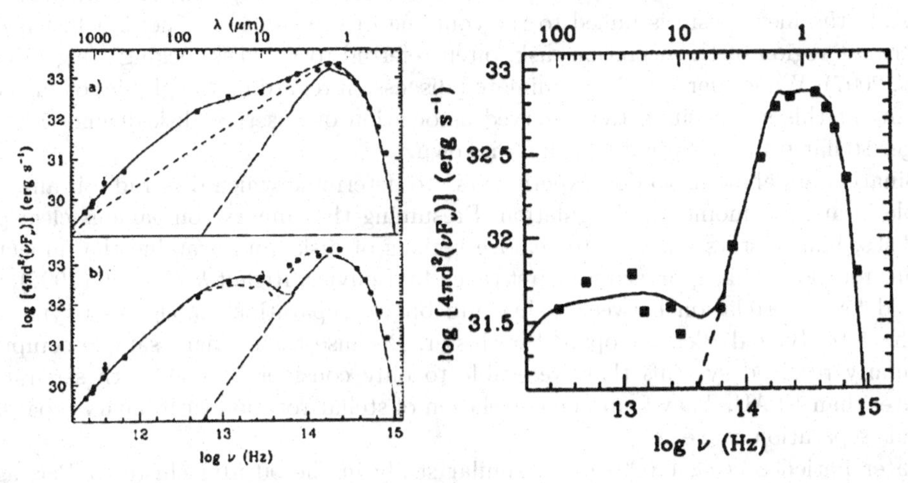

Figure 4. Left panel: The spectral energy distribution of DQ Tau, which is very similar to a canonical Class II spectral energy distribution from a continuous accretion disk. The long-dash line represents the stellar photosphere. The dashed curve in the upper panel represents a reprocessing disk, while the solid line is a best-fit active accretion disk, both continuous from the stellar surface. The solid line in the lower panel is an active accretion disk with a cleared central hole of radius 0.4 AU, while the dashed line represents a similar disk with warm optically thin dust in the central hole. (For more details, see Mathieu *et al.* 1997). Right panel: The spectral energy distribution of V4046 Sgr. Note the markedly different morphology from DQ Tau. The solid line represents an accretion disk with a central hole to the dynamically predicted outer radius of 0.072 AU. (From Jensen & Mathieu 1997.)

In all likelihood, the actual distribution and flow of accreting material is more complex than we have considered. Basri, Johns-Krull & Mathieu note evidence for "complex occultations, ejections, and absorptions near periastron, while the stars look rather placid away from each other." They conjecture that a "nimbus" of material may sit in the circumstellar environment of each star that could contribute to the continuous, lower level accretion. Whatever may be the reality, given this rather different accretion environment it remains all the more remarkable that the spectral signatures of these binaries are indistinguishable from single classical T Tauri stars.

Finally, a digression. While all of the discussion so far in this paper has been in the context of accretion disks, it is worth noting that at periastron it is possible that the stellar magnetospheres of both stars *themselves* interact. At the least we might expect to observe large flare events, such as seen in RS CVn binaries. Basri, Johns-Krull & Mathieu (1997) consider whether such flares might be the proper interpretation of the photometric and spectroscopic events of DQ Tau at periastron, concluding that magnetospheric interactions were not likely to be the source of the photometric brightening or enhanced veiling and emission lines. Nonetheless, if the magnetospheres are as large as 5 stellar radii (as suggested by radio observations of some weak-lined T Tauri stars), then we might expect some interaction. Indeed, as expressed by these authors, it may be notable that DQ Tau is able to pass through roughly a third of its periastron passages *without* observed outbursts.

4. Angular momentum regulation: insights from short-period binaries

An increasing body of observational evidence suggests that early stellar rotation evolution is linked to the presence of accretion disks (see the review by Bouvier in this volume). Possibly the mechanism is linked to the coupling of the stellar magnetic field with the innermost region of the accretion disk, often referred to as "disk locking" (e.g., Herbst *et al.* 2007). While there has been animated discussion regarding the physics of such disk locking at this symposium, the observed association of observed disk diagnostics with longer stellar rotation period stands on its own.

Binary stars allow us to do "experiments" to determine which disk radii, if any, play a role in angular momentum regulation. Presuming that binary companions clear gaps in disks, binary stars allow us to remove regions of disks and examine the impact on stellar rotation. This approach was first taken by Bouvier, Rigaut & Nadeau (1997), who looked for a correlation between $v \sin i$ and binary separation among solar-type stars in the 120 Myr old Pleiades open star cluster. Because their binary sample comprised angularly resolved systems they were able to only consider binaries with separations greater than 10 AU. They found no correlation of stellar rotation with binary separation in this separation range.

Later Patience *et al.* (2002) did a similar study in the 90 Myr cluster α Per, again using a sample of angularly resolved binaries with separations greater than 10 AU. They suggested the presence of an increase in $v \sin i$ for their four binaries with separations of less than 60 AU, with a confidence level of 98%.

The magnetic disk locking mechanism, along with other proposed mechanisms for linking stellar angular momentum evolution to disks, couples the star to innermost regions of the disk, within tenths of an AU from the stellar surface. As such, the binaries expected to have the most direct impact on disk regulation are those with the shortest periods. Thus Meibom, Mathieu & Stassun (2007) recently have repeated these experiments using the spectroscopic binary population among solar-type stars in the 150 Myr open cluster M35.

Specifically, they compared the rotation periods (measured directly through photometric periodicity) of single stars and of primary stars in spectroscopic binaries with semi-major axes of 0.1 AU to 5 AU. Binaries with semi-major axes shorter than 0.1 AU may have primary rotation velocities modified by tidal synchronization (Meibom, Mathieu & Stassun 2006). Primary stars of binaries with separations greater than 5 AU are not detectable with spectroscopic techniques, and thus are included among the "single" star sample. These latter binaries are also wide enough that they are not expected to clear the innermost regions of their circumstellar disks.

The outcome of this study is presented in Figure 5. The primary stars in the shorter period binaries rotate more rapidly than the combined sample of single stars and primary stars in wide binaries. The two distributions are distinct at the 99.9% confidence level.

The results of these three studies, and especially the last, are consistent with mechanisms for stellar angular momentum regulation linked to the innermost regions of accretion disks. That said, it is important to recognize the difference between correlation and causality. Given our almost complete lack of understanding about how such close

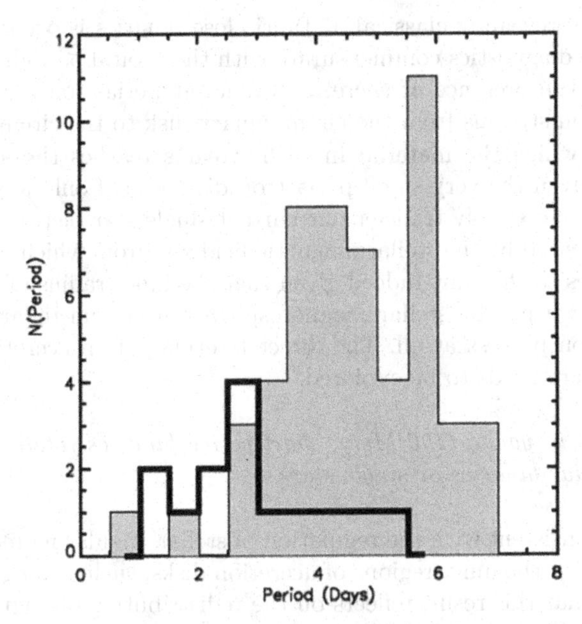

Figure 5. Rotation period distributions for single stars (grey histogram; $\langle P_{rot} \rangle = 4.64$ days) and binary primary stars (solid; $\langle P_{rot} \rangle = 2.95$ days) in the 150 Myr open cluster M35.

binaries form, we must at least entertain the conjecture that rapid stellar rotation and close binaries may be linked through the binary formation process itself.

5. Summary

In one sense, the dynamical interaction of orbiting stars with an accretion disk is one of the most dramatic star-disk interactions, with extensive impact on mass distribution over many AU. Here, however, we have focussed on the implications of close stellar companions for the accretion and outflow processes that have been the primary topic of this symposium.

Three essential findings merit emphasis in this summary:

1. At least 15% of T Tauri stars are binaries with separations within 1 AU.

As observers and interpreters of observational data, we ignore this fact at our own peril. Interesting signatures and behaviors of a supposed single star, and interpreted in the context of single-star–disk interactions, may in fact be associated with the presence of a stellar companion, either through contribution to the combined light or through physical interaction.

2. The presence of close (e.g., $\lesssim$ 0.1 AU) companions does not change the classical T Tauri signatures of accretion and outflow.

Given that we expect such companions to greatly modify the properties of circumstellar disks, if not eliminate them completely, it is remarkable how little impact the presence of a companion has on our standard diagnostics for accretion and outflow. Emission lines, continuum veiling, and spectral energy distributions are all indistinguishable between single classical T Tauri stars and classical T Tauri close binaries.

The one property revealing classical T Tauri close binaries is occasional periodicities in these observable diagnostics commensurate with the orbital period. In the case of DQ Tau in particular, the presence of accretion events at periastron are suggestive of the presence of accretion streams from the circumbinary disk to the circumstellar regions.

The process by which the material in such streams reaches the stellar surfaces remains unknown. Given the very small periastron distances of only a few stellar radii, it is unlikely that we can simply transfer our current single-star picture of a circumstellar accretion disk truncated by the stellar magnetic field and from which material flows onto the stellar field lines to the star. Indeed given that the inner radius of the region cleared by the binary lies within the stellar magnetospheres, a circumstellar disk may not be part of the accretion process at all. The direct interplay of an accretion stream with a stellar magnetosphere needs to be explored.

3. Primary stars in young (150 Myr), short-period binaries rotate more rapidly than primary stars in wide binaries or single stars.

This finding is consistent with the regulation of stellar angular momentum by a mechanism associated with the inner regions of accretion disks, such as magnetic disk locking. It is also possible that this result reflects on the redistribution of angular momentum in the binary formation process itself.

The primary point of this paper is that our observable diagnostics for accretion and outflow associated with classical T Tauri stars are remarkably robust in the presence of very close stellar companions. Arguably, that these diagnostics are so robust is encouraging. Nonetheless, and being intentionally provocative, if our diagnostics are independent of the presence of a close companion that is assuredly greatly modifying the circumstellar environment, perhaps we might wonder to what extent we can in fact use these diagnostics to test models of star-disk interactions around even single classical T Tauri stars.

Acknowledgements

I would like to thank the organizers of this meeting for inviting me, and thereby finally convincing me that studying outflows in young close binaries is as intriguing as studying accretion! The work in this paper with which I have been associated has been done with many friends; I want to particularly acknowledge the contributions of my students Eric Jensen, Søren Meibom, and Keivan Stassun, from whom I have learned so much, and Gibor Basri, whose insights are present throughout this paper. I wish to gratefully acknowledge support of the IAU, and of the National Science Foundation over many years.

References

Alencar, S. H. P., Melo, C. H. F., Dullemond, C. P., Andersen, J., Batalha, C., Vaz, L. P. R., & Mathieu, R. D. 2003, *A&A*, 409, 1037

Artymowicz, P. & Lubow, S. H. 1996 , *ApJ*, 467, L77

Basri, G., Johns-Krull, C. M., & Mathieu, R. D. 1997, *AJ*, 114, 781

Bouvier, J., Rigaut, F., Nadeau, D. 1997, *A&A*, 323, 139

Carr, J. S., Mathieu, R. D., & Najita, J. R. 2001, *ApJ*, 551, 454

Duquennoy, A. & Mayor, M. 1991, *A&A*, 248, 485

Günther, R. & Kley, W. 2002, *A&A*, 387, 550
Hartigan, P., Edwards, S. & Ghandour, L. 1995 *ApJ*, 452, 736
Herbig, G. H. 1977 *ApJ*, 217, 693
Herbst, W., Eislöffel, J., Mundt, R., & Scholz, A. 2007 In *Protostars and Planets* V, B. Reipurth, D. Jewitt, and K. Keil (eds.), University of Arizona Press, Tucson, 297
Hirth, G.A., Mundt, R. & Soif, J. 1997 *A&AS*, 126, 437
Huerta, M., Hartigan, P., & White, R. J. 2005, *AJ*, 129, 985
Jensen, E. L. N., Dhital, S., Stassun, K. G., Patience, J.; Herbst, W. Walter, F. M., Simon, M., & Basri, G. 2007, *AJ*, 134, 241
Jensen, E. L. N., Koerner, D. W.,& Mathieu, R. D. 1996, *AJ*, 111, 2431
Jensen, E. L. N. & Mathieu, R. D. 1997, *AJ*, 114, 301
Martín, E. L., Magazzú, A., Delfosse, X., & Mathieu, R. D. 2005, *A&A*, 429, 939
Mathieu, R. D., Stassun, K., Basri, G., Jensen, E. L. N., Johns-Krull, C. M., Valenti, J. A., & Hartmann, L. W. 1997, *AJ*, 113, 1841
Mathieu, R. D. in *The Formation of Binary Stars*, Proceedings of IAU Symp. 200, Edited by Hans Zinnecker and Robert D. Mathieu, 2001, p. 419.
Meibom, S., Mathieu, R. D., & Stassun, K. G. *ApJ*, in press
Meibom, S., Mathieu, R. D., & Stassun, K. G. 2006, *ApJ*, 653, 621
Patience, J., Ghez, A. M., Reid, I. N., & Matthews, K. 2002 *AJ*, 123, 1570
Prato, L., Simon, M., Mazeh, T., Zucker, S., & McLean, I. S. 2002, *ApJ*, 579, L99
Stempels, H.C. & Gahm, G.F. 2004, *A&A*, 421, 1159

Star-Disk Interaction in Young Stars
Proceedings IAU Symposium No. 243, 2007
J. Bouvier & I. Appenzeller, eds.

doi:10.1017/S1743921307009684

Disc–magnetosphere interactions in cataclysmic variable stars

Coel Hellier

Astrophysics Group, Keele University, Staffordshire ST5 5BG, U.K.
email: ch@astro.keele.ac.uk

Abstract. I review, from an observational perspective, the interactions of accretion discs with magnetic fields in cataclysmic variable stars. I start with systems where the accretion flows via a stream, and discuss the circumstances in which the stream forms into an accretion disc, pointing to stars which are close to this transition. I then turn to disc-fed systems and discuss what we know about how material threads on to field lines, as deduced from the pattern of accretion footprints on the white dwarf. I discuss whether distortions of the field lines are related to accretion torques and the changing spin periods of the white dwarfs. I also review the effect on the disc–magnetosphere interaction of disc-instability outbursts. Lastly, I discuss the temporary, dynamo-driven magnetospheres thought to occur in dwarf-nova outbursts, and whether slow-moving waves are excited at the inner edges of the disc.

Keywords. Accretion, accretion disks, magnetic fields, binaries: close, novae, cataclysmic variables, X-rays: stars.

---

1. Introduction

Given that interactions of an accretion disc with a magnetosphere are widespread in astrophysics, one can ask why it is of particular interest to study those in close binary systems such as the cataclysmic variable stars. One answer is that these systems often show a great range of observational clues. Periodic and quasi-periodic behaviour, often on timescales that are easy to study, is the speciality of cataclysmic variables, making them prime systems for advancing our understanding of accretion.

If you want to see how an emission line varies with the spin-cycle of a magnetic white dwarf in a CV, repeated 10 times for reliability, you need only watch for three hours, and the fossil field of the white dwarf will not have changed. In contrast, the same task for a YSO would take weeks, and the dynamo-driven field might be changing over that time. Further, the field of the white dwarf is more likely to be a simple dipole, and thus easier to model.

In this review I make an observationally led overview of the disc–field interactions in cataclysmic variables. I consider, first, the nature of an accretion flow in the presence of a magnetic field, and whether a disc forms, and then turn to how the disc interacts with the field. For a theoretical account of these topics see Li (1999) or Frank, King & Raine (2002). The definitive review of cataclysmic variable stars is Warner (1995), while for a shorter introduction see Hellier (2001).

One big difference from the situation in YSOs is that in close binaries the accretion originates in a stream from the secondary star through the inner Lagrangian point (L_1). In cataclysmic variables with a highly magnetic white dwarf (exceeding $\sim$30 MG) the white dwarf is phase-locked to the orbit. In such stars (called AM Her stars or polars) the ballistic stream becomes magnetically controlled and is channelled onto a magnetic pole of the white dwarf. Since this situation is least similar to YSOs I will not deal with it here, though for accounts see Wickramasinghe & Ferrario (2000) and Schwope *et al.* (2004).

In a handful of AM Her stars the white dwarf is slightly asynchronous with the orbit, despite a high magnetic field. Perhaps these stars have been knocked out of synchronism by a recent nova eruption (e.g. Schmidt & Stockman 1991). The trajectory of the stream will now change on the 'beat' cycle between the spin and orbital cycles, as the relative orientation of the dipole changes. Typically the accretion stream flips from one pole to the other and back on the beat cycle (10–50 days), though at any one time the star will look like an AM Her with magnetically channelled stream accretion. For accounts of such stars see Ramsay *et al.* (2000), Ramsay & Cropper (2002), Staubert *et al.* (2003) and Schwarz *et al.* (2005).

The main focus of this review will be the 'intermediate polars', which have weaker fields of 1–10 MG. In this regime the field is usually not strong enough to prevent an accretion disc from forming, but is strong enough to carve out the inner disc and for the magnetosphere to dominate the observed characteristics. The hallmark of IPs is X-ray flux from magnetically channelled accretion onto the white dwarf, heavily pulsed at the white-dwarf spin period. In some systems we also see pulsed polarised light, though in many IPs the polarised light is too diluted by other parts of the system to be detected. For an excellent introduction to these stars see Patterson (1994).

2. Disc formation: the case of V2400 Oph

Under what conditions does a disc form around a magnetic white dwarf? If the closest approach of a ballistic stream from the Lagrangian point is further out than the magnetic disruption radius, $R_{\rm mag}$ (at which the magnetic pressure exceeds the ram pressure of the material), then the stream material will accumulate at the circularisation radius (where the angular momentum of the orbit matches that of the L_1 point), and spread inwards and outwards until the inner edge meets $R_{\rm mag}$. However, while this condition applies in some wide binaries such as GK Per, it doesn't hold for the field strengths and orbital periods typical of most IPs. Thus it appears that most IPs form discs even though the trajectory of a ballistic stream would enter the magnetosphere. The case of V2400 Oph gives clues as to how this might occur.

V2400 Oph is the most plausible candidate for an IP which has no disc. We know the white-dwarf spin period from a clear detection of a 927-s pulsation in polarised light (Buckley *et al.* 1997), yet there is no X-ray pulse at this period. Instead the X-rays are pulsed at the 1003-s beat period between the spin and orbital cycles (Fig. 1), a clear indication that the accretion proceeds through a stream which flips between the two magnetic poles as their relative orientation changes. Indeed, an X-ray beat pulse had previously been proposed as the main diagnostic of stream-fed accretion in an IP (Hellier 1991; Wynn & King 1992).

Analysis of how V2400 Oph's spectral lines change with the spin and beat cycles (Hellier & Beardmore 2002) showed a reasonably good match to a simple model of a pole-flipping stream. And yet the low amplitudes of the pulsations and the absence of any orbital modulation indicates that this cannot be the whole story. Hellier & Beardmore (2002) argued for the presence of additional orbiting material that is diluting the emission from the stream. Yet this cannot be an accretion disc, since our experience of these stars suggests that this would produce an X-ray spin-period pulsation.

Instead, we turn to the 'diamagnetic blob' scenario developed by Wynn & King (1995), which suggests that dense 'blobs' of material can orbit quasi-ballistically, but with the addition of a 'magnetic drag' as they cross field lines. Blobs originating in the accretion

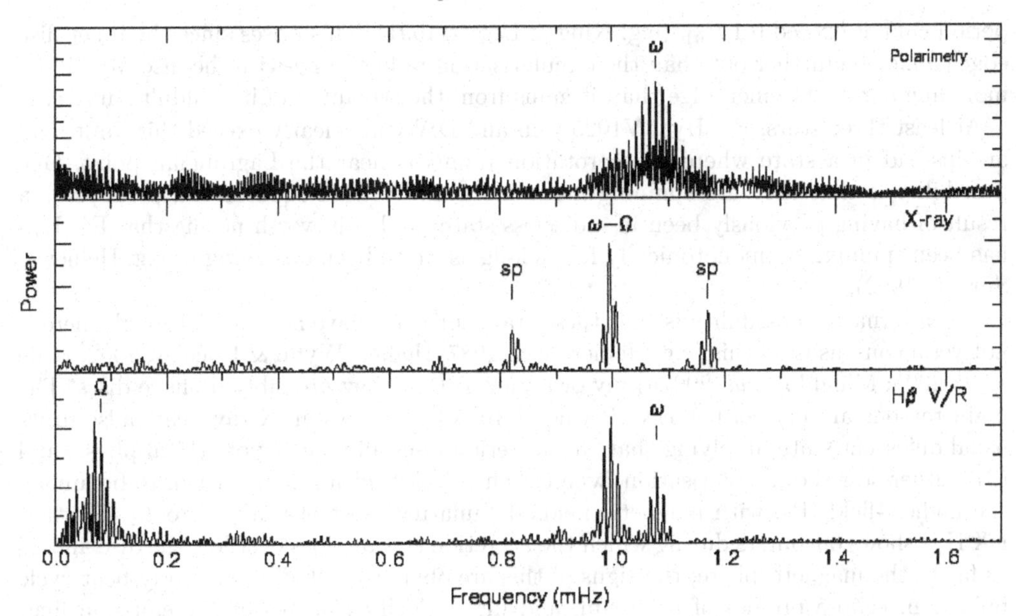

Figure 1. The Fourier transforms of optical polarimetry, X-ray flux and emission line radial velocity for V2400 Oph. The polarimetry reveals the 927-s spin period (denoted ω) while the X-ray flux varies instead at the beat period (ω–Ω, where Ω is the orbital frequency). The emission-line radial velocities vary with all three periods. Figure from Hellier (2002), using data by Hellier and David Buckley.

stream could orbit, lose energy as they cross field lines, spiral inwards and accrete. Yet they would also tend to screen the field from each other. The balance of the two factors would determine whether the blobs accumulate sufficiently to form a disc, and it appears that V2400 Oph is close to this borderline, with sufficient orbiting blobs to carry much of the accretion flow (thus diluting the stream-fed pulsations), yet with the blobs not 'organised' enough to form a disc or to produce an X-ray spin pulse.

V2400 Oph has a field strength (estimated as 9–27 MG; Buckley *et al.* 1995; Väth 1997) that is as high or higher than in any other IP, and this may explain why the orbiting blobs cannot quite form a disc in this star, whereas they do in most IPs.

A possibly similar case is that of TX Col. Again, the main diagnostic of stream-fed versus disc-fed accretion is the ratio of the beat-cycle to the spin-cycle pulsations in the X-ray lightcurve. In TX Col the two pulsations are of comparable magnitude, with the spin pulse being larger at some epochs and the beat pulse at others (Norton *et al.* 1997; Wheatley 1999). So it appears that in TX Col the orbiting material has managed to organise into some sort of disc, yet just as much of the flow is still in the form of a stream flipping between the poles, and the relative proportions of the two fluctuate with time. Interestingly, Mhlahlo *et al.* (2007b) report the detection of high-amplitude quasi-periodic oscillations in the optical light of TX Col, at a period of 6000 s that is unrelated to the spin or beat periods. Understanding these pulsations could be an important clue to an accretion flow on the verge of disc formation.

3. EX Hya-like IPs

For the magnetosphere to be spinning in equilibrium with a disc (meaning that the corotation velocity is close to the Keplerian velocity at the inner disc edge) the spin period cannot exceed 0.1 $P_{\rm orb}$ (e.g. King & Lasota 1991). This arises since the inner disc edge cannot be further out than the circularisation radius, otherwise the disc would lose more angular at its inner edge than it gains from the stream, and it couldn't survive.

At least three stars, EX Hya, V1025 Cen and DW Cnc, clearly exceed this limit, and are instead in a state where the corotation radius is near the Lagrangian point. One possibility is that they are spinning much more slowly than equilibrium, perhaps as a result of having previously been in a discless state (and it is worth noting that EX Hya has been spinning up monotonically for as long as we've been observing it; e.g. Hellier & Sproats 1992).

The alternative possibility is that these stars currently have no disc. Although there is not yet a consensus on this (e.g. Hellier *et al.* 1987; Hellier, Wynn & Buckley, 2002; Belle *et al.* 2005; Mhlahlo *et al.* 2007a) my own view is that they probably do have discs. The main reasons are (1) neither EX Hya nor V1025 Cen shows an X-ray beat pulse in its usual quiescent state, implying that the accretion loses all memory of orbital phase, and (2) neither star shows polarisation, whereas the discless idea requires them to be among the highest-field IPs, with magnetic fields dominating most of the way to L_1. Further, EX Hya shows outbursts during which the accretion stream does overflow the disc and go as far as the magnetosphere; the signs of this are obvious, including an X-ray beat-cycle pulsation, eclipse profiles of a stream, and high-velocity features in the emission lines (e.g. Hellier *et al.* 2000), which increases our confidence that accretion is purely disc-fed in quiescence when these features are not seen.†

Why do these three stars with longer-than-expected spin periods cluster at short periods? One explanation might be that in smaller, shorter-period binaries the synchronisation torques (thought to be caused by interaction of the magnetic fields of the primary and secondary) are stronger. Thus, if a low state of no mass transfer leads the disc to dissipate, a shorter-period system might relatively quickly head for synchronism whereas a longer-period system would not. Once mass-transfer and a disc are re-established the systems would spin back up, but they would still have a higher probability of being found with long, non-equilibrium spin periods.

4. The disc-fed IPs

In Fig. 2 I plot the spin and orbital periods of the known IPs. I have been conservative in choosing what to include, placing emphasis on detections of a periodicity in multiple data sets, preferably including X-ray data.

The line at $P_{\rm spin}$= 0.1 $P_{\rm orb}$ indicates the expected location for equilibrium rotation of a discless accretor, such as V2400 Oph. Systems above this line cannot both possess discs and be in equilibrium. Other than the short-period systems just discussed, most IPs sit on or below this line, compatible with disc-fed accretion. Further interpretation of the $P_{\rm spin}/P_{\rm orb}$ value is hard without knowledge of the magnetic moment or the radius of the inner disc edge, neither of which we have in most IPs. Norton, Wynn & Somerscales (2004) attempt to extract such information from the $P_{\rm spin}/P_{\rm orb}$ values, but their method

† Mhlahlo *et al.* argue that the magnetic influence does go as far as the *outer* edge of the disc in EX Hya in quiescence. The main argument for this is the detection of emission with the low velocity typical of the outer disc which nevertheless appears to circle with the spin cycle. However, EX Hya is peculiar in having a spin period very close to 2/3rds of the orbit; thus orbital-cycle variations do not smear out when folded on the spin cycle (see Hellier *et al.* 1987). The feature seen by Mhlahlo *et al.* is simply the usual *orbital* S-wave from the edge of the disc, which contaminates the plots of data folded on the spin cycle.

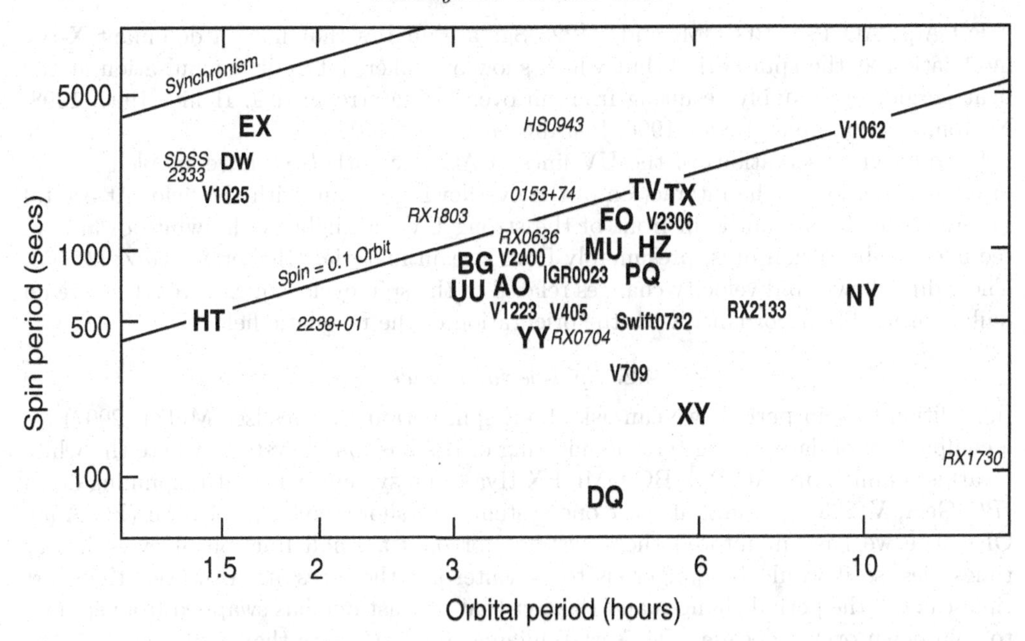

Figure 2. The spin and orbital periods of the intermediate polars, indicated by abbreviated names (some labels have been moved marginally for clarity). Italics indicate that the location on the plot is uncertain, or that the system is not fully secure as an IP, perhaps lacking confirmation from multiple datasets or authors. For details of each system see the compilation by Koji Mukai at the 'IP home page' http://lheawww.gsfc.nasa.gov/users/mukai/iphome/iphome.html

is predicated on a magnetically dominated, non-disc accretion flow, and so would not be applicable to systems with discs.

Observationally, the evidence is that most of these systems do have discs, which is deduced primarily from the fact that the dominant X-ray pulsation is at the spin period (though see Hellier 1991 for a discussion of other indicators). It is also worth noting that none of the IPs appear to show the 'soft X-ray excess' that is sometimes seen in AM Her stars (see Ramsay & Cropper 2004 for AM Her stars and Evans & Hellier 2007 for IPs). The interpretation of soft excesses in AM Hers is that blobbiness of the accretion flow can survive as far as the white-dwarf surface, and that such blobs would not undergo a hard-X-ray-emitting shock but would penetrate the white-dwarf surface and thermalise to produce soft-X-ray emission (e.g. Kuijpers & Pringle 1982). The absence of this effect in IPs suggests that any such blobs are destroyed, being shredded in an accretion disc, or at least during multiple orbits of the white dwarf.

4.1. *Stream-overflow in disc-fed stars*

Possession of an accretion disc does not guarantee that all the accretion flows through it. As pointed out by theorists (e.g. Lubow 1989) the scale height of the stream from the L_1 point is likely to exceed that of the outer disc edge, such that part of the flow continues quasi-ballistically. Around the same time, observers were seeing direct indications of this (e.g. Hellier *et al.* 1989). In many cataclysmic variables the stream will continue inward to some extent, eventually being subsumed into the disc, but if the accretion stream flows as far as the magnetosphere it will produce a beat-cycle X-ray pulsation. The relative power in the beat- and spin-cycle pulsations then presumably gives an indication of the fractions of material accreting by the two paths.

FO Aqr, AO Psc, BG CMi and V1223 Sgr are all IPs that have a dominant X-ray modulation at the spin period, but which show a weaker, intermittent pulsation at the beat period, presumably resulting from an overflowing stream (e.g. Hellier 1993; 1998; Norton, Beardmore & Taylor 1996; Beardmore *et al.* 1998).

In addition, observations of the UV lines of AO Psc with *HST* have enabled a more direct observation of the interaction of the overflowing stream with the field. At orbital phases when the stream is in front of the strong UV backlight of the white dwarf, we see narrow absorption dips, presumably from stream material (Hellier & van Zyl 2005). These dips show rapid velocity changes related to the spin cycle, which reveal the stream flailing around in accordance with the orientation of the magnetic field.

4.2. *Spin-period changes*

In addition to spin periods we can ask about spin-period changes [see Mukai (2007) for a compilation of data]. Long term monitoring of IPs has found systems where the white dwarf is spinning up (AO Psc, BG CMi, EX Hya), and systems where it is spinning down (PQ Gem, V1223 Sgr), and at least one system that shows episodes of both (FO Aqr). Of course we have monitored these systems for only a small fraction of evolutionary timescales, so it would be dangerous to overinterpret these results. However, the short timescales of the period changes and the fact that at least one has swapped from spin up to spin down over a decade (FO Aqr; Williams 2003) suggests that a typical IP below the line in Fig. 2 is hovering about an equilibrium spin period.

A decade ago, Patterson (1994) described this issue as “murky” and suggested that a treasure-trove of $\dot{P}$ data awaited anyone with a good enough theory to interpret it. We are still waiting, and few authors have been brave enough to go beyond Patterson's review.

5. The disc–field connection

One of the hardest questions is what does the connection between the disc and the field look like? What is the radial extent over which disc material feeds onto field lines? What is the vertical extent? What does the vertical section look like?

We have very few observational clues, but one approach is to use eclipses, where the occulting knife edge of the secondary can give spatial information on the X-ray-emitting accretion regions. By tracing these back along field lines, we can deduce information about the disc–magnetosphere interface.

So far, this technique is only possible in XY Ari, which is the only known IP with deep X-ray eclipses. A study by Hellier (1997b) found that the X-ray eclipse egresses occurred in a short enough time that the accretion footprints covered a linear extent of less than 0.1 white-dwarf radii. This suggests that the feeding onto field lines is restricted to a relatively small azimuthal range. That is, at least in quiescence; see § 6 for the changes wrought by an outburst.

Another, more widely applicable technique is to deduce the shapes of the accretion footprints on the white dwarf by interpreting X-ray spin-pulse profiles.

A useful star is HT Cam, which appears to have a very simple X-ray spin pulse, dependent only on whether or not the accretion footprint is on the visible face of the white dwarf. It shows no phase-varying absorption, being devoid of the prominent absorption dips, produced when the accretion flow passes in front of the white dwarf, that are so characteristic of IPs.

Evans & Hellier (2005b) showed that one of the simplest possibilities adequately models HT Cam's spin pulse. In the model, which used a centred dipole, the intensity of the

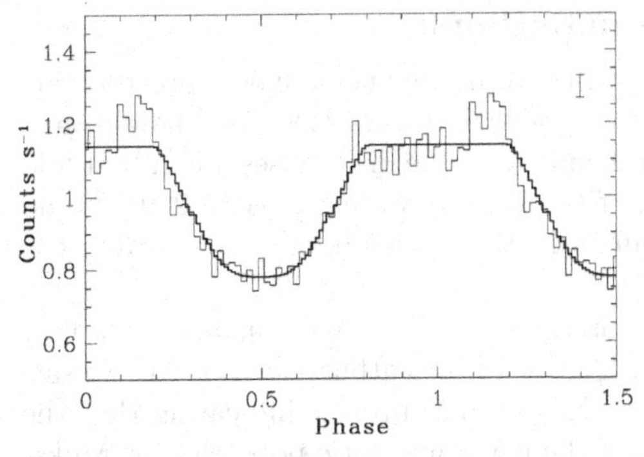

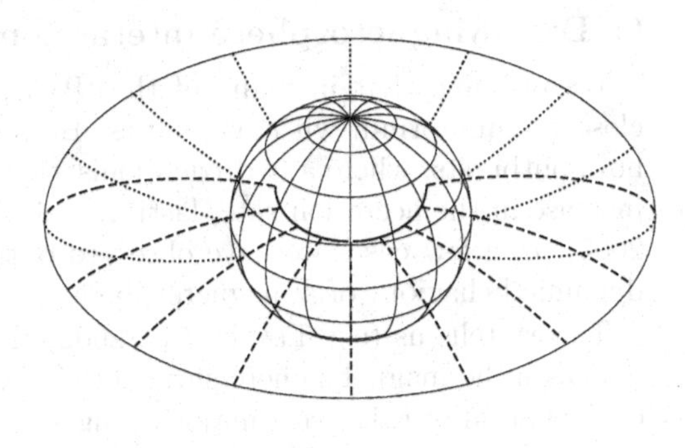

Figure 3. The 'simplest' X-ray spin pulse of HT Cam. The left-hand panel shows the X-ray pulse profile and a fitted model, which is simply a geometric projection of the accretion regions illustrated in the right-hand panel. Figures from Evans & Hellier (2005b).

footprint as a function of magnetic latitude (which presumably maps to accretion rate as a function of azimuth at the inner edge of the disc) was a $\cos^2$ function which peaked at the azimuth to which the magnetic pole pointed, falling to zero 90° away, and then doing the same for the other pole.

V405 Aur, however, shows a more complex, double-peaked spin pulse. In order to explain this Evans & Hellier (2004) suggested that the magnetic dipole was off-centred from the white dwarf, and that the magnetic axis is highly inclined to the spin axis.

All such work is bedevilled by the number of parameters that one could vary in a model (e.g. using multipole, non-symmetric fields) so it should be cautioned that the proposed models may not be unique, with more complex possibilities also possible.

A similar technique to using X-ray pulse profiles is to used spin-pulse profiles of polarised optical light (e.g. Potter *et al.* 1997; 1998). The idea here is that pulsed polarised light originates near to the white dwarf, and so maps to the accretion footprints, whereas unpolarised optical light, even that pulsed on the spin cycle, likely originates from 'accretion curtains' much bigger than the white dwarf. Where both X-ray and polarimetric data combine we can have some confidence in our interpretations, even in complex cases such as PQ Gem, discussed next.

5.1. *Twisting of the field lines*

The 'simplest' IP spin pulse of HT Cam is compatible with field lines that are undistorted. However, detailed analysis of the complex spin pulse of PQ Gem, using X-ray data, optical spectroscopy, and polarimetry, suggests that the accreting field lines are distorted by accretion torques, and that the accreting field lines are 0.1 cycles ahead of the magnetic pole (Mason 1997; Potter *et al.* 1997; Hellier 1997a; Evans, Hellier & Ramsay 2006).

Similar analysis of FO Aqr yields the opposite: the accreting field lines appear to lag the magnetic pole by a quarter of a cycle (Evans *et al.* 2004).

An obvious question is whether these twists are related to the torques on the white dwarf, as shown by period changes. Indeed, PQ Gem, with field lines swept ahead of the pole, is found to be spinning down (Mason 1997). FO Aqr, with field lines swept behind the pole, is spinning up—at least at some epochs. But it has also shown a period of spin down (Williams 2003); and thorough investigation of the X-ray spin pulse over the different epochs (e.g. Beardmore *et al.* 1998) show no obvious shifts in the phases of absorption dips that would imply that the field lines had swapped from trailing to leading.

Thus there is no simple interpretation of the current information on field-line twists and period changes; as yet we have such information on too few systems to discern patterns.

6. Disc–Magnetosphere interactions in outburst

We presume that in many of the IPs the white-dwarf magnetospheres are rotating close to equilibrium with their discs. However, cataclysmic variables also show dwarf-nova outbursts, where a hydrogen-ionisation instability in the disc causes a hundred-fold increase in the accretion rate, lasting typically for several days (e.g. Lasota 2001; Osaki 2005). In a few cases we have observed outbursts in IPs, which allows us to watch the dynamic behaviour of a magnetosphere.

Theory tells us (e.g. Frank *et al.* 2002) that as the ram pressure of the accretion flow increases the magnetosphere shrinks (as $r \propto \dot{M}^{-2/7}$). In an outburst of XY Ari we saw confirmation of this, with major changes in the X-ray pulse profiles indicating that the disc had pushed inwards and blocked the view of the lower magnetic pole (Hellier, Mukai & Beardmore 1997). It took about a day for the disc to push inwards from $\sim 9\ R_{\rm wd}$ to $\sim 4\ R_{\rm wd}$, and eclipse timings indicate that, when it had done so, the flow of material onto field lines was no longer restricted in azimuthal extent, but now flowed from all parts of the inner disc, to fill a complete ring of magnetic longitude around the magnetic poles. It is unclear whether this change is due more to the non-equilibrium rotation (presumably after the disc has pushed inward the Keplerian velocity at its inner edge would now exceed the speed at which the magnetosphere rotates) or whether it relates to the greater scale height of the disc during outburst.

GK Per has an exceptionally long orbital period, and thus a very large disc with long-lived outbursts that are thus easy to study. Analysis of the X-ray pulse profiles suggests that GK Per behaves similarly to XY Ari. The pulse-profiles are explainable by the disc pushing inwards and cutting off the view of the lower pole. Again, the accretion appears to feed from all disc azimuths at the height of outburst, causing complete rings of accretion at the magnetic poles (Hellier, Harmer & Beardmore 2004; see also Vrielmann, Ness & Schmidt 2005).

Thus it is clear that observations of IPs in outburst can reveal important observations of the dynamic interaction of a disc and magnetosphere. However, the fact that such outbursts are occasional, short-lived and unpredictable makes such data hard to obtain. In addition to XY Ari and GK Per, YY Dra (Szkody *et al.* 2002) and HT Cam (Ishioka *et al.* 2002) are among the few systems where this has been achieved. Some IPs (EX Hya, V1223 Sgr & TV Col) show even shorter-lived outbursts, lasting only a day, that are even harder to study. So far it is unclear whether these are disc-instability outbursts (shorter-lived and lower-amplitude owing to the magnetic truncation of the disc) or whether they are episodes of mass transfer from the secondary star (see, e.g., Hellier *et al.* 2000 and references therein).

7. Temporary magnetospheres: DNOs

A long-studied feature of dwarf-nova outbursts is the occurrence of 'dwarf-nova oscillations' in the lightcurves. These are semi-stable periodicities with periods of 6 to 50 secs, which are presumably related to the white-dwarf spin. The puzzle has been why, if they indicate a spinning magnetosphere, are they seen only in outburst in what are otherwise considered to be non-magnetic systems? Surely the magnetic field should be most manifest in quiescence, when the accretion rate is least? The resolution proposed in the series of papers Warner & Woudt (2002), Woudt & Warner (2002), Warner, Woudt & Pretorius (2003) is that the field is only present in outburst, being generated by a dynamo caused by an equatorial belt which is spun up by the extra accretion in outburst

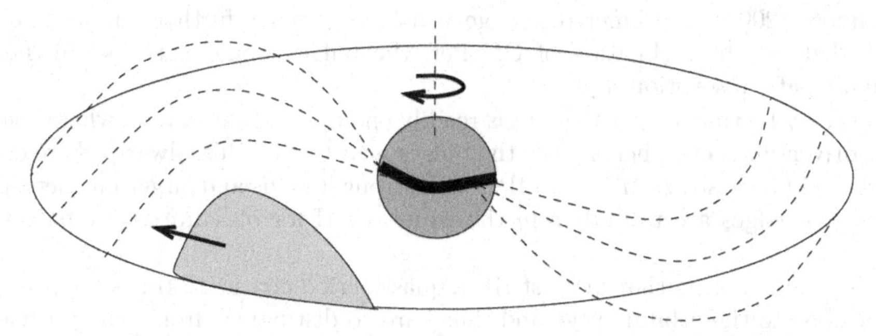

Figure 4. A schematic illustration of a dwarf nova in outburst. While these systems are non-magnetic is quiescence, the increased accretion of outburst spins up an equatorial belt. This slips over the white dwarf, forming a dynamo and generating a temporary magnetosphere. At the inner edge of the disc slow-moving bulges are produced by prograde travelling waves.

and is slipping over the body of the white dwarf. The low-coherence of the oscillations is then explained by the low inertia of the equatorial belt.

The phenomenology of DNOs is too vast to summarise here (see Warner 2004, Pretorius, Warner & Woudt 2006 and references therein), but an important topic is their link to the 'quasi-periodic oscillations' sometimes seen in dwarf novae.

8. QPO waves in the inner disc?

Quasi-periodic oscillations are longer-period than DNOs and less coherent (for an account of the phenomenology see Warner 2004). Although they have been known for decades, it is only relatively recently that a plausible explanation has been proposed. Warner & Woudt (2002) suggest that QPOs are caused by travelling waves at the inner edge of the disc, excited by the interplay of the dynamo-generated field with the inner disc. The waves are slow-moving, prograde bulges that modulate the observed light by periodically obscuring the white dwarf as they circle. Empirically, they appear to circle ≈ 15 times more slowly than the DNO period, which is presumably the rotation period of the temporary, dynamo-driven magnetosphere.

Such bulges can also produce a phenomenon by re-processing the light from the DNO, leading to a quasi-periodicity at the beat period of the DNO and the QPO (see Woudt & Warner 2002; Marsh & Horne 1998).

An interesting question is whether these travelling waves are peculiar to the conditions of a dwarf-nova outburst, or whether they are widespread in disc–magnetosphere interactions. Warner *et al.* (2003) present empirical evidence that the $P_{\rm QPO} = 15 P_{\rm DNO}$ relation holds in a wide range of objects, from neutron stars to black holes. We can thus ask whether the phenomenon is also present in IPs, with their permanent, fossil fields.

One IP almost certainly showing this phenomenon is GK Per, which in outburst shows large-amplitude absorption dips, recurring quasi-periodically with a 5000-s timescale, 15 times longer than the 351-s spin period (Watson, King & Osborne 1985).

One proposed model (e.g. Morales-Rueda, Still & Roche 1999) suggests that the 5000-s timescale is the beat period between the 351-s spin period and the Keplerian velocity at the inner disc edge. However Hellier & Livio (1994) show that, phenomenologically, the modulation must be caused by periodic obscuration by structure circling at 5000 s. Such structure is then best explained by the Warner & Woudt (2002) idea.

Are the travelling-wave bulges there only in the non-equilibrium conditions of outburst, or are they present also in quiescence? It is hard to tell. As discussed in Hellier, Harmer

& Beardmore (2004), the inner disc edge will be $\sim$5 times further out in quiescence, meaning that, at the inclination of GK Per, the bulges would not obscure the white dwarf and create absorption dips.

Thus, it may be that we see the QPOs readily enough in dwarf novae, where the weak, dynamo-driven magnetospheres place the bulges near to the white dwarf, where obscuration effects will be easily seen. But in IPs, with stronger fields and larger magnetospheres, the inner disc edges are too far from the white dwarf for obscuration by bulges to be obvious.

It is also worth noting that in most IPs a quasi-periodicity at 15 times the spin period would be close to the orbital cycle, and thus hard to distinguish from orbital-cycle modulations, which are often prominent (e.g. Hellier, Garlick & Mason 1993; Parker, Norton & Mukai 2005).

Also, given that QPO effects are usually subtle, they would require trains of data of 10 cycles or more to detect, and this is a timescale (30 hrs) on which it is very hard to obtain continuous data trains. Thus for practical reasons, the question of how widespread this phenomenon is in IPs is very hard to answer.

9. Concluding remarks

I have shown above that in many cases observations of magnetic cataclysmic variables give robust clues to how an accretion disc interacts with a magnetosphere. For completeness, I briefly mention two further scenarios. First, there is the possibility of a magnetosphere spinning sufficiently fast to expel material, forming a 'propeller' system. AE Aqr is thought to be in this state, with expulsion of material powered by a rapid spin-down of the white dwarf (e.g. Wynn, King & Horne 1997; Meintjes & de Jager 2000; Meintjes & Venter 2005).

Another possibility is that of a white dwarf with a misaligned spin axis, which is precessing. This would introduce a further periodicity at the precession period into the panoply of cataclysmic-variable periodicities. This has been invoked for stars such as FS Aur and HS 2331+3905 that show periodicities which are otherwise very hard to explain (e.g. Tovmassian *et al.* 2003; Araujo-Betancor *et al.* 2005; Tovmassian, Zharikov & Neustroev 2007).

One notable feature, however, is that owing to the diversity of phenomena even within the intermediate polars, we are often interpreting observational features seen in only one or two systems, rather than analysing patterns seen throughout the class.

There is only one propeller system (AE Aqr); FS Aur and HS 2331+3905 are both unique; only one system (V2400 Oph) is clearly discless; only two systems (EX Hya and GK Per) are well studied in outburst, and both are very different from each other. We have only one system with deep X-ray eclipses (XY Ari, hidden behind a molecular cloud where it cannot be studied in the optical) and one grazing eclipser (EX Hya). Studies of accretion-curtain twists and their relation to spin-period changes are available for only a couple of systems (FO Aqr and PQ Gem); studies of the formation mechanism of the X-ray spin pulses are available for half a dozen stars, but they show a wide range of behaviours, some being single pulsed at the white-dwarf spin period and others being double pulsed (e.g. Evans & Hellier 2005a).

Thus, progress in understanding disc–magnetosphere interactions in these stars is likely to come from studying sufficient systems in detail to look for patterns encompassing the class. Assisting this is the fact that, in the past 5 years, fully a dozen objects have been added to the number we can plot on Fig. 2. Most of these are only lightly studied, and thus the prospects for learning much more about disc–magnetosphere interactions from this class of objects are bright.

References

Araujo-Betancor, S., *et al.* 2005, *A&A*, 430, 629

Beardmore, A.P., Mukai, K., Norton, A.J., Osborne, J.P., & Hellier, C. 1998, *MNRAS*, 297, 337

Belle, K.E., Howell, S.B., Mukai, K., Szkody, P., Nishikida, K., Ciardi, D.R., Fried, R.E., & Oliver, J.P. 2005, *AJ*, 129, 1985

Buckley, D.A.H., Sekiguchi, K., Motch, C., O'Donoghue, D., Chen, A-L., Schwarzenberg-Czerny, A., Pietsch, W., & Harrop-Allin, M.K., 1995, *MNRAS*, 275, 1028

Buckley, D.A.H., Haberl, F., Motch, C., Pollard, K., Schwarzenberg-Czerny, A., & Sekiguchi, K. 1997, *MNRAS*, 287, 117

Evans, P.A., & Hellier, C. 2004, *MNRAS*, 353, 447

Evans, P.A. & Hellier, C. 2005a, in J.-M. Hameury, J.-P. Lasota (eds), *The Astrophysics of Cataclysmic Variables and Related Objects*, ASP Conf. Proc., 330, 165.

Evans, P.A., & Hellier, C. 2005b, *MNRAS*, 359, 1531

Evans, P.A., & Hellier, C. 2007, *ApJ*, 663, 1277

Evans, P.A., Hellier, C., & Ramsay, G. 2006, *MNRAS*, 369, 1229

Evans, P.A., Hellier, C., Ramsay, G., & Cropper, M. 2004, *MNRAS*, 349, 715

Frank, J., King, A.R., & Raine, D. 2002, *Accretion Power in Astrophysics*, (Cambridge: CUP)

Hellier, C. 1991, *MNRAS*, 251, 693

Hellier, C. 1993, *MNRAS*, 265, L35

Hellier, C. 1997a, *MNRAS*, 288, 817

Hellier, C. 1997b, *MNRAS*, 291, 71

Hellier, C. 1998, *Adv. Sp. Res.*, 22, 973

Hellier, C. 2001, *Cataclysmic Variable Stars* (Springer-Praxis: Chichester)

Hellier, C. 2002, in B.T. Gänsicke, K. Beuermann, K. Reinsch (eds), *The Physics of Cataclysmic Variables and Related Objects*, ASP Conf. Proc., 261, 92.

Hellier, C., & Beardmore, A.P. 2002, *MNRAS*, 331, 407

Hellier, C., Garlick, M.A., & Mason, K.O. 1993, *MNRAS*, 260, 299

Hellier, C., Harmer, S., & Beardmore, A. P. 2004, *MNRAS*, 349, 710

Hellier, C., Kemp, J., Naylor, T., Bateson, F.M., Jones, A., Overbeek, D., Stubbings, R., & Mukai, K. 2000, *MNRAS*, 313, 703

Hellier, C., & Livio, M. 1994, *ApJ*, 424, L57

Hellier, C., Mason, K.O., Rosen, S.R., & Cordova, F.A., 1987, *MNRAS*, 228, 463

Hellier, C., Mason, K.O., Smale, A.P., Corbet, R.H.D., O'Donoghue, D., Barrett, P.E., & Warner, B., 1989, *MNRAS*, 238, 1107

Hellier, C., Mukai, K., & Beardmore, A. P. 1997, *MNRAS*, 292, 397

Hellier, C., & Sproats, L.N. 1992, *IBVS*, 3724

Hellier, C., Wynn, G.A., & Buckley, D.A.H. 2002, *MNRAS*, 333, 84

Hellier, C., & van Zyl, L. 2005, *ApJ*, 626, 1028

Ishioka, R., *et al.* 2002, *PASJ*, 54, 581

King, A.R., & Lasota, J.-P. 1991, *ApJ*, 378, 674

Kuijpers, J.; & Pringle, J. E., 1982, *A&A*, 114, L4

Lasota, J.-P. 2001, *New Astron. Revs*, 45, 449

Li, J. 1999, in C. Hellier, K. Mukai (eds.), *Annapolis Workshop on Magnetic Cataclysmic Variables*, ASP Conf. Ser., 157, 235

Lubow, S.H., 1989, *ApJ*, 340, 1064

Marsh, T.R., & Horne, K. 1998, *MNRAS*, 299, 921

Mason, K.O. 1997, *MNRAS*, 285, 493

Meintjes, P. J., & Venter, L. A. 2005, *MNRAS*, 360, 573

Meintjes, P. J., & de Jager, O. C. 2000, *MNRAS*, 311, 611

Mhlahlo, N., Buckley, D.A.H., Dhillon, V.S., Potter, S.B., Warner, B., & Woudt, P.A. 2007a, *MNRAS*, 378, 211

Mhlahlo, N., Buckley, D.A.H., Dhillon, V.S., Potter, S.B., Warner, B., Woudt, P., Bolt, G., McCormick, J., Rea, R., Sullivan, D.J.. & Velhuis, F. 2007b, *MNRAS*, in press (arXiv0705.3259)

Morales-Rueda, L., Still, M.D., & Roche, P., 1999, *MNRAS*, 306, 753

Mukai, K. 2007, IP home page, http://lheawww.gsfc.nasa.gov/users/mukai/iphome/iphome.html

Norton, A.J., Beardmore, A.P., & Taylor, P. 1996, *MNRAS*, 280, 937

Norton, A.J., Hellier, C., Beardmore, A.P., Wheatley, P.J., Osborne, J.P., & Taylor, P. 1997, *MNRAS*, 289, 362

Norton, A.J., Wynn, G.A., & Somerscales, R.V. 2004, *ApJ*, 614, 349

Osaki, Y. 2005, *Proc. Japan Acad., Ser. B: Physical and Biological Sciences*, 81, 291

Parker, T.L., Norton, A.J., & Mukai, K. 2005, *A&A*, 439, 213

Patterson, J. 1994, *PASP*, 106, 209

Potter, S.B., Hakala, P.J., & Cropper, M. 1998, *MNRAS*, 297, 1261

Potter, S.B., Cropper, M., Mason K.O., Hough, J.H., & Bailey, J.A. 1997, *MNRAS*, 285, 82

Pretorius, M.L., Warner, B., & Woudt, P.A. 2006, *MNRAS*, 368, 361

Ramsay, G., & Cropper, M. 2002, *MNRAS*, 334, 805

Ramsay, G., & Cropper, M. 2004, *MNRAS*, 347, 497

Ramsay, G., Potter, S., Cropper, M., Buckley, D.A.H., & Harrop-Allin, M.K. 2000, *MNRAS*, 316, 225

Schmidt, G.D., & Stockman, H.S. 1991, *ApJ*, 371, 749

Schwarz, R., Schwope, A.D., Staude, A., & Remillard, R.A. 2005, *A&A*, 444, 213

Schwope, A. *et al.* 2004, in S. Vrielmann, M. Cropper (eds), *Magnetic Cataclysmic Variables*, ASP Conf. Proc., 315, 92

Staubert, R., Friedrich, S., Pottschmidt, K., Benlloch, S., Schuh, S.L., Kroll, P., Splittgerber, E., & Rothschild, R. 2003, *A&A*, 407, 987

Szkody P., *et al.* , 2002, AJ, 123, 413

Tovmassian, G. et al. 2003, *PASP*, 115, 725

Tovmassian, G.H., Zharikov, S.V., & Neustroev V. V. 2007, *ApJ*, 655, 466

Väth, H. 1997, *A&A*, 317, 476

Vrielmann, S., Ness, J.-U., & Schmitt, J.H.M.M. 2005, *A&A*, 439, 287

Warner, B. 1995, *Cataclysmic Variable Stars* (CUP: Cambridge)

Warner, B. 2004, *PASP*, 116, 115

Warner, B., & Woudt, P.A. 2002, *MNRAS*, 335, 84

Warner, B., & Woudt, P.A. 2006, *MNRAS*, 367, 1562

Warner, B., Woudt, P.A., & Pretorius, M.L. 2003, *MNRAS*, 344, 1193

Watson, M.G., King, A.R., & Osborne, J., 1985, *MNRAS*, 212, 917

Wheatley, P.J. 1999, in C. Hellier, K. Mukai (eds), *Annapolis Workshop on Magnetic Cataclysmic Variables*, ASP. Conf. Ser., 157, 47

Wickramasinghe, D.T., & Ferrario, L. 2000, *New Astron. Revs*, 44, 69

Williams, G. 2003, *PASP*, 115, 618

Woudt, P.A., & Warner, B. 2002, *MNRAS*, 333, 411

Wynn, G.A., & King, A.R. 1995, *MNRAS*, 275, 9

Wynn, G.A., & King, A.R., 1992, *MNRAS*, 255, 83

Wynn, G.A., King, A.R., & Horne, K. 1997, *MNRAS*, 286, 436

Star-Disk Interaction in Young Stars
Proceedings IAU Symposium No. 243, 2007
J. Bouvier & I. Appenzeller, eds.

doi:10.1017/S1743921307009696

The inner gaseous accretion disk around a Herbig Be star revealed by near- and mid-infrared spectro-interferometry

S. Kraus, Th. Preibisch and K. Ohnaka

Max-Planck-Institut für Radioastronomie, Auf dem Hügel 69, 53121 Bonn, Germany
email: skraus@mpifr-bonn.mpg.de

Abstract. Herbig Ae/Be stars are pre-main-sequence stars of intermediate mass, which are still accreting material from their environment, probably via a disk composed of gas and dust. Here we present a recent study of the geometry of the inner (AU-scale) circumstellar region around the Herbig Be star MWC 147 using long-baseline interferometry. By combining for the first time near- and mid-infrared spectro-interferometry on a Herbig star, our VLTI/AMBER and VLTI/MIDI data constrain not only the geometry of the brightness distribution, but also the radial temperature distribution in the disk. The emission from MWC 147 is clearly resolved and has a characteristic physical size of $\sim$ 1.3 AU and $\sim$ 9 AU at 2.2 μm and 11 μm respectively. This increase in apparent size towards longer wavelengths is much steeper than predicted by analytic disk models assuming power-law radial temperature distributions. For a detailed modeling of the interferometric data and the spectral energy distribution of MWC 147, we employ 2-D frequency-dependent radiation transfer simulations. This analysis shows that passive irradiated Keplerian dust disks can easily fit the SED, but predict much lower visibilities than observed, so these models can clearly be ruled out. Models of a Keplerian disk with emission from an optically thick inner gaseous accretion disk (inside the dust sublimation zone), however, yield a good fit of the SED and simultaneously reproduce the observed near- and mid-infrared visibilities. We conclude that the near-infrared continuum emission from MWC 147 is dominated by accretion luminosity emerging from an optically thick inner gaseous disk, while the mid-infrared emission also contains strong contributions from the passive irradiated dust disk.

Keywords. Stars: formation, pre–main-sequence, individual (MWC 147), accretion, accretion disks, techniques: interferometric, radiative transfer.

---

1. Introduction

The spatial structure of the circumstellar material around Herbig Ae/Be stars, i.e. intermediate-mass pre-main sequence stars, is still a matter of debate. Until recently, the spatial scales of the inner circumstellar environment (a few AU, corresponding to $\lesssim$ 0.1 arcsecond) were not accessible to optical and infrared imaging observations, and conclusions drawn on the spatial distribution of the circumstellar material were, in most cases, entirely based on the modeling of the spectral energy distribution (SED). However, as demonstrated by Men'shchikov & Henning (1997) and others, these SED model fits are highly ambiguous, and only the combination of SED modeling with high-resolution imaging can provide crucial constraints on the real geometry of the circumstellar matter. Since these ambiguities concern not only the disk geometry, but also the general nature of the disk (e.g. actively accreting vs. passive irradiated disks), high angular resolution observations are urgently required.

In recent years, near-infrared (NIR) interferometry could make important contributions for a better understanding of the structure of the circumstellar material around

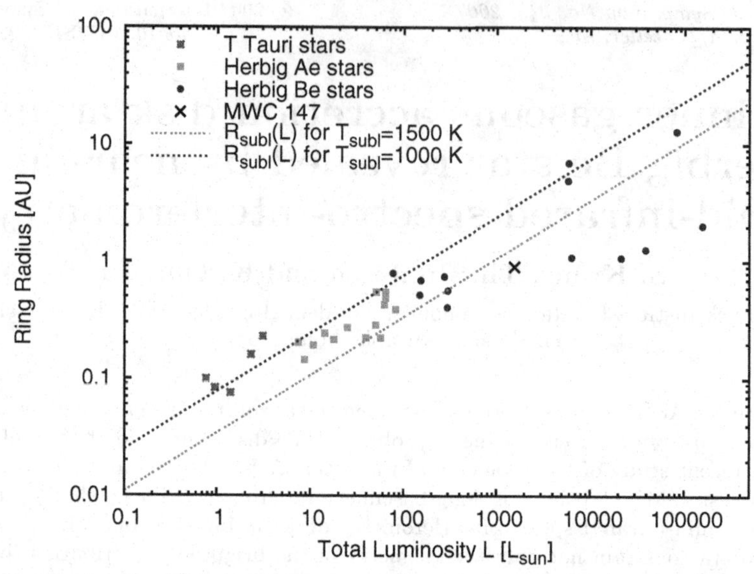

Figure 1. Size-luminosity relation for NIR long-baseline interferometric measurements on YSOs (adopted from Millan-Gabet *et al.* 2007). While the measured sizes for most T Tauri and Herbig Ae stars agree to the dust sublimation radius $R_{\rm subl}$ expected for sublimation temperatures between 1000 and 1500 K (dotted lines), some Herbig Be stars (including MWC 147, marked as cross) show significantly smaller diameters.

these stars (see also the contribution by F. Malbet for a review). A survey performed by Millan-Gabet, Schloerb & Traub (2001) with the IOTA interferometer showed that the near-infrared continuum emission is best described by ring-like or spherical geometries rather than with the "classical" optically thick, geometrically thin accretion disk models (e.g. Lynden-Bell & Pringle 1974; Hillenbrand *et al.* 1992). Later, a correlation between the NIR size and the stellar luminosity was found (Fig. 1; see Monnier & Millan-Gabet 2002), which suggests that the NIR continuum emission mainly traces hot dust at the inner sublimation radius, missing any shielding by optically thick gas inside of the sublimation radius. This observational result also stimulated theoretical work, especially for passive irradiated circumstellar disks, e.g. by Natta, Prusti, Neri, *et al.* (2001) and Dullemond, Dominik & Natta (2001). These models reproduce the ring-like morphology by introducing pronounced, hot dust emission at the inner rim of the circumstellar disk.

However, for the more luminous Herbig Be stars, deviations from the simple $R \propto L^{1/2}$ size-luminosity relation were found (see Fig. 1), indicating that for these hot stars, the inner rim lies closer to the star than expected from the stellar luminosity. Monnier & Millan-Gabet (2002) suggested that this might be due to the presence of an inner gaseous disk, which shields the dust disk from the strong stellar ultraviolet (UV) radiation. Since this shielding would be most efficient for hot stars, it would allow the inner rim of the dust disk around B-type stars to exist closer to the star. Several subsequent studies favor classical accretion disk models (e.g. Eisner *et al.* 2004; Monnier *et al.* 2005), in which the infrared emission contains contributions from the thermal emission of optically thick gas in the innermost disk regions.

Whereas the majority of the existing infrared interferometric studies on YSOs were performed at NIR wavelengths (H and K-band), a smaller number of studies investigated also their mid-infrared (MIR) (N-band) emission (e.g. Leinert *et al.* 2004; Preibisch *et al.* 2006). The latest generation of interferometric instruments provides not only the high spatial resolution achievable with long-baseline interferometry, but also some spectroscopic capabilities. In these cases, the interferograms are spectrally dispersed, such that the fringe contrast (visibility) can be measured for individual spectral channels.

Considering a simple model where each disk annulus radiates as a black-body of different temperature, it is clear that different spectral channels probe different physical regions within the disk. Since spectro-interferometry allows us to measure the spatial geometry (in first approximation the object extension) in each spectral channel, these measurements constrain simultaneously the geometry and also the radial temperature distribution in the disk. In general, one expects the NIR emission to originate mainly from the hottest dust located close to the dust sublimation radius, with significant contributions from the stellar photosphere and/or from scattered light. MIR wavelengths trace also slightly colder dust located further out in the disk. Whereas simple analytic models for the disk temperature distribution provide some understanding for the general dependences, more sophisticated radiative transfer simulations are required to include dust scattering and optical depth effects.

2. Interferometric observations of MWC 147

MWC 147 is a well studied intermediate-mass pre-main sequence star located at a distance of 800 pc in Monoceros. For our modeling, we adopted the stellar parameters by Hernández, Calvet, Briceño *et al.* (2004), namely a spectral type of B6, $L_\star = 1\,550\ L_\odot$, $M_\star = 6.6 M_\odot$, and $R_\star = 6.63\ R_\odot$. The object shows a strong infrared excess of about 6 mag at mid-infrared wavelengths, demonstrating the presence of circumstellar material. Numerous recent observational results strongly suggest the presence of a massive circumstellar disk around MWC 147. Hillenbrand, Strom, Vrba, *et al.* (1992) fitted the spectral energy distribution of MWC 147 with a model assuming a massive accretion disk and estimated an accretion rate of $10^{-5}\ M_\odot\,\mathrm{yr}^{-1}$.

First infrared interferometric observations by Akeson, Ciardi, van Belle, *et al.* (2000) revealed that the K-band size is surprisingly compact (2.28 mas = 0.7 AU, uniform disk diameter). We observed MWC 147 with the interferometric instruments MIDI and AMBER at the ESO Very Large Telescope Interferometer (VLTI); all details are described in Kraus, Preibisch, & Ohnaka (2007). From the resulting seven MIDI and one AMBER measurements, we derive wavelength-dependent visibilities, which can either be fitted directly to model images or used to compute the characteristic size of the emitting region (assuming a certain intensity profile, see Fig. 2). For our fits, we included also additional K-broadband data from the PTI archive, and the upper limit on the H-band size derived by Millan-Gabet, Schloerb & Traub (2001).

As a first step of analysis we compared the interferometric data to analytic disk models. Models of both, passive irradiated circumstellar disks as well as of active viscously accreting disks (Lynden-Bell & Pringle 1974), predict that the radial temperature profile of YSO disks should follow a simple power-law $T(r) \propto r^{-\alpha}$. Most studies infer a power law index of $\alpha = 3/4$ (e.g. Millan-Gabet *et al.* 2001) or $1/2$ (e.g. Leinert *et al.* 2004).

Using the assumption that each disk annulus radiates as a blackbody and a temperature of $T_{\mathrm{subl}} = 1\,500$ K at the inner disk truncation radius, we can compute the wavelength-dependence of the disk size corresponding to these analytic model. Adjusting the resulting wavelength-dependent size to fit the NIR size measured on MWC 147 (see Fig. 2), we find that these analytic models can not reproduce the measured NIR and MIR-sizes simultaneously. Therefore, we performed a more detailed radiative transfer modeling.

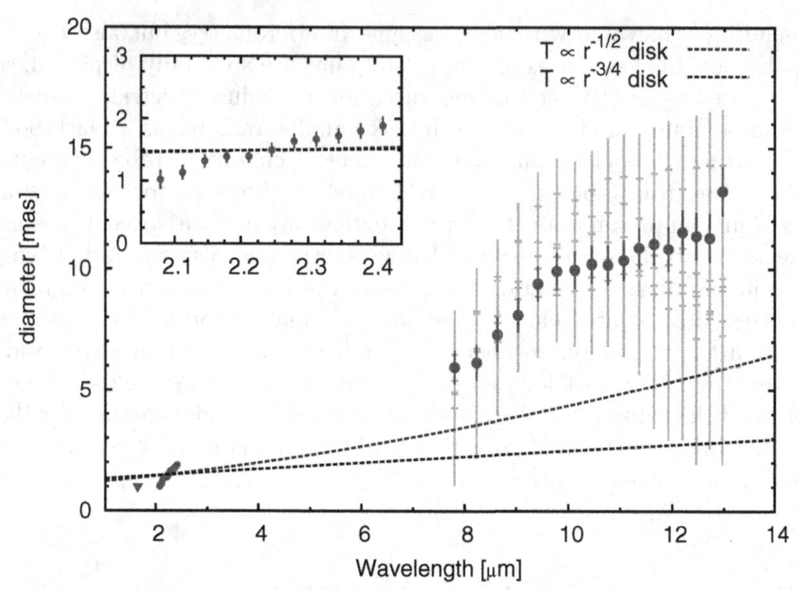

Figure 2. Wavelength-dependence of the measured characteristic size over the H-, K- and N-band. Gaussian intensity profiles were assumed. For comparison, we show the wavelength--dependent size as predicted by simple analytic disk models. As the scaling of these models is arbitrary, we normalize them to the measured NIR size. It is evident that these analytic models can not describe the measured wavelength-dependent size well.

3. 2-D radiative transfer modeling

For our modeling of the interferometric data of MWC 147 we employ the radiative transfer code *mcsim_mpi* (Ohnaka *et al.* 2006), which solves the radiative transfer problem self-consistently using a Monte Carlo approach. For each radiative transfer model, we first check the agreement with the SED of MWC 147, which we constrain using an archival *Spitzer*-IRS spectrum as well as photometric data from the literature. Then, a ray-trace program is used to compute synthetic images for any wavelength of interest. Finally, visibilities are computed from the simulated images for the points of the *uv*-plane covered by the data. Both the disk inclination as well as the orientation of the disk on the sky are treated as free parameters, which we fit in order to find best agreement with the spectro-interferometric visibilities. All details of the modeling are described in Kraus, Preibisch, & Ohnaka (2007).

The dust density distribution of the accretion disk in our models resembles a flared, Keplerian rotating disk with puffed-up inner rim, as parameterized by Dullemond, Dominik & Natta (2001) and extends from the dust sublimation radius (at $\sim$ 2.7 AU, assuming $T_{\rm subl}$ = 1500 K) to 100 AU. The radial density distribution was chosen according to $\rho(r) \propto r^{-3/2}$ and the disk flaring index $\beta = 1.175$. In order to reproduce the shape of the SED, we found that, in addition to the disk, an extended envelope is required, for which we use a density distribution of the form $\rho(r) \propto r^{-1/2}$. The presence of such an envelope is also supported by mid-infrared imaging by Polomski, Telesco, Piña, *et al.* (2002), which revealed an extended, elongated structure.

Fig. 3 shows model images, the SED, as well as the visibilities corresponding to the model of a passive, irradiated accretion disk. We find that the predicted model visibilities (especially in the NIR, but also, to a lower degree, in the MIR) are always much smaller than the measured visibilities. We conclude that although passive irradiated circumstellar

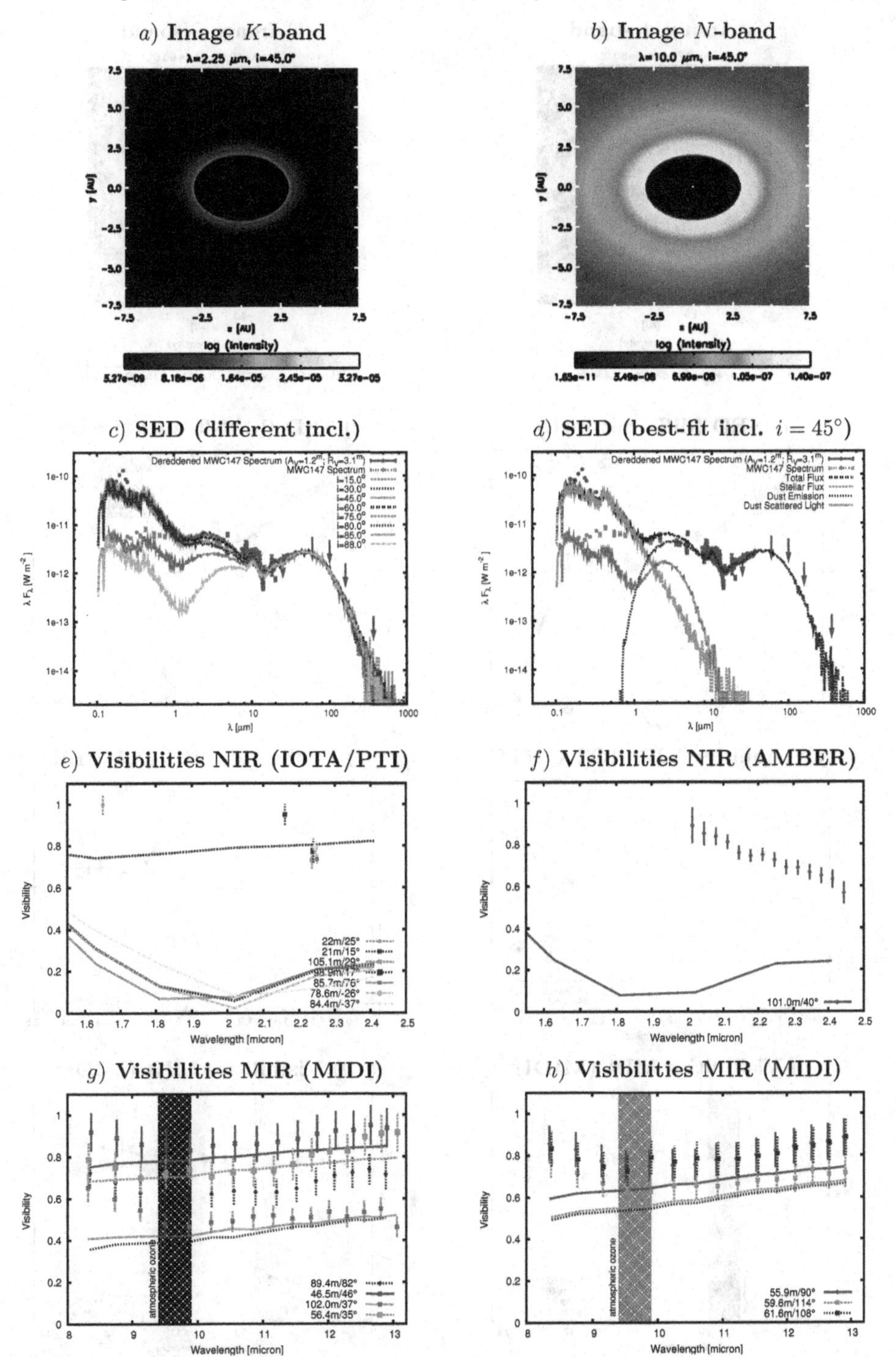

Figure 3. Best-fit radiative transfer model assuming a flared Keplerian disk geometry. This model results in a poor χ_r^2 of 42. Panels *a*) and *b*) show the ray-traced images for two representative wavelengths (2.25 μm and 10.0 μm), *c*) shows the SED for various inclination angles, whereas *d*) gives the SED for the best-fit inclination angle (45°). Panels *e*) to *h*) show the observed and the model NIR and MIR visibilities.

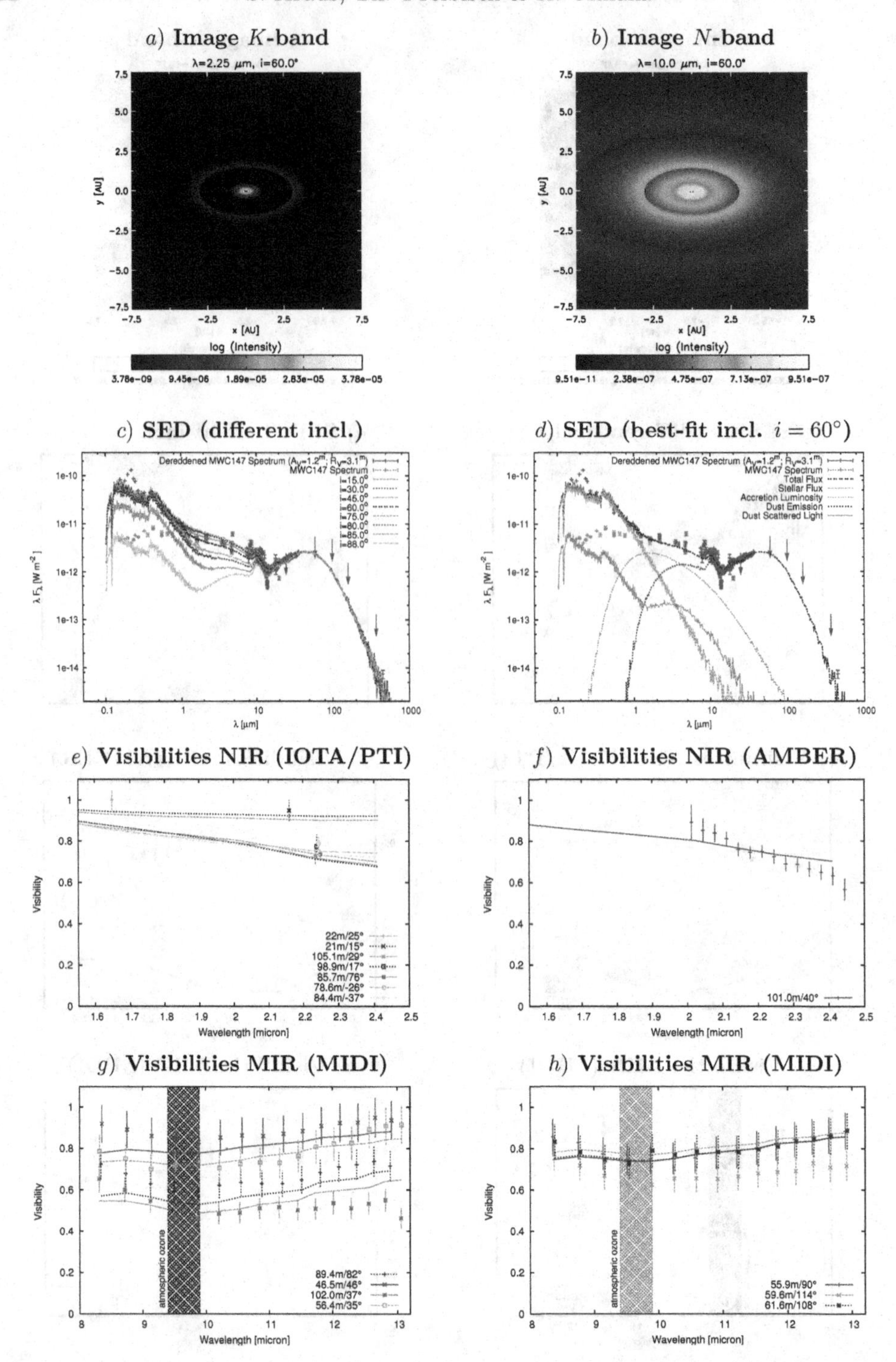

Figure 4. As Fig. 3, but for our best-fit model of a flared Keplerian disk with optically thick gas located inside the dust sublimation radius. With $\chi_r^2 = 1.26$ (inclination 60°), this model fits our spectro-interferometric data much better than the model without optically thick gas emission (Fig. 3).

disk models are able to reproduce the SED of MWC 147, these models are in strong conflict with the interferometric measurements, resulting in $\chi_r^2 \approx 43$.

In passive circumstellar disks, the infrared emission is generally assumed to originate almost entirely from dust; the emissivity of the inner, dust-free gaseous part of the disk, at radii smaller than the dust sublimation radius, is negligible. In an actively accreting disk, on the other hand, viscous dissipation of energy in the inner dust-free gaseous part of the accretion disk can heat the gas to high temperatures and give rise to significant amounts of infrared emission from optically thick gas. The inner edge of this gas accretion disk is expected to be located a few stellar radii above the stellar surface, where the hot gas is thought to be channeled towards the star via magnetospheric accretion columns. While the magnetospheric accretion columns are too small to be resolved in our interferometric data (3 $R_\star$ correspond to 0.09 AU or 0.12 mas), infrared emission from hot gas between the dust sublimation radius and the stellar surface should be clearly distinguishable from the thermal emission of the dusty disk due to the different temperatures of these components and the resulting characteristic slope in the NIR- and MIR-visibilities.

As MWC 147 is a quite strong accretor ($\dot{M}_{\rm acc} \approx 10^{-5}\ M_\odot\,{\rm yr}^{-1}$; Hillenbrand *et al.* 1992), significant infrared emission from the inner gaseous accretion disk is expected. Muzerolle, D'Alessio, Calvet, *et al.* (2004) found that, even for smaller accretion rates, the gaseous inner accretion disk is several times thinner than the puffed-up inner dust disk wall and is optically thick (both in radial as well as in the vertical direction).

In order to add the thermal emission from the inner gaseous disk to our radiative transfer models, we assume the radial temperature power-law by Pringle (1981). Including the accretion luminosity from an inner gaseous disk to the model improves the agreement between model predictions and observed visibilities strongly. With a flared disk geometry and an accretion rate of $\dot{M}_{\rm acc} = 9 \times 10^{-6}\ M_\odot\,{\rm yr}^{-1}$, both the SED and the interferometric visibilities are reasonably well reproduced ($\chi_r^2 = 1.26$, see Fig. 4).

4. Conclusions

Our infrared long-baseline interferometric observations of MWC 147, constrain, for the first time, the inner circumstellar environment around a Herbig Be star over the wavelength range from 2 to 13 μm. We measure a strong increase of the apparent size with wavelength, with a slope much steeper than predicted by the commonly used analytic temperature power-law models. To test whether more realistic physical models of the circumstellar dust environment yield better agreement, we employed 2-D radiative transfer modeling. These models include a dust disk embedded in an extended dust envelope, and were used to simultaneously fit the SED and the spectro-interferometric observables. While models of passive irradiated disks are able to reproduce the SED, they are in strong conflict with the interferometric observables, significantly overestimating the size of both the NIR and MIR emission. Adding an inner gaseous accretion disk component to the model solves the discrepancies and provides good agreement with the interferometric data. The best-fit was obtained with a flared Keplerian disk which is seen under an inclination of $\sim 60°$, extending out to 100 AU and exhibiting a mass accretion rate of $9 \times 10^{-6}\ M_\odot\,{\rm yr}^{-1}$. Our study suggests that the spectro-interferometric capabilities of the latest generation of long-baseline interferometric instruments are particularly well suited to reveal the contributions from active accretion processes taking place close to the star.

References

Akeson, R. L., Ciardi, D. R., van Belle, G. T., Creech-Eakman, M. J. & Lada, E. A. 2000, *ApJ* 543, 313
Dullemond, C. P., Dominik, C. & Natta, A. 2001, *ApJ* 560, 957
Eisner, J. A., Lane, B. F., Hillenbrand, L. A., Akeson, R. L. & Sargent, A. I. 2004, *ApJ*, 613, 1049
Hernández, J., Calvet, N., Briceño, C., Hartmann, L., & Berlind, P. 2004, *AJ* 127, 1682
Hillenbrand, L. A., Strom, S. E., Vrba, F. J. & Keene, J. 1992, *ApJ* 397, 613
Kraus, S., Preibisch, Th., & Ohnaka, K. 2007, *ApJ*, submitted
Lynden-Bell, D. & Pringle, J. E. 1974, *MNRAS*, 168, 603
Leinert, C., van Boekel, R., Waters, L. B. F. M., *et al.*, 2004 *A&A*, 423, 537
Men'shchikov, A. B., & Henning, T. 1997, *A&A*, 318, 879
Millan-Gabet, R., Schloerb, F. P., & Traub, W. A. 2001, *ApJ* 546, 358
Millan-Gabet, R., Malbet, F., Akeson, R., Leinert, C., Monnier, J., & Waters, R. 2007, Protostars and Planets V, eds. B. Reipurth, D. Jewitt, K. Keil, University of Arizona Press, p. 539
Monnier, J. D. & Millan-Gabet, R. 2002, *ApJ* 579, 694
Monnier, J. D., Millan-Gabet, R., Billmeier, R., *et al.*, 2005, *ApJ* 624, 832
Muzerolle, J., D'Alessio, P., Calvet, N. & Hartmann, L. 2004, *ApJ* 617, 406
Natta, A., Prusti, T., Neri, R., Wooden, D., Grinin, V. P., & Mannings, V. 2001, *A&A*, 371, 186
Ohnaka, K., Driebe, T., Hofmann, K.-H., Leinert, C., Morel, S., Paresce, F., Preibisch, T., Richichi, A., Schertl, D., Schöller, M., Waters, L. B. F. M., Weigelt, G., Wittkowski, M. 2006, *A&A* 445, 1015
Preibisch, Th., Kraus, S., Driebe, T., van Boekel, R., & Weigelt, G. 2006, *A&A*, 458, 235
Pringle, J. E. 1981, *ARA&A*, 19, 137
Polomski, E. F., Telesco, C. M., Piña, R., & Schulz, B. 2002, *AJ*, 464, 1

Discussion

ARDILA: Do you include in your modeling absorption by gas located inside the dust sublimation radius?

KRAUS: We agree that full radiative transfer modeling, including gas absorption, would provide a more complete picture. However, including dust and gas opacities in one model would dramatically increase the complexity of the radiative transfer modeling, which was out of the scope of our investigation. Based on the results of the theoretical work by Muzerolle, D'Alessio, Calvet, *et al.* (2004), which predict that the gaseous disk is optically thick and geometrically thin, we expect gas absorption also to be a secondary effect.

CABRIT & NATTA: What are the reasons to include an extended envelope? Would it be possible to fit the data with a pure disk geometry?

KRAUS: We required an extended envelope to successfully reproduce the pronounced shape of the MWC 147 mid- to far-infrared SED and in particular the sharp drop at 15 μm in the *Spitzer*/IRS spectrum. We were not able to reproduce the SED equally well assuming only a disk component. The presence of an extended envelope is also supported by the mid-infrared imaging presented by Polomski, Telesco, Piña, *et al.* (2002), revealing an elongated diffuse emission component, extending ~ 6 arcseconds around MWC 147.

Star-Disk Interaction in Young Stars
Proceedings IAU Symposium No. 243, 2007
J. Bouvier & I. Appenzeller, eds.

doi:10.1017/S1743921307009702

Star-disk interaction in brown dwarfs: implications for substellar formation

Subhanjoy Mohanty
Harvard-Smithsonian CfA, 60 Garden Street, M.S.42, Cambridge, MA 02138, USA
email: smohanty@cfa.harvard.edu

Abstract. I review the current state of knowledge regarding disk accretion in young brown dwarfs (BDs), and the interaction of the disk with the central object. In particular, I discuss (1) observations of accretion/outflow phenomena in BDs; (2) techniques for measuring accretion rates ($\dot{M}_{acc}$); (3) the dependence of $\dot{M}_{acc}$ on the central mass from stars to brown dwarfs; (4) the temporal evolution of $\dot{M}_{acc}$; and (5) observations of variability in the accretion line profiles. I then examine the implications of these issues for the formation mechanism of BDs, and discuss new observations that can further constrain substellar origins.

Keywords. Circumstellar matter, planetary systems, stars: formation, stars: low-mass, brown dwarfs, stars: pre-main-sequence.

1. Introduction

Brown dwarfs (BDs) are by definition substellar objects: with a mass $\lesssim$ 80 M_{Jup} (0.08 $M_{\odot}$), they are incapable of sustaining stable hydrogen fusion, and thus, after an initial period of Deuterium fusion, simply continue to cool down and grow fainter with time. Over the last decade, astronomers have discovered hundreds of these diminutive bodies, both in the field as well as in star-forming regions and young clusters, with masses ranging down to nearly (or perhaps even below) the planetary mass boundary ($\sim$ 12 M_{Jup}). While some BDs are companions to stars, the vast majority are isolated bodies.

The formation mechanism of these ultra-low mass objects presents a challenge to theorists. The central dilemma is easily framed: since the average thermal Jeans mass in molecular clouds is usually of order a solar mass, how does one form free-floating objects ten to a hundred times less massive? Two main competing hypotheses have emerged. The 'ejection' scenario posits that BDs are in fact 'stellar embryos', born within molecular cloud cores but subsequently flung out from the cores by dynamical interactions with their neighbours before accumulating enough gas to become full-fledged stars (Reipurth & Clarke 2001; Bate *et al.* 2003). A key prediction here is that disks around young BDs should be severely truncated by these interactions, and hence far smaller than those girdling newborn stars. As a corollary, the classical T Tauri (CTT) phase of disk accretion that is ubiquitous in young stars may also be shortlived and rare in BDs.

In the alternative 'turbulent fragmentation' picture (Padoan & Nordlund 2004), turbulent shocks in molecular clouds create gravitationally bound cores with a range of masses: from large ones that form stars, to ultra-low mass ones – far below the cloud's mean thermal Jeans mass, but *locally* unstable – that collapse directly into BDs. In this view, there is no fundamental distinction between the formation mechanism of stars and BDs: both form quietly within their own cores. Extended disks, and an associated CTT phase, should then be as common in BDs as in stars.

These two formation mechanisms have key implications for low-mass star formation in general as well. In the first case, gravitational interactions also set the final stellar masses, by limiting the amount of material that can accrete onto nascent protostars: through competitive accretion, repeated truncation of the primordial disks, and outright ejection from the core. In the second case, the mass spectrum of star-forming cores is ultimately regulated by cloud turbulence. Thus, discriminating between these substellar formation scenarios is also vital for understanding the physics governing the *stellar* initial mass function (IMF).

The search for disks, and a CTT-like disk-accretion phase, around young BDs has consequently been a major area of research in the last few years. Disk studies have focussed on excess dust emission at near-infrared to sub-mm/mm wavelengths, while the identification of disk-accretion relies primarily on optical and near-infrared spectroscopy. Surveys have now firmly established that disks are very common in young BDs, just as in stars. In this paper, I review the accretion results.

It is worth pointing out at this juncture that simply the *presence* of disks and accretion, while providing constraints on BD formation, is *not* sufficient to unambiguously distinguish between the two formation scenarios discussed above. I return to this subject in §5 and 6, where I review the constraints implied by current data, and if and how future observations combined with current ones may finally settle the issue of BD formation.

2. Accretion and outflow signatures in brown dwarfs

In solar-type stars, the most straightforward way of identifying ongoing disk accretion (i.e., the CTT phase) is through the presence of optical and/or UV 'veiling' (filling in of photospheric absorption lines), which arises due to excess continuum emission from the hot accretion shock region. Unfortunately, extremely few BDs manifest measurable veiling; as we shall see in §3 and 4.1, this is due to their very small accretion rates.

Another widely used accretion diagnostic in solar-type stars is the equivalent width (EW) of the $H\alpha$ emission line, which is expected to arise both in the accretion funnel flow as well as in the accretion shock: EW[$H\alpha$] > 10Å is usually adopted as signature of ongoing accretion in these stars, while smaller values signify only chromospheric emission. However, this cutoff cannot be extended to arbitrarily late spectral types such as young BDs (SpT $\gtrsim$ M6 at ages of a few Myr). The underlying photospheric continuum at $H\alpha$ (6563Å) lies in the Wien part of the spectrum in the M types, and thus declines drastically with decreasing temperature (later type). Thus even a small *chromospheric* $H\alpha$ flux can easily exceed 10Å in emission EW for very low-mass stars (VLMS) and BDs. To counter this problem, Martín (1998), Barrado y Navascués & Martín (2003) and White & Basri (2003) have proposed spectral-type dependent EW[$H\alpha$] cutoffs for identifying accretion, based on the saturation EW of purely chromospheric $H\alpha$ emission observed as a function of spectral type.

White & Basri (2003) also pointed out a much more straightforward way of identifying accretion, based on the *profile* of the $H\alpha$ emission. They noted that empirically, in high-resolution spectra, the full-width of the $H\alpha$ line at 10% of the emission peak ($\equiv$ FW10) was always $\gtrsim$ 270 kms^{-1} in accretors regardless of spectral type, from solar type stars down to the substellar limit (SpT $\sim$ M6–M7). In subsequent work, Jayawardhana *et al.* (2003) lowered this cutoff to FW10 $\gtrsim$ 200 kms^{-1} for the BD regime (SpT > M6). The simple explanation for this behaviour is that the accretion flow arises from the disk edge at a few stellar radii, and thus attains nearly free-fall velocities in travelling to the central star/BD. Even without other line-broadening mechanisms (e.g., Stark), this implies full-widths $\gtrsim$ 200–250 kms^{-1} for the Hα line wings in low mass stars and BDs (Jayawardhana *et al.* 2003, Mohanty *et al.* 2005). One caveat here is that for the low accretion rates seen in BDs (§3, 4.1), Stark broadening of the line is not expected to be a strong factor, and the infall velocity appears to be the primary driver of the $H\alpha$ line-width (Muzerolle *et al.* 2000; Jayawardhana *et al.* 2003). Consequently, inclination effects can have a significant effect on the full-width: FW10 may fall below the 200 kms^{-1} cutoff even in accreting BDs if the system is observed close to pole-on (Muzerolle *et al.* 2003; Mohanty *et al.* 2005). Nevertheless, combined with other line diagnostics such as permitted emission in HeI (6678Å), OI (8400Å), CaII (8662Å) and forbidden emission in e.g. [OI] (6300Å) (see below), FW10 $\gtrsim$ 200 kms^{-1} provides a good *conservative* diagnostic for identifying accreting VLMS and BDs (Mohanty *et al.* 2005).

Various teams have now obtained high-resolution optical spectra of more than a hundred VLMS and BDs (SpT $\sim$ M5–M9.5: mass $\sim$ 0.1–0.015 $M_\odot$) in nearby star-forming regions (d $\lesssim$ 400pc, age: $\lesssim$1–10 Myr), covering a major fraction of the known BDs in these regions (e.g., Muzerolle *et al.* 2003, 2005; Mohanty *et al.* 2003, 2005; Barrado y Navascués *et al.* 2004). Using the FW10[$H\alpha$] diagnostic above, scores of these sources have been clearly identified as accretors. Some of the salient qualitative features observed are (see Figs. 1, 2, 3):

(1) accretion is observed over the full range of BD masses, down to the lowest mass objects studied ($\sim$ 15 M_{Jup}), approaching the planetary mass boundary;

(2) the broad line-wings of the $H\alpha$ accretion line-profiles indicate free-fall infall velocities, implying the same paradigm applicable to CTTs: magnetospherically channeled accretion flows from an inner disk edge at a few stellar radii;

(3) the $H\alpha$ line profiles in VLMS and BD accretors often evince asymmetries and time-variability analogous to those seen in higher-mass CTTs (see §4.3);

(4) while some emission lines, such as HeI (5876Å), two of the CaII infrared triplet lines (8498, 8542Å) and of course $H\alpha$ itself, appear in both accreting as well as simply chromospherically active VLMS and BDs, others – HeI (6672Å), OI (8446Å) and one of the CaII infrared triplet lines (8662Å) – appear to be present only in accreting VLMS and BDs (though not in *all* the latter), and are thus good secondary indicators of accretion when they are present, in addition to the $H\alpha$ FW10 diagnostic described above;

(5) forbidden emission lines such as [OI] (6300Å), indicative of mass outflow in jets and winds, are observed in a few accreting BDs (see further below).

Accretion signatures are observed in VLMS and BDs in near-infrared spectra as well. In particular, accreting VLMS and BDs often exhibit Paβ and Brγ in emission, while merely chromospherically active VLMS and BDs do not (Fig. 2; Natta *et al.* 2004, 2006; Gatti *et al.* 2006). It is true that these lines are not as sensitive to accretion as $H\alpha$: their smaller optical depth results in their absence in some low mass accretors that still exhibit broad $H\alpha$. Nevertheless, these near-infrared lines are especially important accretion diagnostics in regions of high extinction (e.g., ρ Oph), where high-resolution optical spectroscopy is prohibitively time-consuming.

Finally, accretion signatures in CTTs are often accompanied by those of mass outflow in jets and winds. Indeed, from a theoretical perspective, accretion and outflow are inextricably linked, with the latter carrying away excess angular momentum from the infalling material. In the VLMS and BD regime, jet/wind indicators, such as emission in forbidden lines of [OI], [NII] and [SII] and blue-shifted absorption in $H\alpha$, have been identified in only a handful of sources (Figs. 1, 3; e.g., Barrado y Navascués *et al.* 2004; Whelan *et al.* 2005, 2007; Muzerolle *et al.* 2005; Mohanty *et al.* 2005). This is not too surprising: if the outflow rates are comparable to the infall ones (as theory suggests), then the very small accretion rates in the VLMS/BD regime would generally result in outflow signatures near/below current sensitivity limits. Indeed, in the one BD in which both accretion and outflow rates have been estimated from high-resolution spectra (LS-RCrA-1: Barrado y Navascués *et al.* 2004; Mohanty *et al.* 2005), the outflow estimates

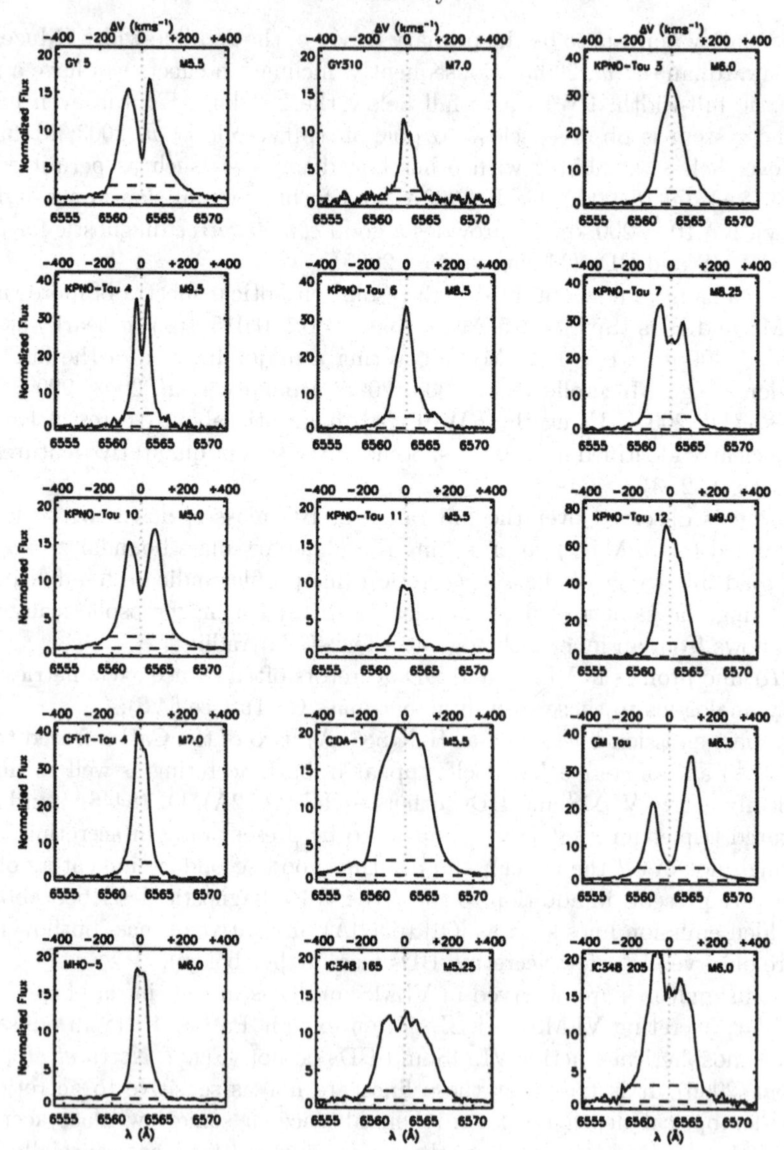

Figure 1. $H\alpha$ profiles in a sample of accreting BDs, exhibiting the broad line wings (except in a few pole-on cases), and line asymmetries due to geometrical and wind effects, associated with CTT-like magnetospheric accretion. From Mohanty *et al.* (2005).

range from an order of magnitude lower to comparable to the accretion rate, roughly consistent with theory. The jets in two BDs (ρ Oph 102 and 2MASS 1207-3932) have now also been spectro-astrometrically resolved (Whelan *et al.* 2005, 2007); the data suggest scaled-down versions of CTT jets, in agreement with the idea that the outflows scale with accretion. However, no BD jet has yet been spatially resolved in imaging.

3. Measuring the accretion rates

Accretion rates ($\dot{M}$) in CTTs are derived most straightforwardly from the accretion luminosity (L_{acc}), as measured from the excess continuum emission in the optical and

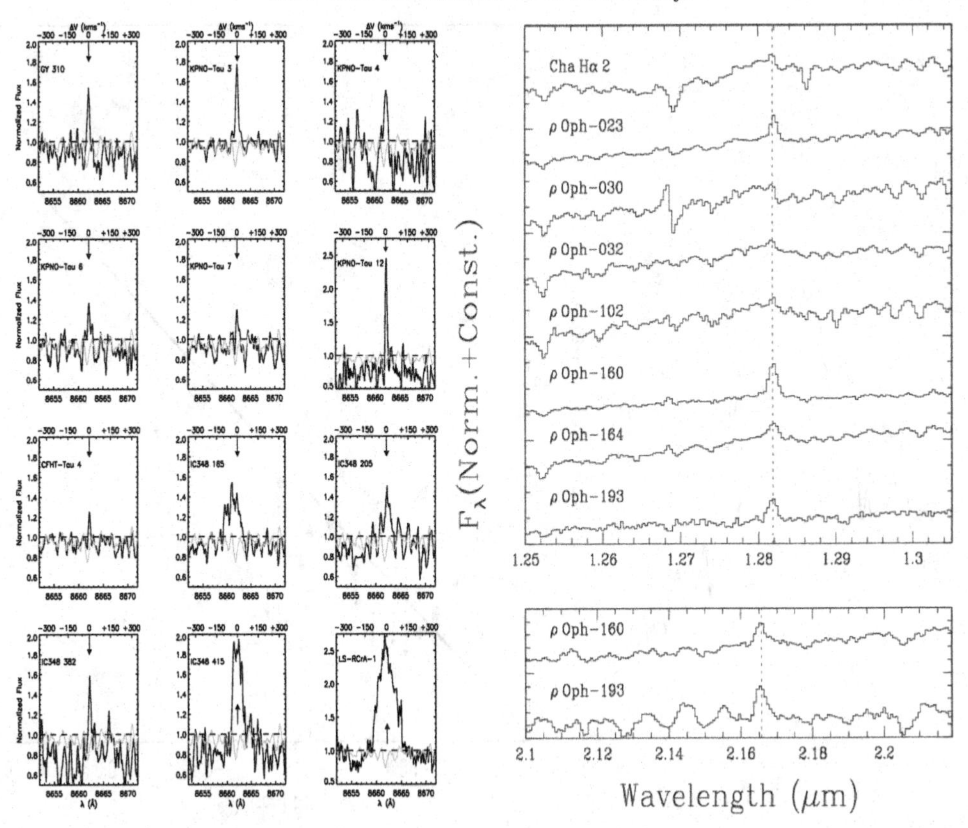

Figure 2. *Left panels*: CaII 8662Å emission (position marked with arrow) in a sample of BD accretors. Horizontal lines indicate the normalized continuum. The superimposed grey lines show the CaII region in non-accreting BDs of the same spectral type as the accretors, for comparison: the non-accretors do not evince any CaII 8662 emission. From Mohanty *et al.* (2005). *Right panels*: Paβ (top) and Brγ (bottom) emission detected in a sample of BDs. The dotted vertical lines show the emission position. From Natta *et al.* (2004).

UV. Unfortunately, as mentioned earlier, the infall rates in most BDs are too small to produce significant excess (Muzerolle *et al.* 2000 calculate that $\dot{M} < 10^{-9}$ $M_{\odot}yr^{-1}$ would be too small for any appreciable excess and veiling). Another technique is to model the $H\alpha$ line profile in detail (Muzerolle *et al.* 1998a). This is the method currently used most widely to directly measure $\dot{M}$ in VLMS and BDs (Muzerolle *et al.* 2000, 2003, 2005). In addition, two secondary methods have come into usage. The first relies on the empirically observed correlation between $\dot{M}$ and the emission flux in CaII infrared triplet lines. The correlation was first noted by Muzerolle *et al.* (1998b) for CTTs, with $\dot{M}$ measured from L_{acc}; Mohanty *et al.* (2005) then showed that it extended into the BD regime as well (with $\dot{M}$ now measured from modeling the $H\alpha$ line), for the 8662Å line of the triplet. The other secondary technique relies on the analogous correlation observed between $\dot{M}$ and the flux in the Paβ and Brγ lines (Natta *et al.* 2004, 2006; Gatti *et al.* 2006). The advantage of these secondary methods, in the VLMS/BD regime, is that once the functional form of the correlation has been established, $\dot{M}$ can be derived from a simple measurement of line fluxes instead of a detailed and time-consuming modeling of the $H\alpha$ profile. The infrared Paβ and Brγ correlations moreover allow $\dot{M}$ to be measured for very extincted sources.

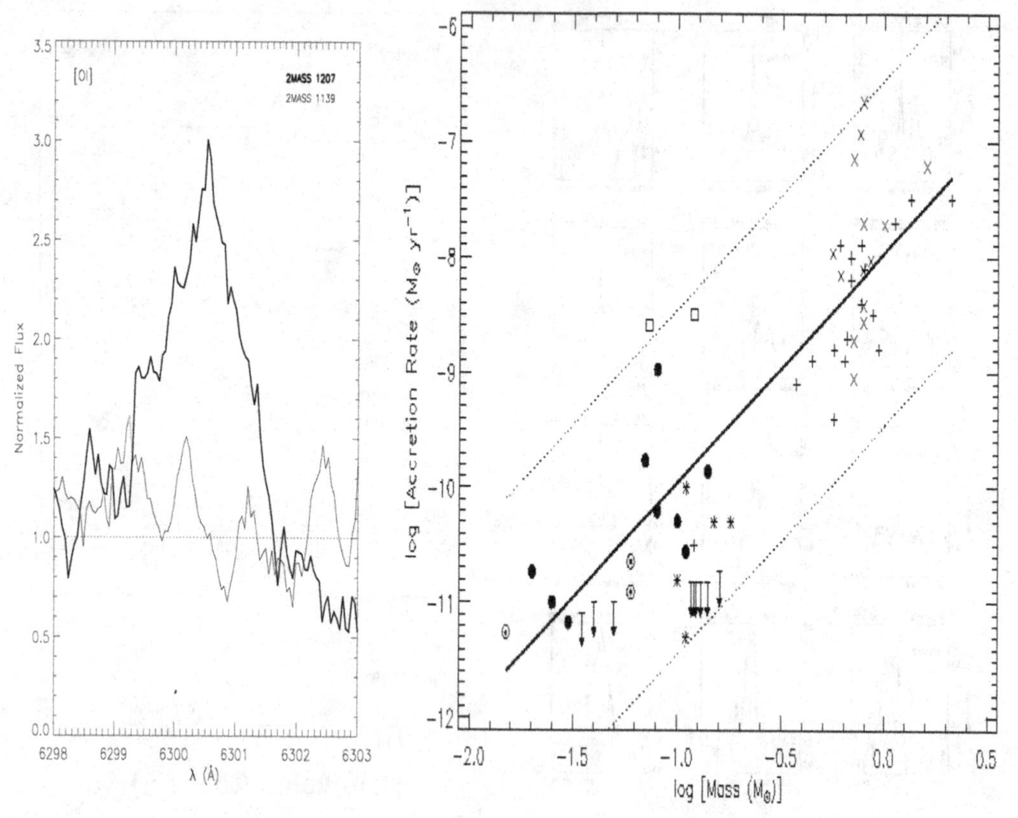

Figure 3. *Left panel*: Forbidden [OI] 6300Å emission from the BD 2MASS1207-3932 (thick line), indicating the presence of an outflowing wind/jet. Thin line is the same spectral region from the non-accreting BD 2MASS1139, without any [OI], for comparison. *Right panel*: $\dot{M}$ vs M_*, from solar-mass CTTs down to ~15 M_{Jup} BDs. Symbols indicate the various studies in the compilation (see Mohanty *et al.* 2005 and references therein). Solid line is the formal fit to the data (excluding $\dot{M}$ upper limits), and follows almost exactly $\dot{M} \propto M_*^2$. Dashed lines are this fit vertically offset by ±1.5 dex, to denote the upper/lower envelope of the trend. Both plots from Mohanty *et al.* (2005).

The drawbacks are threefold. First, from a theoretical standpoint, the basis of the correlations are not completely understood. In particular, the CaII, Paβ and Brγ lines in CTTs appear to arise predominantly in the infalling funnel flow, while in BDs they seem to be produced only in the shock (Muzerolle *et al.* 1998b; Mohanty *et al.* 2005; Gatti *et al.* 2006). Thus, while the lines do probe accretion in both CTTs and BDs, it is not clear why the functional form of the relationship between $\dot{M}$ and the line fluxes should be similar in the two mass regimes. Nevertheless, this does not detract from the usefulness of these secondary techniques, since the correlations are empirically observed.

Second, the CaII, Paβ and Brγ lines are somewhat less sensitive to accretion than $H\alpha$: they are absent for $\dot{M} \lesssim 10^{-11}$ $M_\odot yr^{-1}$, while $H\alpha$ continues to be a viable diagnostic down to $\sim 10^{-12}$ $M_\odot yr^{-1}$. Third, and most important, the viability of these secondary techniques is ultimately based on the accuracy of the $\dot{M}$ from $H\alpha$ modeling in the BD regime (which is how the correlations between $\dot{M}$ and line fluxes are established in the first place). While the modeling has become increasingly sophisticated, it has not been adequately tested against the most direct method – deriving $\dot{M}$ from L_{acc} – in BDs, due to the lack of significant continuum excess in the latter. The modeling also does not currently account for winds, which can affect the $H\alpha$ profile. Consequently, the $\dot{M}$ inferred at present for VLMS and BDs, either from $H\alpha$ modeling or the secondary techniques based on the latter, must be regarded as being somewhat imprecise, though probably correct to within a factor of ∼5 (Muzerolle *et al.* 2003, 2005). In any case, this is the best that can be achieved at present. Using these methods, $\dot{M}$ have now been derived for scores of VLMS and BDs. The accretion rates are in the range 10^{-10}–10^{-12} $M_{\odot}yr^{-1}$: 2–4 orders of magnitude lower than the average in CTTs.

4. Behaviour of the accretion rates

4.1. *Dependence of $\dot{M}_{acc}$ on central mass*

The accretion rates derived so far point to a clear correlation between $\dot{M}$ and the mass of the central object (M_*), all the way from solar-mass stars down to the lowest mass BDs (Fig. 3). In particular, it appears that roughly, $\dot{M} \propto M_*^2$ (albeit with considerable scatter), over ∼4 orders of magnitude in $\dot{M}$ and 2 orders in M_* (Muzerolle *et al.* 2003, 2005; Natta *et al.* 2004, 2006; Mohanty *et al.* 2005).

The reasons for this relationship are unclear, though a few explanations have been offered. Padoan *et al.* (2005) argue that it results from Bondi-Hoyle accretion frrom the surrounding cloud onto the disk. In the latter picture, $\dot{M}_{BH} \propto M[\rho_\infty/(c_\infty^2 + v_\infty^2)]$, where $\dot{M}_{BH}$ is the accretion rate onto the star+disk system from the surrounding medium, M is the total mass of star+disk, and ρ_∞ and c_∞ are respectively the gas density and sound speed in the surrounding medium and v_∞ the velocity of the medium relative to the star+disk system. This directly gives the correlation observed between infall rate and central mass, and explains the observed scatter in terms of the variability of stellar velocity and conditions in the cloud. The first drawback, however, is that Bondi-Hoyle accretion refers to infall from the cloud onto the star+disk, while the observed accretion rates are through the disk onto the star. It is not clear why the two should be related, except perhaps in a time-averaged sense. Moreover, there does not appear to be any systematic difference in $\dot{M}$, for stars of similar mass, between regions very with large differences in cloud properties and stellar velocities, whereas the Bondi-Hoyle scenario would predict a sytematic variation of a few orders of magnitude (e.g., Trumpler 37 versus Taurus: Sicilia-Aguilar *et al.* 2005; Hartmann *et al.* 2006).

Dullemond *et al.* (2006) instead suggest that the relationship arises due to the rotational properties of the original star-forming cores (specifically, they assume that all cores regardless of size are rotating at the same fraction of breakup velocity), combined with properties of viscous accretion disks. Their adopted conditions essentially result in two relationships – $M_{disk} \propto M_*^2$ and $\dot{M} \propto M_{disk}$ – which in combination produce the observed $\dot{M} \propto M_*^2$ correlation. However, the data appear substantially inconsistent with the $\dot{M} \propto M_{disk}$ relation they expect. On the other hand, current disk masses are very insecure, due to large uncertainties in grain opacities; it is hence possible that this explanation for the $\dot{M}$–M_*^2 relation will appear more viable with improved disk masses.

Finally, Hartmann *et al.* (2006) suggest that the relationship arises due to a combination of factors that change with the mass regime: perhaps the lowest mass stars have the least massive disks, that can be thoroughly magnetically active with accretion driven throughout by the magneto-rotational instability, while higher mass stars have more massive disks which have magnetically inactive regions (dead zones) where gravitational instabilities may drive accretion. The viability of this hypothesis, like that of Dullemond *et al.* (2006) above, depends on obtaining more robust disk masses. It also predicts that disk evolution should proceed much more rapidly in the lowest mass sources, with $\dot{M}$ dropping off significantly faster with age in BDs than in higher-mass CTTs. The sample of VLMS and BD accretors as a function of age is too limited at this stage to test this claim (see §4.2).

Finally, a proposed bias that is often brought up in this context deserves mention. Is it possible that the apparent $\dot{M}$-M_*^2 correlation results from an observational bias, wherein VLMS and BDs with very high accretion rates are not identified as such? Specifically, low mass sources with very large $\dot{M}$ would be heavily veiled and masquerade as higher mass stars, thereby escaping inclusion in the VLMS/BD sample. Including such sources would reduce the steepness of the accretion rate/mass relationship, and may remove the correlation altogether. However, this suggestion is probably not viable. A large fraction of young VLMS and BDs known today were originally identified in infrared surveys, and the effects of accretion-related veiling should be much smaller at such wavelengths than in the optical. Thus, sources with high veiling should be apparent by a significant mismatch between the optical and infrared spectral types. No such objects have been identified, implying that the $\dot{M}$ in VLMS and BDs is truly much lower than in CTTs, and that the observed $\dot{M}$–M_*^2 correlation does not result from a bias against low-mass sources with very large accretion rates. On the other hand, it is quite possible that sources with the lowest $\dot{M}$ in all mass regimes are not identified as accretors due to sensitivity limits, and the observed correlation thus represents the upper envelope of the true distribution.

4.2. *Temporal evolution of $\dot{M}_{acc}$*

The number of VLMS and BDs observed in any given star-forming region still remains somewhat small. Nevertheless, some statistical conclusions have become apparent (Mohanty *et al.* 2005). First, in any given region, the fraction of VLMS and BDs that are accreting is comparable to that of CTTs. That is, there does not appear to be any sharp falloff in accreting objects in the VLMS/BD regime. Second, the fraction of VLMS/BDs with measurable accretion appears to decline with age, just as in solar-type stars, with much fewer accretors at $\gtrsim$ 5 Myr than at $\sim$ 1 Myr. Finally, accretion can continue in BDs over timescales comparable to that in CTTs: the oldest known BD accretor, the ~25 M_{Jup} source 2MASS 1207-3932, resides in the $\sim$ 10 Myr-old TW Hydrae Association, which also harbors the oldest known CTTs (e.g., the eponymous star TW Hydra).

The evolution of the accretion *rates* in BDs is less well constrained. On the one hand, the $\dot{M}$ inferred for the oldest BD accretors, at 5–10 Myr, are among the lowest measured in the BD regime ($\sim 10^{-12}$ $M_\odot yr^{-1}$). On the other hand, equally small $\dot{M}$ have also been measured in much younger regions ($\sim$ 1 Myr, e.g., Taurus), and the sample of 5–10 Myr BD accretors is too small to make a meaningful comparison based on average $\dot{M}$ values. At younger ages (<1–3 Myr), there does not appear to be a clear decline in average $\dot{M}$ with time; however, the sample at any given age is still somewhat too small to draw robust conclusions.

4.3. *Variability of line profiles and $\dot{M}_{acc}$*

Accreting VLMS and BDs exhibit both photometric and spectroscopic (emission-line) variability, similar to that seen in CCTs, resulting from both stellar rotation and temporal changes in accretion rate/geometry (Scholz & Eislöffel 2005; Mohanty *et al.* 2003; Scholz *et al.* 2005). In the most detailed study of variability in an accreting BD so far, Scholz *et al.* (2005) show that 2MASS 1207-3932 evinces quasi-periodic variability over timescales of hours to months, with a red-shifted absorption component in the $H\alpha$ line that is modulated by the rotation period of the star. The data suggest a change of about an order of magnitude in $\dot{M}$ over several weeks, and a close to edge-on system with an accretion funnel flow that rotates in and out of the line-of-sight. Extending such detailed studies to large samples of BDs in the future can clarify the physics and geometry of accretion flows in very low mass sources.

5. Implications for brown dwarf formation

To summarize, all the evidence above points to a very similar infancy for BDs and higher-mass stars. Disks and a disk accretion phase are equally ubiquitous in the stellar and substellar regimes, and the phenomenology and physics of the star-disk interaction – magnetospherically channeled inflow from an inner disk edge at a few (sub)stellar radii, the presence of outflowing winds/jets along with inflow, a relationship between the accretion rate and central mass - are also the same in both regimes. Does this point to a similar formation process for BDs as for stars, in agreement with the 'turbulent fragmentation' scenario and in contrast to the 'ejection' hypothesis? While it is tempting to conclude so, the evidence for this is not yet ironclad.

In particular, the optical and infrared accretion and disk diagnostics arise within disk radii of <1 AU. While the presence of disks and disk-accretion does rule out the most severe 'ejection' picture, wherein BD disks are completely sheared away, it does not rule out dynamical interactions which still allow some surrounding material to remain. In particular, current simulations imply that 'ejection' can still permit BD disks up to 10-20 AU in radius (Bate *et al.* 2003). Given the fator of 10^2–10^4 smaller $\dot{M}$ in BDs compared to CTTs, this would also allow disk accretion in BDs to continue as long as in stars. What is required therefore is a measure of the true size of BD disks. Recent sub-mm/mm measurements do imply BD disk radii of at least ~10 AU (they *may* be much larger, but the data cannot determine whether this is so), which is nominally near the limit of the 'ejection' picture (Scholz *et al.* 2006). However, even if the disks were much smaller (~1 AU) initially, they would viscously expand to ~10 AU over Myr timescales. Only if BD disks are on average significantly larger than 10-20 AU can 'ejection' be ruled out confidently. Evidence for this does not yet exist. However, various investigations, some ongoing and others planned for the future, may help settle the question.

6. Further formation constraints: recent and future observations

One of the first tests proposed to test formation scenarios was to measure space velocities and spatial distribution: the initial 'ejection' model (Reipurth & Clarke 2001) suggested that young BDs should have larger mean velocities than higher mass stars, and should therefore also be more spatially dispersed than the latter in star-forming regions. Various surveys (e.g. Luhman 2006) have shown that this is not the case: BDs have similar space velocities and distributions as stars. However, subsequent simulations showed that BD velocities would remain comparable to the average stellar velocity dispersion even for 'ejection' (Bate *et al.* 2003), so this is no longer a stringent test.

The existence of wide binaries is another possible test: the same dynamical interactions that would eject BDs and truncate their disks would also disrupt wide binaries. A few wide separation (~ 40-250 AU) BD binaries have now been discovered (e.g., Chauvin *et al.* 2004; Luhman 2004); at least for these systems, the 'ejection' scenario appears highly unlikely. The discovery of a large sample of such systems would argue against 'ejection' being the primary mode of BD formation. On the other hand, there is increasing evidence that the binary component separation decreases continuously as one moves to lower masses, and that BDs simply continue this trend (e.g., Kraus *et al.* 2006; Basri &

Reiners 2006). A lack of frequent wide binaries among BDs might therefore simply reflect binary formation mechanisms in general, without shedding light on the viability of the 'ejection' scenario for BDs in particular.

Measuring the true size of BD disks provides another means of discriminating between formation scenarios: as mentioned earlier, BD disks much larger than 10–20 AU would strongly argue against 'ejection'. Current modeling, based on fitting spectral energy distributions to observed sub-mm/mm dust emission, already suggests that the disks are at least as large as $\sim$10 AU. However, the number of BDs with such measurements is few (Scholz *et al.* 2006), and the modeling is also dependent on very uncertain dust opacities. Meoreover, as discussed, viscous spreading alone can produce $\sim$ 10 AU disks even if they started out much smaller. A much better test would be to directly resolve the disks. The unprecedented sensitivity and spatial resolution of ALMA will be a great advantage in this area. However, even ALMA might be insufficient: while disks larger than $\sim$10 AU in the nearest star-forming regions ($\sim$150 pc) will nominally be resolved, material beyond $\sim$30 AU in BD disks may be too cold to yield detectable sub-mm/mm emission, even with ALMA (Natta & Testi 2006). Thus ALMA may not be able to decide whether BD disks are significantly larger than implied by ejection combined with viscous spreading.

Finally, the discovery of Class-0 BDs, i.e., proto-BDs embedded in their own isolated cores just like protostars, would provide strong support for the 'turbulent fragmentation' picture and argue against 'ejection'. Indeed, *Spitzer* has recently identified a number of very low luminosity objects (VELLOs), embedded within cores previously thought to be starless. Initial modeling suggests that these Class-0 objects have masses squarely in BD regime (e.g., Young *et al.* 2004; Huard *et al.* 2006; Bourke *et al.* 2006). These masses are rather insecure, however, since they depend on deconvolving the stellar, disk and envelope luminosity contributions; moreover, the accretion rates onto the central objects are poorly constrained. Studies are now underway to get more precise masses by obtaining infrared spectral types for the VELLOs, and also constrain the accretion rates by measuring the Paβ and Brγ emission fluxes. Naively, one expects that if the central masses are currently in the BD regime, and the measured accretion rates are also too low to allow them to eventually reach stellar masses over the normal lifetime of a core, then these must be true proto-BDs. However, reality may not be so simple. While the VELLO masses are themselves very low, the cores they are embedded in are of order a solar mass, i.e., not particularly small ones. It is difficult to imagine why a solar mass core would produce a BD-mass object, and not a low-mass star. Instead, it is possible that the accretion onto the central object is episodic, with long periods of low accretion punctuated by short bursts of intense accretion (as suggested by FU Orionis outbursts); most of the mass accumulated would be during the short bursts. In this case, a BD mass allied with a very low accretion rate would perhaps be the usual state of a low mass star during much of the Class-0 phase, with high accretion rates being statistically rare. Hence currently known VELLOs may not necessarily be proto-BDs, or be able to adjudicate between BD formation scenarios.

What the 'turbulent fragmentation' scenario really says is that BDs form out of gravitationally bound substellar-mass cores. Recent observations already indicate that cores with BD masses are indeed present in star-forming regions (e.g., Walsh *et al.* 2007; Greaves 2005). However, the bound nature of these cores is still uncertain; while the line-width data do suggest that some are gravitationally bound, the conclusion is not robust due to sensitivity issues. ALMA will provide a huge advance in this regard: it should be able to identify substellar-mass cores with ease, and even spatially resolve the nearest ones (Natta & Testi 2006). The firm detection of isolated gravitationally bound BD-mass cores would be a clear indication that BDs can form just like stars, in agreement with the 'turbulent fragmentation' scenario. The further discovery of a proto-BD within one of these cores would of course be the final proof.

Acknowledgements

I acknowledge the support of the Spitzer Fellowship for this work.

References

Barrado y Navascués, D. & Martín, E., 2003, *AJ*, 126, 2997
Barrado y Navascués, D., Mohanty, S. & Jayawardhana, R., 2004, *ApJ*, 604, 284
Basri, G. & Reiners, A., 2006, *AJ*, 132, 663
Bate, M. R., Bonnell, I. A., Bromm, V., 2003, *MNRAS*, 339, 577
Bourke, T. *et al.* , 2006, *ApJ Letters*, 649, L37
Chauvin, G. *et al.* , 2004, *A&A Letters*, 425, L29
Dullemond, C., Natta, A., Testi, L., 2006, *ApJ Letters*, 645, L69
Gatti, T., Testi, L., Natta, A., Randich, S., & Muzerolle, J., 2006, astro-ph/0609291
Greaves, J. S., 2005, *Astronomische Nachrichten*, 326, 1044
Hartmann, L., D'Alessio, P., Calvet, N., & Muzerolle, J., 2006, *ApJ*, 648, 484
Huard,T. *et al.* , 2006, *ApJ*, 640, 391
Jayawardhana, R., Mohanty, S., & Basri, G. 2003, *ApJ*, 592, 282
Kraus, A., White, R., & Hillenbrand, L., 2006, *ApJ*, 649, 306
Luhman, K., 2004, *ApJ*, 614, 398
Luhman, K., 2006, *ApJ*, 645, 676
Martín, E. L., 1998, *AJ*, 115, 351
Mohanty, S., Jayawardhana, R., & Barrado y Navascués, D., 2003, *ApJ Letters*, 593, L109
Mohanty, S., Jayawardhana, R., & Basri, G., 2005, *ApJ*, 626, 498
Muzerolle, J., Calvet, N., & Hartmann, L., 1998a, *ApJ*, 492, 743
Muzerolle, J., Hartmann, L., & Calvet, N., 1998b, *AJ*, 116, 455
Muzerolle, J. *et al.* , 2000, *ApJ Letters*, 545, L141
Muzerolle, J., Hillenbrand, L., Calvet, N., Briceño,C., & Hartmann, L., 2003, *ApJ*, 592, 266
Muzerolle, J., Luhman, K., Briceño, C., Hartmann, L., & Calvet, N., 2005, *ApJ*, 625, 906
Natta, A., Testi, L., Muzerolle, J., Randich, S., Comerón, F., & Persi, P., 2004, *A&A*, 424, 603 [N04]
Natta, A., Testi, L., & Randich, S., 2006, *A&A*, 452, 245
Natta,A. & Testi,L., 2006, *Proceedings of "Science with ALMA: A new era for astrophyisics"*, http://www.oan.es/alma2006/contributions/Natta.pdf
Padoan, P. & Nordlund,Å., 2004, *ApJ*, 617, 559
Padoan, P., Kritsuk, A., Norman, M., & Nordlund,Å., 2005, *ApJ Letters*, 622, L61
Reipurth, B. & Clarke, C., 2001, *AJ*, 122, 432.
Scholz, A. & Eislöffel, J., 2005, *A&A*, 429, 1007
Scholz, A., Jayawardhana, R., & Brandeker, A., 2005, *ApJ Letters*, 629, L41
Scholz, A., Jayawardhana, R., & Wood, K., 2006, *ApJ*, 645, 1498
Sicilia-Aguilar, A., Hartmann, L., Hernández, J., Briceño, C., & Calvet, N., 2005, *AJ*, 130, 188
Walsh, A. *et al.* , 2007, *ApJ*, 655, 958
Whelan, E. *et al.* , 2005, *Nature*, 435, 652
Whelan, E. *et al.* , 2007, *ApJ Letters*, 659, L45
White, R. J. & Basri, G., 2003, *ApJ*, 582, 1109
Young, C. *et al.* , 2004, *ApJS*, 154, 396

Star-Disk Interaction in Young Stars
Proceedings IAU Symposium No. 243, 2007
J. Bouvier & I. Appenzeller, eds.

doi:10.1017/S1743921307009714

Outflow activity in brown dwarfs

Emma T. Whelan[1], Thomas P. Ray[1], Francesca Bacciotti[2], Sofia Randich[2], Ray Jayawardhana[3], Antonella Natta[2], Leonardo Testi[2] and Subu Mohanty[4]

[1]Dublin Institute for Advanced Studies
email: ewhelan@cp.dias.ie

[2]Osservatorio Astrofisico di Arcetri

[3]Department of Astronomy and Astrophysics, University of Toronto

[4]Harvard-Smithsonian Center for Astrophysics

Abstract. While numerous studies have been aimed at understanding the properties of young brown dwarfs relatively little exploration of their potential as drivers of outflows has occurred. Forbidden emission lines are important probes of outflows from young stellar objects, as they trace the shocks which form as an outflow interacts with the ambient medium of its driving source. While forbidden emission was identified in the spectra of young brown dwarfs, indicating the presence of outflows, these lines were weak and confined to the brown dwarf continuum position. Hence their origin in an outflow could not be confirmed. Our approach to this problem, is to analyse the forbidden line regions of brown dwarfs using spectro-astrometry. Spectro-astrometry is a novel technique which allows the user to recover spatial information from a spectrum beyond the limitations of the seeing of the observation. Using this technique we have found two brown dwarf outflows to date. In this chapter we outline this technique, describe our results for the brown dwarfs ρ-Oph 102 and 2MASS1207-3932 and discuss our future plans.

Keywords. 2MASSWJ1207334-393254, stars: low mass, brown dwarfs, stars: formation, ISM: jets and outflows.

---

1. Introduction

The study of young brown dwarfs (BDs) in star forming regions has a special importance as clearly the first question that must be asked about these objects is how are they formed? More and more BDs are being discovered and analysed in nearby stellar nurseries such as Taurus (see Briceño *et al.* 1998; Luhman *et al.* 2003; Muzerolle *et al.* 2005; Grosso *et al.* 2007), Ophiuchius (Wilking *et al.* 1999; Allers *et al.* 2007) or Orion (Lucas & Roche 2000). It is a reasonable first step towards understanding BD formation to compare them with low mass YSOs, in particular the classical T Tauri stars (CTTSs). Specific questions are, do BDs undergo accretion and have accretion disks? and if they are accreting material do they drive outflows?

Many studies have addressed the question on the similarity between young BDs and CTTSs and it is now apparent that BDs and CTTSs bear a strong resemblance. The presence of accretion disks has been confirmed from a study of the spectral energy distributions (SEDs) of a number of BD candidates (Natta *et al.* 2002; Jayawardhana *et al.* 2003a) and Jayawardhana *et al.* (2003b) demonstrated that high resolution optical spectra point to the existence of a T Tauri like accretion phase for BDs. While BDs have been shown to be active accretors (Scholz & Jayawardhana 2006) relatively little is known about their outflow activity. As outflows in CTTSs are directly related to magnetospheric infall (Hartigan *et al.* 1994; Königl & Pudritz 2000), it is feasible that BDs demonstrating strong accretion will drive outflows.

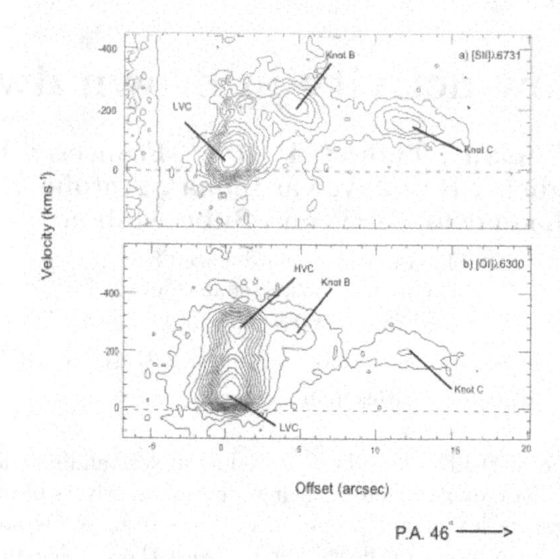

Figure 1. Continuum-subtracted position velocity diagrams of the optical forbidden emission lines from DG Tau taken from Whelan *et al.* (2004). This Figure illustrates the importance of the FELs as tracers of outflows from CTTSs. Note the comparison between the DG Tau jet and the PV diagram of the [OI]λ6300 emission from the BD 2MASS1207-3932 presented in Figure 5.

Early indications that BDs drove outflows came from the discovery of forbidden emission lines (FELs) e.g. [OI]λλ6300, 6363 in their optical spectra (Fernández & Comerón 2001; Muzerolle *et al.* 2003). FELs are important coolants of interstellar shocks and thus are strong tracers of outflow activity (see Figure 1). In the case of BDs however, the FEL regions were weak and not extended. Hence their formation in an outflow while likely could not at first be confirmed. Using the novel technique of spectro-astrometry we have achieved this confirmation and have thus far discovered two BD outflows. Below we shall describe in detail the spectro-astrometric technique and the results we have achieved thus far in the study of young BDs. In the final section we shall outline our future plans.

The discovery that BDs have outflows while relevant to our overall understanding of BD formation is perhaps primarily relevant to the study of the outflow phenomenon. That BDs can launch outflows emphasises the robustness of the outflow mechanism over an enormous range of masses (up to 10^{8-9} $M_\odot$ in Active Galactic Nuclei). Indeed the second BD found to have an outflow, 2MASS1207-3932, is only a $24M_{JUP}$ object thus it is now the lowest mass galactic object that is actively driving an outflow.

2. Spectro-astrometric analysis of the BD FELs

As stated the FELs found in the spectra of young BDs were weak and confined to the source position. Our aim was to recover the position of the emission regions with respect to the BD position using spectro-astrometry. In order to spectrally resolve the BD FEL regions with a usuable S/N, observations were made using the UV-Visual Echelle Spectrometer (UVES) on the VLT. High resolution (R=40,000) cross-dispersed spectra (CD3 disperser, spectral range 4810-6740 Å) were obtained on the ESO VLT UT2.

Spectro-astrometry simply translates as a measurement of the position of the centroid of a flux distribution, as a function of wavelength (hence velocity for a line), producing a plot of offset versus velocity. The profile of a star is smeared by atmospheric turbulence

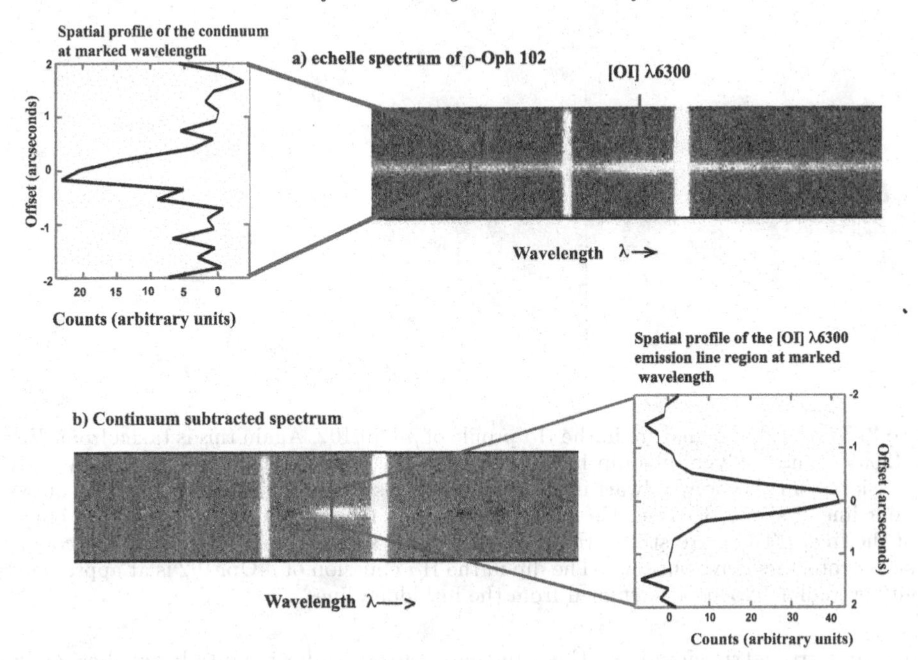

Figure 2. The principles underlying the technique of spectro-astrometry illustrated for the brown dwarf ρ-Oph 102. In the FEL spectrum of a young brown dwarf (pure emission line region originating in the outflow plus BD continuum emission) the contribution from the continuum can be removed by interpolation across the line. The emission centroid of the FEL will be shifted from the continuum centroid (forbidden line emission is quenched close to the star where densities far exceed the critical density.) This shift is accuately measured through Gausian fitting of the spatial profile and through exploring the shifts/offsets with respect to the continuum, one can recover spatio-kinematic information about the flow at milliarcsecond scales.

to appear Gaussian (at least to a first approximation) rather than point-like. Hence, the centroid of the emission is measured by fitting a Gaussian to the spatial profile extracted from the 2D spectrum, at each wavelength or pixel (see Figure 2). If an emission line e.g. a FEL is actually formed in an outflow, positional shifts (weighted by the line intensity) with respect to the continuum centroid are measured in the direction of the outflow (see Whelan *et al.* 2004). Figure 2 illustrates the process of applying spectro-astrometry to [OI]λ6300 region in the spectrum of ρ-Oph 102. Note that before the position of the pure line emission can be mapped the continuum must be removed. Contamination by the continuum emission will tend to drag the line position back towards the source.

While the width of the spatial profile is determined by the seeing, how accurately one can determine the centroid of the emission is, in theory (for fixed seeing), limited only by the strength of the observed signal to noise ratio. The spectro-astrometric accuracy is given by $\sigma_{centroid}$ = Seeing/ $2.355\sqrt{N_p}$ where N_p is the number of detected photons (see Bailey *et al.* 1998). Hence increasing the total number of detected photons increases the astrometric accuracy so that milli-arcsecond precision is possible. Although the BD forbidden emission was relatively very weak (compared to the average CTTS say) the BD continuumm emission was also weaker or comparable to the forbidden emission. This is evident from Figure 2. In order to properly compare the line and continuum our approach was to bin the line and continuum in such a way that the accuracy to which we measured the position of each was the same. Lastly an important consideration when using spectro-astrometry is the possibility of spectro-astrometric artifacts (see Brannigan *et al.* 2006). In our analysis lines like HeIλ6678 and LiI 6708 were found to have

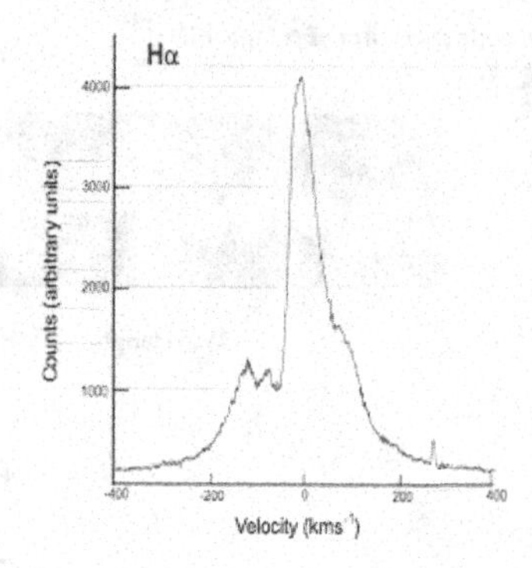

Figure 3. The outflow signature in the Hα profile of ρ-Oph102. Again this is taken from Whelan *et al.* (2005). The P Cygni-like dip in the line profile is a strong signature of outflow activity. Hα emission from the brown dwarf is absorbed as it passes through material moving outwards along our line of sight. Because this material is moving towards us, the dip is on the blueward side of the line. CTTSs are strong Hα emitters, and P Cygni Hα profiles originally confirmed that such protostars drive outflows. The dip in the Hα emission of ρ-Oph102 is at approximately the outflow radial velocity determined from the forbidden lines.

no spectro-astrometric signal. As these lines are chromospheric and photospheric respectively no offset is expected. The lack of positional displacement in such lines rules out the possibility of spectro-astrometric artifacts.

3. Rho Oph 102

The $\sim$ 60 M_{JUP} BD ρ-Oph 102 was the first BD to be confirmed to have an outflow and these first exciting results were presented in Whelan *et al.* (2005). The UVES spectra of ρ-Oph 102 obtained in May 2003 revealed the [OI]$\lambda\lambda$6300, 6363, [NII]λ6583 and [SII]$\lambda\lambda$6716, 6731 FELs to be present (Natta *et al.* 2004). Our analysis put the average velocity of the FELs at $\sim$ -45 kms^{-1} suggesting mass outflow. Moreover the blue-shifted asymmetry mirrored what is seen in CTTSs, and pointed to the presence of a disk. A second indication that an outflow was present came from the Hα line (see Figure 3). Its profile is clearly asymmetrical, i.e., the blue-shifted wing of the line appears to be absorbed in a P-Cygni like fashion at a similar velocity to the FELs ($\sim$ -80 kms^{-1}). However a classical P-Cygni profile, i.e., one that dips below the continuum, is not observed but such a profile is, in any event, a rare occurrence even amongst the CTTSs.

Of the five FELs detected only the [OI]$\lambda\lambda$6300, 6363 and [SII]λ6731 lines were strong enough for any spectro-astrometric analysis. This analysis is presented in Figure 4. Overall the centroids of all the FELs are displaced to the south, i.e. have negative offsets with respect to the continuum and these offsets reach a maximum of 0."08-0."1 at a blue-shifted velocity of $\sim$ -41 kms^{-1}. The scale of this blue-shifted offset would suggest a minimum (projected) disk radius of 0."1 ($\geqslant$ 15 AU at the distance of the ρ-Ophiuchi cloud) in order to hide any red-shifted component. There is no clear spatial offset in Hα even though its higher signal to noise potentially would allow one to measure even smaller offsets than observed in the forbidden lines. This is in agreement with the idea that most of the Hα emission arises from accretion on much smaller scales than are being probed here (see Natta *et al.* 2004). Lastly, no offset is measured in the Liλ6708 and HeIλ6678 lines. As explained above this lack of positional displacement rules out the existence of spectro-astrometric artifacts.

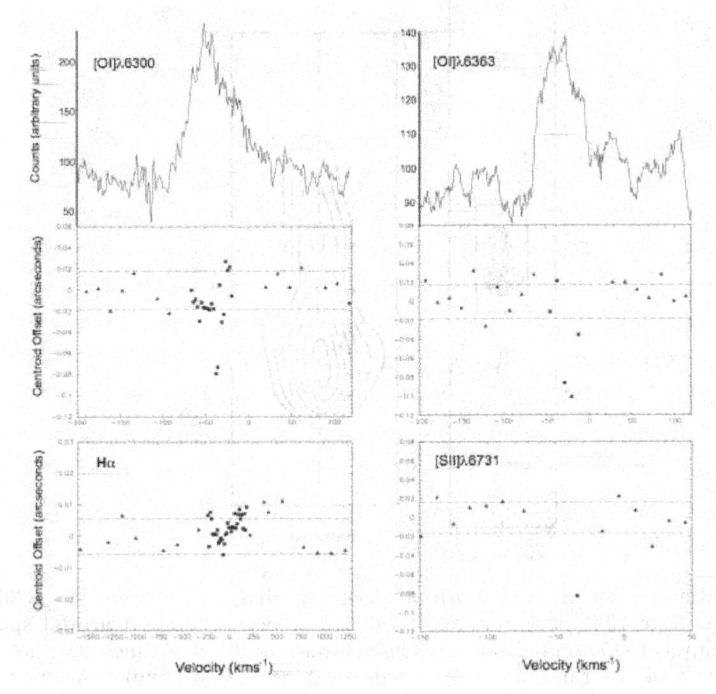

Figure 4. Line profiles (top row) and spectro-astrometric plots (middle row) for the [OI]λ6300 and [OI]λ6363 doublet and spectro-astrometric plots (bottom row) for the Hα and [SII]λ6731 lines. These results were published in Whelan *et al.* (2005). Continuum and line offset points are represented by black triangles and squares respectively. All velocities are systemic and spatial offsets are in the north-south direction (in arcseconds) with negative offsets to the south. Dashed lines delineate the $\pm$ 1σ error envelope. For Hα, note the much smaller offset scale. The [SII] line is blue-shifted to around -40kms^{-1}.

4. 2MASS1207

Following our recent discovery of a bipolar outflow from the $\sim$ 24 M$_{JUP}$ BD 2MASS1207-3932, this BD is now the lowest mass galactic object known to drive an outflow (see Whelan *et al.* 2007). The nature of this BD as a strong accretor was clearly evident from early observations. Mohanty *et al.* (2003) reported bright asymmetric Hα emission with a full width at 10% of the peak flux, of > 200 kms^{-1}. Also Mohanty *et al.* (2005) were the first to observe the [OI]λ6300 line in the spectrum of 2MASS1207 suggesting it could be driving an outflow. Our UVES spectra were obtained in May 2006 and the [OI]$\lambda\lambda$6300, 6363 lines were the only "traditional" FEL tracers seen in the spectrum. A position velocity (PV) diagram of the smoothed [OI]λ6300 spectrum is presented in Figure 5. For this BD the 2D UVES spectra were smoothed using an elliptical Gaussian filter of FWHM 0.12 Å× 0."22. The Gaussian smoothing improved the signal to noise and effectively decreased the spectral and spatial resolution to R $\sim$ 8000 and 0."7 respectively. The PV diagram shows blue and red-shifted emission at velocities of $\sim$ -8 kms^{-1} and $\sim$ + 4kms^{-1} and a relative displacement between the two parts of the line is apparent. This displacement is recovered using spectro-astrometry. Note that the relatively very small radial velocity of the outflow is consistent with the near edge-on disk hypothesis for 2MASS1207-3932.

The results of a spectro-astrometric analysis of the smoothed spectrum are shown in Figure 6. Here the continuum and [OI]λ6300 line emission were smoothed so that the spectro-astrometric error in both regions are the same. The blue and red-shifted

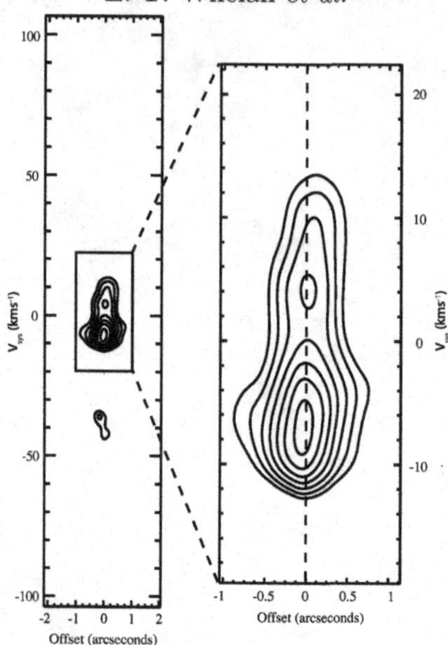

Figure 5. Continuum subtracted Position Velocity diagram of the [OI]λ6300 line in 2MASS1207-3932. Here the spectrum was smoothed in both the spectral and spatial directions using an elliptical Gaussian function. The contours begin at 3 times the r.m.s noise and increase in intervals of the r.m.s noise. The red and blue-shifted components to the line are obvious and an opposing offset in these components is suggested. The offset is recovered using spectro-astrometry.

components to the [OI]λ6300 line are found to be offset in opposing directions to an absolute distance of $\sim$ 80 mas (see Figure 6). Hence it is clear that the [OI]λ6300 emission originates in a faint bipolar outflow driven by 2MASS1207. The 1-σ error in the measurements of 18 mas, is marked by the dashed line in Figure 6. Other lines of interest are [OI]λ6363, Hα and HeIλ6678. The [OI]λ6363 is $\sim$ 3 times fainter than the [OI]λ6300 line and hence is just below the detection limit in the raw spectrum. By smoothing the spectrum using a large gaussian filter (FWHM in dispersion direction = 0.35 Å) the [OI]λ6363 is revealed at a similar velocity to the [OI]λ6300 line, although detailed kinematic data is lost. The primary origin of the Hα and HeIλ6678 lines is in the accretion flow. As this occurs on a very small scale we expect to measure no spectro-astrometric offset and, as can be seen in Figure 6 none is found, again ruling out the presence of spectro-astrometric artifacts.

5. Future Work

While it is generally accepted that accretion activity in BDs is simply scaled down from infall in CTTSs (Jayawardhana *et al.* 2003b), the presumption that BDs also drive scaled-down versions of T Tauri jets and outflows while credible has yet to be generally confirmed. The evidence gathered so far supporting the continuation of the CTT paradigm for outflow activity, into the young BD mass regime, includes the blue-shifted nature of the FELs (mimicking what is seen in the majority of CTTSs) and that the measured offsets lie withing the range estimated for a BD (Whelan *et al.* 2005, 2007 for further details). We plan to develop the work described in this chapter by firstly increasing the sample of BDs known to drive outflows and secondly by working towards capturing the first images of BD outflows. Observing time has been granted on the VLT

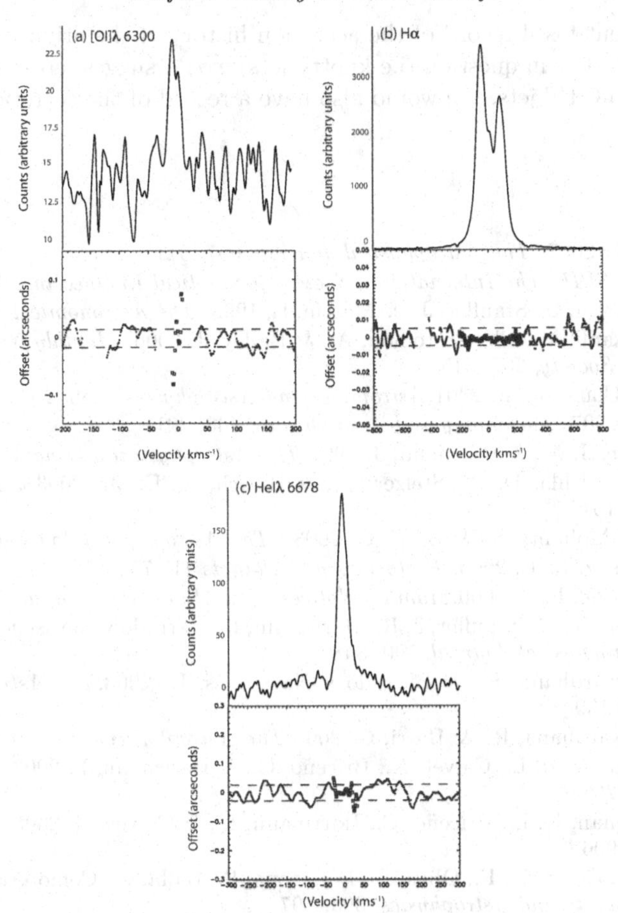

Figure 6. Offset velocity diagrams in the vicinity of the [OI]λ6300, Hα and He I lines. The green dashed lines delineate the $\pm$ 1 σ error envelope for the centroid position of the continuum. The bipolar offset in the [OI]λ6300 reveals the presence of an outflow. No offsets are measured in the Hα and He I lines as expected ruling out the possibility of spectro-astrometric artifacts.

to observe the optical spectra of four more BDs (known to be active accretors) and to image the outflows from ρ-Oph 102 and 2MASS1207-3932. Both these studies will allow us to further investigate how BD outflows compare to outflows driven by CTTSs.

In particular, questions that can be answered from images of the outflows from ρ-Oph 102 and 2MASS1207-3932 are relevant to the collimation, opening angles and morphology of BD outflows. Are BD outflows comparable to those of CTT jets or are they more or less collimated? Are BD outflows knotty? Current models of jet launching and collimation describe how infalling material from the surrounding envelope is re-launched out along the magnetic field lines (from the disk/star) and then collimated by magnetic hoop stresses. A measurement of the degree of collimation in the BD outflows will answer the question as to whether with spectro-astrometry we are detecting a jet or a much wider outflow and also address the role of magnetic fields in the formation of BD outflows. Indeed the fact that we see a bipolar outflow in 2MASS1207 already suggests that we are not detecting a spherically symmetric outflow but one that is axisymmetric. In addition models of jet launching link outflow directly to infall and direct observational evidence exists to support a strong correlation between infall in CTTSs and outflow activity (Hartigan *et al.* 1994). Each Herbig-Haro knot is related directly to an sudden burst in accretion activity hence jets are an excellent fossil record of the accretion history of the driving source. Episodic emission from the BDs in question (i.e knotty jets) would suggest episodic accretion and a similar origin to CTT jets. We would also have a record of the accretion history of the BD.

References

Allers, K. N., *et al.* 2007, *The Astrophysical Journal*, 657, 511

Bailey, J. A. 1998, *SPIE–The International Society for Optical Engineering*, 3355, 932

Briceño, C., Hartmann, L., Stauffer, J., & Martín, E. 1998, *The Astronomical Journal*, 115, 2074

Brannigan, E., Takami, M., Chrysostomou, A., & Bailey, J. 2006, *Monthly Notices of the Royal Astronomical Society*, 367, 315

Fernández, M., & Comerón, F. 2001,*Astronomy and Astrophysics* , 380, 264

Grosso, N., *et al.* 2007, *Astronomy and Astrophysics*, 468, 391

Hartigan, P., Morse, J. A., & Raymond, J. 1994,*The Astrophysical Journal* , 436, 125

Jayawardhana, R., Ardila, D. R., Stelzer, B., & Haisch, K. E., Jr. 2003a, *The Astronomical Journal*, 126, 1515

Jayawardhana, R., Mohanty, S., & Basri, G. 2003b,*The Astrophysical Journal*, 592, 282

Königl, A., & Pudritz, R. E. 2000, *Protostars and Planets IV*, 759

Lucas, P. W. & Roche, P. F. 2000,*Monthly Notices of the Royal Astronomical Society*, 314, 858

Luhman, K. L., Briceño, C., Stauffer, J. R., Hartmann, L., Barrado y Navascués, D., & Caldwell, 2003,*The Astrophysical Journal*, 590, 348

Mohanty, S., Jayawardhana, R., & Barrado y Navascués, D. 2003,*The Astrophysical Journal Letters*, 593, L109

Mohanty, S., Jayawardhana, R., & Basri, G. 2005,*The Astrophysical Journal*, 626, 498

Muzerolle, J., Hillenbrand, L., Calvet, N., Briceño, C., & Hartmann, L. 2003,*The Astrophysical Journal*, 592, 266

Muzerolle, J., Luhman, K. L., Briceño, C., Hartmann, L., & Calvet, N. 2005,*The Astrophysical Journal*, 625, 906

Natta, A., Testi, L., Comerón, F., Oliva, E., D'Antona, F., Baffa, C., Comoretto, G., & Gennari, S. 2002,*Astronomy and Astrophysics*, 393, 597

Natta, A., Testi, L., Muzerolle, J., Randich, S., Comerón, F., & Persi, P. 2004,*Astronomy and Astrophysics*, 424, 603

Scholz, A., & Jayawardhana, R. 2006,*The Astrophysical Journal*, 638, 1056

Whelan, E. T., Ray, T. P., & Davis, C. J. 2004,*Astronomy and Astrophysics*, 417, 247

Whelan, E. T., Ray, T. P., Bacciotti, F., Natta, A., Testi, L., & Randich, S. 2005,*Nature*, 435, 652

Whelan, E. T., Ray, T. P., Randich, S., Bacciotti, F., Jayawardhana, R., Testi, L., Natta, A., & Mohanty, S. 2007,*The Astrophysical Journal Letters*, 659, L45

Wilking, B. A., Greene, T. P., & Meyer, M. R. 1999,*The Astronomical Journal*, 117, 469

Star-Disk Interaction in Young Stars
Proceedings IAU Symposium No. 243, 2007
J. Bouvier & I. Appenzeller, eds.

doi:10.1017/S1743921307009726

Summary and concluding remarks

Immo Appenzeller
University of Heidelberg, Landessternwarte, Königstuhl, D 69117 Heidelberg, Germany
email: I.Appenzeller@lsw.uni-heidelberg.de

Abstract. This symposium was characterized by an intense exchange of new information and by lively discussions on many aspects of the formation and early evolution of low-mass stars. The observational data presented at this meeting, obtained in spectral regions ranging from X-rays to submm waves, were found to be remarkably consistent with the current magnetospheric disc-accretion paradigm for young stellar objects. But there remain open questions, and a full understanding of the star-formation process will require much additional work.

Keywords. Stars: formation, stars: pre–main-sequence, ISM: jets and outflows

---

Agreeing to present a meeting summary comes with the obligation to listen carefully to all the talks and to study all the posters. This is not always an easy or pleasant task. At this conference listening to all the presentations was a real pleasure. Among the many conferences which I attended, this IAU symposium definitely was one of the highlights and one the most rewarding events.

As indicated by the title of the symposium, the central topic of this meeting has been the interaction of very young stars with their accretion disks. The conference was dedicated to Claude Bertout who provided decisive ideas and important details to this subject. From the table of contents of these proceedings it is evident that the themes actually discussed at this meeting went well beyond the central topic and included many different aspects of low-mass star formation. Among these themes were basic concepts of the star-formation process, questions related to the formation and evolution of planetary systems, and new results on young stars of intermediate mass and of very low mass. In addition to to highly informative invited and contributed talks we enjoyed many attractive and well organized posters.

Compared to earlier star formation conferences there was a surprising amount of consensus and agreement on many of the questions which were disputed in the past. This seems to indicate that some basic facts are by now understood and well established. These issues, which, according to the results presented here, appear to be firmly established will be summarized the first section of this report. In a second part of my summary I will list the open questions, the gaps in our present understanding, and potential ways to achieve further progress.

1. What we (seem to) know

1.1. *The observational evidence*

IAUS 243 started with two excellent reviews of the current magnetospheric disk-accretion paradigm of young low-mass stars. First Claude Bertout described the basic model and its interesting history. Then Gibor Basri discussed the spectroscopic evidence which led to our present concepts. Most of the following talks reported new observational results, new models, or new numerical simulations concerning details of the disk-accretion scenario. Remarkably, practically all the new data supported or confirmed the current paradigm.

The magnetospheric disk-accretion model was initially developed on the basis of optical line profiles and NIR photometry. As shown by Tim Harries and his co-workers, improved radiative transfer models result in even better agreement between the computed and the observed Balmer line profiles. On the other hand, it has been known since a long time, that the Balmer line profiles can be ambiguous and that such agreements do not necessarily prove that the models are correct. Therefore, it will be important to extend the new profile calculations to lines which are formed in well defined regions of the accretion-flow, disk, and wind systems. An example of such lines is He I 10830, for which Suzan Edwards showed us very interesting new results. Originating from a high, metastable state of He I, this line is formed in a low-density high-excitation environment. Also particularly valuable in this context are the observations of the UV "transition region" lines, formed in the accretion-shock cooling zones (see David Ardila's contribution), and the profiles of the IR lines of CO, formed in the cool inner disks (described by John Carr).

Among the highlights of this meeting was new and detailed data on the X-ray emission from young stars of different mass. Although there may still be uncertainties concerning the exact location of the X-ray emitting regions, the talks by Thierry Montmerle and Manuel Guedel showed that the observed X-ray properties of T Tauri stars are well explained by the current models. From the present data it seems that the accretion shocks as well as hot magnetospheric coronae, and possibly shocks in the inner jets contribute the observed X-ray flux.

As pointed out by several authors, and as demonstrated with convincing examples by Silvia Alencar, time variability observations provide particularly good tests of model details. Obviously more such observations will be highly valuable and well worth the significant amount of observing time which is needed for such work.

From the talks by Jerome Bouvier and by Jochen Eislöffel we learned that there is at least basic agreement between the statistics of the observed rotational velocities of young stars and the model predictions. The fact that some stars seem not to follow the expectations has to be investigated with a closer look at the corresponding stellar samples.

In spite of the impressive progress in modeling jets and outflows from young stars, there remain open questions concerning the jet physics (see § 2). But, as discussed in the talks of Sylvie Cabrit and Sean Matt (and from other contributions) it is clear that the observed outflows cannot be consistently explained without the presence and the action of large-scale magnetic fields. Thus, although the acceleration of the observed outflows may not yet be fully understood, their properties can be explained *only* in the context of the magnetospheric accretion models.

All the observational results mentioned above support or are consistent with the current disk-accretion models. No observation reported at this meeting appeared to be in serious contradiction with this paradigm.

1.2. *Magnetic fields*

At one point during the discussions Frank Shu happily noted that at this symposium magnetic fields – in the past often regarded as an unwelcome complication of star formation – was one of the major topics. Obviously, magnetic fields are an indispensable ingredient of the current models. When the scenario of the magnetospheric disk accretion was developed, the observational techniques were far from being able to measure the fields predicted by the theory. For many years these fields remained an unproven theoretical prediction. As described in Johns-Krull's excellent review, this situation has changed dramatically. Not only do we observe the fields, but important details could be established.

Most of the magnetic field measurements available at present are based on the profiles of low-excitation absorption lines originating in the cool photosphere of the observed T Tauri stars. In these lines we see fairly strong small-scale fields, which in CTTSs may cover a significant fraction of the stellar surface. No or only a small contribution of a dipole or ordered field is seen in the lines formed in the cool photospheres. However, as reported by Christopher Johns-Krull, at least in some CTTSs, a weaker, ordered large-scale field is observed in the profiles of the He I 5876 emission line, which is thought to be formed in the cooling zones of the magnetic accretion shocks. These results are in good agreement with at least part of the present magnetic disk-accretion models. Obviously, magnetic field measurements in lines produced in the accretion flow provide a particularly valuable test for these models. Therefore, more field measurements based on such lines will be particularly important for a comparison with the theory.

1.3. *Accretion flow models*

Among the most impressive results presented at this meeting were the 2-D and 3-D MHD simulations of the star-disk interaction shown by Marina Romanova and Akshay Kulkarni. Although these calculations still have to assume various physical simplifications, the computed models contain a surprising amount of detail and provide data on the structure as well as on the time evolution of the magnetospheres and accretion flows. The results of these simulations are in surprisingly good agreement with predictions made from the initial, much more simplified magnetospheric accretion models. They again seem to confirm that the basic picture is correct.

2. Open questions and future work

There seems to be general agreement that the accretion flows of young stellar objects are controlled by the common magnetosphere of the disk and the central star. However, the exact topology of these magnetospheres is far from clear. Closely related to the topology of the magnetosphere is the question of whether we always (or only sometimes, or perhaps never) have "disk-locking", and the question where the winds and the outflows from CTTSs originate. In spite of the extensive discussions of this issue, no clear answer could be given at this meeting. On the other hand, the interrelation of these questions became clearer. Progress on this question should become possible from profile observations of lines originating in the inner regions of the outflows. Such observational data will eventually allow us to decide between the different magnetospheric models.

Another question which remained open at this meeting is the origin of the observed X-rays. It seems plausible that much of the hard X-ray flux is produced in hot magnetospheric coronae. But, as discussed (e.g.) by Manuel Guedel, the origin of the soft component is less clear. This components seems to be produced in shocks. The accretion shocks of CTTSs, in principle, provide a natural explanation for this component. However, shocks are also associated with the observed jets and the current models and data do not yet lead to firm conclusions.

Detailed and extensive new MHD simulations presented to us by Christian Fendt, Jonathan Ferreira, Tom Ray and others demonstrated and confirmed that accretion disks and their magnetospheres can naturally produce collimated outflows with properties similar to those of the observed jets from young stars. A problem here is, that similar jet properties can be produced with rather different assumptions. And, although much has been learned about the physical structure of the jets, more information on the jet acceleration regions are needed to distinguish between the different models. Moreover, although the jet models usually start with dipole-like fields of the central objects, the jet evolution leads to quite different topologies. It will be important to develop magnetospheric models which simultaneously describe the development of a collimated outflows and the central fields responsible for the magnetic accretion from the disk to the central star.

At this meeting we saw beautiful examples of computed star-disk systems and their evolution with time. However, at present all such simulations start from some assumed initial conditions. In the real world these initial conditions are the result of the fragmentation and internal interactions in collapsing turbulent interstellar gas clouds. An important task for the future will be to clarify the correct initial conditions by modeling the formation of star-disk systems and their magnetospheres from 3-D collapse calculations, starting with the collapse of molecular clouds, or perhaps even with the formation of these clouds. A step in this direction are the models presented by Frank Shu and by Shu-Ichiro Inutsuka, which start with a rotationally symmetric magnetized cloud core. True 3-D evolutionary computations of turbulent magnetic clouds have been initiated by various groups (e.g., Heitsch *et al.* 2006, Jappsen *et al.* 2005), and present and future computers and modern numerical methods should make it possible to achieve progress in this field in the near future.

During the past years new large optical interferometers have made it possible to directly resolve the inner regions of accretion disks and mass flows around young stars. Examples of such observations have been shown at this meting by Fabien Malbet, Stefan Kraus and Tom Ray. This data provides valuable new information. However, in most cases, little more than visibilities (as a function of wavelength) or aspect ratios have been reported. Obviously, such data arouse our appetite for more detailed information. Modern NIR and MIR interferometers have the potential of producing much more detailed data and – eventually – true images. Thus, there is hope that at future meeting on this topic we will see interferometric images of the star-disk systems which are as detailed and as well resolved as the theoretical models which we have been shown at this symposium.

3. Some historical remarks

Claude Bertout ended his introductory review of this conference with a quotation from George Herbig's summary of the observational part of the IAU Symposium No. 75 ("Star Formation"), which took place in 1976 in Geneva. In his final words George Herbig predicted that "with the perspective of the years, all that we do today will certainly be seen to have been either wrong, or irrelevant , or obvious" (Herbig 1977). By now, more than 30 years later, it seems appropriate to ask, whether this prediction has become true. Although I dislike to disagree with our distinguished colleague George Herbig, according to my assessment, he was only partially correct.

Reading the proceedings of the 1976 IAU symposium, one indeed finds statements and ideas which turned out to be completely wrong or irrelevant. However, in my opinion, none of the issues discussed in 1976 appear (as seen from today) obvious. In view of the complexity of star formation, the details of this process probably will never appear obvious. And some of the information presented at the Geneva meeting was quite correct and was important for the development of our present theories.

At the time of the 1976 meeting Merle Walker's 1972 paper (Walker 1972), suggesting the presence of accretion discs in (part of) the CTTSs, was already well known, and accretion models (and new evidence supporting them) were discussed at the 1976 symposium. But, the chromospheric models of the T Tauri emission spectra were clearly more popular at that time, in part since chromospheric emission was better understood than the complex physics of accretion. Only about three years later, at the 1979 UC

Santa Cruz Astrophysics Summer Workshop, accretion models became to be regarded as a viable alternative.

The recollection of the discussions at the Geneva symposium obviously raises the question of how much of the results of the present symposium will survive the next 30 years.

A look into the old papers shows that much of the less relevant discussions of the 1976 symposium can be traced to the incomplete and often poor observational data available at that time. In 1976 observations of very young stars were restricted to the optical wavelength range and high-resolution spectroscopy existed only for the exceptionally bright objects. Today we have much better and much more comprehensive observational data, which provide more and significantly more reliable constraints. Therefore, I suspect that there is a good chance that somebody who reads the proceedings of IAU Symposium 243 in 30 years will find that the basic features of the magnetospheric disk-accretion model of young stellar objects has been basically correct.

Finally...

Among the participants of this symposium there seems to be a strong consensus that this has been a particularly successful and pleasant meeting. This would not have been possible without the efforts of Jerome Bouvier and the very efficient LOC of this conference. Therefore, I would like to end with a big "MERCY BEAUCOUP" to Jerome Bouvier and to the members of the Grenoble LOC. You did a brilliant job and you were great hosts of a memorable event!

References

Heitsch, F., Hartmann, L., Slyz, A. D., Devrient, J. E., Burkert, A. 2006, *BAAS* 209, 1715

Herbig, G. H. 1977, in: T. de Jong & A. Maeder (eds.), in: Proc. IAU Symposium 75 on *Star Formation*, (Dordrecht: D. Reidel) p. 283

Jappsen, A.-K., Klessen, R. S., Larson, R. B., Li, Y., Mac Low, M.-M. 2005, *A&A* 435, 611

Walker, M.F. 1972, *ApJ* 175, 89

Author Index

Object Index

Subject Index

Printed in the United States
by Baker & Taylor Publisher Services